KICKS AND BLOWOUT CONTROL

SECOND EDITION

KICKS AND BLOWOUT CONTROL

• • • • • • • • • • • • •

SECOND EDITION
(Revised edition of *Well Control Problems and Solutions*)

NEAL ADAMS
LARRY KUHLMAN

PENNWELL PUBLISHING COMPANY
TULSA, OKLAHOMA

Copyright © 1980, 1994 by
PennWell Publishing Company
1421 South Sheridan/P.O. Box 1260
Tulsa, Oklahoma 74101

Library of Congress Cataloging-in-Publication Data

Adams, Neal.
    Kicks and blowout control / Neal Adams, Larry Kuhlman.–2nd ed.
    p. cm.
    Rev. ed. of: Well control problems and solutions. ©1980.
    Includes index.
    ISBN 0-87814-419-6
    1. Oil wells–Blowouts. I. Kuhlman, Larry. II. Adams, Neal.
Well control problems and solutions. III. Title.
TN871.2.A334   1994   94–1884
622' .3382–dc20   CIP

All rights reserved. No part of this book may be
reproduced, stored in a retrieval system, or
transcribed in any form or by any means, electronic
or mechanical, including photocopying and
recording, without the prior written permission of
the publisher.

Printed in the United States of America

1 2 3 4 5   98 97 96 95 94

To N.J. and Marjorie Adams
& Ollie Mae Barrett

—Neal Adams

and to

Henry Kuhlman, my father, who gave
44 years to the drilling industry

—Larry Kuhlman

CONTENTS

PREFACE ix

ACKNOWLEDGMENTS xi

CHAPTER 1 WELL CONTROL EQUIPMENT 1

Fluid Density Control 1
Blowout Preventers 5
Pressure/Flow Control Equipment 30
Accessory Equipment 41
Drill Pipe Blowout Preventers 46
Downhole Blowout Preventers 50
Blowout Preventer Stack Design 53
Choke Devices 59
Choke Manifolds 63
Blowout Preventer Testing 66
Auxiliary Equipment 73
Problems 85
Solutions 88
References 89

CHAPTER 2 WELL CONTROL PROCEDURES AND PRINCIPLES 91

Introduction to Kicks 92
Nonconventional Well Control Procedures 141
Problems 152
Solutions 159
References 161

CHAPTER 3 SPECIAL PROBLEMS 163

Deepwater Well Control 163
Kicks While Running Casing 173

vii

Kicks While Cementing 175
Stripping and Snubbing Operations 179
Kicks While Pipe Is Out of the Hole 190
Handling Drill Pipe Under Pressure 203
Drill Stem Testing 205
Drill String Impairment 207
Well Control with MWD in Drill String 210
Well Control with Downhole Motor in Drill String 211
Well Control with Downhole Motor in Drill String 211
Drill String Shearing 215
Well Control with Hydrate Buildups 218
Well Control in Horizontal Wells 227
Bullheading 235
Hydrogen Sulfide 242
Shallow Gas Handling 258
Kicks with Oil-Base Muds 277
Ultra-Slim Hole Well Control 278
Problems 280
Solutions 282
References 283

CHAPTER 4 BLOWOUTS 287

Blowout Contingency Planning 289
Well Capping 319
Relief Wells 330
Underground Blowouts 407
What Causes Blowouts 420
Problems 422
Solutions 424
References 425

APPENDIX 431

INDEX 469

PREFACE

T he original 1980 release, *Well Control Problems and Solutions*, was the most advanced well control document of its time. It was the basis for the first well control school ever certified by a regulatory authority under current guidelines.

The many well control and blowout control achievements over the last 15 years necessitated the publishing of this second edition. *Kicks & Blowout Control* is the most complete book available on kicks, blowouts, and related well control topics. It contains state-of the-art kick handling procedures and is the most advanced and complete reference on blowouts. No other book in today's industry offers the comprehensive nature of this text.

The book is comprehensive on well control issues. Some included are as follows:

- Basic and advanced kick control
- BOP and related equipment
- Deepwater well control (to 4000 m)
- Kicks while running casing or cementing
- Stripping and snubbing
- Kicks with pipe out of the hole
- Handling drill pipe under pressure
- Well control with MWDs or motors in the drill string
- Hydrates
- Horizontal well control
- Bullheading
- Hydrogen sulfide kicks
- Shallow gas handling
- Kicks with oil base muds
- Ultra-slim hole well control
- Shearing drill pipe
- Blowout contingency planning
- Well capping
- Relief wells
- Underground blowouts
- What causes blowouts

Also, each chapter is complete with solved example problems and unsolved problems at the end of the chapter. (Solutions provided in the Appendix.)

The authors provide a unique blend of experience, education and plain know-how to get a job done. These talents from their nearly 50 years of experience have been focused to produce this book. They have handled severe kicks, land and offshore blowouts, underwater gas flows, high pressure snubbing jobs and every other type of well control situation. They are known for being able to handle the toughest blowout and well control problems.

ACKNOWLEDGMENTS

Many individuals contributed to this second edition of the book. These include family members, daily working associates, members within the oil industry, and friends worldwide. We express our thanks to each who has provided their time and efforts.

Specific acknowledgments must be made to a few individuals. Holly Adams and Donna Foret lost a daddy along the way while Neal Adams was collecting experiences worldwide to share with the industry through this book. Jane, Rhonda, Karl and Debbie, the Kuhlmans, had to put up with many nights of their family leader being at the office or at sites around the world.

The office staff of Neal Adams Firefighters must be acknowledged for their efforts. DeAnne Lofthouse worked diligently on manuscript editing, illustration scanning and all the thousands of other little things that we tend to overlook. Paula Middleton and Nancy Rego made the daily office run while Neal and Larry were focusing on other, more visible projects. The firefighting team all provided valuable experience that has been translated into words in this book.

A silent tribute is given to Louis Records, Sr. for the original encouragement given to Neal Adams many years ago and for developing well control procedures as the current industry knows it today.

CHAPTER 1

WELL CONTROL EQUIPMENT

Many facets of well control must be mastered to control kicks and prevent blowouts. One important part is proper selection and utilization of equipment to control the well. This encompasses not only the surface blowout preventers, but also other items such as mud, mud monitoring equipment, degassers, and mud mixing systems. When all systems are functioning satisfactorily, proper procedures can be executed simply to maintain control of the well and prevent blowouts.

FLUID DENSITY CONTROL

The primary well control tool is hydrostatic pressure of the drilling fluid in the well. The drilling fluid (mud) is used to prevent a well from kicking. If a kick occurs, mud is used to kill the kick and regain control of the well. Therefore, drilling fluid density control is a prime concern of the drilling supervisor.

"Hydrostatic pressure" is defined as the pressure exerted by a column of fluid. In the drilling industry, the fluid is generally considered to be mud, but it could also be brine, gas, air, foam, or water. The formula to calculate hydrostatic pressure is:

$$\text{Hydrostatic pressure} = 0.052 \times \text{mud density} \times \text{depth} \quad (1.1)$$

Where:
hydrostatic pressure = pounds per square inch, psi
0.052 = psi/ft/ppg
mud density = pounds per gallon, ppg
depth = TVD, ft

Example 1.1

Calculate the hydrostatic pressure for each of the following systems.

1. 10,000 ft of 12.0 ppg mud
2. 12,000 ft of 10.5 ppg mud
3. 15,000 ft of 15.0 ppg mud

Solution

Hydrostatic pressure = 0.052 × mud density × depth

1. 0.052 × 12.0 ppg × 10,000 ft = 6,240 psi
2. 0.052 × 10.5 ppg × 12,000 ft = 6,552 psi
3. 0.052 × 15.0 ppg × 15,000 ft = 11,700 psi

A mud gradient can also be used to calculate hydrostatic pressures. "Mud gradient" is defined as the hydrostatic pressure for each vertical foot of mud, and it is calculated using the following formula:

$$\text{Mud gradient} = 0.052 \times \text{mud density} \qquad (1.2)$$

Where:

Mud gradient = psi/ft

0.052 = psi/ft/ppg

mud density = pounds/gallon, ppg

The hydrostatic pressure would be written as:

$$\text{Hydrostatic pressure} = \text{mud gradient} \times \text{depth} \qquad (1.3)$$

Equation 1.1 is a combination of Equations 1.2 and 1.3. Example 1.2 illustrates the mud gradient.

Example 1.2

Use mud gradients to calculate the hydrostatic pressure exerted by 15,000 ft of 15.0 ppg mud.

Solution

1. Using Eq. 1.2,

$$\text{Mud gradient} = 0.052 \times \text{mud density}$$
$$= 0.052 \times 15.0 \text{ ppg}$$
$$= 0.780 \text{ psi/ft}$$

2. Using Eq. 1.3,

$$\text{Hydrostatic pressure} = \text{mud gradient} \times \text{depth}$$
$$= 0.780 \text{ psi/ft} \times 15{,}000 \text{ ft}$$
$$= 11{,}700 \text{ psi}$$

Drilling fluid density is controlled by varying the concentration of high specific gravity solids within the fluid. The fluid density is increased by adding these solids. Density is decreased by either removing the solids or by adding a low density fluid to dilute the solids' concentration. Table 1–1 lists some common materials used to increase fluid density.

Barite is the most commonly used density control material. Its relatively high specific gravity and inert properties make it ideal to use in the mud system. Caution must be taken when using barite from uncertain sources. Poor quality control in some mines may yield a product that is mixed with hydratable clays, which when introduced into the mud system will cause increased mud viscosity properties.

Galena (lead sulfide) is occasionally used for density control in special applications. Its specific gravity of 6.8 generates mud weights that attain

TABLE 1–1 • Fluid density control additives

Additive	Specific Gravity	Maximum Fluid Density (ppg)	Remarks
1. Clay	2.3–2.5	11.5	Viscosity effects control the upper density limit
2. Barite (regular)	4.2–4.3	22.0	Barium sulfate
3. Barite (coarse grind)	4.2–4.3	22.0	Removed with 80-mesh screen
4. Galena	6.8	32.0	Lead sulfide: Special applications only
5. Calcium carbonate	2.7	12.0 (water muds) 11.5 (oil muds)	Ground limestone
6. Sodium chloride	–	10.0	Density is temperature dependent.
7. Calcium chloride	–	11.7	Density is temperature dependent.
8. Zinc chloride	–	17.0	Extremely corrosive
9. Calcium bromide	–	15.0	–
10. Zinc chloride/ calcium chloride	–	14.0	–
11. Zinc bromide	–	19.2	–

high hydrostatic pressures over relatively short columns of fluid. Use of galena muds has been confined to special well control applications, due to the problem of maintaining suspension of the high specific gravity solids in the mud system.

Proper well planning requires a sufficient quantity of barite be maintained on the drilling location to kill a kick. To calculate this volume of barite properly, many operators have established a 1 ppg safety margin, which means that barite volumes will be maintained at a level sufficient to increase the present mud density by 1 ppg. This safety margin is based on well control statistics that show the average kick requires a one-half (0.5) ppg increase in mud weight or less. Thus, the 1 ppg margin incorporates a safety factor of 2 relative to the average kick. The following equation can be used to calculate required barite volumes:

$$\text{Pounds/barrel} = \frac{1490(W_2 - W_1)}{35.4 - W_2} \tag{1.4}$$

Example 1.3 illustrates Eq. 1.4.

Example 1.3

A well is being drilled with 15.0 ppg mud. The hole volume is 850 bbl and the surface pit volume is 350 bbl. How many sacks of barite should be maintained on the drilling location? (Assume that one sack contains 100 lbs of barite.)

Solution

1. Using Eq. 1.4,

$$\text{lbs/bbl} = \frac{1490(W_2 - W_1)}{35.4 - W_2}$$

$$W_1 = 15.0 \text{ ppg}$$

$$W_2 = W_1 + 1.0 \text{ ppg} = 16.0 \text{ ppg}$$

$$\text{lbs/bbl} = \frac{1490(16.0 - 15.0)}{35.4 - 16.0}$$

$$= \frac{1490}{19.4} = 76.8 \text{ lbs/bbl}$$

2. $(850 + 350) \text{ bbl} \times 76.8 \text{ lbs/bbl}$

    $= 92{,}160 \text{ lbs of barite}$

3. $\dfrac{92{,}160 \text{ lbs}}{100 \text{ lbs/sack}} = 921.6 \text{ sacks}$

WELL CONTROL EQUIPMENT

BLOWOUT PREVENTERS

When primary well control has been lost, it becomes necessary to seal the well to prevent an uncontrolled flow (i.e., blowout) of formation fluids. The equipment that seals the well is called a **blowout preventer.** The equipment consists of inside blowout preventers (drill pipe blowout preventers), drill pipe safety valves designed to stop the flow through the drill pipe, and annular or ram preventers designed to stop flow in the annulus. Since there are numerous types of elements in a blowout preventer stack, each will be described with some of its special design features presented.

Many blowout preventer manufacturers and models exist in the industry. The following discussions are designed to give an overview and some details. Equipment operations vary among models, pressure ratings, and manufacturers. Manufacturers should be consulted for additional details on various models or equipment not discussed. (Some manufacturers are not discussed. It is not possible to provide details on all equipment sources within the constraints of a general-purpose kick and blowout control book.)

Annular/Ram Blowout Preventers

The **blowout preventer stack** is designed to control the flow of fluids in the annulus, and it may be a composite of several types of blowout preventer elements. Some, but not all, components include annular preventers, ram preventers (e.g., pipe, blind, shear and variable), and drilling spools. Each element will be discussed with blowout preventer stack design criteria presented in later sections. (A listing of various blowout preventer component specifications is presented in the Appendix).

Annular Preventers

The first preventer normally closed when shut-in procedures are initiated is the **annular preventer.** The four basic segments of the annular preventer are the head, body, piston, and steel-ribbed packing element. (See Figure 1-1.) When the preventer's closing mechanism is actuated, hydraulic pressure is applied to the piston, causing it to slide upward. This forces the packing element to extend into the wellbore around the drill string. The preventer element is opened by applying hydraulic pressure to slide the piston downward, and it allows the packing to return to its original position.

Caution should be taken when closing the annular preventer to insure that it is closed only when the drill string is in the hole. When pipe is not present and the annular preventer is closed on open hole, the rubber packing element is subjected to high stresses. The overall life of the element is shortened by open hole closures. When emergencies exist, most

FIGURE 1-1 • Annular closing mechanism (Courtesy of Hydril)

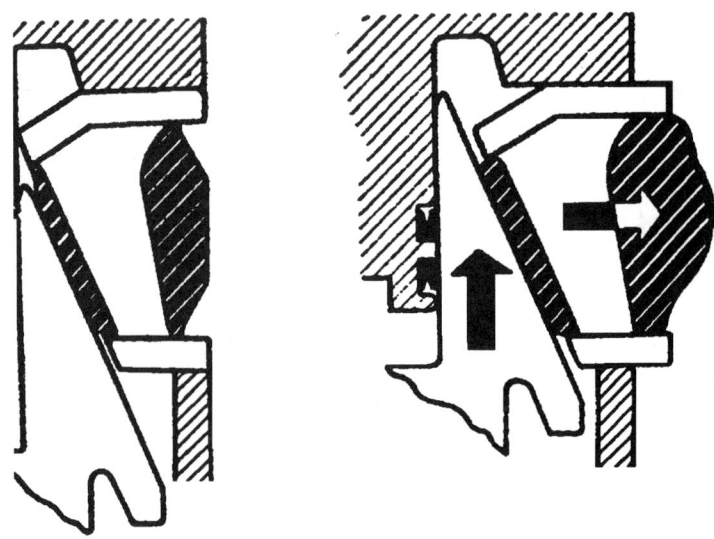

annular preventers will seal on open hole with an element in good condition. However, the pressure rating on open hole closure is one half the normal pressure rating.

The initial recommended hydraulic closing pressure for most annular preventers is 1,500 psi (10 MPa). However, this varies with models and pressure ratings. (A listing of recommended closing pressures for most types of preventers and valves used in well control is presented in the Appendix.)

Under certain conditions, hydraulic pressure should be reduced to minimize damage to the rubber portion of the element and possibly to large diameter tubulars. Several manufacturers have recommended closing pressures that depend on the kick casing pressure after the well is shut-in and for stripping operations. (Figures and Tables for these pressures are presented in following sections in the chapter.)

While well killing procedures are in progress, it is not always necessary to exert hydraulic pressure on the preventer in excess of the kick pressure. Some annular preventers, as well as many ram-type preventers, are designed to utilize wellbore pressures to aid in maintaining closure. In some cases, it may be observed that the preventer will remain closed even when virtually no hydraulic pressure is applied.

Many annular preventers have elements that can be changed without removing the pipe if the element becomes damaged during well killing operations. Some annular models, however, are relatively difficult to

Well Control Equipment

disassemble and reassemble under field conditions. When the preventer becomes damaged, the pipe rams below the annular should be closed and locked. The top plate, or cover, of the preventer must be removed, and the rubber element must be lifted out with the rig hoist line.

With the element out of the body, use a knife and split the rubber from the pipe. Cut the new rubber between the ribs and install the element in reverse order from the removal sequence. (See Figure 1–2.) After the top plate is replaced, the annular can be used again in the kill operations. This procedure is obviously not applicable for subsea preventers. If the hydraulic closing unit is positioned significantly above the preventers, it may be necessary to drain the hydraulic fluid line to the annular before the cover can be removed.

FIGURE 1–2 • Annular element replacement (Courtesy of Hydril)

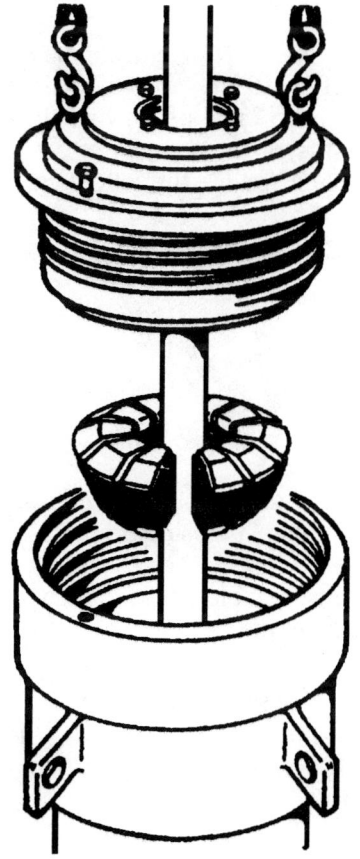

A special design feature of the annular preventer is that it will allow stripping operations to be carried out while maintaining a seal during pipe/tool joint passage.

The annular packing element is generally considered superior to the rams for stripping purposes due to its abrasion resistance and capability to flex as tool joints pass. Although the accumulator pressure regulator will maintain constant hydraulic pressure on the packing element, caution must be exercised because the slow response of regulators requires tool joints to be moved slowly through the preventer in order to avoid damage to the packing element. All manufacturers have recommended stripping practices that should be followed. A surge bottle should be installed on the annular hydraulic closing line as close to the unit as possible.

Hydril Corp.® manufactures several models of annular preventers with different packing elements available for specific types of service. Hydril's line of annular blowout preventers is based on more than 40 years of experience.

A standard annular BOP will strip drill pipe and tool joints or close off the annulus or open hole to full working pressure. The universal seal-off feature permits closure and seal-off on virtually anything in the bore—drill pipe, kelly, tool joints, or tubing.

Hydril's annular BOPs include the following:

- GX®
- GK®
- GL®
- MSP®
- GS®
- GKS®
- RS®

The GS is a snubbing preventer.

Table 1-2 lists the available packing elements, and Tables 1-3, 1-4, and 1-5 give the recommended closing pressures for the MSP, GK, and GL models, respectively. When the GL model is used in subsea applications, special closing pressure tables, dependent upon the water depth, are available from the manufacturer.

Hydril GX operating features (See Figure 1-3.)

1. The GX is designed for both surface and subsea applications. Manufacturer's recommended model for deepwater applications.
2. The operating piston is balanced with equally sized opening and closing chambers. This equalizes the hydrostatic forces

Well Control Equipment

TABLE 1-2 • Hydril packing elements (Courtesy of Hydril)

Packing Type	Color Code	Letter Code	Manufacturer's Recommended Usage
Natural rubber	Black	N	Water-base muds with less than 5% oil; operating temperatures –30° to 225°F (–35°C to 107°C); applicable for H_2S service.
Nitrile rubber	Red	NBR	Oil-base muds with aniline points between 165°F and 245°F; applicable for H_2S service; operating temperatures 32°F to 190°F (0° to 88°C).
Neoprene	Green	N	Oil-base muds with operating temperatures between –30°F to 170°F (–35° to 77°C), applicable for H_2S service.

TABLE 1-3 • "MSP" closing pressures (Courtesy of Hydril)

	Preventer size (in.–psi)			
Pipe OD, in.	7 1/16–2,000	9–2,000	11–2,000	21 1/4–2,000
5 1/2	–	–	350	500
4 1/2	350	400	450	550
3 1/2	400	500	550	600
2 7/8	400	550	650	650
2 3/8	500	650	750	700
1.90	600	750	850	800
1.65	700	850	850	900
CSO	1,000	1,050	1,150	1,100

exerted on the piston by the control fluid in subsea conditions. It results in constant closing pressure regardless of water depth or mud weight.
3. Replaceable wear plate eliminates metal-to-metal contact between the packing unit inserts and the BOP head.
4. H_2S serviceable.
5. Latched head for easy access to packing unit and seals.
6. Available in a GX Annuflex model for applications in subsea drilling systems. This model combines a GX annular preventer with a subsea riser flex joint.

TABLE 1-4 • "GK" closing pressures (Courtesy of Hydril)

Pipe OD, in.	7 1/16-3M	7 1/16-5M	7 1/16-10M	7 1/16-15M	7 1/16-20M	9-3M	9-5M	9-10M	11-3M	11-5M
6 5/8										350
5									450	450
4 1/2	350	400	350	2,100	2,200	400	450	350	450	450
3 1/2	400	450	550	2,100	2,200	500	600	380	550	525
2 7/8	400	450	750	2,100	2,200	550	650	570	650	800
2 3/8	500	500	850	2,100	2,200	650	750	760	750	900
1.90	600	600	900			750	850	860	920	
1.66	700	700	1,000			850	950	950	950	
CSO	1,000	1,000	1,150			1,050	1,150	1,000	1,150	1,150
								1,150		

	11-10M	13 5/8-3M	13 5/8-5M	13 5/8-10M	16-2M	16-3M	16 3/4-5M	18-2M
		200	600	700	350	450		500
	500	800	650	700	400	500		550
	800	900	650	1,200	500	500	600	600
	900	1,000	700	1,400	600	600	650	650
	1,000	1,100	750	1,400	700	700	750	700
	1,100	950	800	1,500	800	800	850	750
		1,000	1,000	1,500	900	950	950	850
			1,000		1,000	1,000	1,050	950
	1,200	1,200	1,150	2,200	1,150	1,150	1,150	1,150

Manufacturer's note: The pressures above are a guideline. Maximum packing until life will be realized by use of the lowest closing pressure that will maintain a seal. For subsea applications, see the appropriate operator's manual for computation of best closing pressure.

Well Control Equipment   11

TABLE 1-5 • "GL" closing pressures (Courtesy of Hydril)

	Preventer size, (in.–psi)								
	13 5/8–5,000 Well pressure (psi)			16 3/4–500 Well pressure (psi)			18 3/4–5,000 Well pressure (psi)		
Pipe OD, in.	2,000	3,500	5,000	2,000	3,500	5,000	2,000	3,500	5,000
7	900	950	1,100	700	825	950	700	825	950
5	900	1,000	1,100	750	850	1,000	800	900	1,000
3 1/2	1,200	1,200	1,200	800	925	1,050	1,000	1,050	1,100
CSO	1,400	1,500	1,500	1,400	1,500	1,500	1,500	1,500	1,500

Note: These are closing pressures from the secondary chamber to the opening chamber for surface installations. Consult the manufacturer for subsea applications.

FIGURE 1-3 • Hydril "GX" preventer (Courtesy of Hydril)

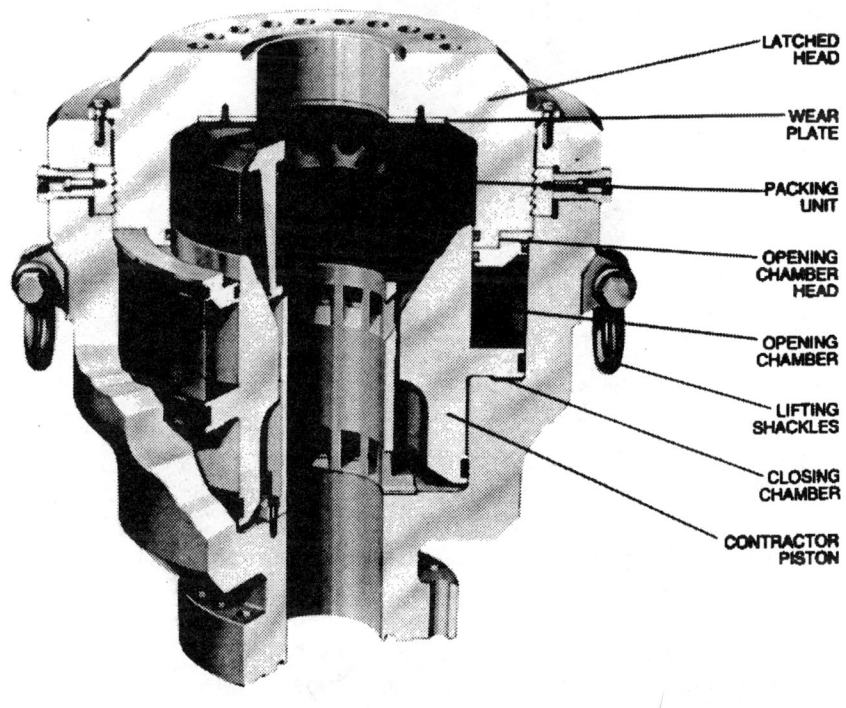

FIGURE 1-4 • Hydril "GK" preventer (Courtesy of Hydril)

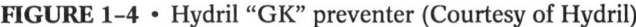

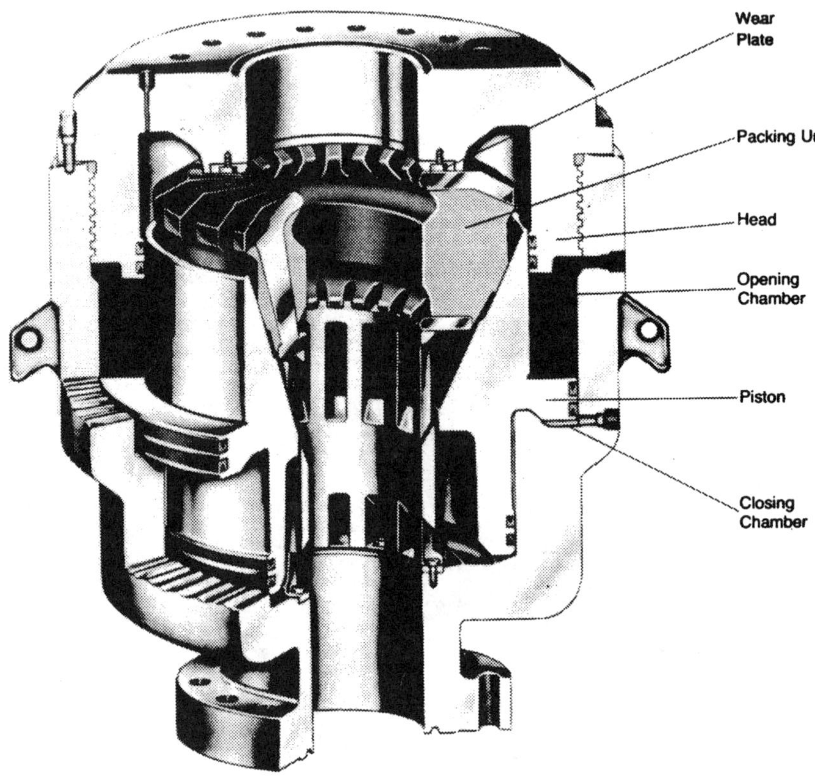

Hydril GK operating features (See Figure 1-4.)

1. Full range of bore sizes and pressure ratings.
2. Available with flanges of higher pressure rating than the preventers to connect with high pressure, ram-type preventers.
3. Most GK annulars have a screwed head design. Latched head is available on units 13 5/8", 5,000, and larger.
4. Sealing assistance is gained from the well pressure after initial seal off (except for GK 7 1/16", 15,000 and 20,000).
5. Designed for use on surface installations, but it can also be used subsea. Requires high accumulator pressures when used in subsea installations.
6. Has provisions to measure piston travel to gauge packing element wear.

Well Control Equipment

Hydril GL operating features

1. Designed and developed for both surface and subsea installations.
2. Closing pressure must be maintained during all seal off operations. Closing pressure must be increased as well pressure increases.
3. Available in large bore models. Only available in 5,000 psi (34 MPa) working pressure rating. Available in an integral dual BOP unit.
4. Latched head design for fast access.
5. Has a secondary chamber designed to balance riser hydrostatic pressure in subsea systems. Secondary chamber can be connected two ways to optimize operations for different effects—minimize closing/opening fluid volumes or reduce closing pressure.

Hydril MSP operating features

1. Low pressure service—500 to 2,000 psi (3–14 MPa).
2. Available in bore sizes from 7 1/16" through 30".
3. A primary usage is in diverter systems, but it can be used as a BOP on surface or subsea installations.
4. Sealing assistance is gained from the well pressure except in the MSP 29 1/2" – 500, which is not wellbore assisted.

Hydril also offers a snubbing annular BOP. The GS is available in 10,000 and 15,000 psi models. According to the manufacturer, the packing unit is designed specifically for continuous stripping and snubbing. Its operational design is similar to other Hydril units.

The GKS annular BOP/stripper is an annular BOP with application to workovers. This BOP/stripper is applicable to stripping, snubbing, stuffing, and wireline operations.

Hydril offers the RS tubing stripper for moving long strings of tubing or drill pipe in or out of the well under pressure. It is automatic, and it is always in contact with the tubing. It is available with a removable slip bowl and slip assembly.

Shaffer®, a Varco company, manufactures annular preventers operating on the same general principle as the Hydril annulars. (See Figure 1–5.) The Shaffer annular preventer, called the Spherical®, is widely used, and it has undergone rigorous testing with good results. In particular, its stripping capability has a long-term life according to testing and field results. Under controlled stripping testing, the Spherical element lasted two to four times longer than the others tested.

Table 1–6 describes the packing elements available for the Shaffer unit. New high temperature annular elements are being developed.

The unit is available in standard units, lightweight units for airlifting, and for Arctic service. All units close and open with 1,500 psi (10.3 MPa) accumulator pressure for most applications (7" (178 mm) OD or less,

FIGURE 1-5 • Shaffer "Spherical®" preventer (Courtesy of Shaffer, Inc.)

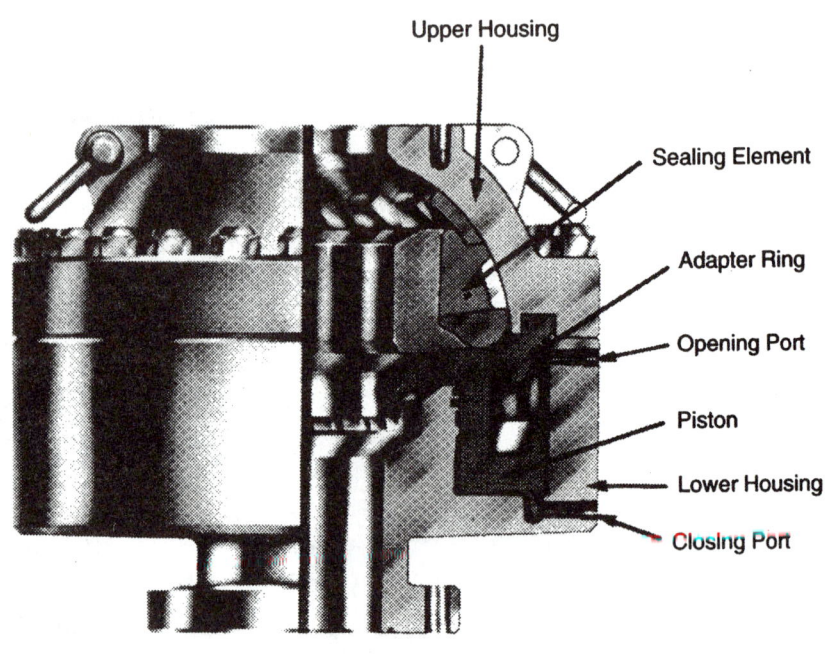

TABLE 1-6 • Shaffer packing elements (Courtesy of Shaffer, Inc.)

Packing Type	Color Code	Manufacturer's Recommended Usage
Natural rubber	Red	Low temperature operations in water-base muds; stripping abrasion resistance. Temperature range −20° to 170°F (−29° to 77°C).
Buna (nitrile)	Blue	Oil- and water-base muds; H_2S service. Temperature range 40° to 170°F (4° to 77°C).
Ultra Temp™		High-temperature prolonged operations at 250°F (121°C). Intermittent operations to 350°F (177°C). Currently available only for ram preventers. Annular element is under development.

stationary). Figure 1-6 provides recommended spherical closing pressures for large pipe. Shaffer operation manuals should be consulted for stripping operations closing pressures.

Shaffer spherical operating features

1. Full range of bore sizes and pressure ratings.
2. Will close on open hole.

Well Control Equipment 15

FIGURE 1-6 • Shaffer annular recommended closing pressures on large pipe (Courtesy of Shaffer, Inc.)

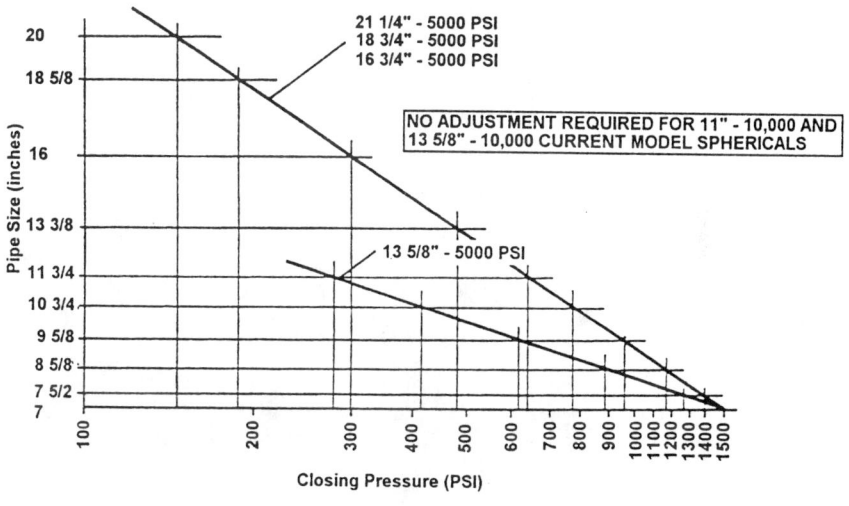

3. Applicable for surface and subsea service.
4. Slight sealing assistance is gained from the well pressure.
5. Good stripping service.
6. Available in lightweight, dual, and Arctic models.
7. Low operating pressure.

Cooper Oil Tools (formerly Cameron® Iron Works) manufactures the preventer shown in Figure 1-7. The model D®, which is a re-engineered version of the original A model, has received extensive field and laboratory testing since 1971, and it has good well control operating characteristics. The functional mechanism is a hydraulic-operated, sliding piston-type.

The "D" annular BOP closing pressure forces the operating piston and pusher plate upward, which displaces the solid elastomer donut and forces the packer to close inward. As the packer closes, steel reinforcing inserts rotate inward to form a continuous steel support ring at the top and bottom of the packer. The inserts remain in contact with each other whether the packer is open, closed on pipe, or closed on open hole. Refer to the "D" annular operation manual for details on closing pressure requirements.

The "D" annular-packing element is available in CAMULAR™ material for most sizes/pressure ratings. CAMULAR provides longer service life, excellent diesel mud resistance, and can operate to 180°F (82°C) or intermittently to 200°F (93°C). The "DL" annular packer can be split for installation while pipe is in the hole.

FIGURE 1-7 • Cooper "D" preventer (Courtesy of Cooper Oil Tool)

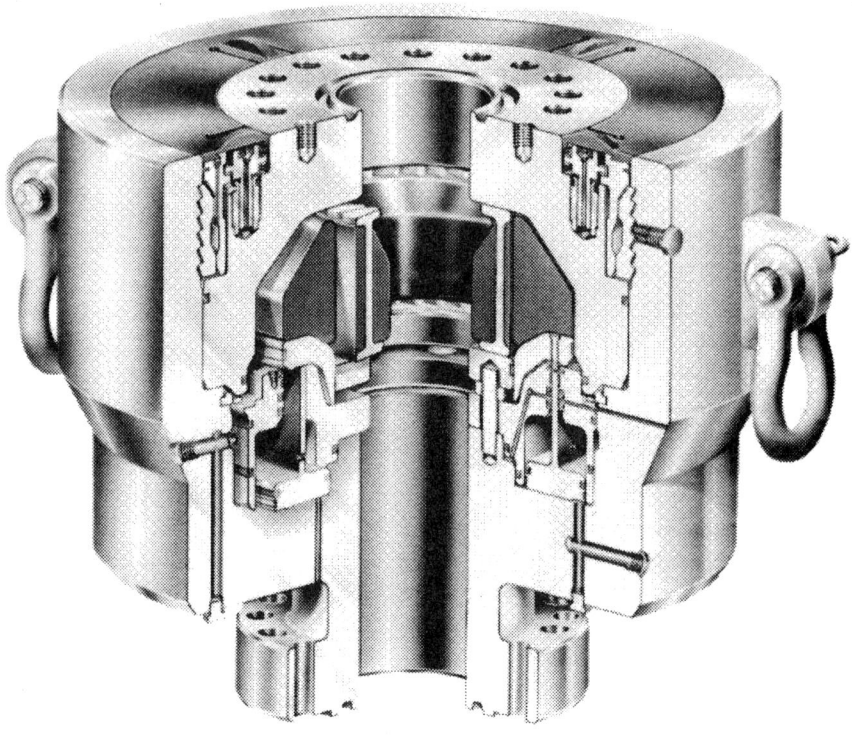

Cooper "D" operating features

1. Full range of bore sizes and pressure ratings.
2. Will close on open hole.
3. Sealing assistance is gained from well pressure.
4. Quick release top latch allows fast element change.
5. Low vertical height.
6. Low element weight.
7. Requires less fluid to open and close the element than comparable models, but at a higher pressure.

Ram-Type Preventers

Unlike the operation of the annular preventer, ram preventers seal the annulus by forcing two elements to make contact with each other in the annular area. These elements have rubber packing seals that affect the complete closure. Other than the sealing mechanism, ram-type blowout preventers differ greatly from annular preventers because each type and size of ram has one function and cannot be used in a variety of applications. (The exception is the variable bore ram).

For example, ram bodies with a set of 4 1/2 in. (114 mm) pipe rams will seal on 4 1/2 in. pipe and will not seal with any other size of pipe, nor will they seal without pipe in the well. Ram-type preventers are generally considered to be more reliable in high pressure service, as well as more easily serviceable and shorter. Ram types that warrant discussion are pipe, blind, variable bore, and shear rams.

Several design features of all ram preventers should be understood by the drilling supervisor. One feature is the direction of the pressure seal. Most ram preventers are designed to hold pressure from the lower side, which means that (1) the preventer will not secure the well if it is installed upside down, and (2) that the ram will not pressure test from the top down. The last consideration is important when designing a blowout preventer stack arrangement and the manner in which it will be pressure tested.

Another special design feature is the secondary rod sealing action that is available for most types of ram preventers. Due to routine wear, the primary rod sealing mechanism may begin to leak. The secondary rod sealing mechanism is used to provide an additional measure of protection in sealing the area around the rod used to actuate the preventer.

Most ram preventers are manufactured with a self-feeding action for the rubber sealing section. (See Figure 1–8.) As the rubber wears, the small extrusion plates are forced into the recessed area, which allows additional rubber to extend past the ram face and aid in securing a seal. If the rams are used improperly, the self-feeding action will cause the rubber seal to extend an excessive distance into the wellbore, which will cause over stressing and rapid deterioration of the element. Because of this, pipe rams should not be closed routinely if pipe is not in the hole.

Ram bodies are universal because they will accept either blind, shear/blind, or pipe ram elements. Also, units are available that are comprised of single, double, or triple ram bodies. In the multiple-unit ram bodies, any combination of pipe and blind ram elements may be used. (See Figure 1–9.)

Figure 1–10 shows a pipe ram assembly with the ram block, rubber seals, and pipe guides that center the pipe during closure.

Care must be taken in ram size selection when aluminum drill pipe is being used. This type of pipe has a tube outside diameter in the middle section slightly smaller than the tube outside diameter near the tool joint. Regular 4 1/2 in. (114 mm) pipe rams will seal on the middle tube section of 4 1/2 in. aluminum pipe but not near the tool joint, which can be done with steel pipe. The shut-in procedures must be planned accordingly to account for this irregularity.

FIGURE 1-8 • Ram packer self-feeding action (Courtesy of Hydril)

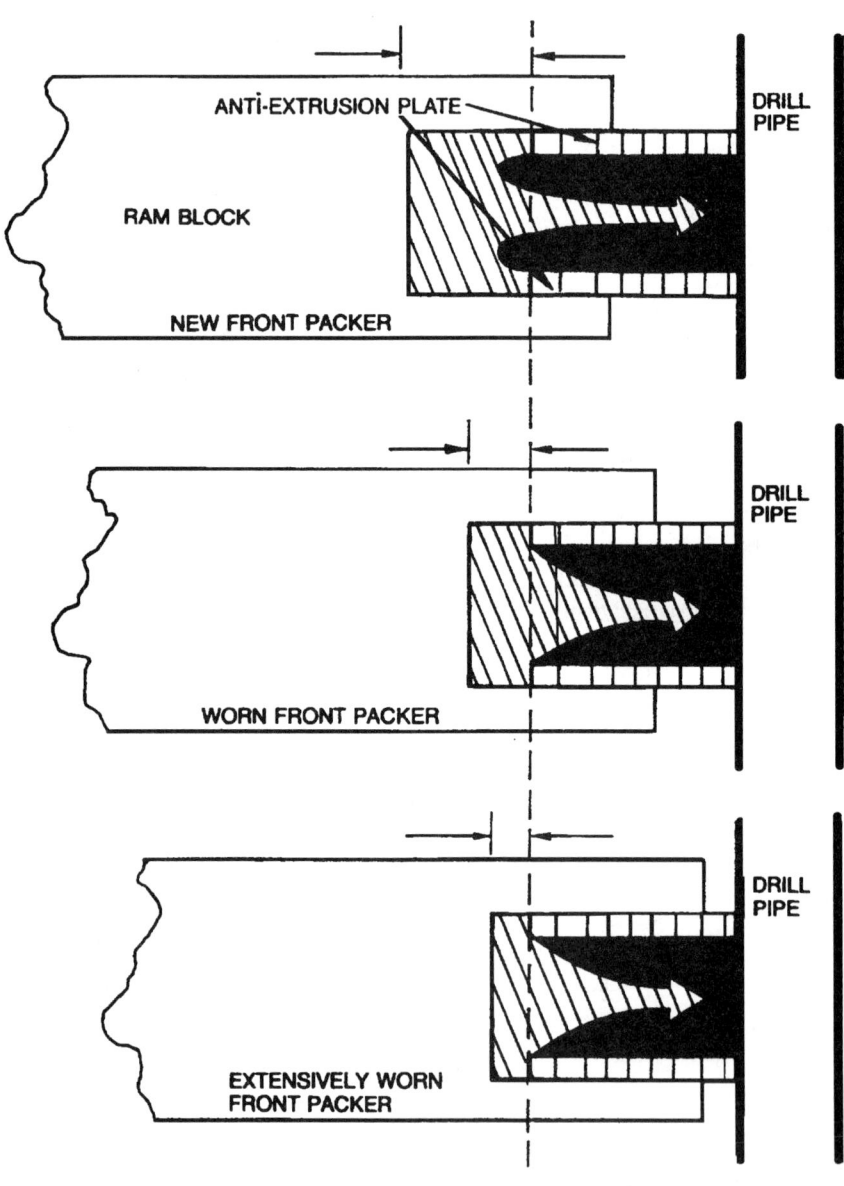

Well Control Equipment 19

FIGURE 1-9 • Ram types (Courtesy of Cooper Oil Tool)

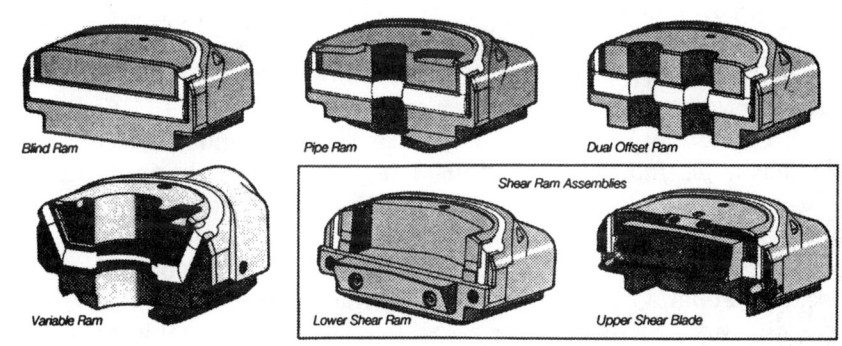

FIGURE 1-10 • Pipe ram assembly (Courtesy of Cooper Oil Tool)

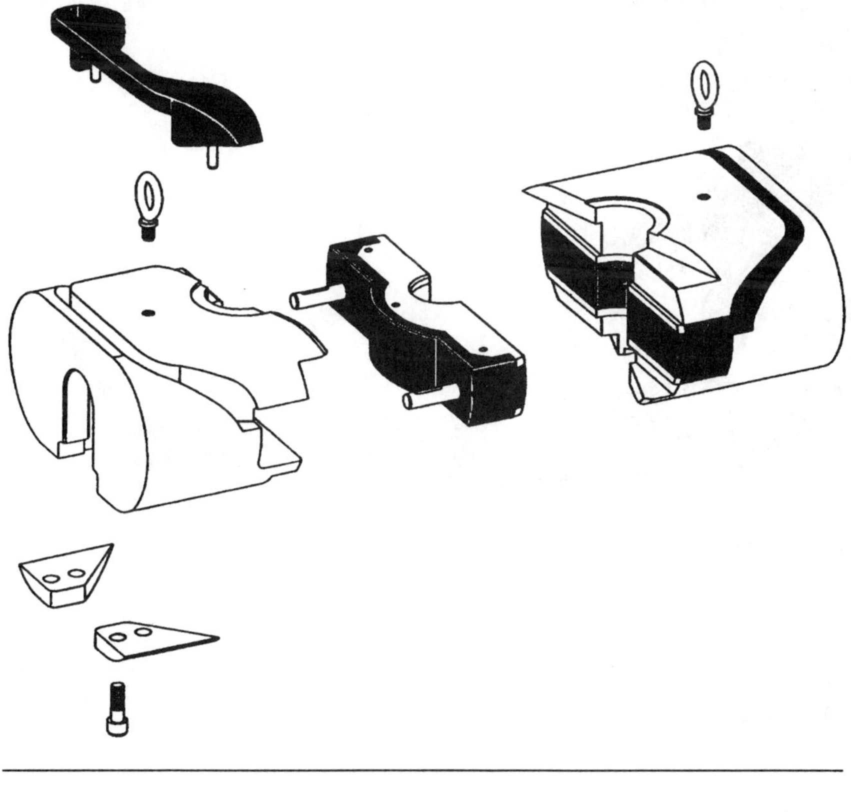

One special design feature of some pipe ram elements is that when closed and locked, the ram can support the weight of the drill string, if necessary, by hanging a tool joint on the ram. This feature is useful when storm conditions exist or blowouts are impending. However, this usage is not recommended under normal conditions. If this practice is routine, special hardness ram elements can be purchased for this function.

Blind rams are designed to seal the well if pipe is not in the hole. The element is flat-faced and contains a rubber section. The rams are not designed to effect a seal when pipe is in the hole; although, occasionally the pipe will be cut if the blind rams are accidentally closed. Thus, precautions should be taken with the blowout preventer control panel to ensure that blind rams cannot be accidentally closed.

Most manufacturers offer a variable bore ram, or multi-ram. (See Figure 1–11.) It has the capability to seal on several pipe sizes. Some claims are made that these rams can seal on hexagonal kellys. Manufacturers should be consulted for pipe size sealing ranges for variable bore rams. An additional feature is that a BOP stack with variable rams and a tapered pipe string in the well can use fewer preventer elements in some cases.

FIGURE 1–11 • Variable bore ram assembly (Courtesy of Cooper Oil Tool)

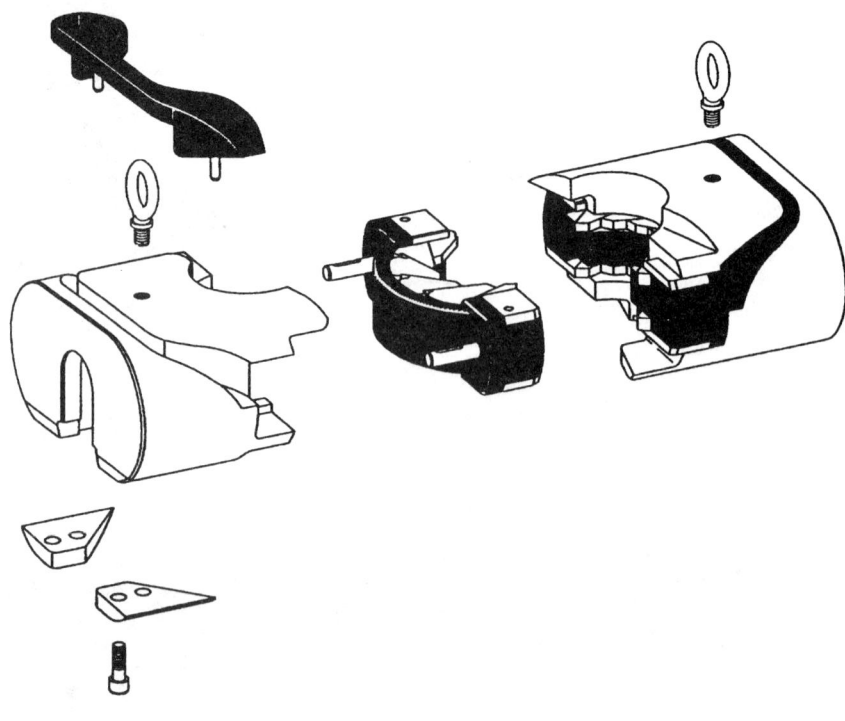

WELL CONTROL EQUIPMENT

Some variable bore rams may not be able to support as much pipe hang-off weight as pipe rams, or none at all. The manufacturer should be consulted.

All manufacturers offer shear/blind rams. These have the capability to shear certain sizes of pipe and casing and to provide a seal on the open wellbore. (See Figure 1–12.) Shearing dynamics vary with manufacturers. Also, shearing capability varies with pipe manufacturer and size/weight, BOP type and size, and position of rams in the stack. It is suggested that operators develop a shearing policy to account for well conditions and BOP types. The precautions mentioned for blind rams also apply to shear/blind rams.

Ram rubbers have been developed with good wear and temperature resistance. Significant improvements have been made in elastomer technology. This allows stripping/snubbing in hot environments or situations where the well must be flowed while running pipe in or out of the well. (Note: This is a special services operation and not to be attempted routinely without training and proper equipment.) Most of the newer models are suitable for H_2S service.

The following discussion on available ram bodies does not encompass every model nor does it completely cover the models presented. The manufacturers should be consulted for complete details. A listing of various specifications of ram-type preventers is presented in the Appendix.

FIGURE 1–12 • Shear/blind ram assembly (Courtesy of Shaffer, Inc.)

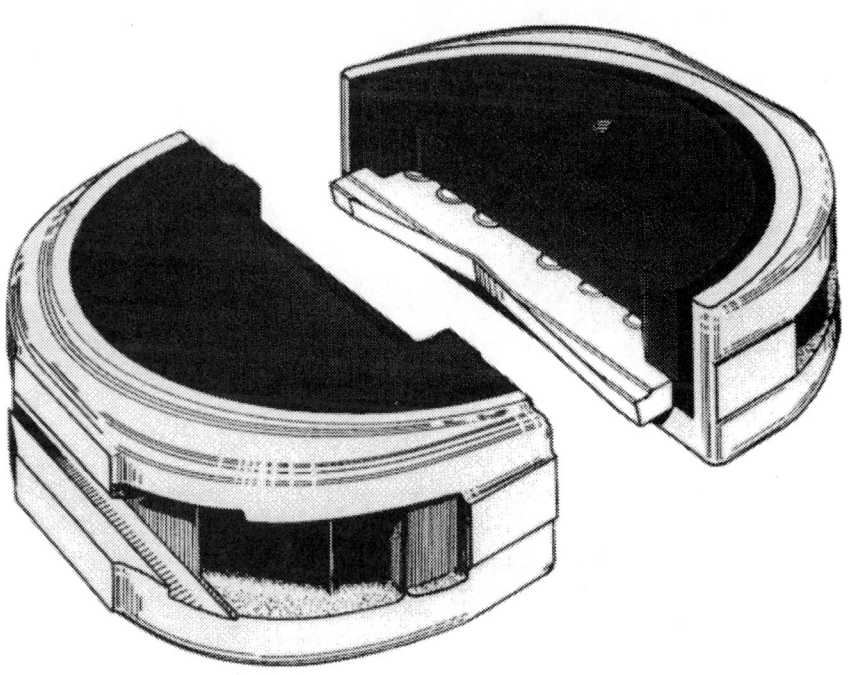

Cooper Oil Tools (formerly Cameron Iron Works) has manufactured several models of ram preventers. These include the following:
- F™
- SS™
- QRC™
- U™
- U-II™
- T™

Although the types F, SS, and QRC are no longer manufactured, some are still used in the industry.

Cooper offers CAMRAM™ BOP packers and top seals that have a high resistance to diesel drilling muds and H_2S corrosion inhibitors, increased stripping life, 250°F (120°C) sustained service, and 300°F (149°C) intermittent service. CAMRAM 350 packers and top seals can operate in temperatures to 350°F (177°C) and H_2S concentrations to 35% without detrimental effects.

Cooper type "F" operating features

1. Low-closing pressures required due to low closing ratios.
2. Well pressure aids in maintaining rams closed.
3. Rams can be changed and repaired in the field.
4. Relatively lightweight.
5. Only one screw required for manual locking.
6. The hydraulic operator can be replaced while the preventer is locked in pressure service.

Cooper type "SS" (Space Saver) operating features

1. Low vertical height. (It was originally designed for rigs with low substructures.)
2. Ram position can be determined by external observation of the locking screws.
3. Well pressure assists in maintaining rams closed.
4. Rams can be changed and repaired in the field.

Cooper type "QRC" (Quick Ram Change) operating features

1. Rams can be manually locked in the closed position.
2. Well pressure helps to maintain rams closed.
3. Rams can be changed and repaired in the field.
4. Ram position can be determined by exterior observation.
5. Has secondary operating rod seal.

FIGURE 1-13 • Cooper "U" (Courtesy of Cooper Oil Tool)

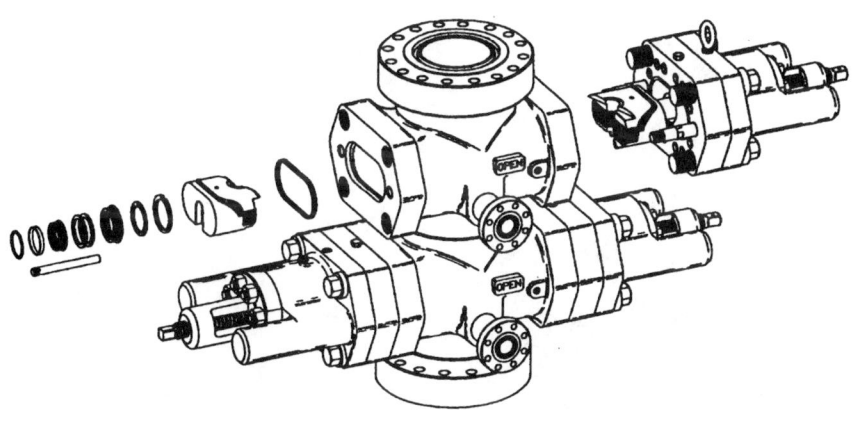

Cooper type "U" operating features (See Figure 1-13.)

1. Designed for both surface and subsea operations.
2. Well pressure helps maintain rams closed and increases the sealing force.
3. Rams can be changed and repaired in the field.
4. Ram position can be determined by exterior observations.
5. Has secondary rod seal.
6. A "wedgelock" mechanism is available to hydromechanically lock the rams. Pressure balance chamber is used for subsea applications.
7. Shear rams are available. Some sizes require special stroke shear bonnets to provide the additional stroke distance and force required for shearing.
8. Hydraulic bonnet bolt tensioning system is available for larger "U" units to reduce maintenance time.
9. The "U" offers the widest available size range in Cooper BOPs.

Cooper type "U-II" operating features (See Figure 1-14.)

1. Short stroke bonnet reduces the opening stroke by about 30%, reduces the BOP overall length, and reduces the weight supported by the ram change pistons compared to the original "U" BOP.
2. The BOP has an internally ported hydraulic bonnet tensioning system.
3. Available in 18 3/4" (476 mm) sizes for 10,000 and 15,000 psi (69–103 MPa).

FIGURE 1-14 • Double U II ram preventer (Courtesy of Cooper Oil Tool)

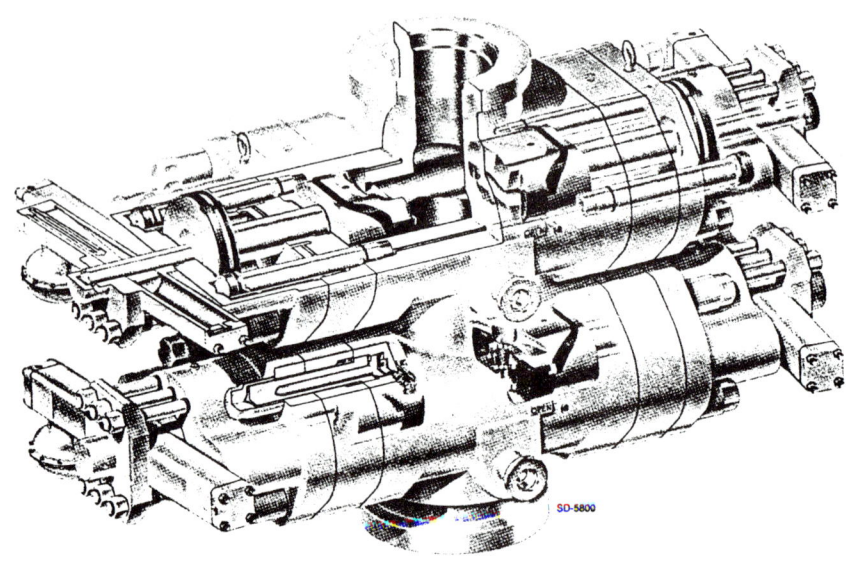

Cooper type "T" operating features (See Figure 1-15.)

1. Fast and easy service. Side ram removal permits reduced stack height and facilitates change-out of rams and packers. Sliding the rams to the side simplifies access and handling. Hydraulic bolt tensioning system used for fast disassembly/assembly and accurate tensioning.
2. Metal-to-metal bonnet gaskets increase safety and reliability in critical service applications.
3. The standard "T" BOP has enough piston area to shear commonly used sizes and grades of drill pipe.
4. All operating system and well bore seals can be replaced with the bonnet ram in the ram-change position.
5. Improved "ST" hydromechanical locking (wedge-type) system.
6. Available in sizes from 1.315" to 13.375" (33–340 mm) and for aluminum sizes ranging from 3.5" to 5" (89–127 mm).

Shaffer, a Varco company, has manufactured several models of preventers. These include the following:

- LWS™
- LWP™
- SL™

Well Control Equipment 25

FIGURE 1-15 • T ram preventer (Courtesy of Cooper Oil Tool)

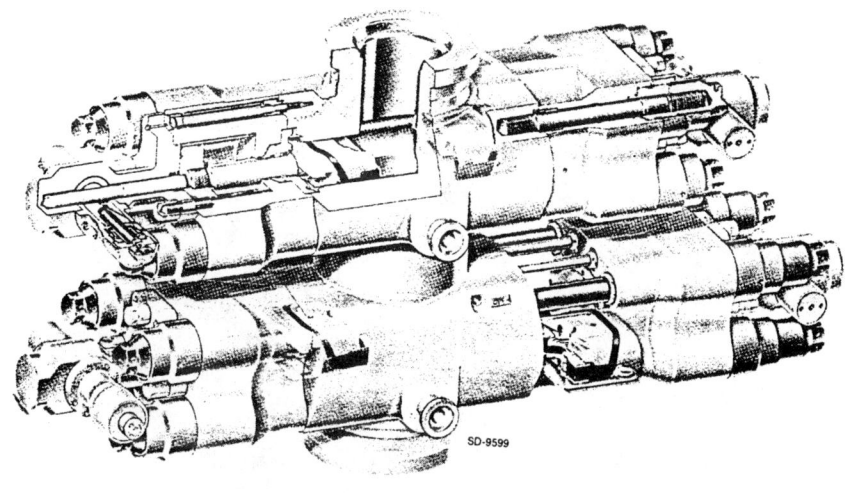

- Sentinel®
- XHP®
- B
- E

Types XHP, B, and E will not be presented in this discussion, but many of the older models are still in operation. (The XHP, B, E, and other older models are no longer discussed in Shaffer company literature.) The new designs are the product of more than 60 years of experience.

Shaffer ram-type preventers have single piston hydraulic operators, self-draining bodies, and minimal maintenance space requirements in-line with the preventer. Special Arctic models are available for low temperature service.

Shaffer "LWS" operating features (See Figure 1-16.)

1. One of the best known Shaffer ram preventers. Has many of the same features as the most advanced SL model.
2. Wellbore pressure helps to maintain rams closed.
3. Rams can be changed and repaired in the field.
4. Relatively lightweight.
5. New models have secondary rod sealing system and older models can have it retrofitted.
6. Ram position can be determined by external observation with manual locking system.

FIGURE 1-16 • Shaffer "LWS™" ram preventer (Courtesy of Shaffer, Inc.)

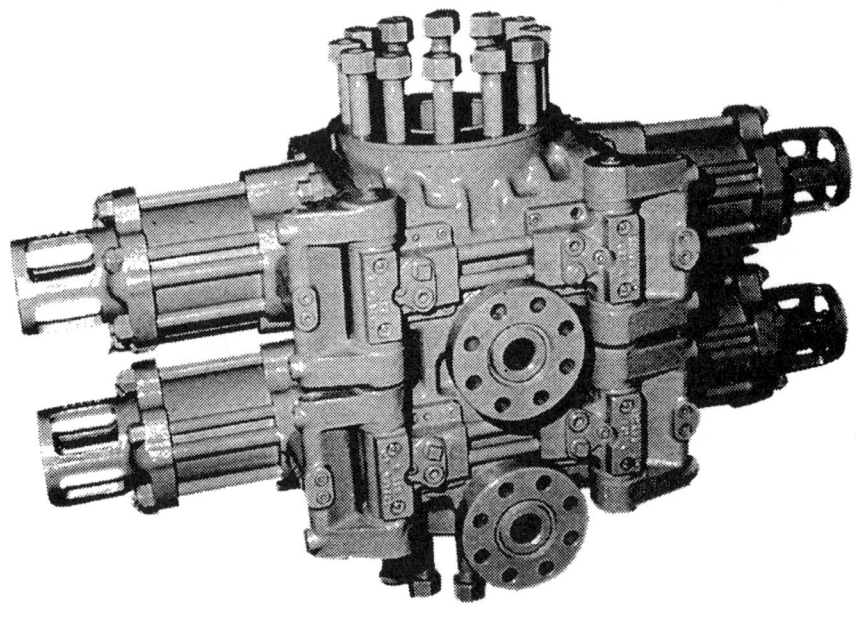

7. External ram locking mechanism or can have an automatic Poslock locking system installed.
8. Swinging hinged door opening mechanism requires some additional lateral space, but less clearance at each end.
9. Applicable for H_2S service.
10. Rams are available suitable for hanging off the drill string up to a maximum of 600,000 lbs (270,000 kg).

Shaffer "LWP™" operating features (See Figure 1-17.)

1. Limited size and pressure ratings. Primarily for workover and well servicing operations.
2. Same basic features as the LWS.
3. Relatively lightweight.
4. Rams can be changed and repaired in the field.
5. Swinging hinged door opening mechanism requires some additional lateral space, but less clearance at each end.
6. Ram position can be determined by external observation.
7. Only manual locking system is available.
8. H_2S trim is optional.

FIGURE 1-17 • Shaffer "LWP™" ram preventer (Courtesy of Shaffer, Inc.)

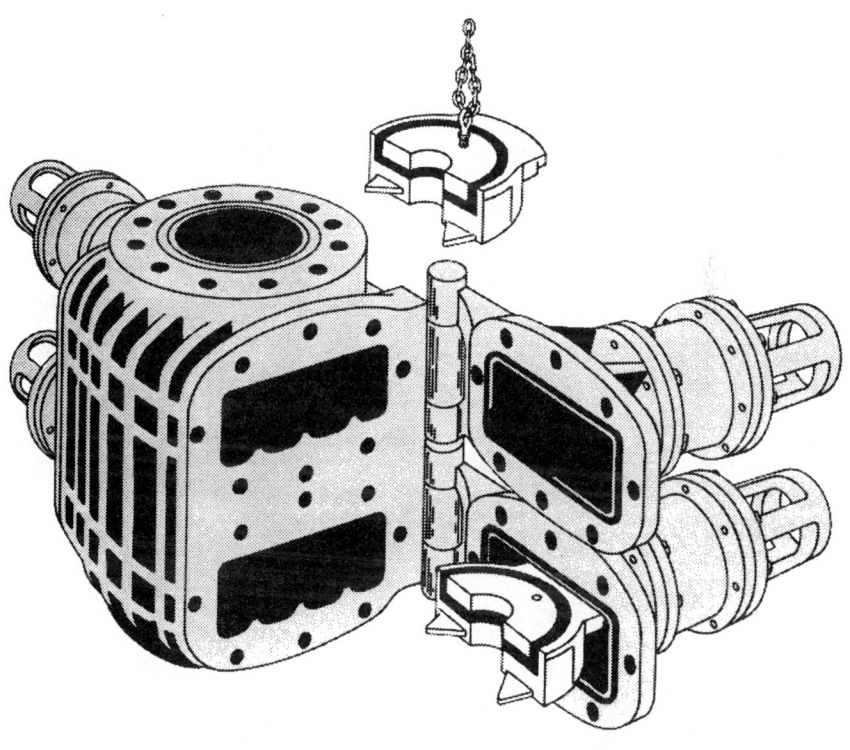

Shaffer "SL" operating features (See Figure 1-18.)
1. Available in a wide range of sizes and pressure ratings.
2. Flat doors simplify ram changes. Swinging hinged door opening mechanism requires some additional lateral space, but less clearance at each end.
3. Rams can be changed and repaired in the field.
4. Door seals on all sizes have a hard backing to minimize extrusion and pinching.
5. H_2S trim is standard. Shear rams are available with sour trim.
6. PosLock or MultiLock system locks the rams automatically each time the rams are closed. Manual locks and the newly developed automatic UltraLock systems are available.
7. Model SL-D rams are available that support a 600,000 lb (270,000 kg) drill string.
8. New UltraTemp packer and seal compounds are available for continuous service in temperatures to 350°F (177°C).

FIGURE 1-18 • Shaffer "SL™" ram preventer (Courtesy of Shaffer, Inc.)

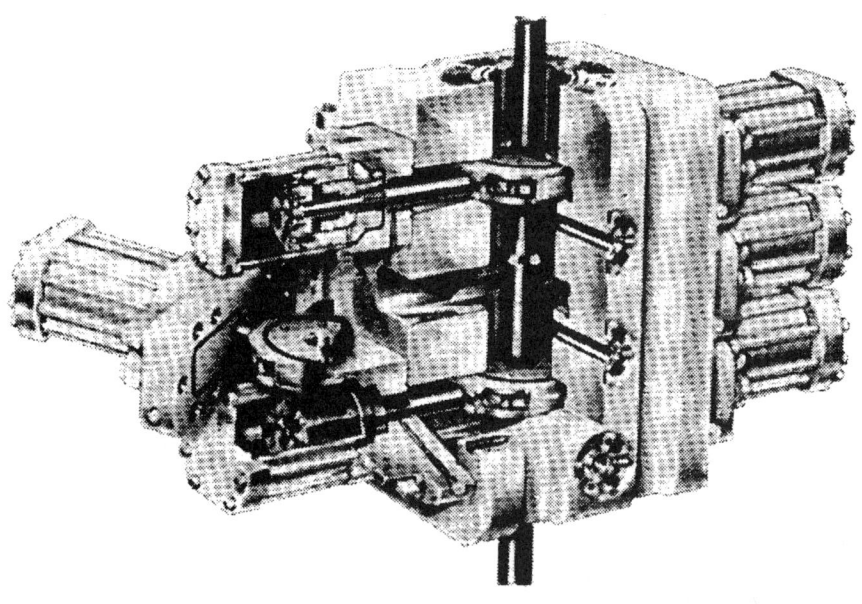

Shaffer "Sentinel" and "Sentinel II" operating features (See Figure 1-19.)

1. Short, compact, and lightweight, which is advantageous for well servicing, workover, fracing, and low pressure drilling.
2. Optional manual chain-drive or hydraulic closing systems.
3. Manual closing system requires only one person to open or close the rams.

Hydril Company manufactures several types of ram preventers. The drilling preventers are available to 15,000 psi (103 MPa) and up to 21 1/4" (540 mm) bores. A snubbing preventer is available in a 4 1/16" (103 mm), 10,000 psi (69 MPa) and 15,000 psi (103 MPa) units. A workover ram preventer is offered in 7 1/16" (179 mm), 5,000 psi (34 MPa). It is shorter and lighter than other workover rams in the industry according to the manufacturer.

Hydril drilling ram operating features (See Figure 1-20.)

1. Offered in full range of sizes and pressure ratings.
2. Available with manual or automatic "Multiple Position Locking (MPL™)" systems.
3. Cylinder liner and upper seal seat are field replaceable or field repairable.
4. Secondary rod sealing system.

WELL CONTROL EQUIPMENT 29

FIGURE 1-19 • Shaffer "Sentinel®" ram preventer (Courtesy of Shaffer, Inc.)

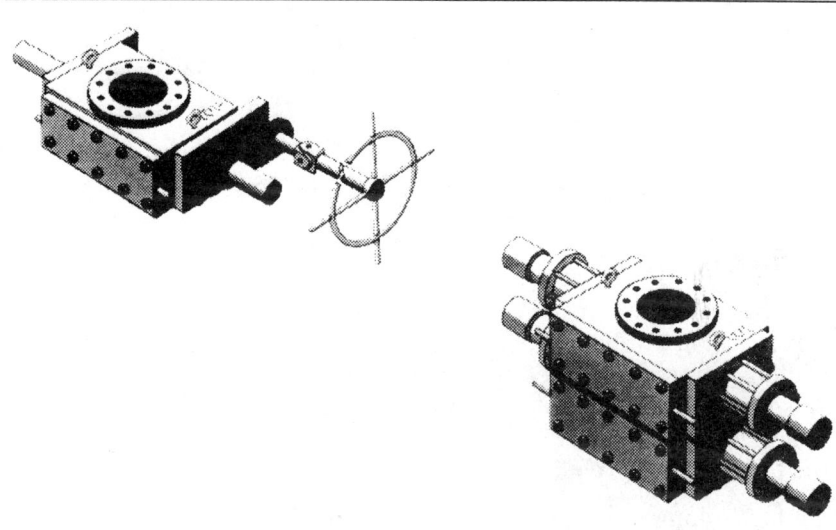

FIGURE 1-20 • Hydril drilling ram preventer (Courtesy of Hydril)

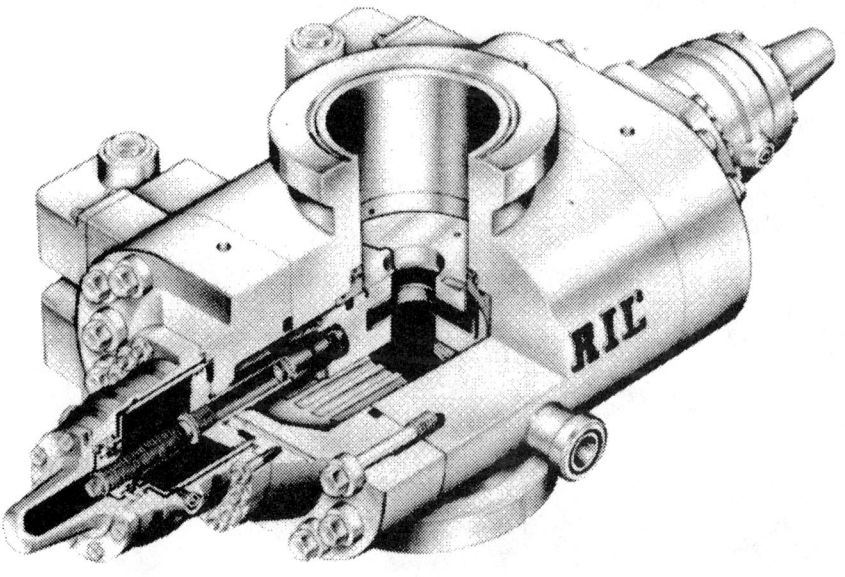

5. Rams can be changed and repaired in the field.
6. Swinging hinged door opening mechanism requires some additional lateral space, but less clearance at each end.
7. Sloped ram cavity is self draining to avoid buildup of mud and sand.
8. Rams are designed to permit drill pipe hang off.

PRESSURE/FLOW CONTROL EQUIPMENT

Drilling Spools

If blowout preventer units without built-in outlets are used, it is necessary to install a drilling spool. It is a cylindrical spacer spool with flanged, studded, or hub/clamp connections on each end and one or more outlets. It is placed within the blowout preventer stack to provide a means for attaching mud lines, termed choke and kill lines. The spool should meet the following API requirements:

1. Working pressure consistent with the blowout preventers.
2. One or more side outlets, no smaller than 2 in. (50 mm) diameter, with a pressure rating consistent with the blowout preventer stack.
3. Vertical bore diameter at least equal to the maximum internal diameter of the innermost casing. If the spool is to pass slips, hangers, or test tools, the bore should be at least equal to the maximum bore of the uppermost casing head, or blowout preventer stack.

Figure 1-21 illustrates a flanged drilling spool with two side outlets.

FIGURE 1-21 • Flanged drilling spool (Courtesy of Cooper Oil Tool)

Well Control Equipment

Casing Head

The basis of all blowout preventer stacks and usually the first component installed is the casing head. The head can be equipped with flanged, slip-on and weld, or threaded connections for attachment to the casing and the preventer stack. It can have threaded or flanged side outlets. The casing head should at least meet the minimum API requirements as follows:

1. Working pressure rating that equals or exceeds the maximum anticipated surface pressure to which it will be exposed.
2. Bending strength equal to or exceeding the outermost casing to which it is attached.
3. End connections with mechanical strength and pressure capacity comparable to corresponding API flanges or to the pipe to which it is attached.
4. Adequate compressive strength to support subsequent casing and tubing weight to be hung therein.

Figure 1–22 is an example of a casing head with threaded lower connection and flanged upper connection.

FIGURE 1–22 • Casing head with threaded lower connection and flanged upper connection (Courtesy of Cooper Oil Tool)

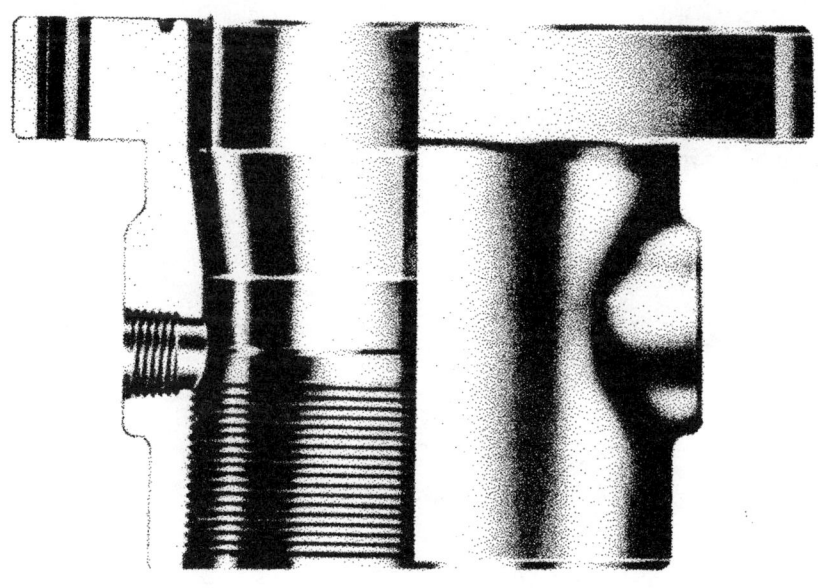

Diverter Systems

In certain cases, proper well control procedures demand that a kick not be shut in, but the well should be allowed to blow out in a controlled manner away from the rig. (Reasons for these procedures will be presented in later sections.)

Blowout diversion procedures do not require a full blowout preventer stack; instead, a special diverter unit, or annular preventer, with a relatively low working pressure is used to contain the vertical flow. One or more horizontal outlets are used to direct the flow through a vent line(s) to a safe distance from the rig.

Figure 1–23 illustrates a purpose-built diverter stack in which an annular preventer is used as a diverter unit. A diverter spool with one or more vent outlets is attached to the annular. The flow line(s) is attached

FIGURE 1–23 • Purpose-built diverter system

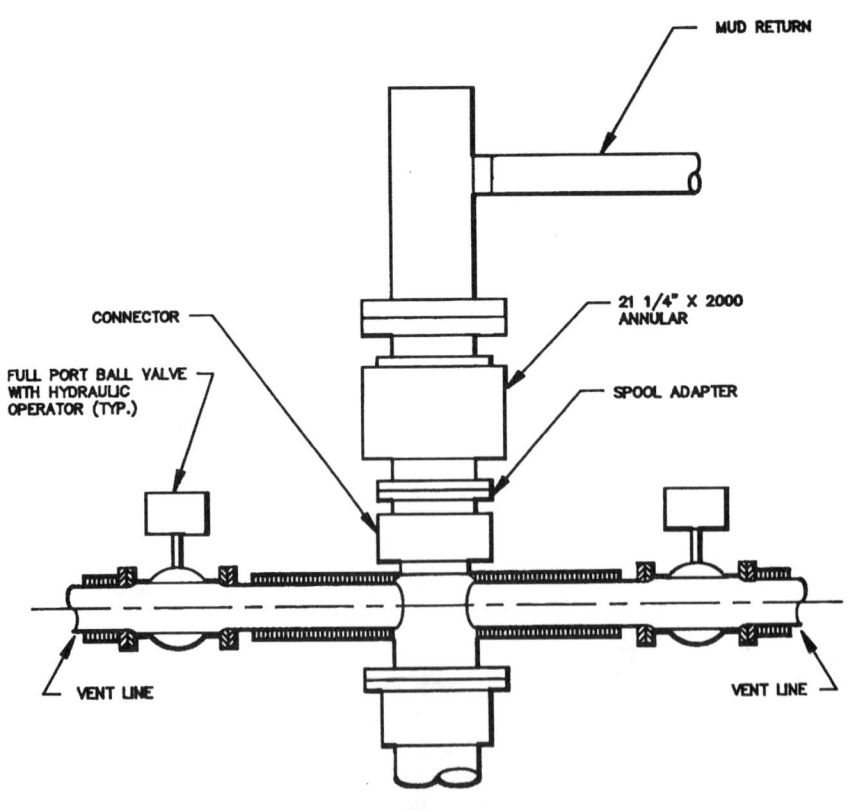

Well Control Equipment

to the outlets. Flow path choice is made with a valve on each vent line. Special erosion-resistant spool pieces have been developed to handle high volume, erosive flows.

Hydril Company offers several diverter units and components for diverter systems including the FSP 28-2000 Diverter BOP™, FS 21-500 Diverter™, SXV/MSP Diverter BOP™, and DS 12-500 Flow Selector™.

The FSP and FS combine a Hydril annular and integral venting through one or more outlets. The SXV/MSP combines an MSP annular with an SXV spool containing an integral vent. On the FSP and FS, the vent outlet(s) opens automatically as the annular piston and valve sleeve move upward while closing the annular. Integral valving eliminates the need for one or more valves on the vent/mud flow lines. The FS marine unit is shown in Figure 1-24.

FIGURE 1-24 • Hydril "FS" diverter unit (Courtesy of Hydril)

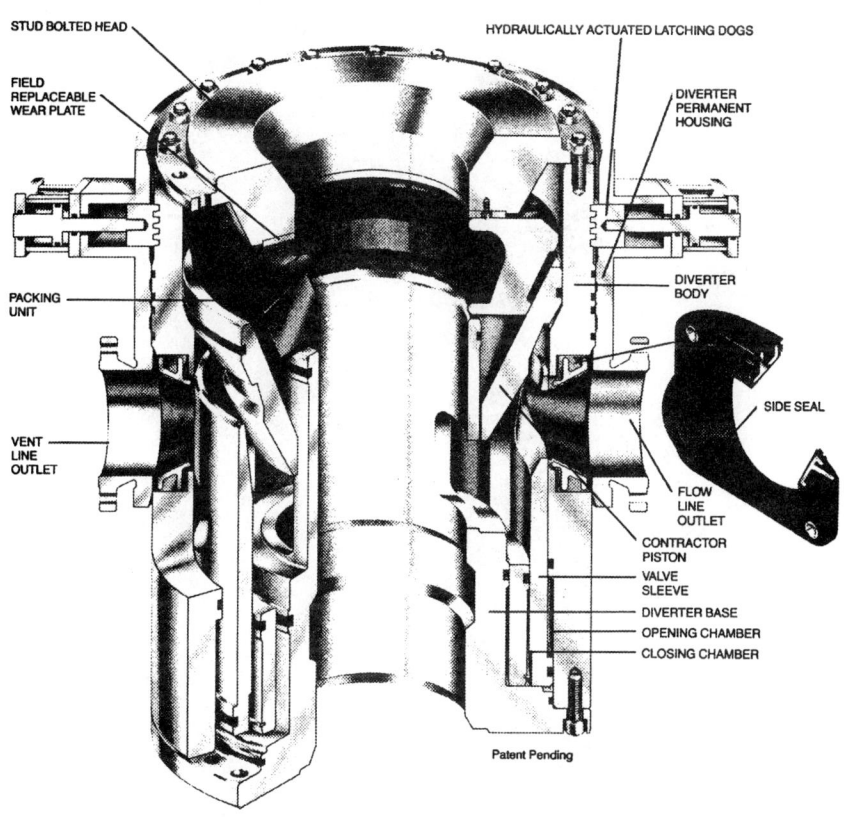

The FSP has a 28" (711 mm) bore and is primarily for bottom-supported offshore rigs. The FS has a 21" (533 mm) bore and is applicable for floating drilling.

The SXV spool can be attached to an MSP annular or separated by a spacer spool. The hydraulic control signal that closes the MSP annular simultaneously opens the SXV valve by using interconnecting hydraulic lines. The SXV/MSP assemblies are used for surface or subsea diverting and are available in a 20" to 30" (508–762 mm) range.

The DS 12–500 Flow Selector is a 3-way valve to provide the capability to switch from one vent line to another without complete shutoff during the process. The deflector, or target plug, is made of hardened steel to resist the erosive effects of the diverted flow.

Shaffer, a Varco company, offers a complete shutoff (CSO) diverter that integrates the Shaffer Spherical 21 1/4" (540 mm) annular with a vent outlet spool piece. The assembly fits into a housing that is securely mounted under the rig floor. (See Figure 1–25.) The unit can close on open hole or a wide range of tubulars.

FIGURE 1–25 • Shaffer diverter unit (Courtesy of Shaffer, Inc.)

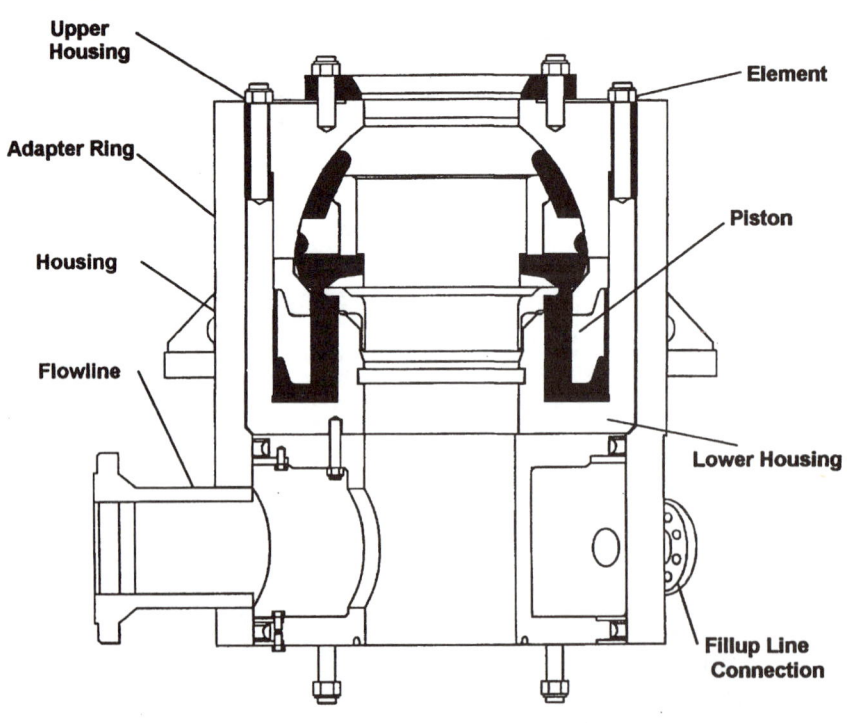

Well Control Equipment

ABB Vetco Gray offers diverter units (initially developed by Regan Forge and Engineering Co.) that operate on a different operating mechanism than the other available diverters. The Vetco KFDJ™ (for bottom-supported offshore rigs) and KFDS™ (for floating rigs) have rubber packers that are actuated by applying hydraulic pressure on the exterior of the elements, thereby forcing them inward to seal on the pipe in the well. This arrangement is also unique because it is designed to allow the insert (i.e., inner rubber packer) to be removed and exchanged without removing the top cover. These units provide rapid closure with low closing fluid volumes. They do not have the capability to close on open hole. The model KFDJ is shown in Figure 1–26.

The Vetco KFDS-CSO™ unit provides open hole shutoff. It is an integrated unit using a Shaffer Spherical annular similar to the Shaffer diverter.

Rotating Head/BOP

The primary function of an annular preventer is to provide pressure control while allowing a small amount of pipe movement. Occasionally, a tool is needed that will provide greater amounts of pipe movement

FIGURE 1–26 • Vetco KFDJ diverter unit (Courtesy of ABB Vecto Gray, Inc.)

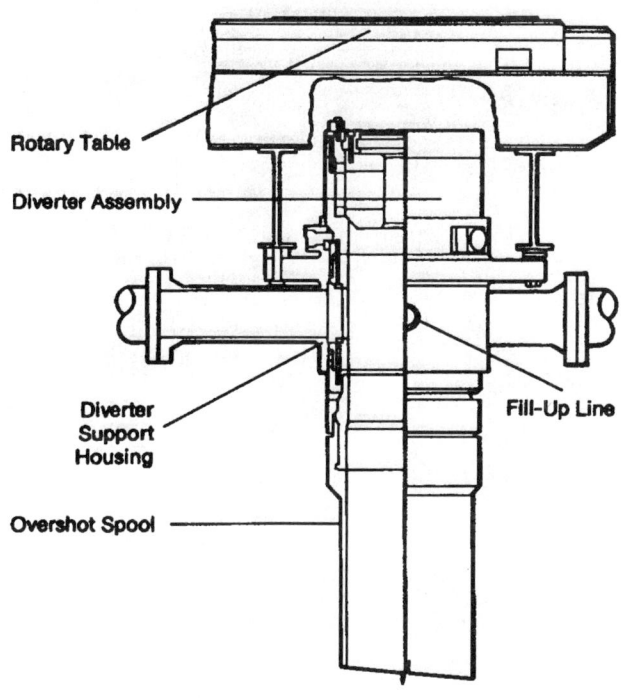

flexibility at lower service pressures. The rotating head/BOP serves this purpose. Rotating heads/BOPs have been used in air and gas drilling, controlled pressure drilling, and in reverse circulation operations with well pressures to 2,000 psi (14 MPa) and at rotating speeds to 150 rpm. When used in controlled pressure drilling, the head allows the use of lighter muds with increased penetration rates and reduced swabbing. The head also maintains the gas under pressure during a kick to reduce its volume. (See Figure 1-27.)

FIGURE 1-27 • Shaffer Type 79™ Rotating BOP (Courtesy of Shaffer, Inc.)

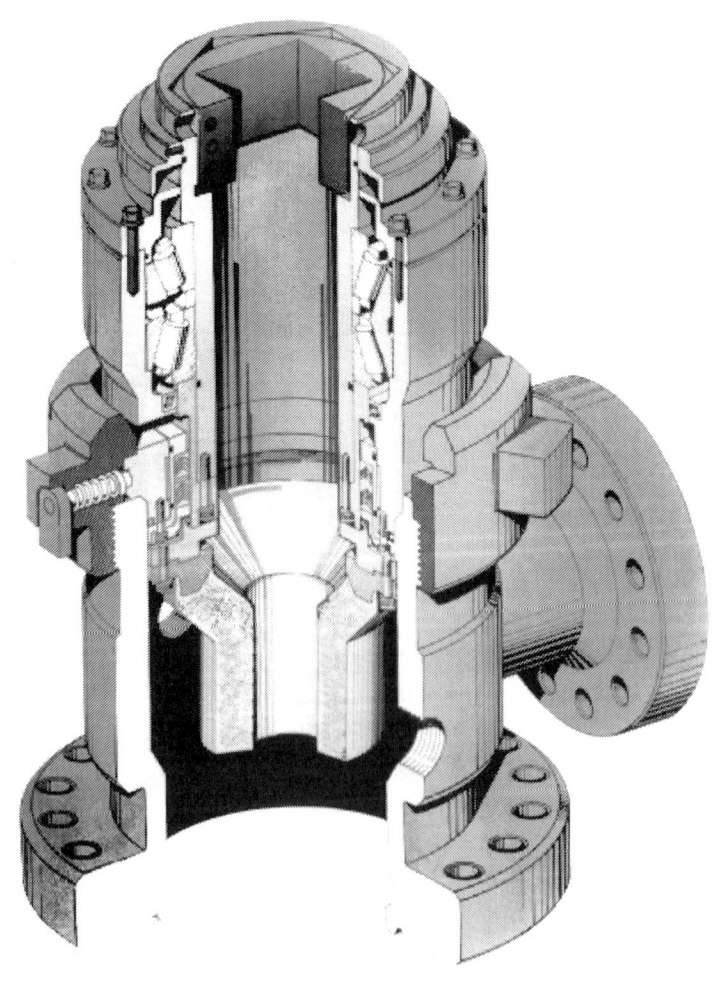

Choke and Kill Lines

In well killing operations, it will generally be necessary to circulate fluid down the drill pipe, up the annulus, and through an exit at the surface. The lines attached to the blowout preventers to provide this exit are termed "choke lines." The choke line carries the mud and kick fluid from the blowout preventer stack to the choke. The primary purpose of the kill line is to serve as a back-up choke line. The kill and choke lines may be used to pump mud directly into the annulus if necessary, although the kill line usually performs this function.

The kill and choke lines may be attached to several members of the blowout preventer stack. These lines could be attached to the outlets of the drilling spool shown in Figure 1-21, or they could be attached directly to the blowout preventer. Only under extreme circumstances (and never preferentially) should the choke and kill lines be attached to the casing head, casing spool, or below the lowermost set of rams. (See the section on preventer stack design for a further explanation.)

The kill and choke lines should meet a number of requirements. Some, but not all, are as follows:

1. Pressure ratings of these lines should be consistent with the blowout preventer stack.
2. The lines should meet all minimum blowout preventer testing requirements.
3. The lines should have a consistent inner diameter to avoid erosion due to diameter changes.
4. Direction changes should be kept to a minimum. If required to make several angular changes between the BOP stack and the choke manifold, it may be advisable to use target tees or crosses to absorb the turbulent erosion effects at these points.

Plug/Ball Valves

The name of this valve type describes its construction. A steel ball (see Figure 1-28) (or plug) has an opening (or orifice) along a line passing through its center. Plug valves and ball valves operate in a similar manner. The valve is open when the ball/plug is oriented so the orifice is in line with the fluid flow path. It is closed when the ball is oriented so the opening is perpendicular to the flow path (i.e., 1/4 turn to fully open or fully close).

The valve can be manually operated or operated remotely by using hydraulic actuators. In practical applications, it will often be necessary to pressurize the downstream side of the valve before opening if high pressure is acting on the upstream side of the valve. This pressurization is

FIGURE 1-28 • Ball valve (Courtesy of Cooper Oil Tool)

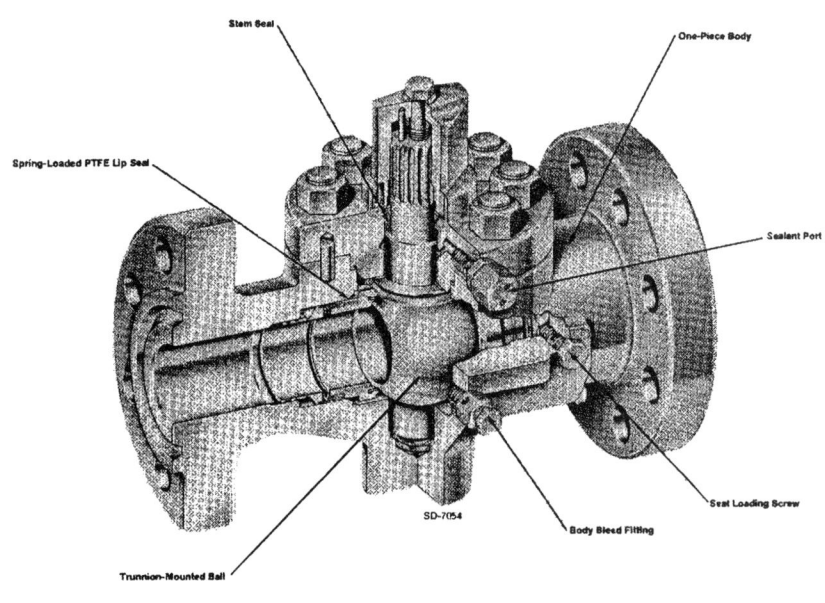

done because (1) the upstream pressure may create a differential pressure sufficient to cause the opening stem of the valve to shear when a sufficient external force is applied and (2) to avoid a sudden pressure surge on the downstream equipment.

Gate Valves

A gate valve utilizes closing and sealing mechanisms different than a ball or plug valve. In the gate valve shown in Figure 1-29, a blank plate is positioned across the flow path to halt fluid flow. When the valve is opened, the plate is moved so a section of the plate containing an orifice is positioned across the flow path to allow fluid movement through the orifice. Similar to the ball valve, the gate valve can be operated either manually or through remote hydraulic controls.

Gate valves, such as the Cooper Type "FLS" gate valve shown in Figure 1-29, have several design features that make them particularly useful in well control operations. They are designed for service with water, oil, gas, or mud as the circulating fluid. They are termed a "WOGM" valve. Metal-to-metal primary seals are utilized. Secondary seals are included for enhanced performance and to assist on low pressure sealing.

FIGURE 1-29 • "FLS" manual gate valve (10,000 psi) (Courtesy of Cooper Oil Tool)

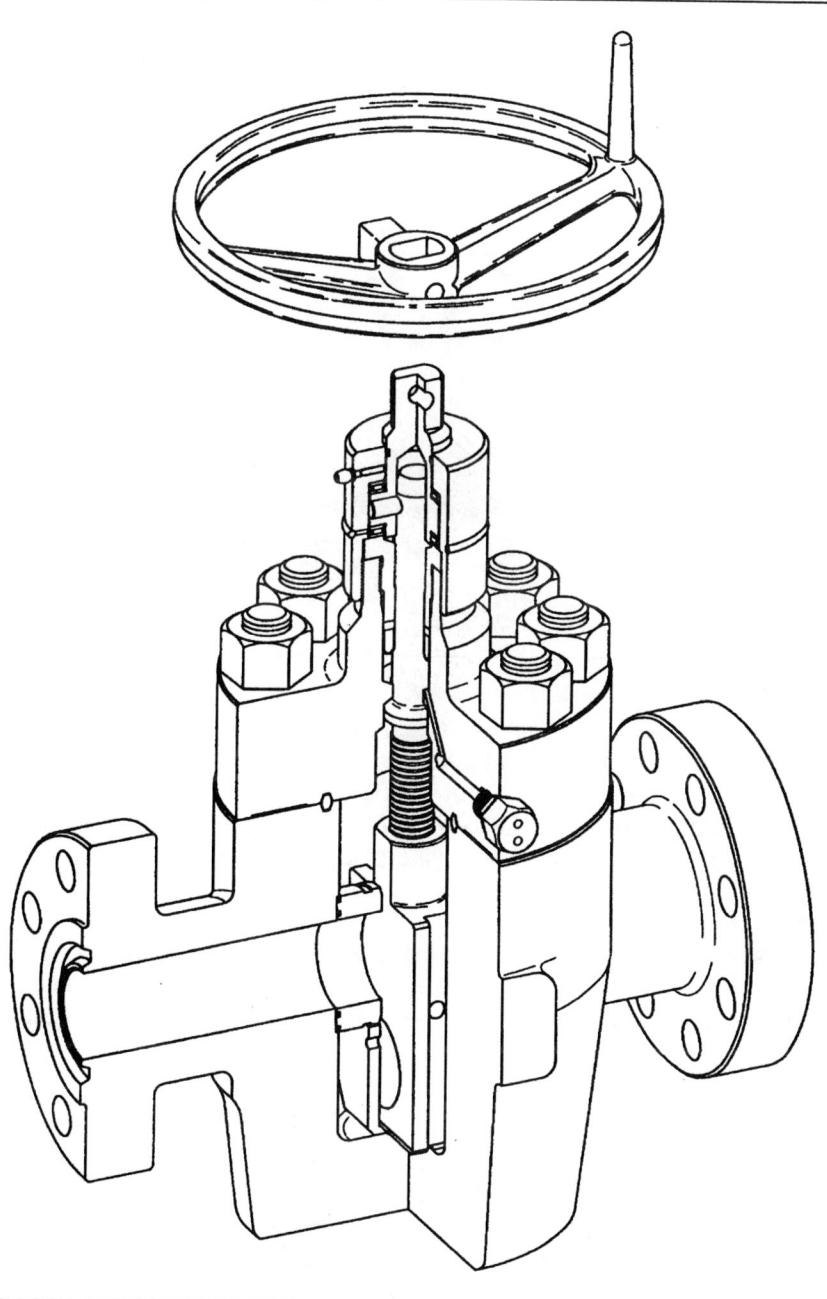

FIGURE 1-30 • Shaffer "HB™" hydraulic-operated gate valve (Courtesy of Shaffer, Inc.)

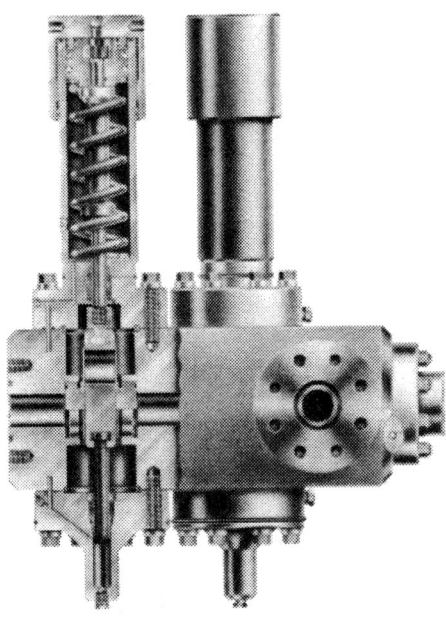

Hydraulic-operated gate valves are used on subsea BOP stacks and some surface stacks. Special versions, such as the Shaffer Type HB™ subsea gate valve shown in Figure 1-30, have hydraulic actuators which hydraulic pressure-assist the valve open or closed. They typically have a spring mounted on the closing side of the actuator piston to assist in maintaining closure after the gate has moved to the closed position in case of control fluid pressure loss.

Neither gate valves nor ball/plug valves are designed to operate under high flow rate conditions. Also, they are not intended for use as a choke device. Special drilling chokes are available to control high rate, abrasive fluid flows.

Check Valve

When well control conditions warrant pumping fluid directly into the annulus, a check valve on the kill line has applications. The valve shown in Figure 1-31 is generally spring-loaded and allows fluid movement in one direction only. Mud can be pumped into the annulus through the valve, but it cannot reverse direction and flow out of the annulus. A special design feature of some models is that the internal elements can be

WELL CONTROL EQUIPMENT 41

FIGURE 1-31 • Check valve (Courtesy of Hydril)

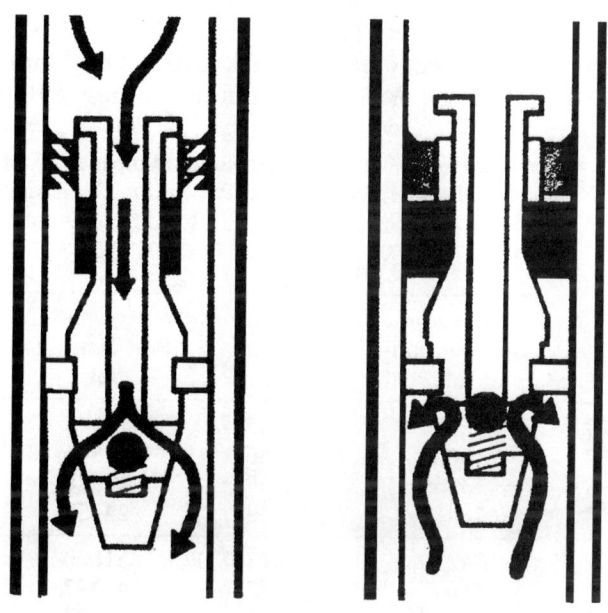

removed when necessary to allow fluid flow in either direction. A check valve is generally identified in schematic notation by placing an arrow immediately above the valve to indicate the fluid flow direction.

ACCESSORY EQUIPMENT

Several pieces of accessory equipment are necessary for proper blowout preventer operation. This equipment includes flange bolts, clamp connectors, H_2S or sour gas trim, and seal rings. Guidelines for the proper use of this equipment are as important as proper stack design or pressure-testing procedures.

Blowout Preventer Connections

When the elements of a blowout preventer stack are connected, they must provide a pressure seal at the connection point consistent with the pressure rating of the preventer elements.

One problem associated with preventers under pressure service is leaking connections. BOP connections, and the make-up and routine monitoring of these connections, are of primary importance in well control. The two types of connections available are flanged (bolted) and clamped hub.

Bolted flanges are the most common type of connection used in the industry. The bolts must be selected so tensile ratings are sufficient to withstand the maximum load that may be imposed. Also, the bolt torque applied must meet certain values to effect the flange pressure seal. Table 1–7 lists the API recommended values for bolt tension

TABLE 1–7 • Recommended flange bolt tension and torque (From "Recommended Practices for Blowout Prevention Equipment Systems for Drilling Wells" API RP 6A, 16th Edition. October 1, 1989.)

Bolt Size	40,000 psi Stress		52,500 psi Stress	
	Bolt Tension lbf	Make-Up Torque ft-lbs	Bolt Tension lbf	Make-up Torque ft-lbs
1/2-13 UNC	5674	45	7448	59
5/8-11 UNC	9026	86	11846	113
3/4-10 UNC	13355	150	17528	196
7/8- 9 UNC	18482	239	24257	313
1 -8 UN	24229	361	31800	474
1 1/8-8 UN	31617	522	41497	686
1 1/4-8 UN	39987	726	52483	953
1 3/8-8 UN	49339	976	64757	1281
1 1/2-8 UN	59672	1277	78320	1676
1 5/8-8 UN	70988	1635	93171	2146
1 3/4-8 UN	83284	2054	109311	2695
1 7/8-8 UN	96563	2538	126739	3331
2 -8 UN	110824	3093	145456	4060
2 1/4-8 UN	142290	4435	186755	5821
2 1/2-8 UN	177683	6116	233209	8028
2 5/8-8 UN	196852	7097	258368	9314
2 3/4-8 UN	217003	8176	284817	10731
3 -8 UN	260250	10653	341578	13982
3 1/4-8 UN	307424	13585	403495	17830
3 3/4-8 UN	418554	20967	542790	27519
3 7/8-8 UN	442541	23157	580834	30393
4 -8 UN	472509	25494	620168	33461
4 1/2-8 UN	602200	36412	790388	47790
4 3/4-8 UN	672936	42879	883229	56289

Notes:
1. Based on API Spec 6A, Supplement 1, October 1991.
2. ASTM A193, Grade B7 bolts (min. yield 105,000 psi – ≤ 2.5", 95,000 psi – > 2.5") and ASTM A194, Grade 2H or 2HM heavy hex nuts.
3. Assumes bolt and nut are lubricated with API 5A2 thread compound.
4. It is recognized that applied torque to a nut member is only one of several ways to approximate tension and unit stress in a stud bolt. Tabulated values are presented for convenience and guidance only.

TABLE 1-8 • Comparison of make-up torque with various lubricants (Courtesy of Shaffer, Inc.)

Lubricant	Coeff. of Friction	% of Effort to Friction	% of Effort to Tension	Relative Torque, ft-lb, 1 3/8" Bolt & 1/2 Min. Yield
Light machine oil, as shipped	0.15	92.5%	7.5%	1,445
Oily machined surfaces, 4140 steel	0.13	91.7%	8.3%	1,290
Graphite with petrolatum	0.10	89.2%	10.8%	998
Tool joint compound	0.08	86.9%	13.1%	820
Select-a-torq. 503 Moly-Graph	0.07	84.7%	15.3%	702
API 5A2 lubricant	0.05	77.9%	22.1%	487

and torque requirements for various bolt sizes under different design conditions.

The lubricant applied to the bolt is very important. An analysis of the screw jack formula shows a major contributing factor to required torque is friction. Thread friction is dependent on lubricant type used on the threads. Table 1-8 compares the effect of six lubricants on the coefficient of friction on the threads. This table points out that the choice of lubricants will determine the difficulty incurred in attaining the proper bolt torque. Additional lubricants are available that offer even greater friction reduction (less make-up torque) than the API 5A2 thread lubricant. (Note: This discussion is directly applicable to drill pipe tool joint make-up and that the makeup torque is somewhat lubricant dependent.)

Clamped hub connections are preferred by some users due to the relative ease of operation and time savings when compared to bolted flanges. The preventers and valves have hub ends that mate with a wraparound clamp connector. An example is shown in Figure 1-32. When the mating hubs are aligned, the clamp is attached and tightened. This reduces the stabbing time and difficulty. The clamped hub connections are designed to withstand the same pressure as bolted flanges.

Preventers should never be welded together. Welding presents problems such as heat effects on the metallurgy and temporarily irreversible connections. Manufacturers of preventers report that many preventer failures are the result of an operator welding on the body without applying necessary heat treatment.

FIGURE 1-32 • Grayloc bolted clamp (From "Recommened Practices for Blowout Prevention Equipment Systems for Drilling Wells," API RP 6A, 16th Edition, October 1, 1989.)

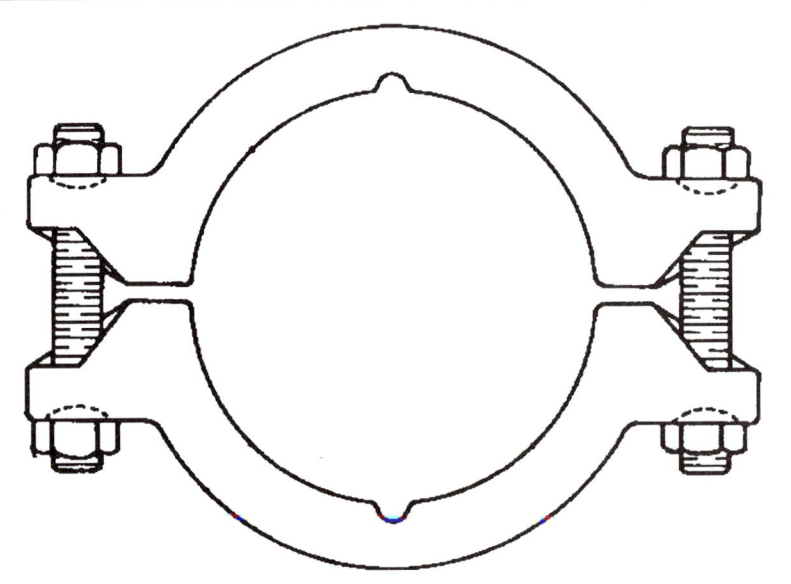

Seal Rings and Ring Grooves

The pressure seal at the flanges is effected by seal rings and seal ring grooves. The most common ring grooves are the API 6B for working pressures of 2,000 to 5,000 psi (14–34 MPa) and the API 6BX for working pressures of 5,000 to 20,000 psi (34–138 MPa). (See Figure 1-33.)

Seal rings for each type of ring groove are also presented in Figure 1-33. The "R" gaskets and "RX" pressure-energized gaskets are used with the 6B flange, and the "BX" is designed for the 6BX flange. Types "RX" and "BX" are pressure-energized rings, which will use well pressure to a degree to create a seal and do not rely completely on flange bolt torque that could result in a leak. Type "R" uses bolt torque exclusively to cause a seal between flanges. The pressure-energizing feature is advantageous since normal preventer vibrations will reduce bolt torque. (Periodic checking and tightening of bolts should be a part of blowout preventer maintenance procedures.)

H_2S Trim

Drilling in H_2S environments causes not only personnel safety problems, but also causes adverse effects on the equipment used in well control. Standard metals with a tensile strength greater than 80,000 psi (550 MPa) are subject to hydrogen embrittlement and sulfide stress cracking, which

Well Control Equipment 45

FIGURE 1-33 • API seal rings and grooves (From "Recommened Practices for Blowout Prevention Equipment Systems for Drilling Wells," API RP 6A, 16th Edition, October 1, 1989.)

API TYPE R RING-JOINT GASKETS
(For Use in 6B Flanges)

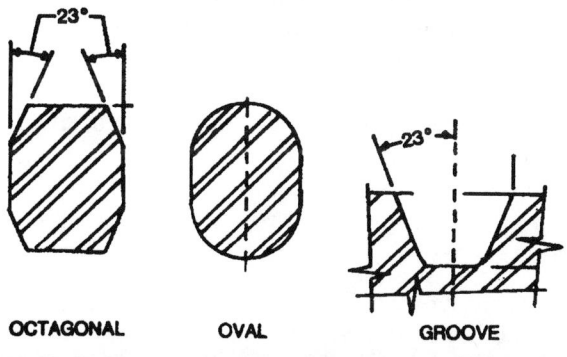

API TYPE RX PRESSURE ENERGIZED RING-JOINT GASKETS
(For use in 6B flanges and segmented flanges)

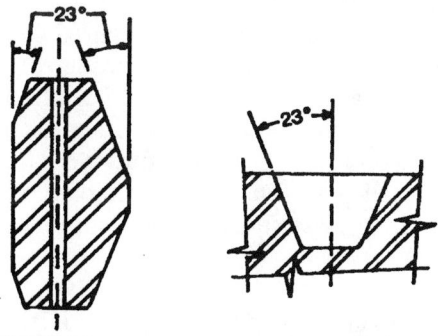

API TYPE BX PRESSURE ENERGIZED RING-JOINT GASKETS
(For use in 6BX flanges)

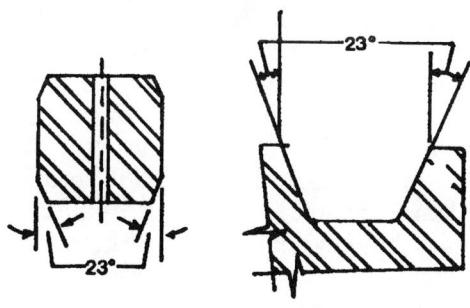

is a result of H_2S exposure. For this reason, preventers must be used that have a tensile strength less than 80,000 psi, or a hardness of 22 or less on the Rockwell "C" scale. To achieve this prerequisite, it may be necessary to use larger amounts of softer (lower yield) metal in order to contain the required pressures of well control service.

Some equipment, as noted, is serviceable in H_2S environments without any alterations. Some components require special trim. For complete details on any equipment used in H_2S service, the manufacturer should be consulted. It may be necessary in some applications to perform remedial work to the equipment such as the addition or replacement of special trim or heat treating to relieve stresses.

Wear Bushing

Routine daily activities with the drill pipe and kelly can cause excessive wear within the blowout preventer stack if certain precautions are not taken. To avoid this wear, bushings are sometimes installed in the wellhead. Wear bushings absorb the rotation and trip wear and can be replaced at a fraction of the cost required to rework blowout preventers or wellheads. (See Figure 1–34.)

The manner in which the bushing is secured should be given some consideration. Some bushings are lowered through the blowout preventer stack into the wellhead and remain unsecured, while others can be locked into place with external locking screws or other locking devices. This is an important consideration. Occasionally, wells have had blowouts because an unsecured wear bushing was blown up inside the BOPs making it impossible to close several of its components.

DRILL PIPE BLOWOUT PREVENTERS

The prevention of blowouts through the drill pipe is an important facet of well control. When a kick occurs, the influx fluid will generally enter the annulus due to the direction of drilling fluid flow during normal drilling circulation. However, if the kick fluid does enter the drill pipe, the shut-in drill pipe pressures will be greater than normal kick conditions because of the vertical column of mud displaced by the relatively small volume of influx fluid. As a result, the selection and utilization of the drill pipe blowout prevention equipment is essential for proper kick control.

Several tools are available to contain drill pipe pressures during kicks. The primary tool is the kelly cock. When the kelly is not in use, drill string valves are necessary to control the pressures. These valves may be automatic or manually controlled valves, and they may be a permanent part of the drill string or installed when the kick occurs.

Well Control Equipment

FIGURE 1-34 • Wear bushing (Courtesy of Cooper Oil Tool)

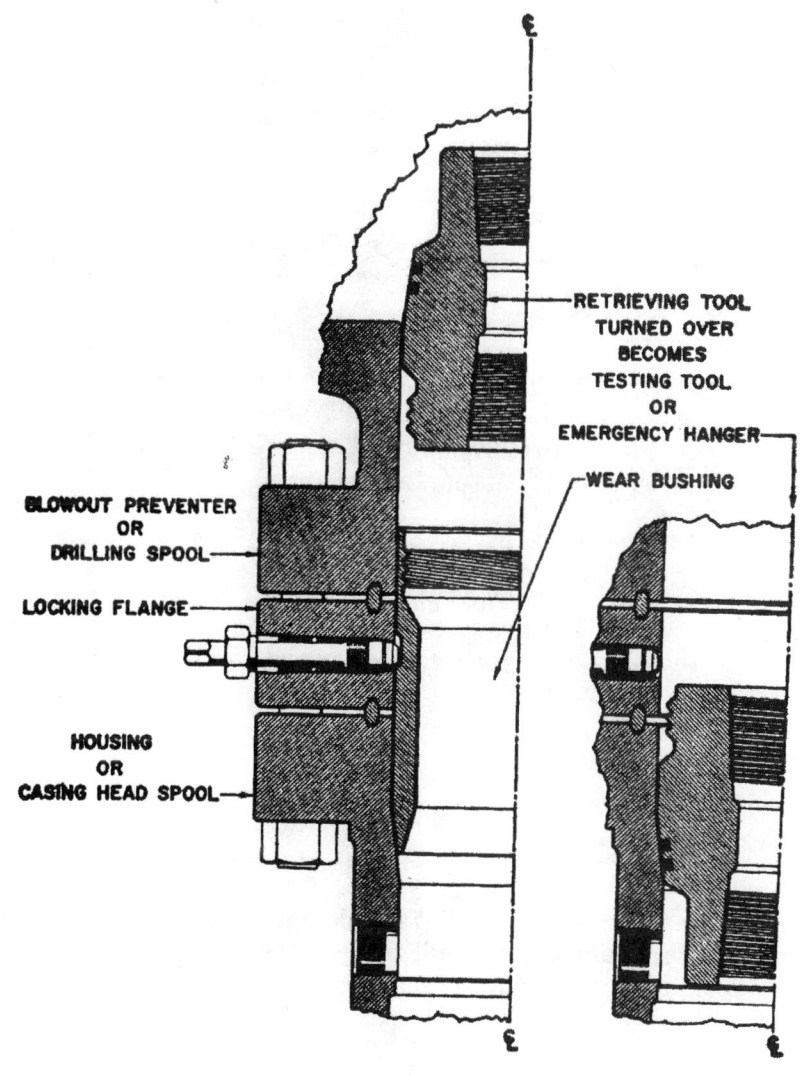

Kelly Cock

The kelly, through which rotary motion is applied to the drill string, is the tubular connection between the drill string and the surface drilling equipment. Valves are generally placed above and below the kelly to provide pressure protection for the kelly and all the surface equipment.

FIGURE 1-35 • Kelly cock (Courtesy of Cooper Oil Tool)

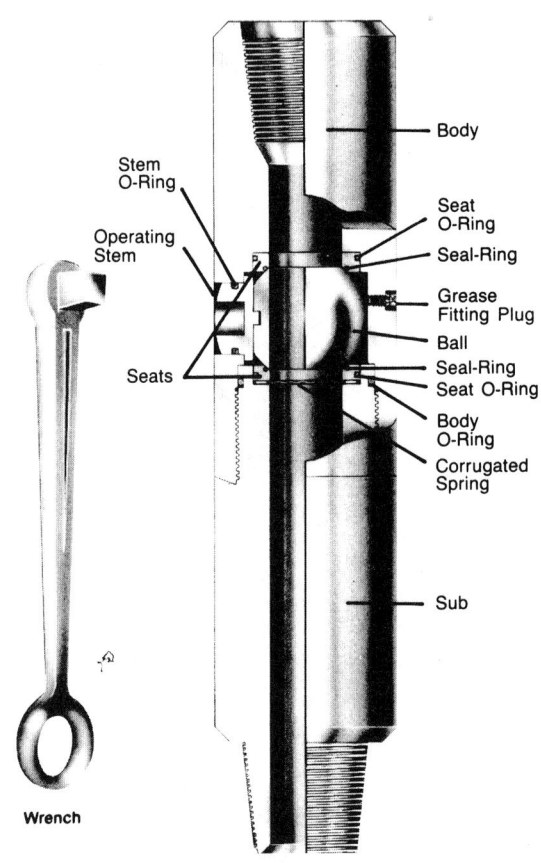

These valves, called "kelly cocks," should have a pressure rating consistent with the remainder of the drill string and be capable of sustaining the wear and hook load required of the hoisting equipment. (Figure 1-35.)

Automatic Valves

An automatic closure, or float valve, in the drill string will generally allow fluid movement down the drill pipe, but it will not allow flow in an upward direction when closed properly. The valve may be the flapper-type, a spring-loaded ball, or the dart-type, and it may be permanent or pump-down installed. Although the valve prevents drill pipe blowouts, it is often used primarily to minimize flow back during connections or to prevent bit plugging.

Well Control Equipment

FIGURE 1-36 • Regular and vented flapper (Courtesy of Hydril)

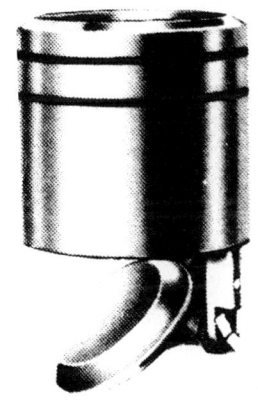
PLAIN FLAPPER

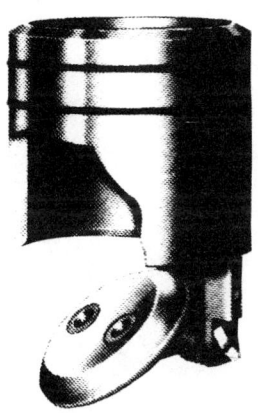
VENTED FLAPPER

Since a direct reading of static drill pipe pressures is not possible with a conventional float valve, alternative and more complicated pressure reading procedures must be implemented. This problem can be circumvented if a flapper-type valve is used with small built-in fluid ports to allow pressure build-up at the surface while still preventing a blowout. Figure 1-36 shows both the conventional and the vented flapper.

Safety Valves

A full-opening safety valve is usually installed on the drill pipe after a kick occurs when the kelly is not in use. The safety valve's advantage is it can be in the open position when stabbed on the drill pipe to minimize the effect of upward moving flow lifting the valve. The flow will pass through the valve during the stabbing and make-up, then the valve can be closed. (See Figure 1-37.)

Some types of automatic valves can be locked in the open position to achieve this stabbing. The valve can be closed after make-up.

The manually operated safety valve possesses one feature that makes it advantageous over the automatic valve in certain applications. When in the open position, the manual valve has a non-obstructed orifice. The automatic valve locked in the open position has the sealing mechanism (flapper, ball, or dart) as an obstruction. The manual valve can be opened to allow passage of wireline tools with a diameter smaller than the valve bore. This cannot be done with the automatic valve.

FIGURE 1-37 • Full opening safety valve (Courtesy of Hydril)

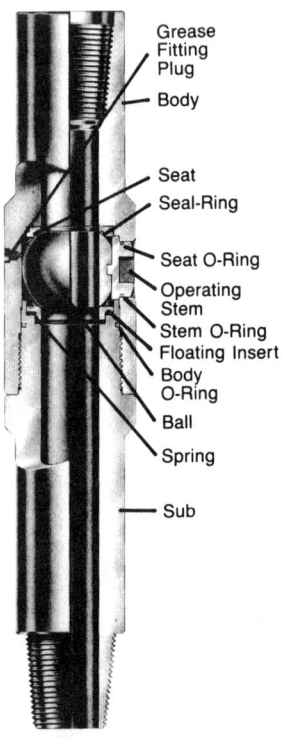

ABB Vetco Gray manufactures a quick-coupling attachment for emergency drill pipe closure. (See Figure 1-38.) The coupling is designed to drop over the pipe with an open valve attached and automatically latch under and seal off around the drill pipe tool joint. After the pressure is bled off, the coupling is released by depressing a ring, or individual dogs, depending on the type of pipe. The coupling will accept either manual or automatic safety valves.

DOWNHOLE BLOWOUT PREVENTERS

Blowout preventers have been thought of as surface equipment designed to contain kicks. Downhole blowout preventers have been designed based on a different concept. The downhole preventer, which is generally an inflatable drilling packer, is designed to seal the annulus, contain the kick fluid below the packer, and allow kill mud to be circulated above the packer. Downhole BOPs were first developed in the early 1960s.

FIGURE 1-38 • ABB Vetco Gray Fast Shut-Off Coupling (Courtesy ABB Vetco Gray, Inc.)

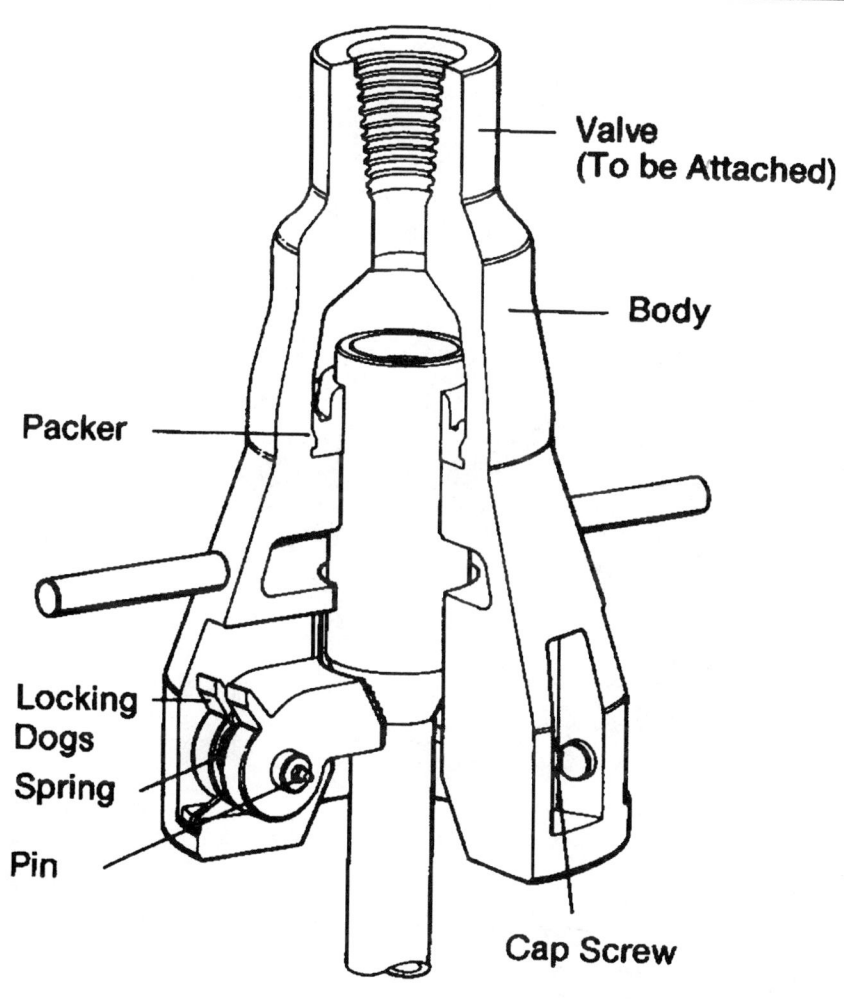

After heavy muds are circulated to the surface, the packer is released, and the kick fluid is circulated from the well. The advantage of this procedure is that the wellbore is not exposed to the full stresses of the kill pressures. Figure 1-39 illustrates the sequence of kill steps using a downhole preventer.

Disadvantages of using downhole preventers are the same basic problems associated with any drilling, or open hole, packer. Among these are the time required to inflate the packer after the kick has occurred, attaining a seat in the open hole, and packer malfunction due to prolonged wear. Downhole preventers have never received prominent use.

FIGURE 1-39 • Downhole packer kill procedure

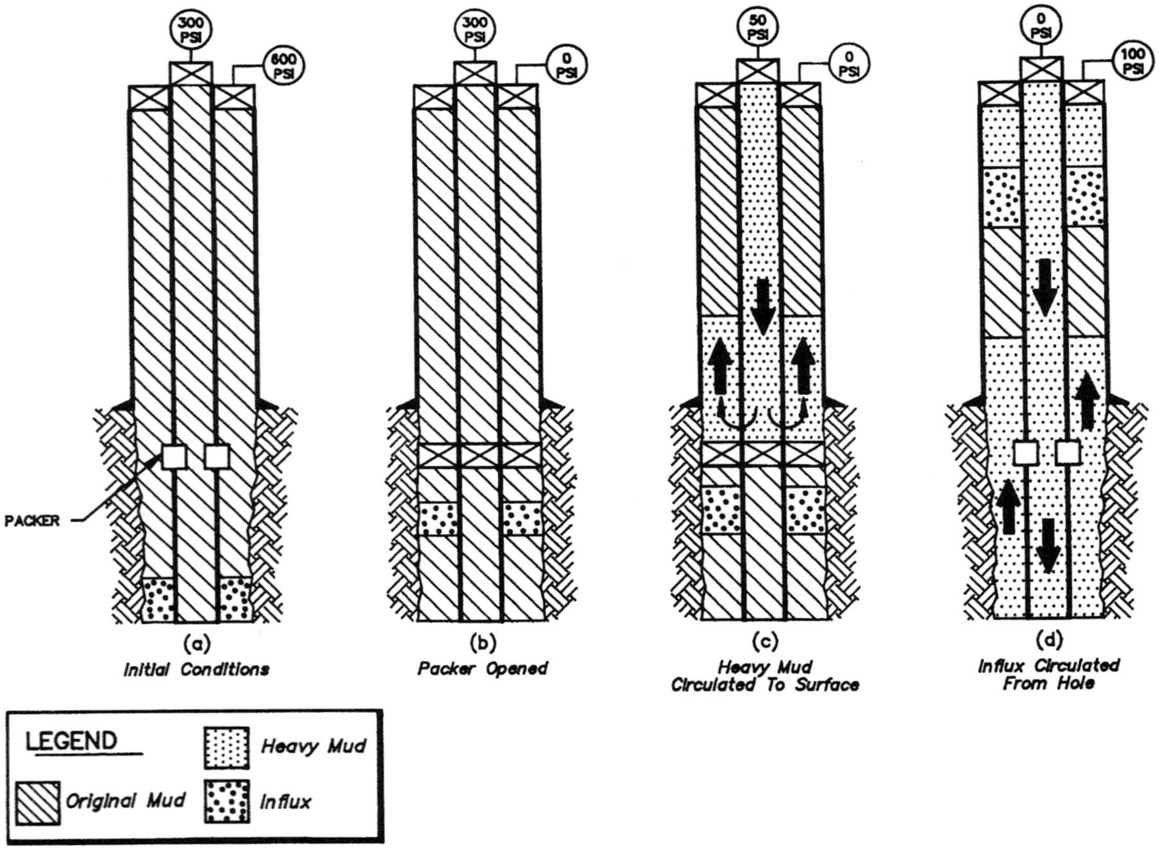

BLOWOUT PREVENTER STACK DESIGN

Considerations in designing an arrangement of annular/ram blowout preventers include pressure design, component selection and arrangement, subsea-related variations, and diverter systems.

Pressure Design

Several well-founded viewpoints relate to the pressure requirements that preventer stacks should meet. Some argue that the working pressure needs to be no greater than the burst strength of the exposed casing string, formation fracture pressure of the shallowest exposed zone, or a predetermined, maximum allowable surface casing pressure. It can be seen these guidelines may present serious problems when applied in severe well control situations.

The most common of the early guidelines is that the preventers need to be no stronger than the casing string to which they are attached. The fallacy is it assumes the casing string has been properly designed to withstand kick-imposed stresses. This quite often is not the case. It would follow that if the casing is improperly designed, the preventer pressure rating is also improperly designed.

The safest procedure for selecting preventer pressure ratings is to insure that the preventers can withstand the worst case conditions. These conditions occur when all drilling fluids have been evacuated from the annulus and only low density formation fluids (gas) remain. This procedure is illustrated in Example 1.4.

Example 1.4

A well is to be drilled to 10,600 ft and has an expected bottomhole pressure (BHP) equivalent to 10.5 ppg mud. What pressure rating should the preventers have? (Assume a gas density of 2.5 ppg.)

Solution

1. Determine the maximum anticipated formation pressure.

    Pressure = 0.052×10.5 ppg $\times 10,600$ ft = $5,787$ psi

2. Determine the gas hydrostatic pressure acting downward on the zone assuming that the mud is evacuated from the hole.

    Pressure = 0.052×2.5 ppg $\times 10,600$ ft = $1,378$ psi

3. The pressure imposed on the preventer would be the difference between the formation pressure and gas hydrostatic pressure.

$$5{,}787 \text{ psi} - 1{,}378 \text{ psi} = 4{,}409 \text{ psi}$$

The preventers must be able to withstand 4,409 psi (30 MPa). Using API designations (Table 1–9), a 5,000 psi/5M (34 MPa) working pressure rating system would be required.

Experience suggests this method should be used in shallow well situations where it is possible to achieve a complete mud evacuation. However, as the well depth increases, it becomes less likely a full mud evacuation will occur. A modification, based on a percentage of the maximum possible pressure load, could be used to determine the required preventer pressure rating. This percentage would depend on the operator's experiences in a particular drilling environment. Example 1.5 illustrates the technique modification for deep wells.

Example 1.5

A North Sea operator wishes to drill an expected bottomhole pressure of 16.0 ppg equivalent at 16,500 ft. The operator's experience dictates that an 80% design factor would account for unexpected eventualities. What pressure rating should the preventers have? (Assume a gas density of 2.0 ppg.)

Solution

1. BHP = 0.052 × 16.0 ppg × 16,500 ft = 13,728 psi
2. Gas pressure = 0.052 × 2.0 ppg × 16,500 ft

   = 1,716 psi
3. Resultant pressure = BHP − gas pressure

   = 13,728 psi − 1,716 psi

   = 12,012 psi
4. Working pressure = resultant pressure × 80%

   = 9,609 psi

Using the API designations in Table 1–9, 10,000 psi/10M (70 MPa) working pressure preventers would be necessary to control the well properly.

Well Control Equipment

TABLE 1-9 • Blowout preventer pressure rating designations* (From "Recommended Practices for Blowout Prevention Equipment Systems for Drilling Wells," 2nd Edition, RP 53, May 25, 1984.)

API Class	Working Pressure psi	Service Condition	Flange Size, in.	Minimum Vertical Bore, in.	Matching Casing Size, OD, in.
0.5 M	500	Diverter	29 1/2	29 1/2	30
2 M	2,000	Light duty	26 3/4	26 3/4	20
			20	21 3/4	20 and 18 5/8
			16	16 3/4	16
3 M	3,000	Low pressure	26 3/4	26 3/4	30
			20	20 3/4	30 and 18 5/8
			12	13 5/8	13 3/8 and 11 3/4
			10	11	10 3/4 and 9 5/8
			8	9	8 5/8 and 7 5/8
			6	7 1/16	7 5/8 to 4 1/2
5 M	5,000	Medium pressure	21 1/4	21 1/4	20
			18 3/4	18 3/4	18 5/8
			16 3/4	16 3/4	16
			13 5/8	13 5/8	13 3/8 and 11 3/4
			10	11	10 3/4 and 8 5/8
			6	7 1/16	7 5/8 to 4 1/2
10 M	10,000	High pressure	21 1/4	21 1/4	20
			18 3/4	18 3/4	18 5/8
			16 3/4	16 3/4	16
			13 5/8	13 5/8	13 3/8 and 11 3/4
			10	11	10 3/4 to 8 5/8
			6	7 1/16	7 5/8 to 4 1/2
			9	9	8 5/8
			7 1/16	7 1/16	7 5/8 to 4 1/2
15 M	15,000	Extreme pressure	13 5/8	13 5/8	13 3/8 and 11 3/4
			11	11	10 3/4 to 8 5/8
			9	9	8 5/8
			7 1/16	7 1/16	7 5/8 to 4 1/2
20 M	20,000	Extreme pressure	7 1/16	7 1/16	7 5/8 to 4 1/2

*API standard bores; however, other bores are used.

Component Design

After the pressure rating for the preventer has been selected, the component arrangement must be considered. The logic will be developed using four components on a surface (land) stack; namely, an annular, pipe rams, blind rams, and drilling spool. Logic for the minimum stack can be extended to any number of components.

Figure 1–40 shows a good arrangement for this four-member stack. If one component fails, there will always be a back-up system. This sequence of operations explains the design:

Step 1. The annular preventer is closed.

Step 2. If the annular fails while killing the well, the lower set of pipe rams are closed.

Step 3. Emergency procedures are exercised. Either the annular is changed, the blind rams are changed to pipe rams, or both annular and blind rams are changed.

Lower pipe rams are not for circulation purposes, but they are simply to close in the well while repairs to the upper members are made. Also, a kill or choke line should never be attached below the lowermost set of pipe rams (i.e., ram outlets or casing head valves).

The valves adjacent to the blowout preventer stack should be arranged based on the back-up principle. The innermost valve next to the stack is for emergency use only, while the next valve outward is for day-to-day actuation. The outer valve is generally a hydraulic-actuated valve for remote control during kick killing procedures.

In deepwater drilling, the blowout preventers are located on the sea floor. This necessitates installing certain built-in safety redundancies in the stack. Since component failure cannot be readily repaired, additional preventer elements must be installed to handle most eventualities.

A typical subsea stack (see Figure 1–41) illustrates the same logic developed in the previous section for a minimum stack by insuring a built-in, back-up system is available. Some back-up systems are shown. In this illustration, they are the two annular preventers, shear/blind rams as the top set of ram preventers to allow for rig emergency/rig departure, two choke/kill lines with the primary line on top and the secondary line on the bottom, dual valves on each choke/kill line, and extra pipe rams.

Many instances occur in shallow hole sections where it will not be possible to shut in a well to control a kick due to an insufficient fracture gradient. When this happens, a blowout must be diverted away from the rig using a diverter system as previously discussed. As soon as the kick is observed, the diverter line(s) is opened, the flow line is isolated, and the diverter unit (or annular preventer) is closed.

Well Control Equipment

FIGURE 1-40 • Basic surface BOP stack arrangement

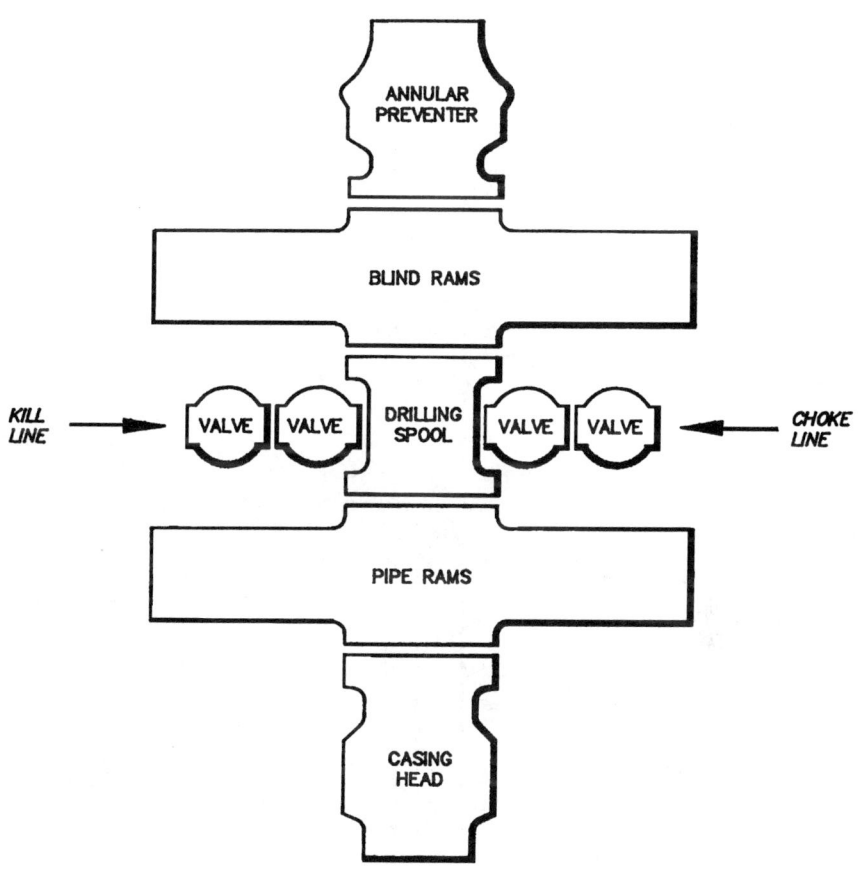

The arrangement shown in Figure 1-42 has several important features recommended for diverter systems. The control panel is designed so movement of a single control lever opens the diverter valve(s), closes the diverter unit, and isolates the flow line. Control lever movement in the opposite direction closes the valve(s), opens the unit, and opens access to the flow line. Diverter line options should take advantage of the wind direction.

The lines should be as large as possible with a suggested minimum ID of 14 in. Angles and bends should be avoided in the diverter lines to prevent plugging, erosion, and back pressure problems (i.e., the lines should be straight). The preventer may be a low pressure annular preventer or some type of diverter unit used to direct the flow into the lines. However, computer modeling and blowout experiences show a minimum recommended rating is 1,000 psi (7 MPa).

FIGURE 1-41 • Typical subsea stack

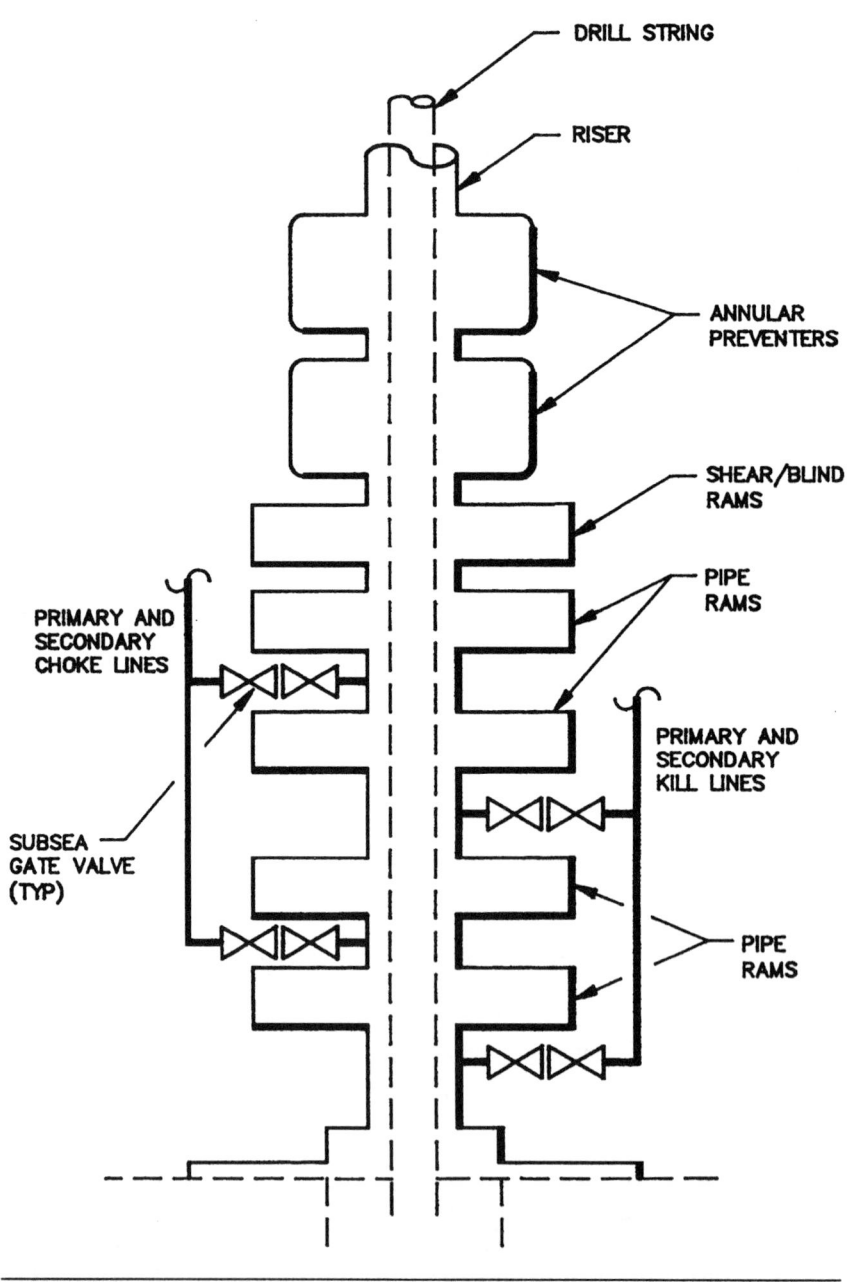

WELL CONTROL EQUIPMENT

FIGURE 1-42 • Typical diverter stack

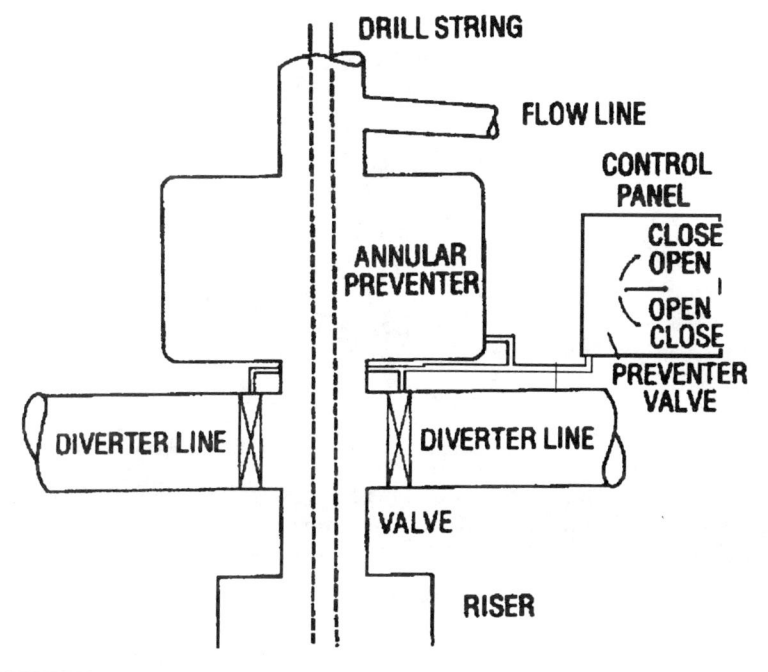

CHOKE DEVICES

A "choke" is used to control flow and pressure. The choke resistance creates a back pressure that controls formation pressure when a kick is circulated from the well.

Many choke types have been used for well control purposes. Among these are positive, or fixed-orifice size chokes, and adjustable chokes manufactured with rubber or steel trim. The primary type of choke used today is the steel, remote-controlled adjustable choke because of its greater durability over the rubber choke and its ability to change orifice sizes quickly.

Manually Adjustable Steel Chokes

The manually adjustable steel choke is an older type that is still used. This type creates a back pressure with a stem and beveled seat mechanism. Fluid is allowed to flow through the seat or orifice. As some alteration in the amount of back pressure is required, the stem is positioned in the seat to create a greater or lesser resistance to flow. Back pressure control is attained by the degree to which the stem is forced into the beveled seat or extracted from it. (See Figure 1-43.)

The main benefit of this choke is its simplicity. Problems with some models have been noted from field usage. The stem and seat mechanism on some models has a tendency toward turbulent erosion, thereby reducing its sealing ability. Shale cuttings build-up and choke plugging has been observed due to the inability of the choke to allow passage of the cuttings. Also, the placement of this choke in the manifold requires the operator to be removed from the main driller's console area during choke operation, which increases the difficulty of well control procedures. Manual chokes are commonly provided as back-up for the remote-controlled, adjustable choke.

Remote-Control, Adjustable Steel Chokes

The advent of the remote-control steel adjustable chokes has improved pressure control in most situations. The improved steel choke components can tolerate all types of kick fluids for long periods at high pressures, if necessary. Two common remote adjustable choke types are discussed.

A common choke type consists of two tungsten carbide plates with "half-moon" orifices that control fluid flow depending on the relative position of the orifices. The orifices are offset in the closed position and, as one plate is rotated with respect to the other, the orifices become aligned, which allows fluid flow through the choke.

Rotating Disc Choke Operating Features

1. The design allows for complete closure.
2. Operates hydraulically by means of a pump operated by rig air, by a manually-operated hand pump, or manually with an attached bar at the choke.
3. Variable choke speed control.
4. Automatic features as an option.
5. 10,000 psi (70 MPa) minimum working pressure rating– 20,000 psi (140 MPa) units available.
6. H_2S service model optional.

Another remote-control choke type utilizes a modified version of the stem and seat design (rod/gate and cylinder system) to develop the desired back pressure. Mud is circulated through the cylinder when the choke is open. (See Figure 1-44.) As the choke is closed, the rod/gate is forced hydraulically into the cylinder to create an obstruction and resistance to flow.

FIGURE 1-43 • Cooper H2 adjustable choke (Courtesy of Cooper Oil Tool)

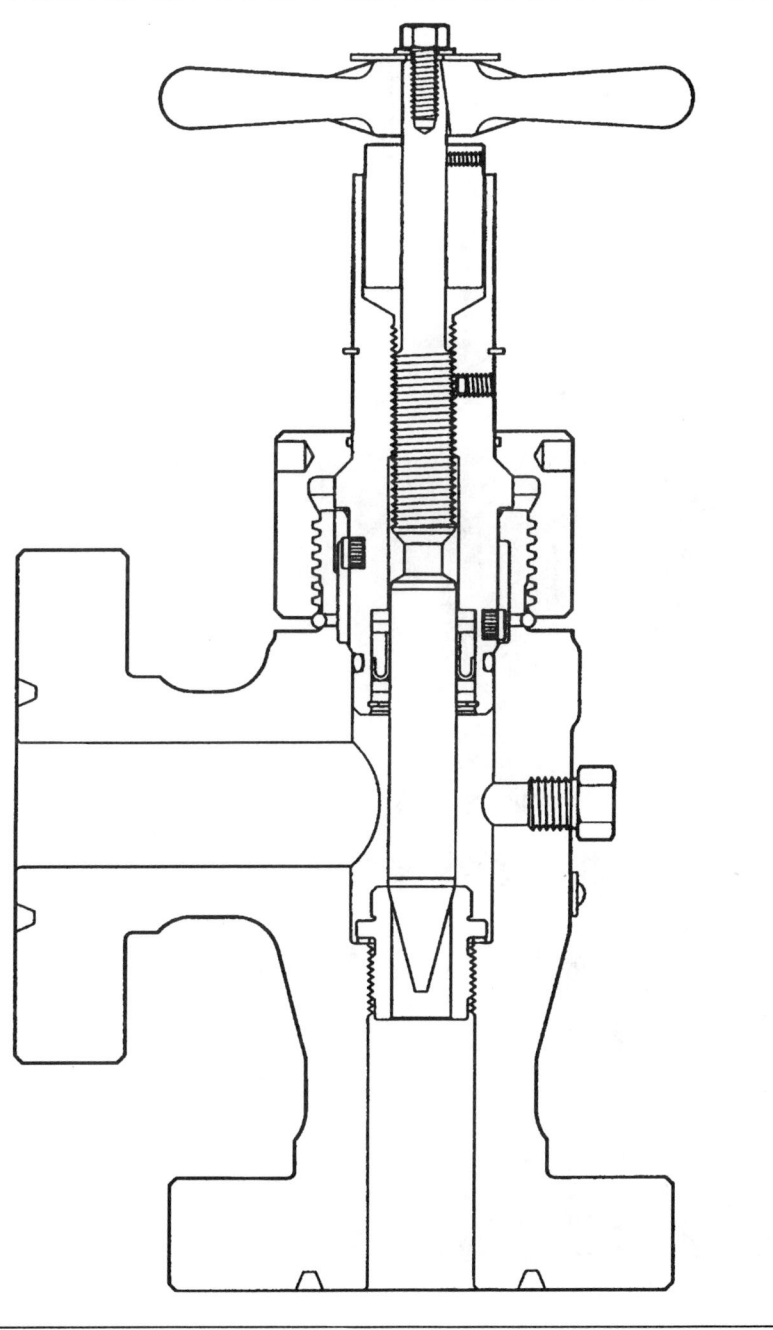

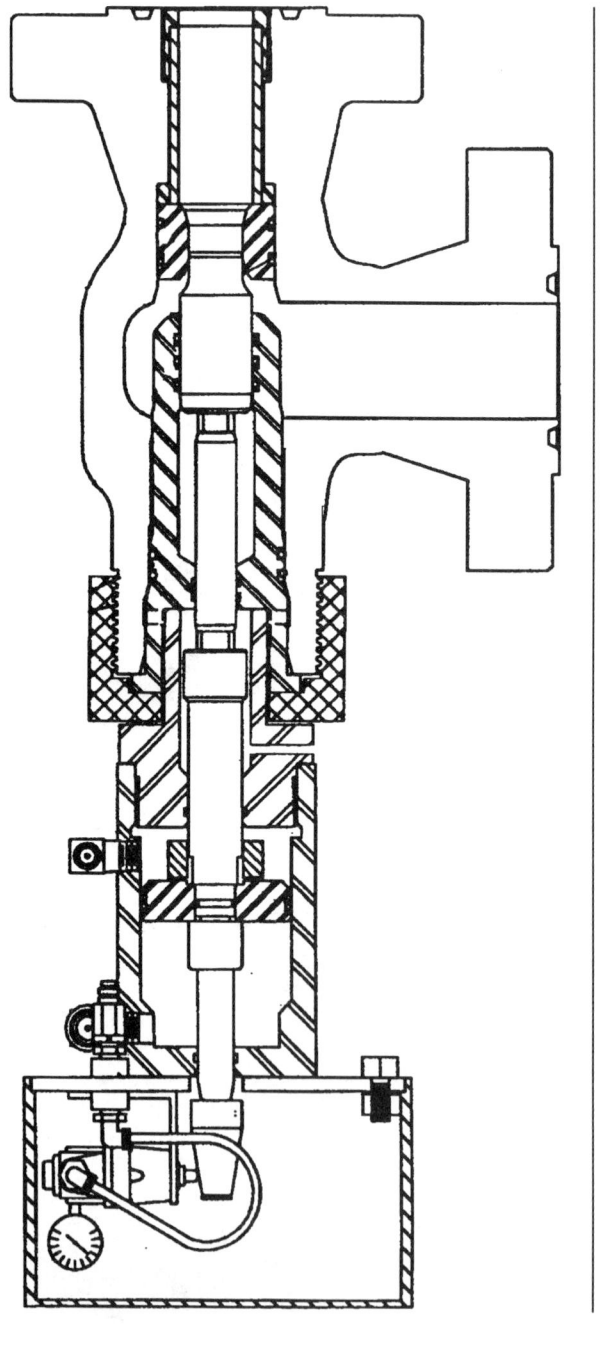

FIGURE 1–44 • Remote-control adjustable choke (Coursty of Power Chokes, Inc.)

Well Control Equipment

Stem and Seat Choke Operating Features

1. Some models may not allow complete closure, resulting in an inability to pressure test with water (except with new trim). The choke will generally seal, however, when mud is used. (Note: Chokes are not normally depended on for complete shut-off. Typically, gate or plug valves are used for that purpose.)
2. Operated hydraulically by means of a pump and rig air, hydraulically with an attached nitrogen bottle, or manually with a hand operated pump.
3. Variable choke speed control.
4. 5,000 psi–20,000 psi (34–140 MPa) working pressure models.
5. Most models are suitable for H_2S service.
6. Stem and seat are made of tungsten carbide.
7. Stem and seat are reversible for double life.
8. "Maximum allowable choke manifold pressure" option is available on some models, which will automatically adjust the choke when the pressure exceeds a preset range.
9. The choke control panel can connect to two separate chokes and alternately operate either with a switch on the panel face.

The control panel for choke operation will generally contain the gauges and controls necessary to monitor the well during the kick killing operation. The panel should contain, as a minimum, accurate drill pipe and casing pressure gauges, cumulative pump stroke counter, and choke control lever (knob). (See Figure 1–45.)

Some optional items for inclusion on the panel are rig air supply monitor, variable choke speed control lever/knob, relative choke position indicator, pump stroke rate indicator, and automatic control switches. The panel should be installed where the operator can communicate with key personnel such as the driller, in close proximity to a BOP control station, and near adequate lighting to facilitate accurate pressure readings.

CHOKE MANIFOLDS

The **choke manifold** is an arrangement of valves, lines, and chokes to control the flow of mud and kick fluids from the annulus during the killing process. Design conditions include a variety of fluids such as mud, oil, water, or gas; high pressures; low upstream flow rates; high downstream velocities; and solids in the produced fluids such as sand, shale, or pipe protector rubbers.

The manifold should control pressures by using one of several chokes. It should be able to divert flow to one of several areas including a burning pit, reserve pit, mud pit, or overboard (if offshore), as applicable. The

FIGURE 1-45 • Single choke remote-control panel (Courtesy of Power Chokes, Inc.)

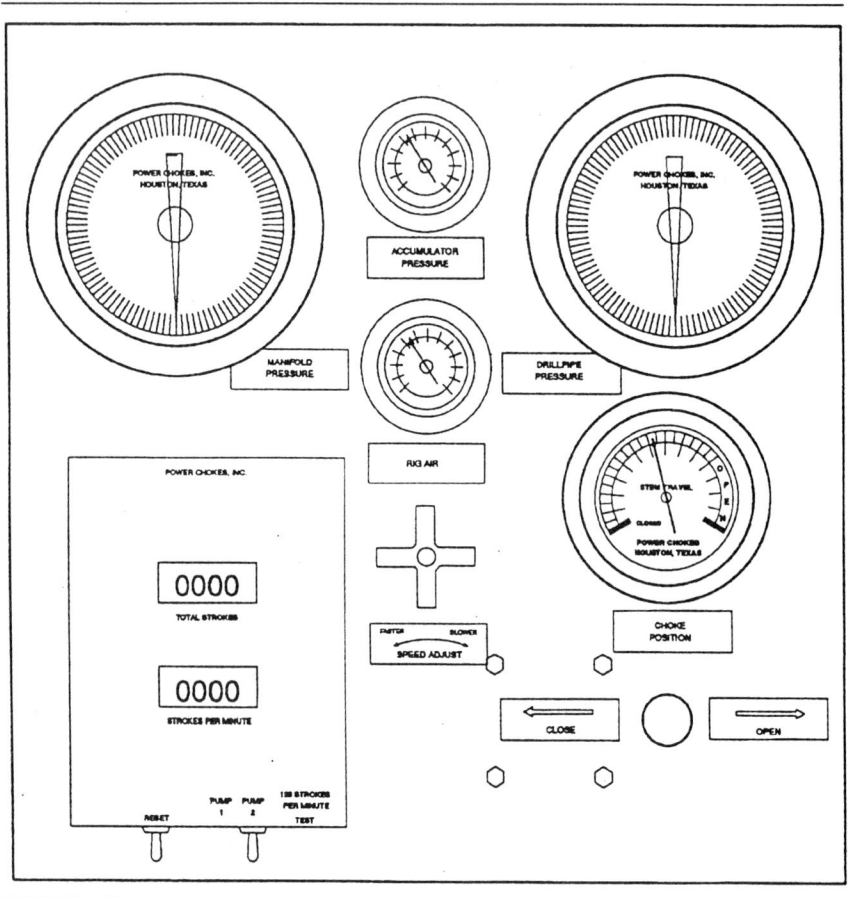

choke manifold components should have pressure ratings at least equal to the preventer stack, and they should meet all pressure testing specifications imposed on the preventers. It should be suitably anchored to prevent movement during the killing operation. The choke manifold should feature easy access to every component. All lines should be constructed as straight as possible. All lines and valves should have consistent bore sizes to minimize turbulent erosion.

Manifold Design

The principle applied to the design of the blowout preventer stack also applies in designing the choke manifold. The procedure is to insure that

Well Control Equipment

a back-up system is available if the primary component fails. Also, it is good practice to start drilling operations using the manifold necessary to reach the total depth to avoid installing a different manifold with each casing setting depth.

Figure 1-46 is a choke manifold suitable for many drilling and well control operations. Note that this design meets the requirements for choke manifolds. Buffer chambers are used at the downstream connections to act as hydraulic cushions and to minimize erosion. Two manually adjustable chokes have been provided due to high stem and seat erosion rates and possible cuttings packoff associated with these chokes. A direct line for gas from the preventer stack to the burn pit has been provided, if it becomes necessary to divert the well temporarily. (Note: This design does not constitute a true diverter system.) Although not shown, additional valved outlets are provided for pressure gauges and pressure sensors for the remote-control choke.

FIGURE 1-46 • Choke manifold (Courtesy of Cooper Oil Tool)

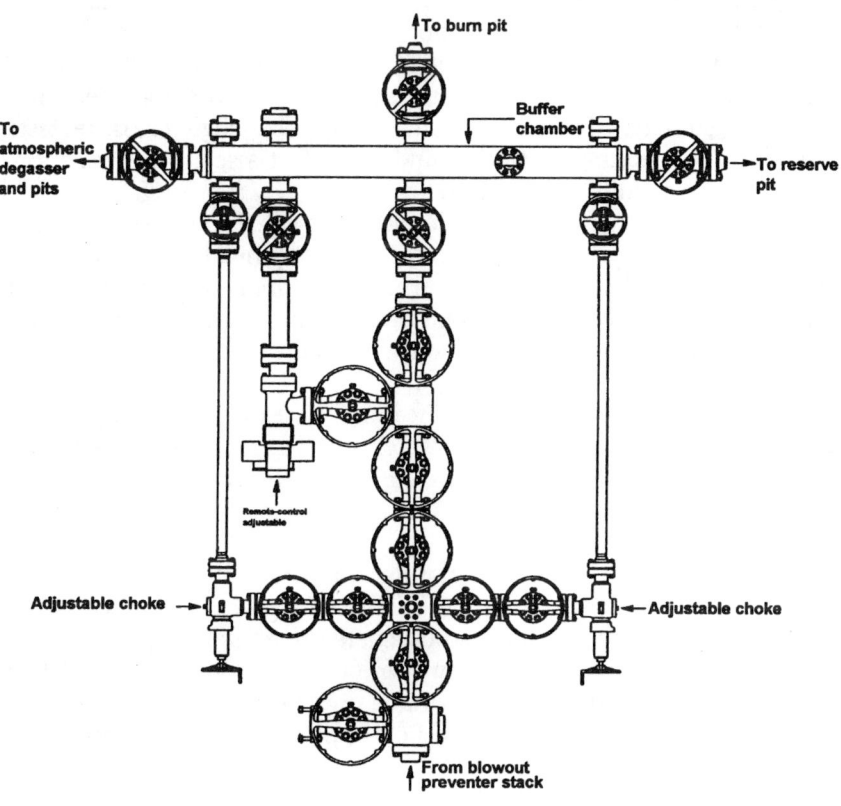

BLOWOUT PREVENTER TESTING

The blowout preventer stack must be pressure tested after assembly/installation to insure that it can control the designed pressures and periodically retested for confirmation of pressure integrity. Important considerations in blowout preventer testing are test fluids, pressures, testing equipment, procedures, and frequency of retesting.

Test Fluids

Clean water is the best test fluid because of availability. It will not plug small leaks as will mud. If high pressure gas wells are to be drilled, some operators test with an inert gas such as nitrogen. Oxygen or hydrocarbon gasses should never be used to pressure test a stack.

Test Pressures

A high and low pressure testing procedure should be employed. A low pressure test in the range of 100 to 300 psi (0.7–2.1 MPa) should be applied to the stack. Although the stack is generally washed and cleaned with water prior to testing, it is difficult to remove existing dried mud particles completely from a potential leak hole. High pressure tests applied to the mud may pack the mud and effect a seal, whereas a low pressure test may allow the leak to be detected.

The initial high pressure test should be directed to the lesser of the rated working pressure of the preventer stack, wellhead, or upper part of the casing string. The preventers' working pressure is the recommended option. The same pressure would be used to test all blowout prevention equipment (e.g., stack, kelly cocks, and manifold). The test should be held for at least three minutes.

Other high pressure tests (after the initial test) should be to a pressure equal to at least 70% of the preventer stack working pressure, but not more than the wellhead-working pressure or 70% of the casing internal yield pressure. The high test pressure should never be less than the maximum expected surface pressure.

Since the overall life of the annular element depends on imposed pressure and number of actuations, it is allowed an exception. After the initial test to its working pressure, subsequent test pressures can be to 50% of the rated working pressure. However, most operators want annulars tested to at least 70% of the rated working pressure. Reduced annular test pressures are justified because actual field applications generally use rams when kick pressures approach test pressures.

Well Control Equipment

Testing Equipment

The pumps used to generate pressures for preventer testing may be any type that is capable of attaining the desired pressures. However, since most test pressures and volumes will be out of the range of rig pumps, a smaller high pressure pump must be used. In many applications, a cementing-type, positive displacement pump is suitable if it is convenient. If a cementing-type pump is not available, several service companies offer preventer testing and provide a small, high pressure, low volume reciprocating pump, manifold, and chart recorder. (See Figures 1–47 and 1–48.)

FIGURE 1–47 • Koomey™ test unit (Courtesy of Shaffer, Inc.)

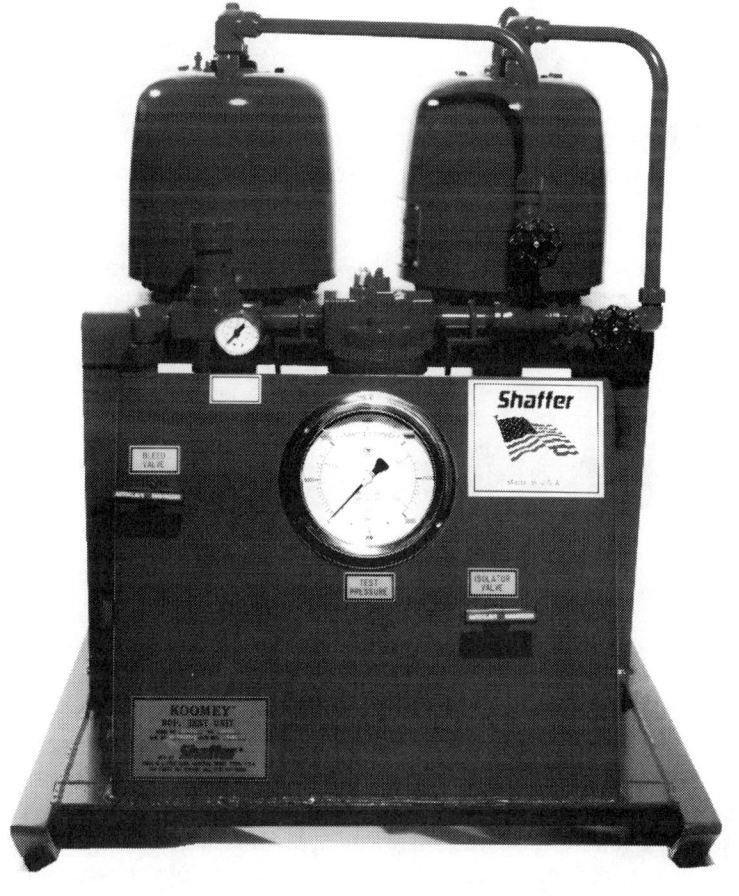

FIGURE 1-48 • Chart recorder (Courtesy of Shaffer, Inc.)

It is generally not desirable to expose the casing and open hole sections to the test pressures used on the preventers. A test plug must be set in the bottom of the preventers or in the wellhead to prevent this occurrence. The plugs most commonly used are the wellhead, or boll-weevil plug, and the cup-type plug.

The boll-weevil (wellhead) plug is designed to seat in the wellhead, and each will generally seal in only one type of head. (See Figure 1-49.) The plug is lowered into the head with a joint of drill pipe or a special test joint to test the pipe rams and annular preventer. The pipe is removed, leaving the plug resting on the head to test the blind rams.

FIGURE 1-49 • Boll-weevil test plug (Courtesy of Cooper Oil Tool)

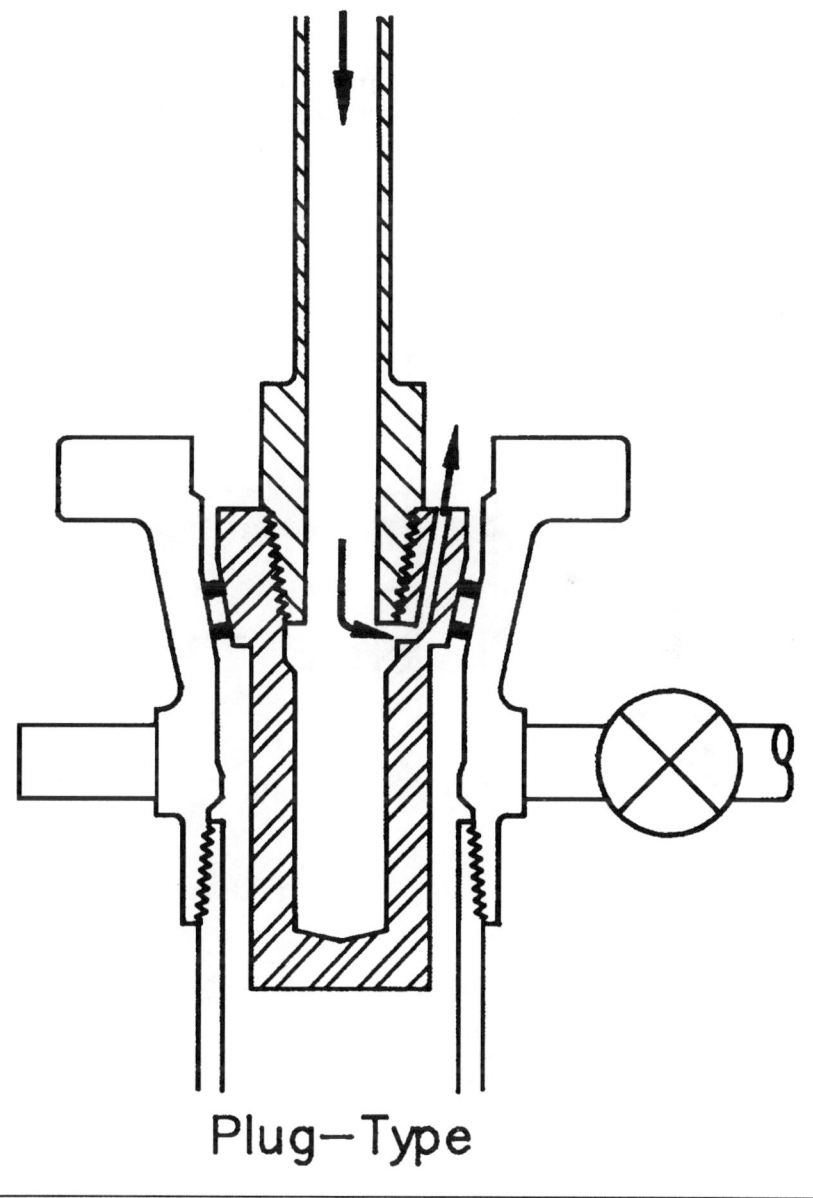

Plug–Type

If a boll-weevil plug is used, care must be taken that the plug is designed for the existing wellhead. Wellheads with the same basic dimensions often require different plugs. As an example, one manufacturer of 7 in. (178 mm) wellheads has 7 different test plugs for this size of head.

Cup-type testers (see Figures 1–50 and 1–51) are more universal because they effect a seat in the casing and not the head. Although the cup

FIGURE 1–50 • Cup tester (Courtesy of Cooper Oil Tool)

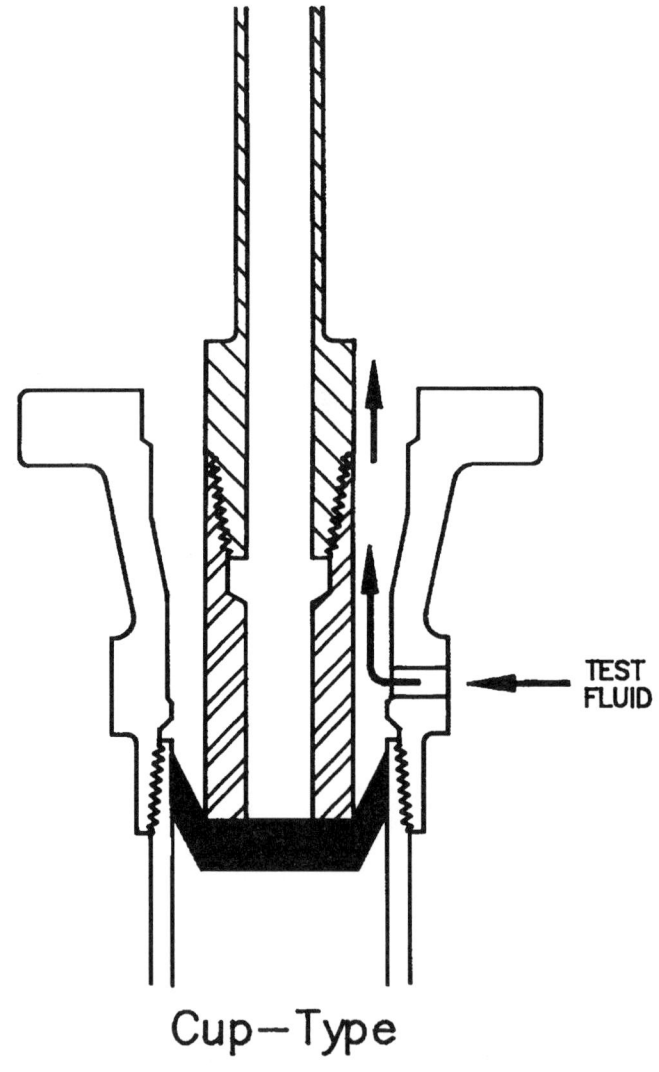

Cup–Type

Well Control Equipment

FIGURE 1-51 • Cooper type "F" cup tester (Courtesy of Cooper Oil Tool)

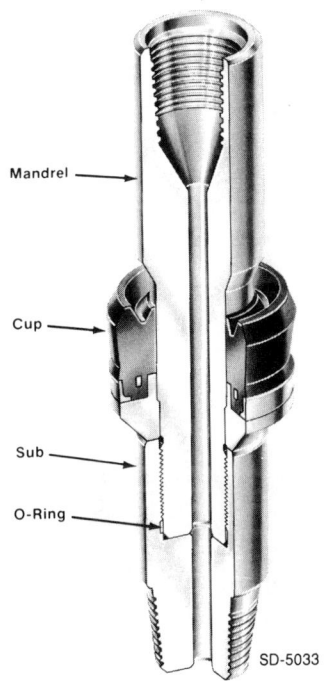

can be placed at any point in the casing, pressure testing specialists recommend positioning it opposite the slips in the casing spool or head.

Since the cup is not supported by the wellhead, the force (cup area × pressure) created by the pressure test must be supported by the drill pipe or test joint. This will often limit the use of drill pipe for testing because its yield strength might be exceeded. The cup cannot be used to test the blind rams.

A combination plug is available that offers the advantages of both the cup and the boll-weevil plug. (See Figure 1-52.) The plug is supported by the head and allows testing of the blind rams while the cup creates a pressure seal.

Testing Procedures

Although the order of element testing may vary with companies, the basic procedures generally remain the same. A schematic representation that provides the order of testing for all BOP elements should be made available for use during testing. The schematics are useful because of their thorough, yet simple, presentation.

FIGURE 1-52 • Combination boll-weevil cup tester (Courtesy of Cooper Oil Tool)

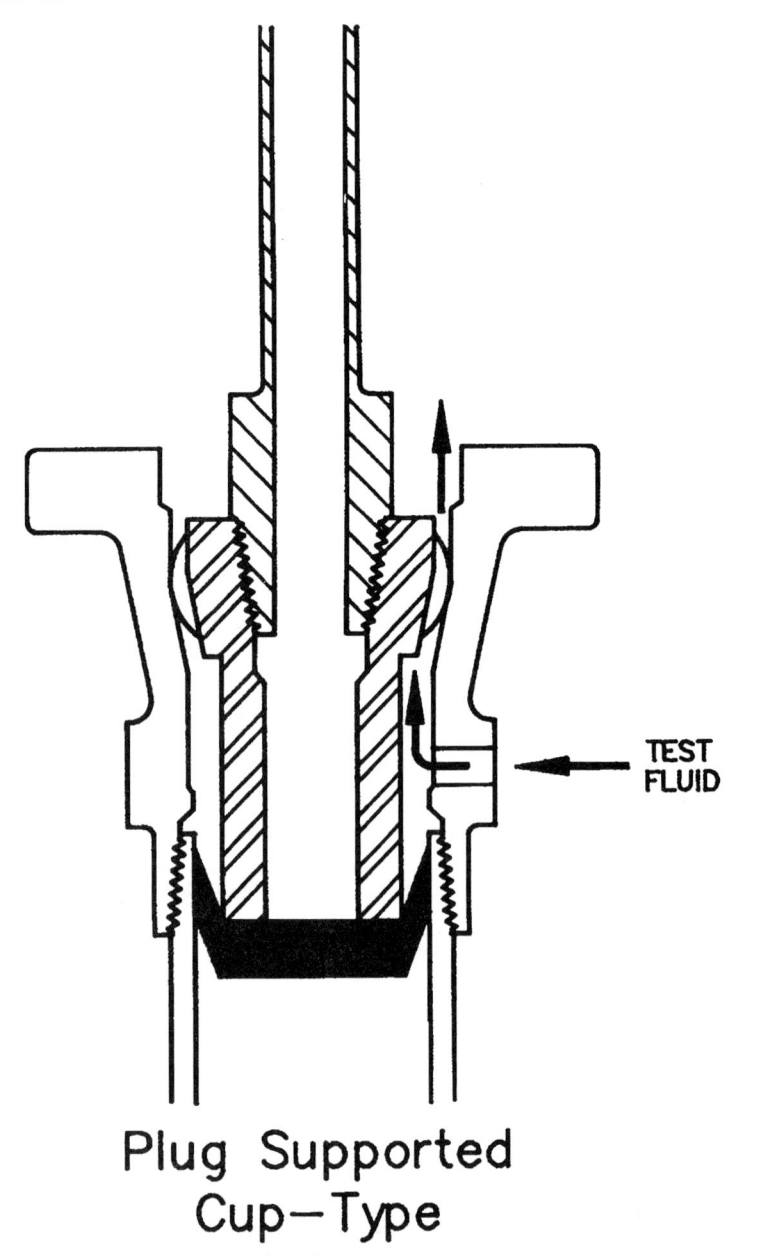

TABLE 1-10 • Checklist for preventer pressure testing

1. Make, size, and type of wellhead equipment.
2. If a reworked head is used, insure that no changes in type were made without appropriate changes in marking.
3. Working pressure rating of the wellhead and preventers.
4. Size of drill pipe or work string.
5. Type of tool joint.
6. If a mixed string is used, list both sizes.
7. Size and weight of casing.

Specialists have observed that problems during pressure testing are usually the result of a few basic causes. Some reasons include the test plug does not fit the wellhead, it will not seat properly because wellhead hanger holddown studs leak, or the test joint/plug threads leak. Packing glands around hold down studs can leak. Rams or ram bodies could be upside down. Ring gaskets or bonnet seal gaskets might be bad, or seal grooves may be damaged. These problems fall into one of three general categories: (1) mistakes in test preparation (wrong plug sizes), (2) equipment installation errors, and (3) lack of maintenance.

If a service company is used to test the preventer, the checklist in Table 1-10 will minimize the problems associated in test preparation.

Testing Frequency

The preventers should be tested on initial assembly and installation. They should be retested weekly, after setting and cementing a casing string, and after any repair that requires breaking a pressure connection. Many operators require a retest prior to entering a transition zone.

Subsea stacks should be tested before running the stack to minimize delays for repairs after installation on the subsea wellhead. Another test is run after the stack is on bottom.

After each test has been completed, it should be entered on the morning report form, and a test report should be completed and entered in the well history file. Figure 1-53 is a sample report form.

AUXILIARY EQUIPMENT

Closing Unit and Control Systems

The closing unit and control system provides closing energy to the blowout preventer stack. This is usually done with a hydraulic system designed and built to provide actuating pressure to the equipment in five

FIGURE 1-53 • Sample BOP test report form

Well:_____ Contractor:_____
BOP test date:_____ Well depth:_____
Last test:_____ Last casing set at:_____
Test fluid:_____

Equipment		Test pressure, psi	Test period, minutes	Remarks
Annular preventer:				
Pipe rams:	upper			
	lower			
Blind ram:				
Kill line:	valve 1			
	valve 2			
	valve 3			
Choke line	valve 1			
	valve 2			
	valve 3			
Manifold:	valve 1			
	valve 2			
	valve 3			
	valve 4			
Choke:	hydraulic			
	manual			
Kelly cock & kelly				
Floor safety valve				

Type Test: Initial_____ Weekly_____ Ram Change_____

seconds or less and to maintain the required pressures as desired. Annular preventer closure time should not exceed 30 seconds.

The working of the closing unit is a function of hydraulic oil stored in accumulators under compressed nitrogen. As hydraulic fluid is forced into an accumulator vessel (bottle) by a small volume output, high pressure pump, the nitrogen is compressed and stores potential energy. When the preventers are actuated, the pressured fluid is released and actuates the preventers. Pressure regulators are used to reduce the stored energy to the required operating pressure. Hydraulic pumps replenish the accumulators with the same amount of fluid as used to work the preventers. Figure 1-54 shows a closing unit that includes the accumulator bottles, pumps, controls, and a hydraulic fluid reservoir tank.

A precharge pressure is generally applied to the nitrogen to ensure that the fluid can be forced from the bottles when necessary. Precharges may range from 750 to 3,000 psi (5-21 MPa) with the desired precharged

Well Control Equipment 75

FIGURE 1-54 • Koomey™ closing unit and control system (Courtesy of Shaffer, Inc.)

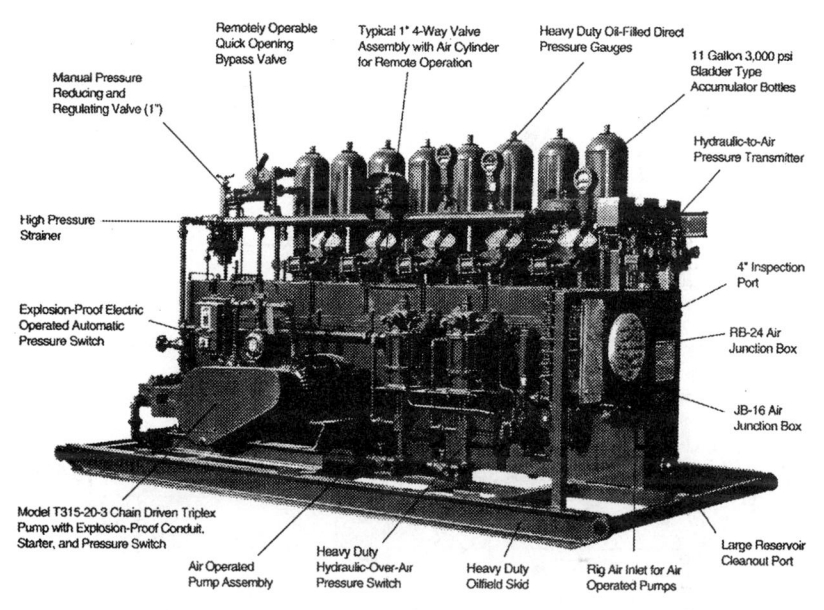

pressure being dependent on the service conditions during fluid drawdown.

Figure 1-55 is a drawdown curve for 3 different precharge pressures, and it is used to select the size and quantity of accumulator bottles with respect to preventer stack actuation pressure and volume requirements.

The closing unit must be equipped with two pressure regulating devices so different pressure levels can be supplied with the unit. As an example, an accumulator pressure of 3,000 psi (21 MPa) is recommended in some cases, but the pressure must be regulated to provide 1,500 psi (10 MPa) to the annular preventer, since this is the maximum recommended closing pressure for most annulars. Accordingly, other stack members may require a different operating pressure. (Recommended operating pressures are listed in the Appendices.) A bypass valve is built into the accumulator if it becomes necessary to use full pressure to close the preventers in emergency conditions.

Another purpose of this hydraulic system is to maintain constant pressures when stripping pipe through the annular preventer. As tool joints are stripped through the packing element, the accumulators must allow the excess fluid volume to move from the annular closing chamber. After the tool joint passes through the packing element, the accumulators

FIGURE 1-55 • Accumulator draw down curves (Courtesy of Shaffer, Inc.)

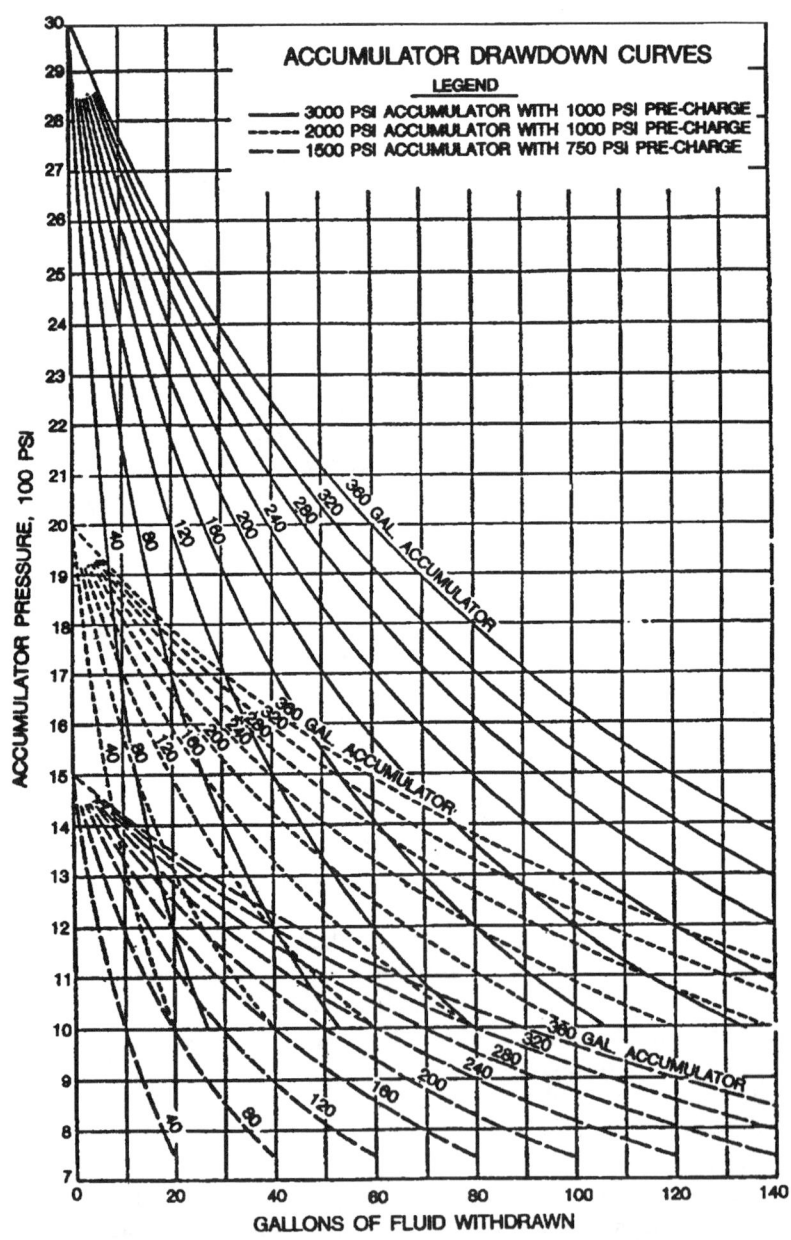

Well Control Equipment

must force additional fluid back into the annular preventer to maintain a constant pressure.

It is strongly recommended by the manufacturers to install a surge bottle on the hydraulic closing line as near as possible to the annular for stripping operations. A surge bottle in the circuit will sharply increase the useful life of the annular element if stripping is required.

Design Procedures

Closing unit and accumulator design specifications vary, but the system should have the ability to close a minimum of three members of the stack (one of which must be the annular) without having to recharge the accumulators.

One specification requires the accumulators to close all members of the stack without recharging and to have 50% of the original fluid remaining as a reserve after the functions have been completed and without the pumps on-line. A minimum final pressure of 200 psi (1.4 MPa) over the precharge (or 1,200 psi for 1,000 psi precharge) is required to ensure that the preventers remain closed. Example 1.6 shows the requirements for the listed stack components.

Example 1.6

What would be the minimum requirements for an accumulator if the following elements are in use? Use the preventer information in the Appendix.

Element	No.	Type	Size (in.)	Pressure rating (psi)
Annular	1	Hydril "GK"	11	10,000
Rams	2	Cameron "U"	11	10,000
Valves	4	Cameron "F"	3 (3 1/8 bore)	10,000

Solution

Part I. Volumetric requirement

Element	Gallons to Close
Annular (1)	25.10
Rams (2)	6.72
Valves (4)	1.12
	32.94 gal

The closing unit accumulator system should have a minimum volumetric capacity of $32.94 \times 2 = 65.88$ gal (or about 70 gallons).

Part II. Pressure requirement

Using the drawdown curves shown in Figure 1–55, several options are available. Two options are an 80 gal or greater accumulator with 3,000 psi charge, or a 120 gal or greater accumulator with a 2,000 psi charge.

It should be noted that the current API documents covering these systems (API RP 53 and API Spec 16A) are undergoing revisions. Also, requirements differ between U.S.A. specifications and regulatory agencies in other countries, such as Norway. Individual operators sometimes have additional requirements. The designer sizing the system should refer to the latest specification revisions to consider possible areas of operations and potential users.

Mud Mixing Equipment

During the kick killing process, it becomes necessary to increase the mud density by adding weight materials (such as barite) to the mud system. It is important to design the mud mixing system so mud density can be increased as quickly as possible to initiate kick killing procedures. Three main components of the mud system are the mixing pumps, the hopper, and the barite.

Mixing Pumps

The mud mixing pump is used to add weight material and chemicals to the mud stream. Although centrifugal pumps are generally used, small reciprocating pumps have performed efficiently. The centrifugal pump consists of an impeller, suction and discharge lines, and a power source. Mud enters the pump from the suction line, accelerates due to the centrifugal action of the impeller, and leaves the discharge line at high velocities.

The size of the centrifugal pump depends on its function. Figure 1–56 shows the required pump flow rates for a range of barite feed rates. For example, if the bulk barite system can feed barite at a rate of 15 sacks (100 lb, 45 kg, per sack) per minute, the centrifugal system should be designed to produce a 1,500 gpm (340 m^3/hr) flow rate to the hopper to increase the density by a design factor of 1.0 ppg (120 kg/m^3). If the centrifugal pumps are called on to do more than mix mud during a kick, either the pump sizes must be increased or more pumps must be used.

Mud Hoppers

The mud hopper is used to add weight material to the mud stream. The hopper consists of inlet and exit mud lines, a jet mixer, and one or more valves. Figure 1–57 illustrates a recommended design for a hopper. The dimensions shown are for optimum efficiency. Variations from this design may result in poor hopper performance.

FIGURE 1–56 • Required mud flow rates for barite addition (Courtesy of Oil and Gas Journal)

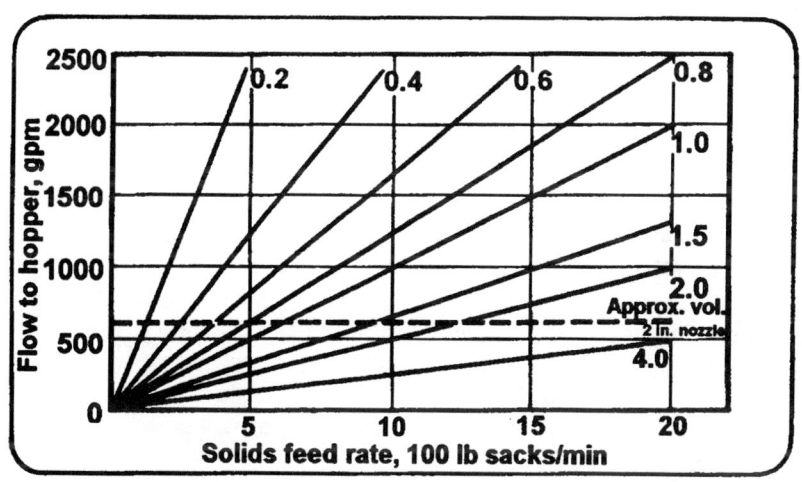

FIGURE 1–57 • Recommended mud hopper design (Courtesy of Oil and Gas Journal)

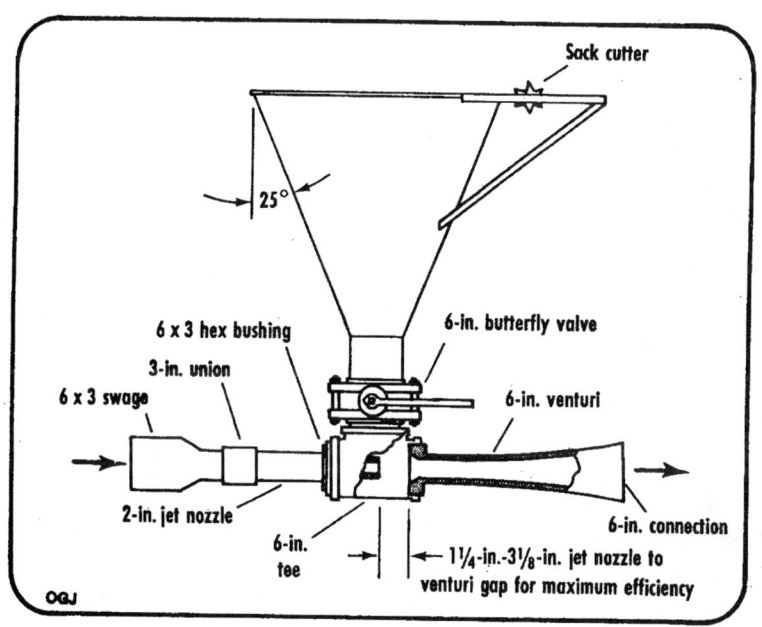

Bulk Mud Systems

Barite is marketed commercially in several forms. The particle size is usually a fine grade for drilling purposes, but it can be obtained in coarse grade, which is removed by the shale shaker during the first circulation. The coarse grade is used when it is not desirable to have solids remain in the mud system. Barite can be obtained in bulk quantities or in 50 or 100 lb (23–45 kg) sacks.

The sack weight material (barite, etc.) is not convenient in well control operations. Since it is usually stacked in 30 sack lots, time and space are consumed in individual sack handling and pallet movement.

Bulk weight material is preferred for well control because fewer crew members are necessary for mixing, and weight-up time is generally much shorter. Bulk tanks with a capacity of 500 to 1,500 sacks can be used with mud hopper equipment already attached. It is necessary to attach mud lines to the hoppers and fill the tanks. Large volumes of barite can be added quickly without cutting sacks or moving pallets.

Mud Pumping Equipment

Several types of rig pumps can be used during normal circulation and kick killing procedures. Two types are the double-acting duplex pump and the single-acting triplex pump. Other available pumps are generally based on the same principles as either the duplex or triplex pumps.

The double-acting duplex pump has two liners with four sets of suction and discharge valves. A set of valves is located on each end of the liner so fluid is pumped when the stroke is in the forward direction as well as in the reverse direction. The pump has the characteristic of moving large volumes of fluid at relatively low stroke rates.

The single-acting triplex pump has three liners. A set of valves is located on each liner. The pump generally strokes at faster rates than the duplex, but the output volume is comparable. Experience has shown the triplex is lighter weight, has better wear resistance, and is easier to maintain.

Pump Crippling

In many critical well control situations, it may be necessary to pump at low rates to avoid fracturing the formation and inducing an underground blowout. Unfortunately, the trend in rig pump design has been to increase the pump size which increases the minimum flow rate. As a result, it may be necessary to pull certain valves or rods that cripple the designed efficiency of the pump.

Well Control Equipment

FIGURE 1–58 • Recommended pump crippling guidelines

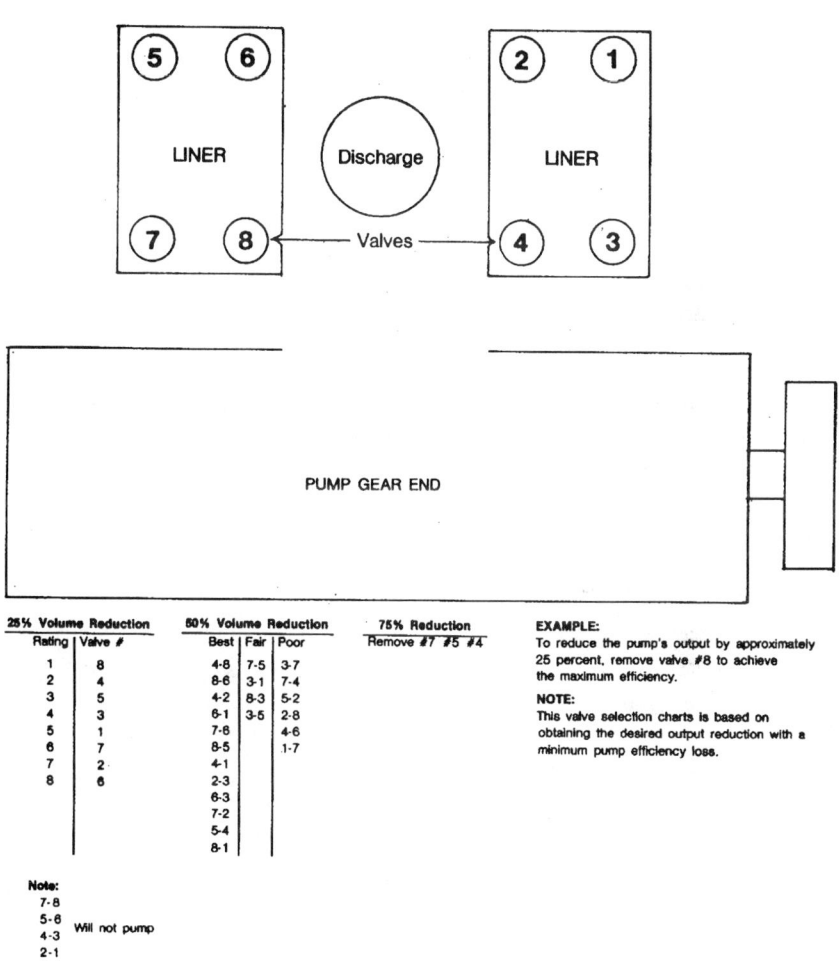

Figure 1–58 points out the valves to pull in order to gain approximate percentage reductions and still maintain maximum efficiency. The chart is only for the duplex, due to its number and arrangement of valves. Crippling of the triplex would be similar to the duplex. SCR systems in use today reduce the need for pump crippling in most cases.

Trip Tank

It is necessary to monitor the amount of mud that exits or enters the hole as the drill string is run in or out. The monitoring, or measurement, can be done either by using the rig pumps and calculating the number of strokes required to fill the hole, or by using a trip tank.

A trip tank is used to measure accurately within ± 1.0 barrel (0.2 m³). As the pipe is pulled from the hole, the mud from the tank is allowed to fill the hole as needed, which at the same time denotes the amount of mud being used. The mud can fill the hole by gravity feed or by a pump with a return line from the bell nipple to the tank. The advantage over pump stroke counting is that it is a continuous fill-up device and does not require as much of the driller's attention.

Degassers

Degassers remove air or gas entrained in the mud system to insure that the proper density mud is recirculated down the drill pipe. If the gas or air is not removed, the mud weight measured in the pits may be misleading. This will result in the addition of unnecessary amounts of weight material, thereby giving true mud densities down the hole greater than desired. Common degassers are the vacuum and atmospheric types.

Gas Busters

The atmospheric separator (i.e., "poor-boy degasser") is the first line of defense on gas removal in most well control operations. A typical unit schematic is shown in Figure 1–59. The mud and gas enters the top and is allowed to separate through gravity segregation. The unit is useful because of its ease of operation, maintenance, and construction, as well as its ability to remove large volumes of gas. It should be noted the vent line should be sufficiently long to insure that gas does not vent near the rig floor (i.e., it should be run to the top of the derrick).

Problems with this unit include: insufficient body size/construction, insufficient liquid seal, small diameter vent lines, or excessive gas flow rates through the degasser that perhaps should be flared at gas-to-surface conditions. Rig explosions have occurred due to plugging or poor-boy degasser failures and subsequent rig gasification.

Vacuum Degassers

The vacuum degasser consists of a vacuum generating tank which, in effect, pulls gas out of the mud. (See Figure 1–60.) Some degassers have a small pump to create a vacuum. Most vacuum degassers, regardless of type, have a minimum required mud throughput for efficient operation.

Several other types of degassers are available, such as the centrifugal spray-type or the pressurized separator. The centrifugal spray-type is easy to install and operate. The pressurized separator is perhaps the best

Well Control Equipment

FIGURE 1-59 • Schematic of typical atmospheric degasser

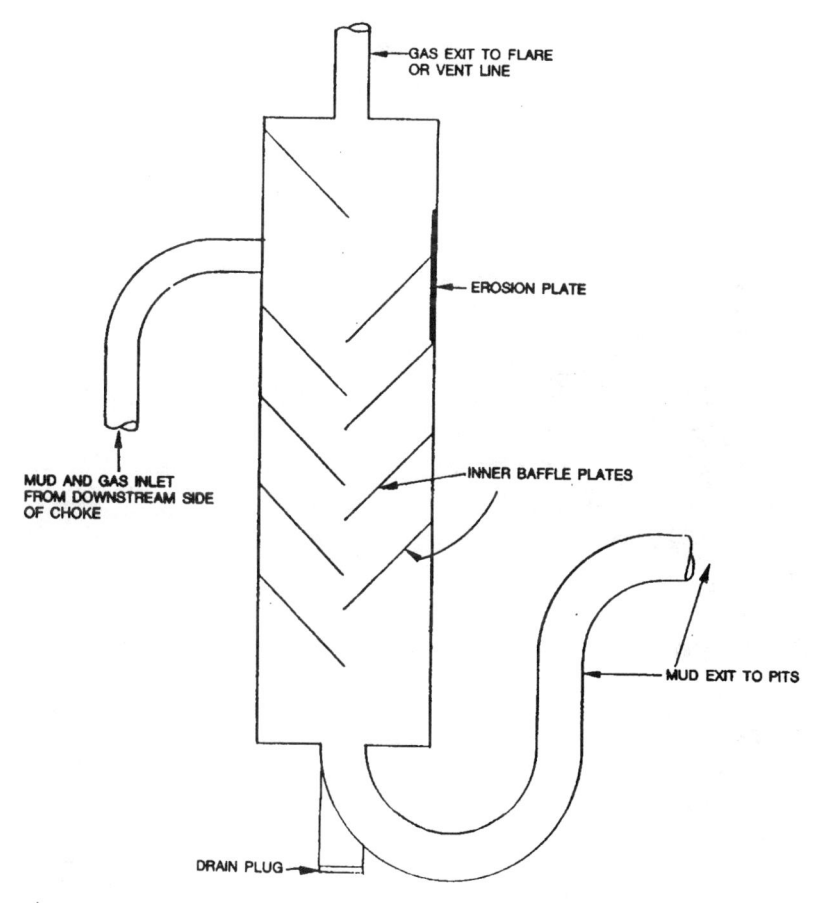

degassing tool for severe gas kick control and has a good service record under these conditions. The unit is complex to operate and maintain, and it is seldom used in current practice.

Mud Monitoring Equipment

Mud system monitoring is key to maintaining safe control of the well. The mud system gives warning signs and indications of kicks that can be used to reduce the severity of the kick by early detection and shut-in before a large influx is taken. If this system is properly monitored, other drilling problems such as lost circulation can be minimized.

FIGURE 1-60 • Welco degasser

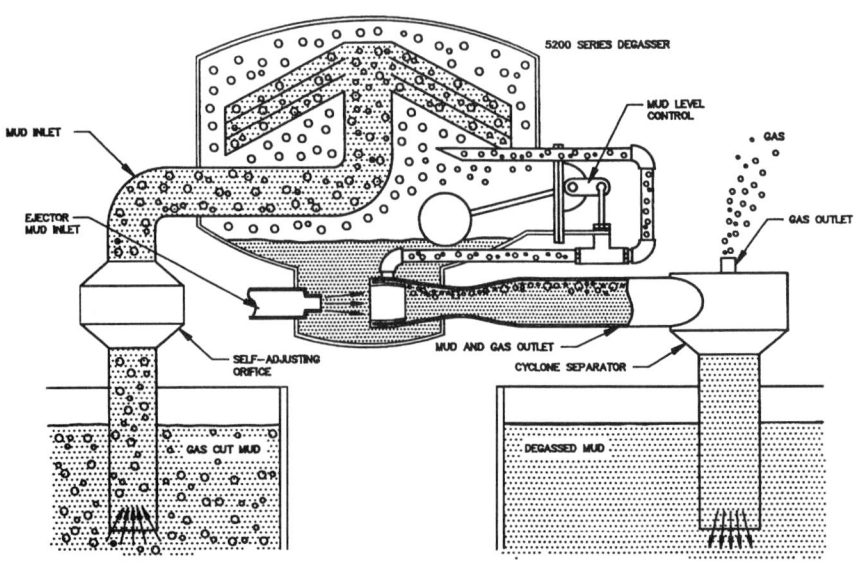

Flow Detectors

When a kick occurs, one primary warning sign will be an increased flow rate leaving the well. A flow monitor gauges mud flow rate and, if any abnormal changes occur, the monitor records the changes and sounds an alarm notifying the crew. The detector also indicates lost circulation if the flow rate decreases.

The most common type is a flapper, which is placed in the flow line. A tension spring is attached to the flapper and adjusted to the warning device. If the flow rate increases, the flapper changes position and creates a new spring tension recorded by the monitor. The reverse is true when lost circulation occurs.

Pump stroke counting is a viable procedure for filling the hole as the pipe is pulled. The flow monitor can be synchronized with the mud pumps to signal mud flowing out of the bell nipple. It can be set up to automatically shut down the pumps and record the pump strokes required to fill the hole.

Pit Monitors

Another key warning sign of a kick is an increased pit volume. As the formation fluid enters the hole, an equal volume of mud is displaced into the pits, which can be recorded by the proper type of detection equipment.

WELL CONTROL EQUIPMENT

The basis of most pit monitoring systems is a float level attached to a calibrated recorder. In many operations, especially floating drilling, the recorder should have a pit volume totalizer (PVT) feature that compensates for pit level changes from vessel heave and roll.

Gas Detectors

Several gas detectors are available that function on different principles. They all generally report the gas content as units of gas in the mud stream. When a certain level of gas has been sensed, an alarm sounds or a light signals the crew. The disadvantages of gas detectors are maintenance problems, the general inability to function in large concentrations of gas, and a misleading nature in kick detection.

PROBLEMS

(*Author's Note:* At the end of each chapter, a set of problems is given similar to those contained within the chapter. Several problems are similar to each example problem.)

Problems denoted with an asterisk (*) are advanced in content and designed to develop a more complete understanding of certain principles. These problems are for more advanced students and should not be assigned routinely to the beginner.

1.1 Calculate hydrostatic pressures for each of the following systems:

    a. 13,500 ft of 14.0 ppg mud

    b. 8,600 ft of 9.0 ppg salt water

    c. 17,000 ft of 18.5 ppg mud

1.2 A well is 15,000 ft deep. It contains 7,500 ft of 15.0 ppg mud and 7,500 ft of 16.0 ppg mud. What are the hydrostatic pressures for each section, and what is the total hydrostatic pressure at 15,000 feet?

1.3 A typical kick situation has developed the following arrangement of fluids in the annulus:

    a. 2,500 ft of 12.0 ppg mud

    b. 2,500 ft of 8.6 ppg saltwater

    c. 3,500 ft of 12.0 ppg mud

    d. 4,000 ft of 13.1 ppg mud

What is the hydrostatic pressure of each interval, and what is the total pressure exerted at the bottom of the hole?

1.4* An equivalent mud weight is often used to convert a combination of pressures into mud weight units of pounds per gallon to compare equivalent systems. The equivalent mud weight (E.M.W.) formula is derived from the hydrostatic pressure equation:

$$\text{E.M.W.} = \frac{\text{Total pressures}}{0.052 \times \text{depth}} \text{ or } \frac{\text{Total pressures} \times 19.23}{\text{depth}}$$

Using the solution from Problem 1.2, what is the equivalent mud weight at 15,000 ft?

1.5* What is the equivalent mud weight of the system developed in problem 1.3 at 2,500 ft? At 5,000 ft? At 8,500? At 12,500 ft?

1.6* A kick situation has developed that has yielded 500 psi on the annulus pressure gauge. The annulus contains 8,000 ft of 10.0 ppg mud above 1,000 ft of 9.0 ppg saltwater. What is the equivalent mud weight at 8,000 ft? At 9,000 ft?

1.7 If the active mud system of the Louisiana Producer No. 14 contains 1,260 bbl of 12.5 ppg mud, what is the number of 100 lb barite sacks necessary to increase the mud weight to 13.5 ppg? Number of tons?

1.8 In problem 1.7, if the mud weight was increased from 17.5 ppg to 18.5 ppg, would the requirements be the same? If not, how much would be required in 100 lb sacks?

1.9 The AMSCO Oil Co. is drilling at 12,675 ft with 13.2 ppg mud. It becomes necessary to increase the bottomhole hydrostatic pressure by 450 psi. What mud weight is required to achieve this increase? If the active mud volume is 975 bbl, how much barite will be required?

1.10* In a special blowout situation, the Dry Hole Oil and Gas Co. Wildcat No. 182 must develop a hydrostatic pressure of 7,400 psi over a 5,000 ft interval. What mud weight will be required to achieve this pressure? Assuming that a galena mud must be used, what volume in tons of galena is required if the mud system presently contains 460 bbl of 14.0 ppg mud. (Hint: See Equation 1.4.)

1.11* The mud density required to kill a particular underground blowout was 26.0 ppg. A total of 500 bbl of 18.0 ppg mud was used as the base fluid. How many tons of galena would be required to weight the mud?

Well Control Equipment

1.12* If the hydrostatic pressure must be increased by 700 psi in a well containing 12,600 ft of 11.5 ppg mud, how much barite in 100 lb sacks is required? If galena were used instead of barite, how many tons of galena would be necessary? The system volume is 1,200 bbl.

1.13 A shallow well is drilled to 5,500 ft in a normally pressured formation (9.0 ppg). In order to safely control the bottom-hole pressures, what should be the preventer pressure rating? (Assume a 2.5 ppg gas density in all problems in this chapter.)

1.14 An intermediate depth well is expected to encounter 10.1 ppg formation pressures at 11,050 ft. Will 3,000 psi working pressure preventers be sufficient if a 100% design factor is used? If not sufficient, what pressure rating should be used?

1.15 The Ocean Tide Exploration Co. will drill a development well to 14,600 ft with an expected formation pressure of 16.9 ppg at that depth. The operator's experience in the area dictates that an 80% design factor should be used in preventer selection. What should be the pressure rating of the preventers?

1.16 An operator elects to design an accumulator system to actuate all stack members with a 50% fluid reserve capacity and a final minimum pressure of 1,500 psi. What are the minimum volumes and initial charge pressures acceptable if the following stack is used? (Use the drawdown curves in the text.)

Element	No.	Type	Size (in.)	Pressure rating (psi)
Annular	1	Shaffer	6	5,000
Rams	3	Shaffer "LWS"	7 1/16	10,000
Valves	3	Cameron "F"	3 (3 1/8 Bore)	10,000

1.17 Using the same requirements as in Problem 1.16, what would be the minimum acceptable accumulator capacities and pressures for the following subsea stack?

Element	No.	Type	Size (in.)	Pressure rating (psi)
Annular	2	Hydril "GK"	11	10,000
Rams	5	Cameron "U"	11	10,000

1.18 A particular governing body rules that the accumulator system must be of sufficient size to close the three stack members with the largest fluid requirements and have a 50% reserve fluid capacity. Also, after activation the accumulator must have a minimum final pressure of

1,200 psi. If a well is drilled in the governed domain and utilizes the following stack, what are the size and initial pressure requirements for the accumulator? What must be minimum operating pressure on the annular preventer if the casing pressure is 1,550 psi?

Element	No.	Type	Size (in.)	Pressure rating (psi)
Annular	1	Regan "KFL"	13 5/8	5,000
Rams	3	Hydril "X"	11	10,000
Rams	1	Cameron "U"	11	10,000

SOLUTIONS

Note: All solutions are approximate depending on user's round off procedures.

1.1 a. 14.0; 9,828 psi

   b. 9.0; 4,025 psi

   c. 18.5; 16,354 psi

1.2 15.0; 5,850 psi

   <u>16.0; 6,240 psi</u>

   12,090 psi

1.3 a. 12.0; 1,560 psi

   b. 8.6; 1,118 psi

   c. 12.0; 2,184 psi

   d. 13.1; 2,725 psi

1.4 15.5 ppg

1.5 2,500 ft; 12.0 ppg

   5,000 ft; 10.3 ppg

   8,500 ft; 11.0 ppg

   12,500 ft; 11.7 ppg

1.6 8,000 ft; 11.2 ppg
   9,000 ft; 11.0 ppg

1.7 857 sacks

   42.9 s. tons

1.8 1,110 sacks

   55.5 s. tons

1.9   13.9 ppg
      47,299 lbs
      473 sacks
      23.6 s. tons

1.10 28.5 ppg
      279 s. tons

1.11 155 s. tons

1.12 1.1 ppg
      barite; 43.1 s. tons
         862 sacks
      galena; 35.6 s. tons
         712 sacks

1.13 2,000 psi

1.14 5,000 psi

1.15 10,000 psi

1.16 92.36 gal (~ 100 gal)
      200 gal accumulator at 3,000 psi precharge

1.17 241.4 gal
      360 gal at 3,000 psi

1.18 160 gal at 2,000 psi

REFERENCES

Adams, N.J. "Well Control Manual," Manual Training Guide. February 1978.

Adams, N. *Well Control Problems and Solutions.* Tulsa, Oklahoma: The Petroleum Publishing Company, 1980.

API Specification 6A (SPEC 6A), "Specification for Wellhead and Christmas Tree Equipment," 16th ed., 1 October 1989.

API Specification 16A (SPEC 16A), "Specification for Drill Through Equipment," 1st ed., 1 November 1986.

API RP 53, "API Recommended Practices for Blowout Prevention Equipment Systems for Drilling Wells," 2nd ed., 25 May 1984.

Lee, H.A. "Good Design Can Improve Mud-Hopper Performance," *Oil & Gas Journal,* 12 December 1977.

CHAPTER 2

WELL CONTROL PROCEDURES AND PRINCIPLES

Well control and blowout prevention have become particularly important topics in the oil industry for a number of reasons. Among these reasons are higher drilling costs, possible loss of life, and waste of natural resources when blowouts occur. One additional reason for concern is the increasing number of governmental regulations and restrictions placed on the oil industry partially as a result of recent, much-publicized well control incidents.

For these and other reasons, it is important that drilling people understand well control principles and the procedures followed to control potential blowouts properly.

Here are some important keys that can be used to control kicks and prevent blowouts. These are presented based on our work as blowout specialists:

- Shut in the well quickly.
- Kicks happen as frequently while drilling as tripping out of the hole.
- When in doubt, shut down and get help. Many small kicks turn into big blowouts because of improper handling.
- Do not hurry and make mistakes. Take your time and get it right the first time. You may not have an opportunity to do it again.

More details are presented in this chapter. Unusual problems occurring during kick-killing are discussed in Chapter 3.

INTRODUCTION TO KICKS

Different drilling problems confront the operator on a day-to-day basis. Among these are lost circulation, stuck pipe, deviation control, and well control. The drilling problem considered in this discussion is well control. Other drilling problems are presented when related to some aspect of well control.

A "kick" can be defined as a well control problem in which the pressure found within the drilled rock is greater than mud hydrostatic pressure acting on the borehole or rock face. When this occurs, the greater formation pressure has a tendency to force formation fluids into the wellbore. This fluid flow is called a kick. If the flow is successfully controlled, the kick has been killed. A "blowout" is the result of an uncontrolled kick.

The severity of a kick depends on several factors. One is the ability of the rock to allow fluid flow. The "permeability" of rock describes its ability to allow fluid movement. The "porosity" measures the amount of space in the rock containing fluids. A rock with high permeability and high porosity has a greater potential for a severe kick than a rock with low permeability and porosity. For example, sandstone is considered to have a greater kick potential than shale because, in general, sand has a greater permeability and porosity than shale.

Another controlling variable for kick severity is the pressure differential involved. "Pressure differential" is the difference between the formation fluid pressure and the mud hydrostatic pressure. If the formation pressure is much greater than the hydrostatic pressure, a large negative differential pressure exists. If this negative differential pressure is coupled with high permeability and porosity, a severe kick can occur.

A kick is labeled in several ways. One label depends upon the type of formation fluid that entered the borehole. Known kick fluids include gas, oil, saltwater, magnesium chloride water, hydrogen sulfide (sour) gas, and carbon dioxide. If gas entered the borehole, the kick is called a "gas kick." Furthermore, if a volume of 20 bbl (3.2 m^3) of gas entered the borehole, the kick could be termed a 20 bbl (3.2 m^3) gas kick.

Another method of labeling kicks is the required mud weight increase necessary to control the well and kill a potential blowout. For example, if a kick required a 0.7 ppg (84 kg/m^3) mud weight increase to control the well, the kick could be termed a 0.7 ppg (84 kg/m^3) kick. (It is interesting to note that an average kick will require about 0.5 ppg (60 kg/m^3) mud weight increase or less.)

An additional important consideration in well control is the pressure the formation rock can withstand without sustaining an induced fracture. This rock strength is often called the "fracture gradient" and is usually expressed in lb/gal equivalent mud weight.

Well Control Procedures and Principles

The "equivalent mud weight" is the summation of pressures exerted on the borehole wall and includes mud hydrostatic pressure, pressure surges due to pipe movement, friction pressures applied against the formation as a result of pumping the drilling fluid, or any casing pressure caused by a kick. For example, if the fracture gradient of a formation is determined to be 16.0 ppg, the well can withstand any combination of the above mentioned pressures that yields the same pressure as a column of 16.0 ppg (1920 kg/m^3) mud to the desired depth. This combination could be (1) 16.0 ppg (1920 kg/m^3) mud, (2) 15.0 ppg (1800 kg/m^3) mud and some amount of casing pressure, (3) 15.5 ppg (1860 kg/m^3) mud and a smaller amount of casing pressure, or (4) other combinations.

Causes of Kicks

Kicks occur as a result of formation pressure being greater than mud hydrostatic pressure, which causes fluids to flow from the formation into the wellbore. In almost all drilling operations, the operator attempts to maintain a hydrostatic pressure greater than formation pressure and thus prevent kicks. However, on occasion (and for various reasons), the formation will exceed the mud pressure and a kick will occur. A study of reasons for this imbalance will explain the causes of kicks. Here are the key causes of kicks:

- Insufficient mud weight
- Improper hole fillup on trips
- Swabbing
- Cut mud
- Lost circulation

More details follow.

Insufficient Mud Weight

Insufficient mud weight is one of the predominant causes of kicks. A permeable zone is drilled while using a mud weight that exerts less pressure than the formation pressure within the zone. Fluids begin to flow into the wellbore, and the kick occurs.

Abnormal formation pressures are often associated with causes for kicks. Abnormal formation pressures are more than under normal conditions. In well control situations, formation pressures greater than normal are the greatest concern. Since a normal formation pressure is equal to a full column of native water, abnormally pressured formations exert more pressure than a full water column. If these abnormally pressured formations are encountered while drilling with mud weights insufficient to control the zone, a potential kick situation has developed. Whether or not the kick occurs depends upon the permeability and porosity of the rock.

TABLE 2-1 • Abnormal pressure indicators

Qualitative Methods
• Paleontology
• Offset well-log analysis
• Temperature anomaly
• Gas counting
• Mud or cuttings resistivity
• Cutting character
• Hole condition
Quantitative Methods
• Shale density
• "d" exponent
• Normalized penetration rate
• Other drilling equations

A number of methods can be used to estimate formation pressures in an effort to prevent this type of kick. Some are listed in Table 2-1.

Kicks caused by insufficient mud weights seem to have the obvious solution of drilling with high mud weights. However, this is not always a viable solution. First, high mud weights may exceed the fracture gradient of the formation and induce lost circulation. Second, mud weights in excess of the formation pressure may significantly reduce the penetration rates. Also, pipe sticking becomes a serious consideration when excessive mud weights are used. The best solution is to maintain a mud weight slightly greater than formation pressure until the mud weight begins to approach the fracture gradient requiring an additional string of casing.

Improper Hole Fill-Up During Trips

Improperly filling the hole during trips is another predominant cause of kicks. As the drill pipe is pulled out of the hole, the mud level falls because the drill pipe steel had displaced some amount of mud. With the pipe no longer in the hole, the overall mud level decreases. The following example illustrates the hydrostatic reduction when pulling drill pipe and drill collars.

Example 2.1

Calculate the hydrostatic pressure reduction when pulling 10, 93 ft stands of drill pipe from the hole without filling the hole. (Use the Appendix.)

Hole size = 8 1/2 in. (casing ID)

Drill pipe = 4 1/2 in., 16.6 lb/ft

Collars = 7 in. OD with 2.5 in ID

Drill pipe displacement = 0.00648 bbl/ft

WELL CONTROL PROCEDURES AND PRINCIPLES

Collar displacement = 0.0415 bbl/ft
Annular capacity (4 1/2 × 8 1/2 in.) = 0.05 bbl/ft
Annular capacity (7 × 8 1/2 in.) = 0.0226 bbl/ft
Drill pipe capacity = 0.01422 bbl/ft
Collar capacity = 0.0061 bbl/ft
Mud weight = 15.0 ppg

Solution

1. What is the total fluid displaced by 10 stands of pipe?

   10 stands × 93 ft/stand × 0.00648 bbl/ft = 6.0264 bbl

2. How many ft does 6.0264 bbl fill?

   Annular capacity plus drill pipe capacity = 0.05 + 0.01422

   = 0.06422 bbl/ft 6.0264 bbl/0.06422 bbl/ft = 93.84 ft

3. Calculate the pressure reduction.

   93.84 ft × 0.052 × 15.0 ppg = 73.2 psi

Example 2.2

Using the information given in Example 2.1, calculate the hydrostatic pressure reduction when pulling only one stand of collars without filling the hole.

Solution

1. 1 stand × 93 ft/stand × 0.0415 bbl/ft = 3.8595 bbl

2. Annular capacity plus collar capacity

   = 0.0226 + 0.0061 = 0.0287 bbl/ft

   3.8595 bbl/0.0287 bbl/ft = 134.47 ft

3. Calculate the pressure reduction.

   134.47 ft × 0.052 × 15.0 ppg = 104.8 psi

In this example, note that pulling collars without filling the hole is 10 times more critical with respect to displacement than pulling drill pipe without filling the hole.

It is necessary to fill the hole with mud periodically to avoid reducing the hydrostatic pressure thereby allowing a kick to occur. Several methods can be used to fill the hole, but all must be able to measure accurately the amount of required mud. It is not satisfactory under any conditions to allow a centrifugal pump to fill the hole continuously from the suction pit, since accurate mud volume measurement is not possible. Two methods most commonly used to monitor hole fill up are a trip tank and pump stroke measurement.

A trip tank has a calibration device to monitor the volume of mud entering the hole. The tank can be placed above the preventer to allow gravity feed into the annulus, or a centrifugal pump may pump mud into the annulus with the overflow returning to the trip tank. The advantages of a trip tank include the hole remains full at all times, and an accurate measurement of the mud entering the hole is possible.

Another method of keeping a full hole is to fill the hole periodically with a positive displacement pump. A flow line device can be installed to measure pump strokes required to fill the hole and will automatically shut off the pump when the hole is full. The following example illustrates use of the rig pump to fill the hole during a trip.

Example 2.3

Calculate the number of pump strokes required to fill the hole if 10 stands of pipe are pulled from the hole. (Use the data from Example 2.1 and the Appendix.)

Pump = double-acting duplex pump, 6 in. liner × 18 in. stroke

Output = 0.1916 bbl/stroke, or 5.2 strokes/bbl

Solution

1. From Example 2.1, 10 stands of pipe displacement is 6.0264 bbl.

2. How many strokes will be required?

    Barrels × strokes/bbl

    = 6.0246 bbl × 5.2 strokes/bbl

    = 31.3 or 32 strokes (per 10 stands)

Swabbing

Swab pressures are created by pulling the drill string from the borehole. Swab pressure is negative and reduces the effective hydrostatic pressure throughout the hole below the bit. If this pressure reduction lowers the effective hydrostatic pressure below the formation pressure, a potential kick has developed. Variables controlling swab pressures are pipe pulling speed, mud properties, hole configuration, and the effect of "balled" equipment. Some effects can be seen in Table 2–2.

Well Control Procedures and Principles

TABLE 2-2 • Swab pressures (in psi) in various hole sizes with several pulling speeds for a 14.0 ppg mud, 4 1/2 in. pipe

Hole Size, in.	Pulling Speeds, seconds/stand					
	15	22	30	45	68	75
8 1/2	267	167	124	98	84	75
6 1/2	589	344	256	192	159	140
5 3/4	921	524	394	289	231	200

Cut Mud

Gas contaminated mud will occasionally cause a kick, although this is rare. The mud density reduction is usually caused by fluids from the core volume cut and released into the mud system. As the gas is circulated to the surface, it expands and reduces the overall hydrostatic pressure sufficient to allow a kick to occur.

Example 2.4

Using the data given below, what is the reduction in mud weight near the surface from core volume cutting?

Hole size = 12 1/4 in.   Pump rate = 10 bbl/min

Depth = 9,000 ft   Mud weight = 9.3 ppg

Drilling rate = 100 ft/hr

Formation pressure = normal

Gas zone = 50 ft, 20% porosity, 25% water saturation

Solution

1. What is the gas volume in the sand?

    $(50 \text{ ft}) (\pi/4) (12.25^2/12^2) (0.20) (1 - 0.25) = 6.1 \text{ ft}^3$

2. What is the expanded volume of gas at surface? (The appropriate Z factors are used.)

$$P_1V_1/T_1Z_1 = P_2V_2/T_2Z_2$$

$[(9,000 \times 0.465) + 15] (6.1)/(637)(.94) = (15)(V_2)/(1)(530)$

$$V_2 = 1,516 \text{ ft}^3$$

3. What volume of mud is mixed with the 1,516 ft³ of gas?

   Pumping time while drilling the sand is 30 minutes.

   Pump rate = 10 bbl/min

   Volume = 30 min × 10 bbl/min = 300 bbl

   300 bbl × 5.615 ft³/bbl = 1,684.5 ft³ of mud

4. 1,516 ft³ of gas will be mixed with 1,684.5 ft³ of mud. The mud weight will be reduced to approximately 4.4 ppg, or half its original weight.

Although the mud weight is cut severely at the surface, the hydrostatic pressure is not reduced significantly, since most gas expansion occurs near surface and not at the hole bottom.

Lost Circulation

Occasionally, kicks are caused by lost circulation. A decreased hydrostatic pressure occurs from a shorter mud column. When a kick occurs from lost circulation, the problem may become severe. A large volume of kick fluid may enter the hole before the rising mud level is observed at the surface. It is a recommended practice to fill the hole with some type of fluid to monitor fluid level.

Warning Signs of Kicks

Warning signs and possible kick indicators can be observed at surface. It is the responsibility of each crew member to recognize and interpret these signs and take proper actions. All signs do not positively identify a kick; some warn of potential kick situations. Key warning signs include the following:

- Flow rate increase
- Pit volume increase
- Flowing well with pumps off
- Pump pressure decrease and pump stroke increase
- Improper hole fill-up on trips
- String weight change
- Drilling break
- Cut mud weight

Here, each warning sign is identified as either primary or secondary relative to its importance in kick detection.

Flow Rate Increase

An increase in flow rate leaving the well while pumping at a constant rate is one primary kick indicator. The increased flow rate is interpreted to mean that the formation is aiding the rig pumps moving fluid up the annulus by forcing formation fluids into the wellbore. (Primary indicator)

Pit Volume Increase

If the pit volume is not changed as a result of surface controlled actions, a pit increase indicates a kick is occurring. Fluids entering the wellbore displaces an equal volume of mud at the flow line and results in a pit gain. (Primary indicator)

Flowing Well With Pumps Off

When the rig pumps are not moving the mud, a continued flow from the well indicates a kick is in progress. An exception is when the mud in the drill pipe is considerably heavier than in the annulus as in the case of a slug. (Primary indicator)

Pump Pressure Decrease And Pump Stroke Increase

A pump pressure change may indicate a kick. Initial fluid entry into the borehole may cause the mud to flocculate and temporarily increase the pump pressure. As the flow continues, the low density influx will displace heavier drilling fluids and pump pressure may begin to decrease. As the fluid in the annulus becomes less dense, the mud in the drill pipe tends to fall and the pump speed may increase. (Secondary indicator)

Other drilling problems may exhibit these same signs. A hole in the pipe, called a "washout," will cause pump pressure to decrease, and a twist-off of the drill string will give the same signs. It is proper procedure, however, to check for a kick if these signs are observed.

Improper Hole Fill Up On Trips

When the drill string is pulled out of the hole, the mud level should decrease by a volume equivalent to the removed steel. If the hole does not require the calculated volume of mud to bring the mud level back to the surface, it is assumed a kick fluid has entered the hole and filled the displacement volume of the drill string. Even though gas or saltwater entered the hole, the well may not flow until enough fluid has entered to reduce the hydrostatic pressure below the formation pressure. (Primary indicator)

String Weight Change

Drilling fluid provides a buoyant effect to the drill string and reduces the actual pipe weight supported by the derrick. Heavier muds have a greater buoyant force than less dense muds. When a kick occurs and low density formation fluids begin to enter the borehole, the buoyant force of the mud system is reduced. The string weight observed at the surface begins to increase. (Secondary indicator)

Drilling Break

An abrupt increase in bit penetration rate, called a "drilling break," is a warning sign of a possible kick. A gradual increase in penetration rate is an abnormal-pressure-detection indicator, and it should not be misconstrued as an abrupt rate increase.

When the rate suddenly increases, it is assumed the type of rock being drilled has changed. It is also assumed the new rock type has the potential to kick as in the case of a sand, whereas the previously drilled rock did not have this potential as in the case of shale. Although a drilling break may have been observed, it is not certain that a kick will occur, but only that a new formation has been drilled that has kick potential. (Secondary indicator)

It is recommended practice when a drilling break is recorded that the driller should drill 3 to 5 ft (1 to 1.5 m) into the sand and stop to check for flowing formation fluids. However, it is industry practice to only stop and check for flow if drilling has been progressing for some interval (i.e., 50 ft or more) in a formation with a potential to seal (shale, etc.) or one which is slow drilling. Flow checks are not generally performed in top hole drilling or if drilling through a series of stringers where repetitive breaks are encountered.

Cut Mud Weight

Reduced mud weight observed at the flow line has occasionally caused a kick to occur. Some causes for reduced mud weight are core volume cutting, connection air, or aerated mud circulated from the pits and down the drill pipe. Fortunately, the lower mud weights from the cuttings effect are found near the surface, generally due to gas expansion, and do not appreciably reduce mud density throughout the hole. Table 2–3 shows gas cutting has a very small effect on bottom-hole hydrostatic pressure.

An important point to remember about gas cutting is that if the well did not kick in the time required to drill the gas zone and circulate the gas to the surface, only a small possibility exists that it will kick. Generally, gas cutting only indicates a formation has been drilled that does contain gas. It does not mean the mud weight must be increased. (Secondary indicator)

Well Control Procedures and Principles

TABLE 2-3 • Effect of gas cut mud on the bottom-hole hydrostatic pressure

Depth, ft	10 ppg mud cut to 5 ppg Pressure Reduction, psi	18 ppg mud cut to 16.2 ppg Pressure Reduction, psi	18.0 ppg mud cut to 9 ppg Pressure Reduction, psi
1,000	51	31	60
5,000	72	41	82
10,000	86	48	95
20,000	97	51	105

Kick Detection and Monitoring with MWD Tools

MWD systems monitor mud properties, formation parameters, and drill string parameters during circulation operations. The system is widely used for drilling, but it also has application for well control including the following:

- Drilling efficiency data such as down-hole weight on bit and torque can be used to differentiate between ROP changes due to drag and those caused by formation strength. Monitoring bottom-hole pressure, temperature and flow through the MWD tool is not only useful for early kick detection, but it can also be valuable during a well control kill operation.
- Formation evaluation capabilities, such as gamma ray and resistivity measurements, can be used to detect influxes into the wellbore, identify rock lithologies, and predict pore pressure trends.
- The MWD tool allows monitoring of the acoustic properties of the annulus for early gas influx detection. Pressure pulses generated by the MWD pulser are recorded and compared at the standpipe and at the top of the annulus. Full scale testing has shown that the presence of free gas in the annulus is detected by amplitude attenuation and phase delay between the two signals. For water-based mud systems this technique has demonstrated the ability to consistently detect gas influxes within minutes before significant expansion has occurred. Further development is currently under way to improve the system's capability to detect gas influxes in oil-based mud.
- Some MWD tools feature kick detection by means of ultrasonic sensors. In these systems, an ultrasonic transducer emits a signal that is reflected off the formation back to the sensor. Small quantities of free gas significantly alter the acoustic impedance of the mud.

Automatic monitoring of these signals permits detection of gas in the annulus. It should be noted these devices only detect the presence of gas at or below the MWD tool.

The MWD tool offers kick detection benefits if the response time is less than the time for surface indicators to be observed. The tool can provide early detection of kicks and potential influxes as well as monitor the kick killing process. Tool response time is a function of the complexity of the MWD tool and the mode of operation. The sequence of data transmission (gamma, resistivity, tool face orientation, etc.) determines the update times of each type of measurement. Many MWD tools allow for reprogramming of the update sequence while the tool is in the hole. This feature can enable the operator to increase the update frequency of critical information to meet the expected needs of the section being drilled. If the tool response time is longer than required for surface indicators to be observed, the MWD only serves as a confirmation source.

Shut-In Procedures

When one or more warning signs of kicks are observed, steps should be taken to shut in the well. If there is any doubt as to whether the well is flowing, shut it in and check the pressures. Also, there is no difference between "just a small flow" and a "full flowing" well because both can very quickly turn into a big blowout.

There has been some hesitation in the past in closing in a flowing well due to the possibility of sticking the pipe. It can be shown that for all types of pipe sticking (differential pressure, heaving shale, and sloughing shale) it is better to close in the well quickly, reduce the kick influx; and as a result, reduce the chances of pipe sticking. The primary concern at this point is to kill the kick safely; and when feasible, the secondary concern is to avoid pipe sticking.

Some concern has been expressed about fracturing the well and creating an underground blowout as a result of shutting in the well when a kick occurs. If the well is allowed to flow, it will eventually become necessary to shut in the well, at which time the possibility of fracturing the well will be greater than if the well had been shut in immediately after the initial kick detection. Table 2–4 shows an example of higher casing pressures as a result of continuous flow.

Initial Shut-In

There has been considerable discussion as to the merits of "hard" shut-in procedures versus "soft" shut-in procedures. The hard shut-in procedure has the annular preventer(s) closed immediately after the pumps are shut down. In soft shut-in procedures, the choke is opened prior to closing the

Well Control Procedures and Principles

TABLE 2-4 • The effect of continuous influx on the casing pressure as a result of failure to close in the well.*

Volume of Gas Gained, bbl	Casing Pressure, psi
20	1,468
30	1,654
40	1,796

*In a 16,000 ft well with typical geometry

preventers, then the choke is closed. Arguments in favor of soft shut-in procedures are (1) it avoids a "water hammer" effect due to stopping fluid flow abruptly, and (2) it provides an alternate means of well control (low choke pressure method) if the casing pressure becomes "excessive."

The water hammer effect has no proven substance. The low choke pressure method is an unreliable procedure.

The primary argument against the soft shut-in procedure is that a continuous influx is permitted while the procedures are executed. For these reasons, only the hard shut-in procedures are presented.

Hard shut-in procedures for well control depend upon the type of rig and the drilling operation occurring when the kick is taken.

1. Drilling—land or bottom-supported offshore rig
2. Tripping—land or bottom-supported offshore rig
3. Drilling—floating rig
4. Tripping—floating rig
5. Diverter procedures—all rigs (when surface pipe is not set.)

Drilling—Land or Bottom-Supported Offshore Rig

These rigs do not move during normal drilling operations. Some are land and barge rigs, jack-ups, and platform rigs.

Shut-In Procedures

1. When a primary kick warning sign has been observed, immediately raise the kelly until a tool joint is above the rotary table.
2. Stop the mud pumps.
3. Close the annular preventer.
4. Notify the company personnel.
5. Read and record the shut-in drill pipe pressure, the shut-in casing pressure, and the pit gain.

Raising the kelly is an important procedure. With the kelly out of the hole, the valve at the bottom of the kelly can be closed if necessary. Also, the annular preventer members can attain a more secure seal on pipe than a kelly.

Tripping—Land or Bottom-Supported Offshore Rig

A high percentage of well control problems occur when a trip is being made. The kick problems may be compounded when the rig crew is preoccupied with the trip mechanics and fails to observe the initial warning signs of the kick.

Shut-In Procedures

1. When a primary warning sign of a kick has been observed, immediately set the top tool joint on the slips.
2. Install and make up a full opening, fully opened safety valve on the drill pipe.
3. Close the safety valve and the annular preventer.
4. Notify the company personnel.
5. Pick up and make up the kelly.
6. Open the safety valve.
7. Read and record the shut-in drill pipe pressure, shut-in casing pressure, and pit gain.

Installing a full opening safety valve in preference to an inside-blowout preventer (float) valve is a prime consideration because of the advantages offered by the full opening valve. If flow is encountered up the drill pipe as a result of a trip kick, the fully opened, full opening valve is physically easier to stab. Also, a float-type inside blowout preventer valve would automatically close when the upward moving fluid contacts the valve. (Assume that a manual lock, float valve is not in use.)

Also, if wireline work such as drill pipe perforating or logging becomes necessary, the full opening valve will accept logging tools approximately equal to its ID, whereas the float valve may prohibit wireline work altogether. After the kick is shut in, an inside blowout preventer float valve may be stabbed on the full opening valve to allow stripping operations.

Drilling—Floating Rig

A floating rig moves during normal drilling operations. The primary types of floating vessels are semisubmersibles and drillships.

Several differences in shut-in procedures apply to floaters. Drill string movement can occur, even with a motion compensator in operation. Also, the blowout preventer stack is on the sea floor. To solve the problem of possible vessel and drill string movement and the resulting wear on the preventers, a tool joint may be lowered on the closed pipe rams. The string weight is hung on these rams. This procedure is not necessary if the rig has a functional motion compensator.

When the stack is located a considerable distance from the rig floor, the problem is to insure that a tool joint does not interfere with the closing of the preventer elements. A spacing out procedure should be executed when the BOP is tested after running the BOP stack. Close the rams, slowly lower the drill string until a tool joint contacts the rams, and record the position of the kelly at that point. Space out should occur so that a tool joint and lower kelly valve are above the rotary table. Spacing should be correlated to tide measuring equipment on the rig floor.

The following procedure could be altered to just using the annular preventer and motion compensator for cases where (1) the SICP and SIDPP are low and close to the same value (indicating oil or water), or (2) the "kick volume" is less than 20–30 bbls and the time to kill the well is less than 2–3 hours. Be sure that the closing pressure on the annular is reduced to a value within the range recommended by the manufacturer for this situation to avoid annular element failure.

Shut-In Procedures

1. When a primary warning sign of a kick has been observed, immediately raise the kelly to the level previously designated during the spacing out procedure (tide adjusted).
2. Stop the mud pumps.
3. Close the annular preventer.
4. Notify the company personnel.
5. Close the upper set of pipe rams.
6. Reduce the hydraulic pressure on the annular preventer.
7. Lower the drill pipe until the pipe is supported entirely by the rams.
8. Read and record the shut-in drill pipe pressure, shut-in casing pressure, and pit gain.

Tripping—Floating Rig

The procedures for kick closure during a tripping operation on a floater is a combination of floating drilling procedures and the immobile rig tripping procedures.

Shut-In Procedures

1. When a primary warning sign of a kick has been observed, immediately set the top tool joint on the slips.
2. Install and make up a full opening, fully opened safety valve in the drill pipe.
3. Close the safety valve and the annular preventer.
4. Notify the company personnel.
5. Pick up and make up the kelly.
6. Reduce the hydraulic pressure on the annular preventer.
7. Lower the drill pipe until it is supported by the rams.
8. Read and record the shut-in drill pipe pressure, shut-in casing pressure, and pit gain.

Diverter Procedures—All Rigs

When a kick occurs in a well with insufficient casing to control a kick safely, a blowout will occur. Since a shallow underground blowout is difficult to control and may cause the loss of the rig, an attempt is usually made to divert the surface blowout away from the rig. This is the common practice on land or offshore rigs that are not mobile. (Diverter equipment is discussed in Chapter 1.) Special attention must be given so the well is not shut in until the diverter lines are opened.

Diverter Procedures

1. When a primary warning sign of a kick has been observed, immediately raise the kelly until a tool joint is above the rotary table.
2. Increase the pump rate to maximum output.
3. Open the diverter line valve(s).
4. Close the diverter unit (or annular preventer).
5. Notify the company personnel.

Recent experiences show that shallow gas flows are difficult to control. Industry philosophy is improving and new handling procedures are being developed. Chapter 4 contains extensive discussions on shallow gas handling and blowout control that goes beyond the scope of "routine" kick killing.

Crew Member Responsibilities for Shut-In Procedures

Each crew member has different responsibilities during shut-in procedures. These are listed according to job classification.

Floorhand (Roughneck)

1. Notify the driller if any warning signs of kicks are observed.
2. Assist in installing the full opening safety valve if a trip is being made.
3. Initiate well control responsibilities after shut-in.

Derrickman

1. Notify the driller if any warning signs of kicks are observed.
2. Initiate well control responsibilities and begin mud mixing preparations.

Driller

1. Shut in the well immediately if any of the primary warning signs of kicks are observed.

Well Control Procedures and Principles

2. If a kick occurs while making a trip, set the top tool joint on the slips and direct the crews in the installation of the safety valve prior to closing the preventers.
3. Notify all proper company personnel.

Obtaining and Interpreting Shut-In Pressures

"Shut-in pressures" are defined as pressures recorded on the drill pipe and casing when the well is closed. Although both pressures are important, the drill pipe pressure will be used almost exclusively in killing the well. The shut-in drill pipe pressure is abbreviated SIDPP. Shut-in casing pressure is SICP. (At this point, assume that the drill pipe does not contain a float valve.)

Reading and Interpreting Pressures

During a kick, fluids flow from the formation into the wellbore. When the well is closed in to prevent a blowout, pressure builds at the surface due to formation fluid entry into the annulus and the difference between the mud hydrostatic pressure and the formation pressure.

Since this pressure imbalance cannot exist for long, the surface pressures will finally build so that the surface pressure plus the mud and influx hydrostatic pressures in the well are equal to the formation pressures. Equations 2.1 and 2.2 express this relationship for the drill pipe and the annular side, respectively.

$$\text{SIDPP} + \text{Drill pipe hydrostatic pressure} = \text{Bottom-hole formation pressure} \quad (2.1)$$

$$\text{SICP} + \text{Annular mud hydrostatic pressure} + \text{Annular influx hydrostatic pressure} = \text{Bottom-hole formation pressure} \quad (2.2)$$

Example 2.5 and Figure 2–1 show how the shut-in pressures are read and interpreted.

Example 2.5

While drilling at 15,000 ft, the driller observed several primary warning signs of kicks and proceeded to shut in the well. After the shut-in was completed, he called the company man and started to record the following pressures and pit gains. The well was shut-in at 6:00 A.M.

Shut-in Time	SIDPP, psi	SICP, psi	Pit Gain bbl
6:00 A.M.	650	950	20
6:05 A.M.	750	1,000	20
6:10 A.M.	775	1,040	20
6:15 A.M.	780	1,040	20

The final shut-in pressures after 15 minutes were recorded as follows:

    SIDPP   —   780 psi
    SICP   —   1,040 psi
    Pit gain   —   20 barrels

Interpretation of Recorded Pressures

An important basic principle can be seen in Figure 2–1. It is observed that formation pressure (BHP) is more than drill pipe hydrostatic pressure by an amount equal to the SIDPP. The drill pipe pressure gauge is a bottom-hole pressure gauge. The casing pressure cannot be considered a direct bottom-hole pressure gauge due to generally unknown amounts of formation fluid in the annulus.

Constant Bottom-hole Pressure Concept

Figure 2–1 can be used to illustrate another important basic principle. It was stated that the 780 psi (5.4 MPa) observed on the drill pipe gauge was the amount necessary to balance mud pressure at the hole bottom with the pressure in the gas sand at 15,000 ft (4,600 m). A basic law of physics states formation fluids travel from areas of high pressure to lower pressures only, and they do not travel between areas of equal pressures, assuming gravity segregation is neglected.

If the drill pipe pressure is controlled so total mud pressure at the hole bottom is slightly greater than formation pressure, then there will be no additional kick influx entering the well. The concept is the basis of the constant bottom-hole pressure method of well control in which the pressure at the hole bottom is kept constant and at least equal to formation pressure.

Effects of Time

In Example 2.5, 15 minutes were used to obtain shut-in pressures. The purpose of this time period is to allow pressures to reach equilibrium, and it will be sufficient to balance formation pressures. The required time will depend on variables such as the type of influx, rock permeability and porosity, and the original amount of pressure underbalance. This may take a few minutes to several hours. The amount of time required is dependent upon conditions surrounding the kick.

FIGURE 2-1 • Hydrostatic pressure

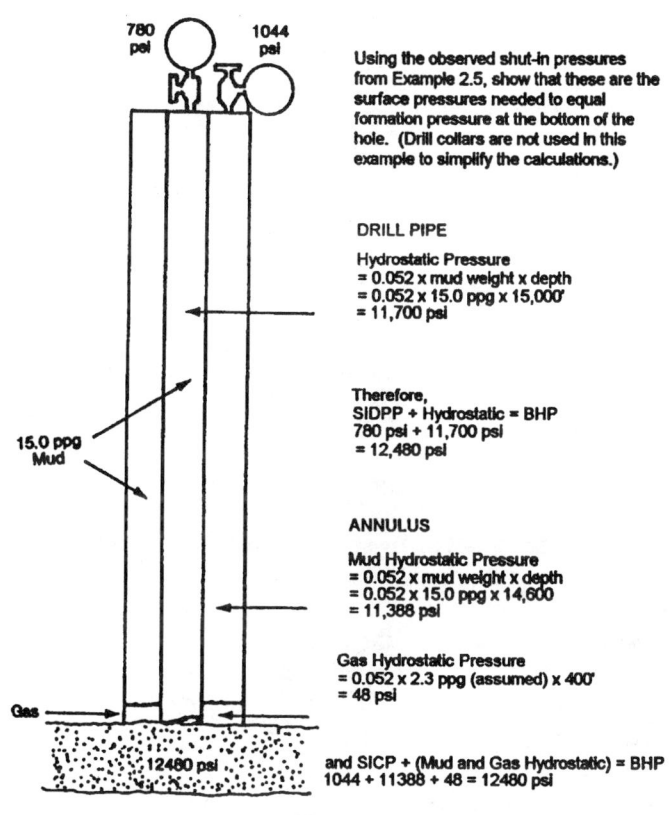

NOTE: In practical situations, the amount or type of influx will not be (exactly) known and therefore the annulus pressure should not be used to calculate inforamtion pressures.

Several other factors affect the time allowed for pressures to stabilize. Gas migration is the movement of low density fluids up the annulus. It tends to build pressure at the surface if time is allowed for migration. Also, the influx may have a tendency to deteriorate the hole stability and cause either stuck pipe or hole bridging. These problems must also be considered when reading the shut-in pressures.

Trapped Pressure

"Trapped pressure" is any pressure recorded on the drill pipe or annulus more than the amount needed to balance bottom-hole pressure. Pressure

TABLE 2-5 • Guidelines to check for trapped pressure

1. When checking for trapped pressure, bleed from the casing side only. The reasons for this include (1) the choke is located on the casing side, (2) this avoids contamination of the mud in the drill pipe, and (3) it avoids the possibility of plugging the bit jets.
2. Use the drill pipe pressure as a guide, since it is a direct bottom-hole pressure indicator.
3. Bleed small amounts (1/4 to 1/2 bbl) of mud at a time. Close the choke after bleeding and observe the pressure on the drill pipe.
4. Continue to alternate the bleeding and subsequent pressure observation procedures as long as the drill pipe pressure continues to decrease. When the drill pipe pressure ceases to fall, stop bleeding and record the true shut-in drill pipe pressure and casing pressure.
5. If the drill pipe pressure should decrease to zero during this procedure, continue to bleed and check pressures on the casing side as long as the casing pressure decreases. (Note: This step normally will not be necessary.)

---

can be trapped in the system in several ways. Common ways are gas migrating up the annulus and tending to expand or closing the well in before the mud pumps have quit running. Using a pressure reading containing trapped pressure results in erroneous kill calculations.

Guidelines help to check when releasing trapped pressure. If they are not properly executed, the well will be much more difficult to kill. These guidelines are listed and explained in Table 2-5.

Since trapped pressure is more than the amount needed to balance bottom-hole pressure, trapped pressure can be bled without allowing any additional influx into the well. However, after the trapped pressure is bled off and if the bleeding is continued, more influx will be allowed into the well and the surface pressures will begin to increase.

Although bleeding procedures can be implemented at any time, it is advisable to check for trapped pressure when the well is shut in initially and to recheck it when the drill pipe is displaced with kill mud if any pressure remains on the shut-in drill pipe.

Example 2.6 illustrates the bleeding procedures.

Example 2.6

A kick was taken and shut in. The SIDPP was read as 525 psi, and the SICP was 760 psi. The company representative checked for trapped pressure by bleeding small amounts from the choke and recording shut-in pressures.

Increment Number	Bleeding Volume, bbl (approximate)	SIDPP, psi	SICP, psi
0	–	525	760
1	1/2	510	745
2	1/2	500	735
3	1/2	490	725
4	1/2	480	715
5	1/2	475	710
6	1/2	475	710
7	1/2	475	715

The true pressures were recorded as:

$$SIDPP = 475 \text{ psi}$$
$$SCIP = 715 \text{ psi}$$

Drill Pipe Floats

A kick can occur when a drill pipe float valve is used. (See Chapter 1 for a discussion of float valves.) Since a float valve prevents fluid and pressure movement up the drill pipe, there will not be a drill pipe pressure reading after the well is shut-in. Several procedures can obtain the drill pipe pressure, and each depends on the amount of information known when the kick occurs.

Table 2–6 describes the procedure to obtain the drill pipe pressure if the slow pumping rate (kill rate) is known. Table 2–7 gives a procedure if the kill rate is not known. Examples 2.7 and 2.8 illustrate these procedures.

TABLE 2–6 • Procedure to establish shut-in drill pipe pressure (SIDPP) if the kill rate is known

1. Shut in the well, record the shut-in casing pressure (SICP) and obtain the kill rate either from the driller or the daily tour report.
2. Instruct the driller to start the pumps and maintain the pumping rate at the kill rate (strokes).
3. As the driller starts the pumps, use the choke to regulate the casing pressure at the same pressure that was originally recorded at shut-in conditions.
4. After the pumps are running at the kill rate with the casing pressure properly regulated at shut-in pressure, record the pressure on the drill pipe while pumping.
5. Shut down the pumps and close the choke.
6. The shut-in drill pipe pressure equals the total pumping pressure minus the kill rate pressure, or

$$SIDPP = \text{Total pressure} - \text{kill rate pressure}$$

TABLE 2-7 • Procedure to establish the shut-in drill pipe pressure (SIDPP) if the kill rate is not known

1. Shut in the well.
2. Line up a low volume, high pressure reciprocating pump on the stand pipe.
3. Start pumping and fill up all of the lines.
4. Gradually increase the torque on the pumps until the pumps begin to move fluid down the drill pipe.
5. The shut-in drill pipe pressure is the amount of pressure required to initiate the fluid movement. This is assumed to be the amount needed to overcome the pressure acting against the bottom side of the valve.

Example 2.7

A kick was taken on a well in which a drill string float was used. The kill rate and pressure taken immediately prior to the kick was 26 spm at 650 psi. The shut-in pressures were 400 psi on the casing (SICP) and zero on the drill pipe. Establish the true SIDPP.

Solution

1. Instruct the driller to run his pumps at 26 spm.
2. Operate the choke to maintain the casing pressure at the initial pressure of 400 psi.
3. After the drill pipe pressure has stabilized, record this value as the total pumping pressure. As an example, assume the total pressure to be 870 psi.
4. From Table 2-6,

$$SIDPP = \text{total pressure} - \text{kill rate pressure} \quad (2.3)$$

$$SIDPP = 870 \text{ psi} - 650 \text{ psi} = 220 \text{ psi}$$

Example 2.8

A trip was made for a new bit in which the jet sizes were changed. The driller was on bottom and had drilled 5 ft when a kick was taken. The kill rate with the new bit was not known, and the drill string contains a float. Establish the shut-in drill pipe pressure.

Solution

Since the kill rate is not known, the procedure listed in Table 2-7 must be exercised.

1. Line up a low volume, positive displacement pump on the stand pipe and fill up all the lines.

WELL CONTROL PROCEDURES AND PRINCIPLES 113

2. Pressures were increased on the stand pipe at a low rate with the following results:

Volume pumped, bbl	SIDPP, psi
0	0
1	0
2	0
3	70
4	150
5	220
6	300
7	300

3. The SIDPP was assumed to be 300 psi. A graph (see Figure 2–2) aids in determining the desired values.

FIGURE 2–2 • Procedure to estabilsh the shut-in pipe pressure on Example 2.8 when the kill rate was not known

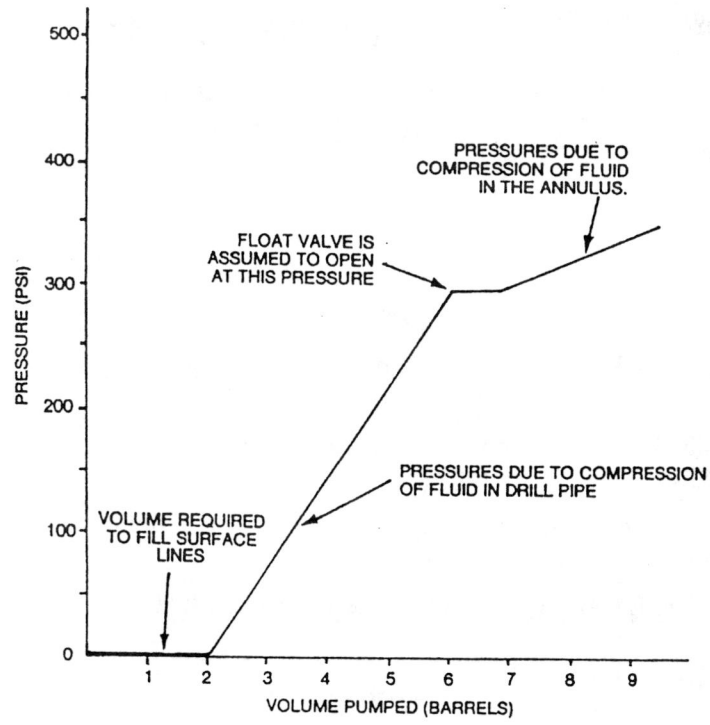

Table 2-6 is important in another application. Suppose a kick was taken and the shut-in drill pipe pressure was known (no float valve). A kill rate had not been established. Step 6 of this table could be modified to read:

$$\text{Kill rate pressure} = \text{Total pressure} - \text{SIDPP} \qquad (2.4)$$

The procedures in Table 2-6 remain the same with the exception that Equation 2.4 would be substituted for Equation 2.3.

Establishing the shut-in drill pipe pressure becomes more complex if the kill rate was not previously established and a float valve in the string prohibits pressure readings at the surface. Table 2-7 must be used initially to determine the SIDPP, after which Equation 2.4 and Table 2-6 must be implemented to establish the kill rate. Example 2.9 illustrates this technique.

Example 2.9

A kick was taken in which the kill rate was not known, and a float valve was used in the string. The SICP is 550 psi. Establish the SIDPP and the kill rate.

Solution

1. After preparing all surface equipment, the company man began pressurizing the drill pipe with the following results:

Volume pumped, bbl	SIDPP, psi
0	0
1	0
2	50
3	150
4	250
5	260
6	265

2. The SIDPP was assumed to be approximately 250 psi.

3. The driller was instructed to bring his pumps to 20 spm (arbitrary value). The company man used the choke to maintain the initial value to 550 psi on the casing.

4. After the pressures stabilized on the drill pipe, the total pressure was 950 psi.

5. Using Equation 2.4, the kill rate was established.

$$\text{Kill rate pressure} = \text{Total pressure} - \text{SIDPP} \qquad (2.4)$$

$$= 950 \text{ psi} - 250 \text{ psi}$$

$$= 700 \text{ psi (at 20 spm)}$$

Well Control Procedures and Principles

TABLE 2-8 • Influx gradient evaluation guidelines

Influx Gradient	Influx Type
0.05 – 0.2	Gas
0.2 – 0.4	Probable combination of gas, oil, and/or saltwater
0.4 – 0.5	Probable oil, or saltwater

Kick Identification

When a kick occurs, it may prove interesting to know the type of influx (gas, oil or saltwater) entering the wellbore. It must be remembered that well control procedures developed here are designed to kill all types of kicks safely. The formula required to make this kick influx calculation is as follows:

$$\text{Influx gradient} = \text{mud gradient (drill pipe)} - (\text{SICP} - \text{SIDPP})/\text{influx height} \quad (2.5)$$

Where:

Influx gradient = formation fluid gradient entering the wellbore, psi/ft

Mud gradient = mud gradient in the drill pipe, psi/ft

SICP = shut-in casing pressure, psi

SIDPP = shut-in drill pipe pressure, psi

Influx height = formation fluid length in the annulus, ft

The influx gradient can be evaluated using the guidelines in Table 2-8.

Although SICP and SIDPP can be determined accurately for Equation 2.5, it is difficult to determine the influx height. This requires knowledge of the pit gain and exact hole size. Example 2.10 illustrates Equation 2.5.

Example 2.10

While drilling, a kick was taken with the following known data. Assume no drill collars in the hole to simplify the example. What type of fluid entered the well?

Depth = 15,000 ft

Mud weight = 15.0 ppg

Drill pipe = 4 1/2 in.

Hole size = 8 1/2 in.

SIDPP = 780 psi

SICP = 1,100 psi

Pit gain = 50 bbl

Solution

1. 1 bbl of fluid occupies 20 vertical ft in a 4 1/2 × 8 1/2 in. annulus.
   50 bbl × 20 ft/bbl = 1,000 ft (influx height)

2. Mud gradient = 0.052 psi/ft/ppg × 15.0 ppg
   = 0.780 psi/ft

3. Influx gradient = mud gradient $- \dfrac{(SICP - SIDPP)}{\text{influx height}}$

   $= 0.780\,\text{psi/ft} - \dfrac{(1{,}100\,\text{psi} - 780\,\text{psi})}{1{,}000\,\text{ft}}$

   $= 0.780\,\text{psi/ft} - 0.320\,\text{psi/ft}$

   $= 0.460\,\text{psi/ft}$

4. From Table 2–8, the influx is probably salt water or oil.

Kill Weight Mud Calculation

It is necessary to calculate the mud weight needed to balance bottom-hole formation pressure. "Kill weight mud" is defined as the amount necessary to balance formation pressure exactly. It will be shown in later sections that it is safer to use the exact required mud weight without variations.

Since the drill pipe pressure has been defined as a bottom-hole pressure gauge, the SIDPP can be used to calculate the mud weight necessary to kill the well. The kill mud formula follows:

$$\text{K.W.M.} = \text{SIDPP} \times 19.23/\text{depth} + \text{O.W.M.} \qquad (2.6)$$

Where:

K.W.M. = kill weight mud, ppg

SIDPP = shut-in drill pipe pressure, psi

19.23 = reciprocal of 0.052, ppg/psi/ft

Depth = true vertical bit depth, ft

O.W.M. = original weight mud in the drill pipe, ppg

Since the casing pressure does not appear in Equation 2.6, a high casing pressure does not necessarily indicate a high kill weight mud. The same is true for pit gain, since it does not appear in Equation 2.6. Example 2.11 uses the kill weight mud formula.

Example 2.11

What will be the kill weight mud for the kick data given below?

> True vertical depth = 11,550 ft
> O.W.M. = 12.1 ppg
> SIDPP = 240 psi
> SICP = 1,790 psi
> Pit gain = 85 bbl

Solution

> K.W.M. = SIDPP × 19.23/depth + O.W.M.
> = 240 psi × 19.23/11,550 ft + 12.1 ppg
> = 0.4 ppg + 12.1 ppg
> = 12.5 ppg

Well Control Procedures

Many well control procedures have been developed over the years. Some have utilized systematic approaches, while others were based on logical, but perhaps unsound, principles. The systematic approaches will be presented in this section. A discussion in later sections will be given to several other methods of well control.

In previous sections, the constant bottom-hole pressure concept was developed in which the total pressures (mud hydrostatic pressure, casing pressure, etc.) at the hole bottom would be maintained at a value slightly greater than formation pressures to prevent further influxes of formation fluids into the wellbore. Also, since the pressure would be only slightly greater than the formation pressure, this minimizes the possibility of inducing a fracture and an underground blowout. This concept can be implemented in three ways.

1. *One circulation, or Wait and Weight method.* After the kick is shut in, weight the mud to kill density, then pump out the kick fluid in one circulation using the kill mud. (An alternate name often applied is the "engineer's method.")
2. *Two circulation, or Driller's Method.* After the kick is shut in, the kick fluid is pumped out of the hole before the mud density is increased.
3. *Concurrent Method.* Pumping begins immediately after the kick is shut in and pressures are recorded. The mud density is increased as rapidly as possible while pumping the kick fluid out of the well.

If applied properly, each method achieves constant pressure at the hole bottom and will not allow additional influx into the well. Procedural and theoretical differences make one procedure more desirable than the others.

Wait and Weight Method

Figure 2-3 describes the one circulation, or Wait and Weight method. At point 1, the shut-in drill pipe pressure is used to calculate the kill weight mud. The mud weight is increased to kill density in the suction pit. As the kill mud is pumped down the drill pipe, the static drill pipe pressure is controlled to decrease linearly, until at point 2 the drill pipe pressure should be zero. The heavy mud has killed the drill pipe pressure.

FIGURE 2-3 • Drill pipe pressure graph of the one circulation method

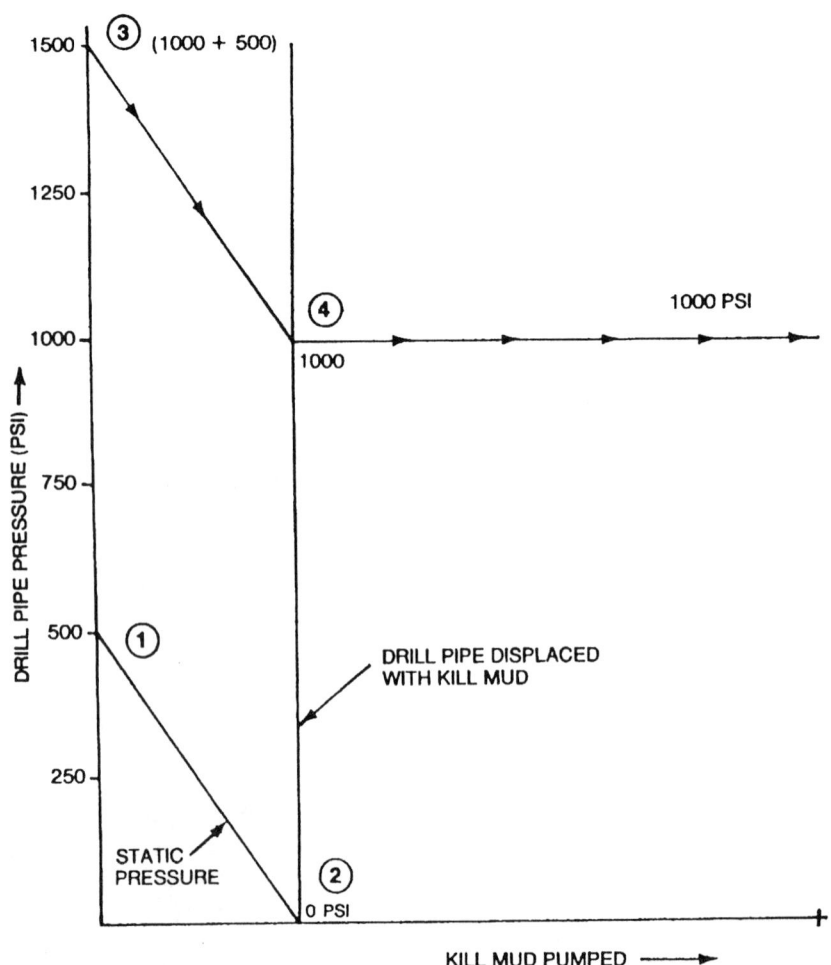

Well Control Procedures and Principles

Point 3 shows the initial pumping pressure on the drill pipe is the total of SIDPP plus the kill rate pressure. While pumping kill mud down the pipe, the circulating pressure decreases until at point 4 only the pumping pressure remains. From the time kill mud is at the bit until it reaches the flow line, the choke is used to control the drill pipe pressure at the final circulating pressure. The driller insures the pump remains at the kill speed.

Driller's Method

In the two circulation, or Driller's Method, the circulation is started immediately. Kill mud is not added in the first circulation. The drill pipe pressure will not decrease during the first circulation. (See Figure 2–4.) The purpose is to remove the kick fluid from the annulus.

In the second circulation, the mud weight is increased and causes a decrease from the initial pumping pressure at 1 to the final circulating pressure at 2. This pressure is held constant while the annulus is displaced with kill mud.

Concurrent Method

This method is the most difficult to execute properly. As soon as the kick is shut in, pumping begins immediately after reading the pressures. The mud density is increased as rapidly as rig facilities will allow. The difficulty is determining the mud density being circulated and its relative position

FIGURE 2–4 • Drill pipe pressure graph of the two circulation method

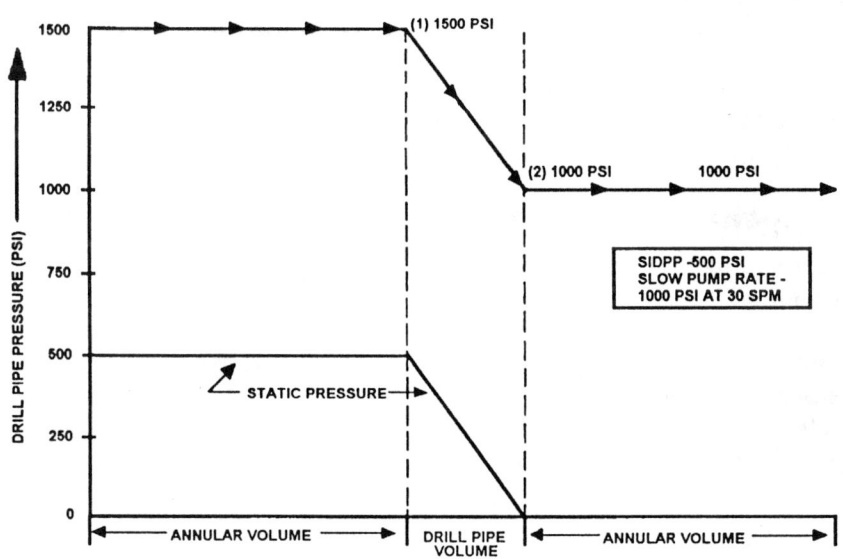

FIGURE 2-5 • Drill pipe pressure graph of the concurrent method

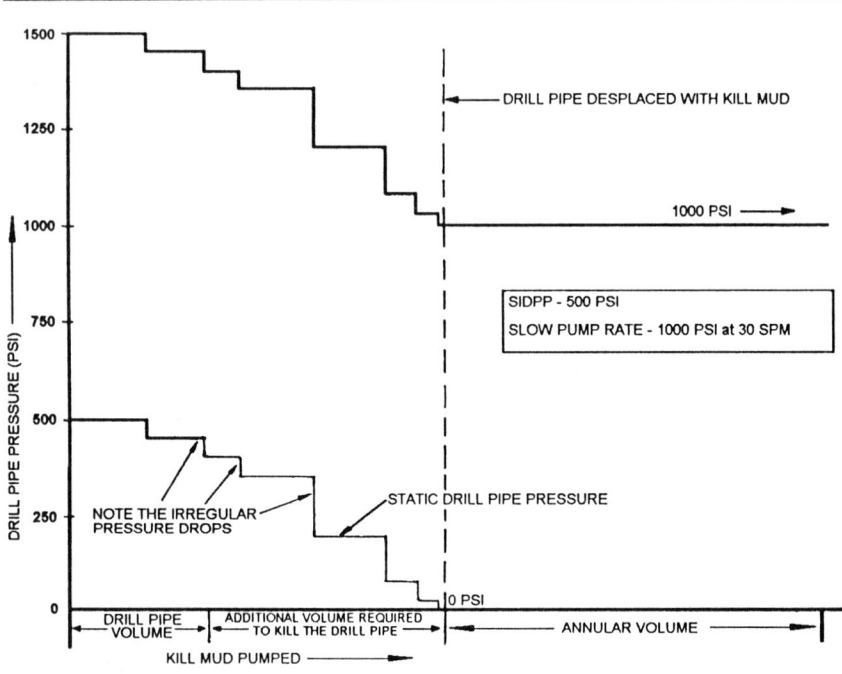

in the drill pipe. Since this position determines the drill pipe pressures, the rate of pressure decrease may not be as consistent as seen in the other two methods. (See Figure 2-5.) As a new density arrives at the bit or some predetermined depth, the drill pipe pressure is decreased by an amount equal to the hydrostatic pressure of the new mud weight increment. When the drill pipe is displaced with kill mud, the pumping pressure is maintained constant until kill mud reaches the flow line.

Constant Bottom-hole Pressure Methods

Determining the best well control method for most situations involves several considerations. Some are (1) time required to execute the kill procedure, (2) surface pressures from the kick, (3) complexity relative to ease of implementation, and (4) down hole stresses applied to the formation during the kick killing process. All points must be analyzed before a procedure can be selected. The following list briefly summarizes the general opinion in the industry:

1. Wait and Weight method should be used in most cases.
2. Driller's Method should be used if a good casing shoe exists, and there is going to be a delay in weighting up the system.

3. Concurrent Method should only be used in rare cases. It would be primarily used for a severe kick (1.5 ppg or greater) with a large influx and a potential problem with developing lost circulation. The pump rate should be kept to a minimum to allow the weight to be raised continuously.

In an analysis of kick killing procedures, emphasis is placed on the one and two circulation methods. Inspection of the procedures will show these are opposite approaches while the concurrent method falls somewhere between.

Time

Two important considerations relative to time are required for the kill procedure. The first is time required to increase the mud density from the original weight to the final kill weight mud. Since a few operators are very concerned with the pipe sticking during this time, the well control procedure is often chosen that minimizes the waiting time required to increase the mud density. Procedures with the least amount of initial waiting time are the concurrent method and the two circulation method. In both procedures, pumping begins immediately after the shut-in pressures are recorded.

An important time consideration, however, is not the initial waiting time but the overall time required for the complete procedure to be implemented. Figure 2-3 shows that the one circulation method requires one complete fluid displacement (drill pipe and annulus), while the two circulation method (see Figure 2-4) requires the annulus be displaced twice in addition to the drill pipe displacement. In certain situations, extra time for the two circulation method may be serious with respect to hole stability or preventer wear.

Surface Pressures

During the course of well killing, surface pressures may approach alarming values. This may be a problem in gas volume expansion near the surface. The kill procedure with least surface pressure required to balance the bottom-hole formation pressure is important.

Figures 2-6 and 2-7 point out the different surface pressure requirements for several kick situations. The first major difference is noted immediately after the drill pipe is displaced with kill mud. The necessary casing pressure begins to decrease from the increased kill mud hydrostatic pressure in the one circulation procedure. This decrease is not seen in the two circulation method, since this procedure does not circulate kill mud initially. In fact, the casing pressure increases as the gas bubble expansion displaces mud from the hole.

The second pressure difference occurs as the gas approaches the surface. The two circulation procedure has higher pressures as a result of

FIGURE 2-6 • Annular pressures for one circulation method vs. two circulation method in a 10,000 ft well

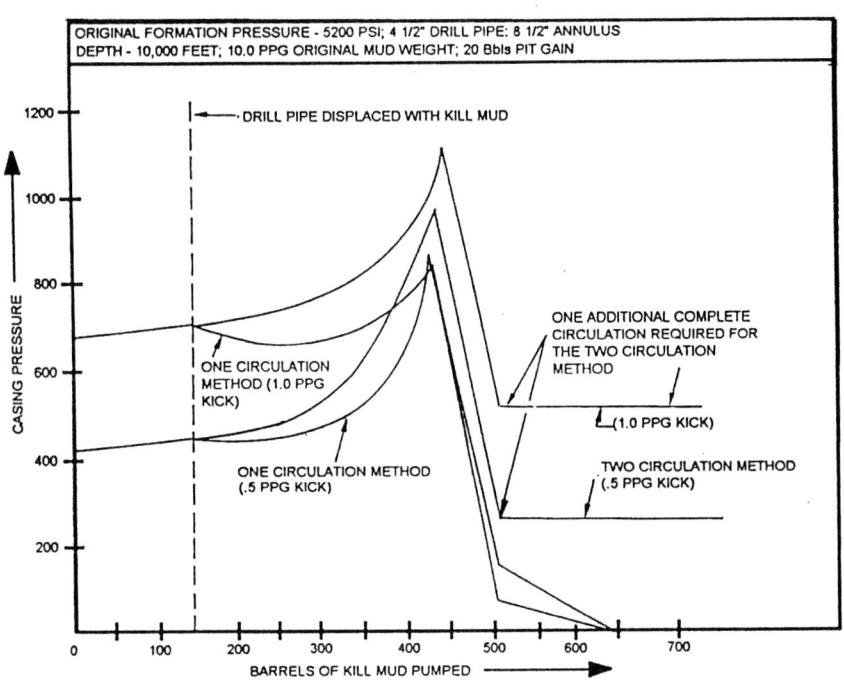

the lower density original mud weight. It is interesting to note these high necessary casing pressures suppress the gas expansion to a small degree resulting in a later arrival of gas at the surface.

Procedure Complexity

Process suitability is partially dependent on the ease with which it can be executed. The same principle holds true for well control. If a kick killing procedure is difficult to comprehend and implement, its reliability is diminished.

The concurrent method falls into the category of reduced reliability from procedure complexity. To perform this procedure properly, the drill pipe pressure must be reduced according to mud weight being circulated and its position in the pipe. This implies that (1) the crew will inform the operator when a new mud weight is being pumped, (2) rig facilities can maintain this increased mud weight increment, and (3) the mud weight position in the pipe can be determined by pump stroke counting. Many operators have discontinued using this complex method.

FIGURE 2-7 • Annular pressures for one circulation method vs. two circulation method in a 15,000 ft well

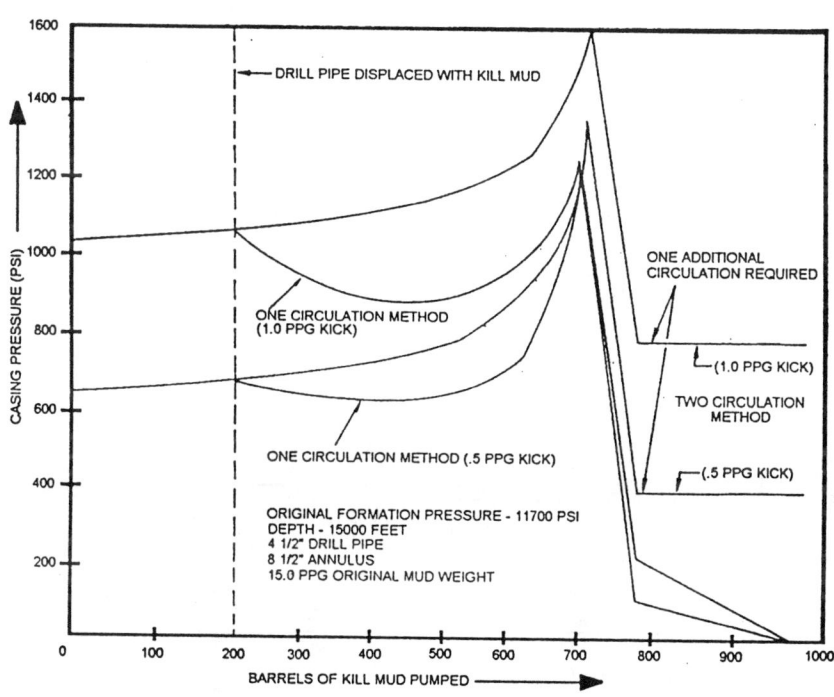

One and two circulation methods are receiving prominent use because of application ease. In both procedures, the drill pipe pressure remains constant for long intervals of time. Also, when displacing the drill pipe with kill mud, the drill pipe pressure decrease is virtually a straight-line relationship and not staggered as in the concurrent method. (See Figure 2-5.)

Downhole Stresses

Although all considerations are important, the primary concern should be the stresses imposed on the borehole wall. If the kick-imposed stresses are greater than the formation can withstand, an induced fracture occurs allowing the possibility of an underground blowout. The procedure that imposes the least downhole stress while maintaining constant pressures on the kicking zone is considered the most conducive to safe kick killing.

Equivalent mud weights are a useful tool to measure downhole stresses. The "equivalent weights" are defined as the total pressures to a depth converted to pounds per gallon mud weight.

$$\text{Equivalent mud weight} = (\text{Total pressures} \times 19.23)/\text{depth} \quad (2.7)$$

FIGURE 2-8 • Equivalent mud weight comparison for the one circulation method vs. the two circulation kill procedure (0.5 ppg kick at 10,000 ft)

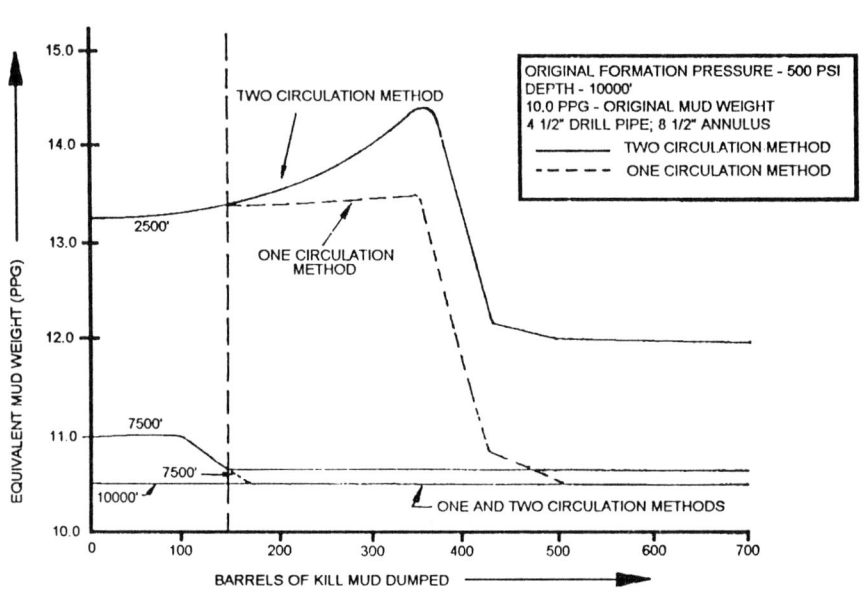

The equivalent mud weights for the systems in Figures 2-6 and 2-7 are presented in Figures 2-8 and 2-9. The one circulation method has consistently lower equivalent mud weights throughout the killing process after the drill pipe has been displaced. The procedures generally exhibit the same maximum equivalent mud weights. They occur from the time the well is shut in until drill pipe is displaced.

Figures 2-8 and 2-9 illustrate an important principle. The maximum stresses occur very early in circulation for the deeper depth and not at the maximum casing pressure intervals. The maximum lost-circulation possibilities will not occur at the gas-to-surface conditions as might seem logical to the casual observer. If a fracture is not created at shut-in, it will probably not occur throughout the remainder of the process. A full understanding of this behavior may calm operator's concerns about formation fracture as the gas approaches the surface.

Variables Affecting Kill Procedures

Although variables that affect kick-killing do not necessitate a change in the basic procedure structure, they may cause an unexpected behavior that can mislead an operator into making bad judgments. The one circulation method will be used to demonstrate the effect of these variables.

FIGURE 2-9 • Equilent mud weight comparison for the one circulation method vs. the two circulation kill procedure (0.5 ppg kick at 15,000 ft)

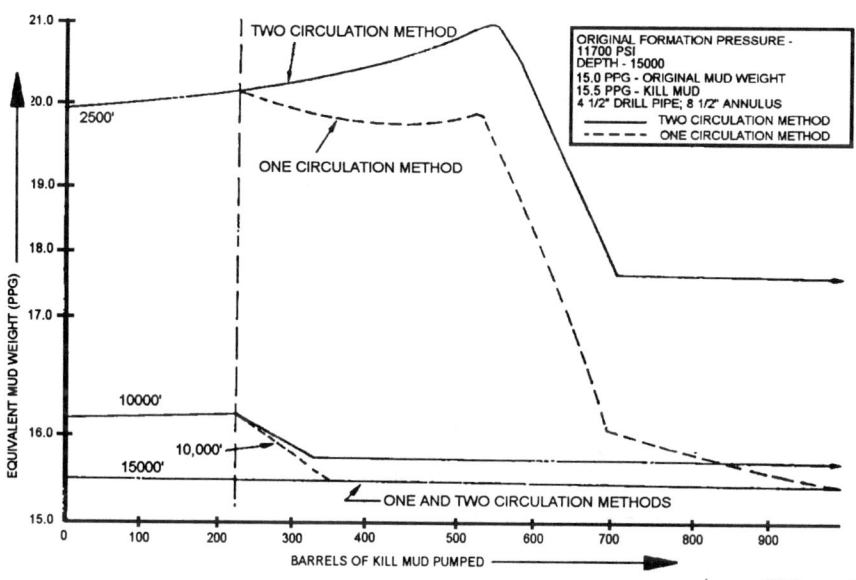

Influx Type

The influx type entering the wellbore plays a key role in casing pressure behavior. The influx can range from a heavy oil to fresh water. The most common is gas or saltwater. Each has a pronounced casing pressure curve and different downhole effects.

Gas kicks are generally more dramatic than other influx types. Some reasons are (1) the rate at which gas enters the wellbore, (2) high casing pressures resulting partially from the low density fluid, (3) gas expansion as it approaches the surface, (4) fluid migration up the wellbore, and (5) fluid flammability. A typical gas kick casing pressure curve is shown in Figure 2-10.

Gas expansion from decreased confining pressures as the fluid is pumped up the wellbore affects the kick killing process. (See Figure 2-10.) As the gas begins to expand, the previously decreasing casing pressure begins to increase at an accelerating rate. This higher casing pressure may give the false impression that another kick influx is entering the well. Immediately after the gas-to-surface conditions, the casing pressure decreases rapidly, which may give the impression that lost circulation has occurred.

Both casing pressure changes are expected behaviors and do not indicate an additional influx or lost circulation. The possibility of lost circulation is less at gas-to-surface conditions than at the initial shut-in conditions. (See Figures 2-8 and 2-9.)

FIGURE 2-10 • Typical gas kick casing pressure curve for the one circulation method

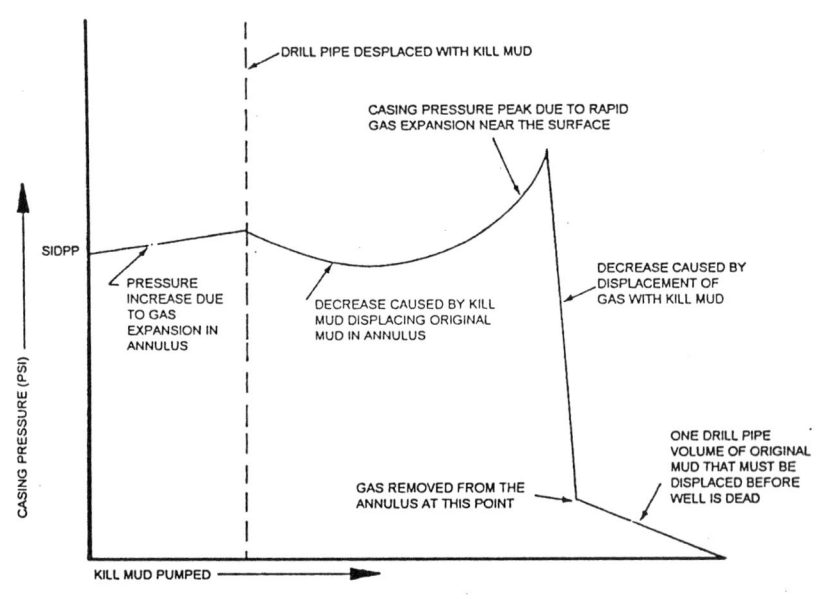

When gas expands, the increased gas volume displaces fluid from the well resulting in a gain. Figure 2-11 shows the pit gain for the problem illustrated in Figure 2-6. This pit gain is in addition to the volume increase from weight materials. Since the pit gains in volume, it is logical to assume the flow rate exiting the well increases. (See Figure 2-12.)

Gas migration may cause special problems. There have been numerous recent studies of gravity segregation phenomena in an effort to quantify a migration rate. Field data from a professional well killing corporation suggests a rate of 7 to 15 ft/min in mud systems. Regardless of the rate, the migration effect must be considered because of gas expansion potential. If the fluid is not allowed to expand properly during the migration period, trapped pressure will be generated at the surface. If unnecessary expansion occurs, additional formation gas will enter the well. Example 2.12 illustrates the gas migration phenomenon with an actual field case.

Example 2.12

While drilling a development well from an offshore platform, a kick was taken. The SIDPP was 850 psi and the SICP was 1,100 psi. Storm conditions forced the tender (barge) to be towed away from the platform to avoid damage to the tender or platform legs. The removal of the tender caused all support services to the platform to be severed, including the mud and pumps.

Well Control Procedures and Principles

FIGURE 2-11 • Pit gain for the 1.0 ppg kick in Figure 2-6

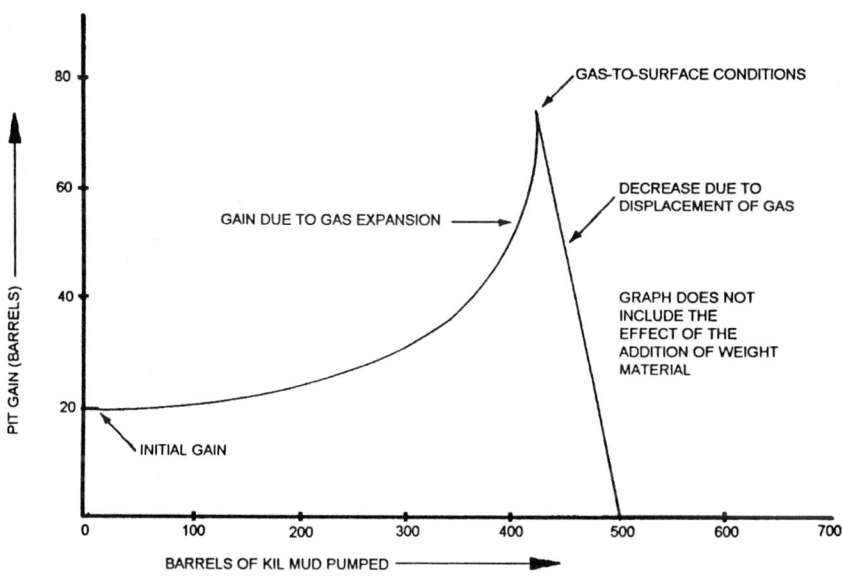

FIGURE 2-12 • Typical representation of flow rates in and out during a kick killing

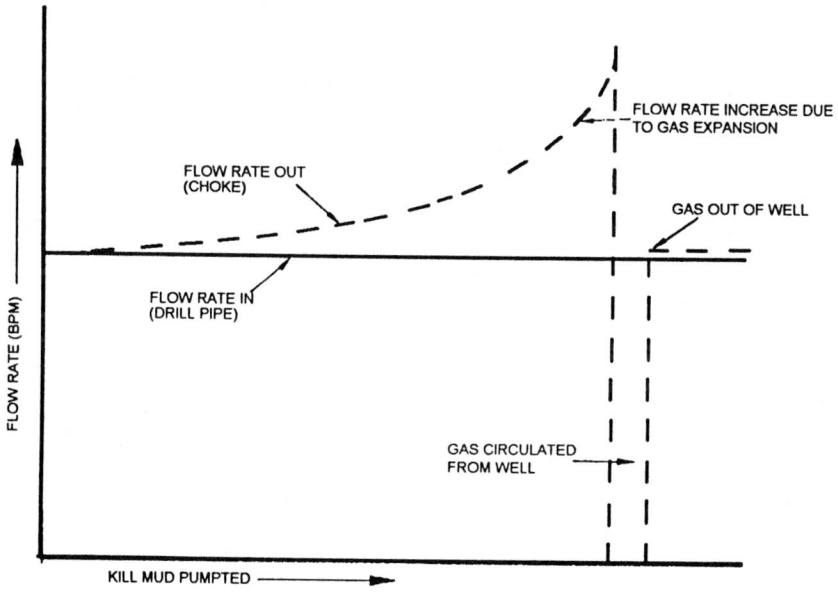

The engineer on the platform knew the kick would become a problem from gas migration up the annulus. To satisfy the situation, he allowed the migration to build the pressure on the drill pipe to 900 psi, which he utilized as a 50 psi safety margin. Thereafter, the migration was allowed to build the SIDPP to 950 psi before he would bleed a small volume of mud from the annulus to reduce drill pipe pressure to 900 psi. Since bottom-hole pressure was still 50 psi more than formation pressure, no additional influx occurred. This procedure was continued until the gas reached the surface, at which time the pressures ceased to increase and remained at 900 psi. After support services were restored to the rig, the gas was pumped from the well and kill procedures were initiated.

This example points out the manner in which gas migration can be safely controlled by utilizing the concept of the drill pipe pressure as a bottom-hole pressure indicator.

Saltwater kicks do not pose the same problems as gas kicks. Volume expansion does not occur. Also, since saltwater is more dense than gas, casing pressures are less than for a comparable volume of gas. (See Figure 2-13.) Shut-in pressures for the 50 bbl (7.9 m^3) saltwater kick are approximately the same as that seen in Figure 2-6 for a 20 bbl (3.2 m^3) gas kick under the same conditions.

FIGURE 2-13 • Typical saltwater kick casing pressure curve

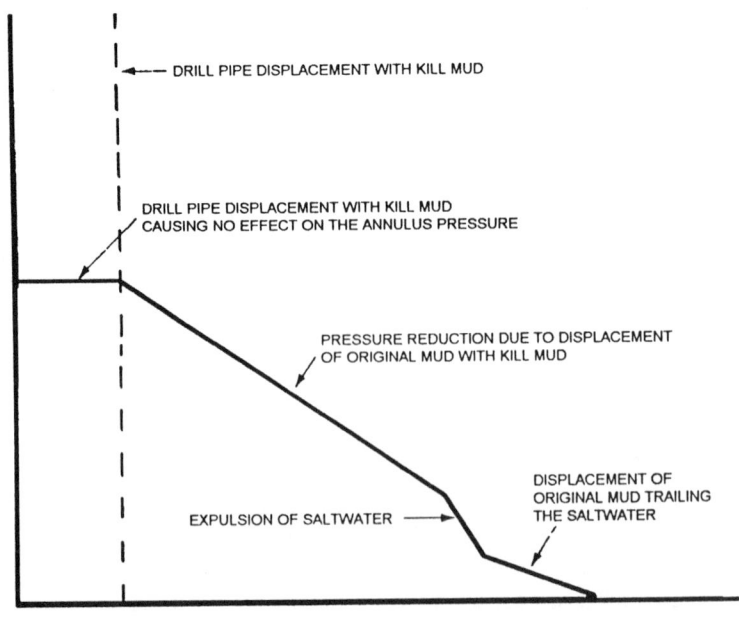

Hole stability and pipe sticking are generally more severe with a saltwater kick than gas. The saltwater fluid causes a fresh water mud filter cake to flocculate and create pipe sticking tendencies and unstable hole conditions. The severity increases with large kick volumes and with extended waiting periods before the fluid is pumped from the hole.

Volume of Influx

The fluid volume entering the well is a controlling variable on the casing pressure throughout the kill process. Increased influx volumes give rise to higher initial SICP as well as greater pressure differences at the gas-to-surface conditions. (See Figure 2–14.) This representation points out the importance of quick closure rather than hesitation.

Kill-Weight Increment Variations

The original mud density must be increased in most kick situations to kill the well. The incremental density increase has some effect on casing pressure behavior. (See Figure 2–15.) The gas-to-surface pressure condi-

FIGURE 2–14 • Comparison of casing pressure curves for 10, 20, and 50 bbl kick volumes

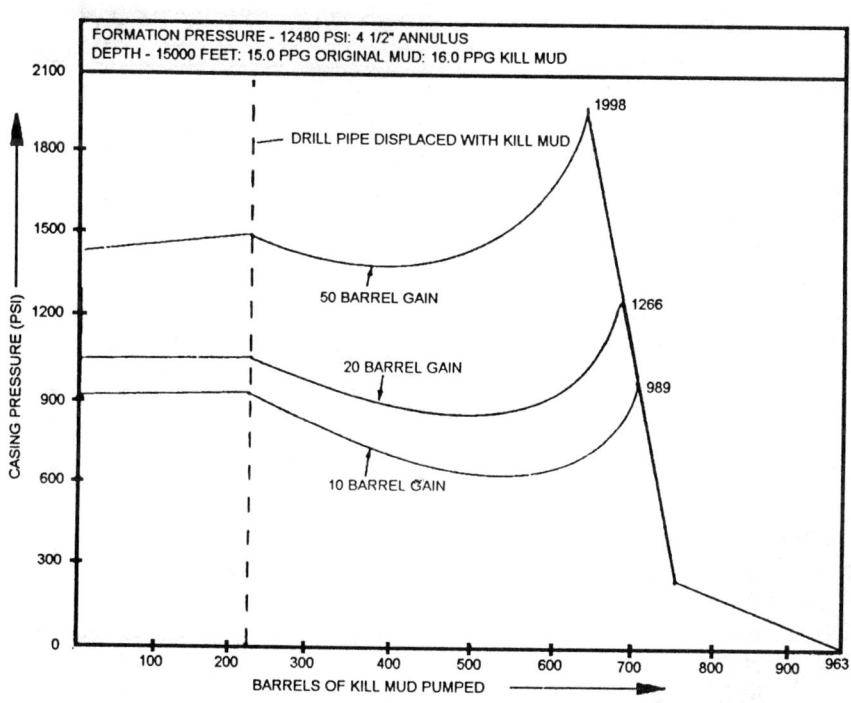

FIGURE 2-15 • Comparison of several kill mud weight increments

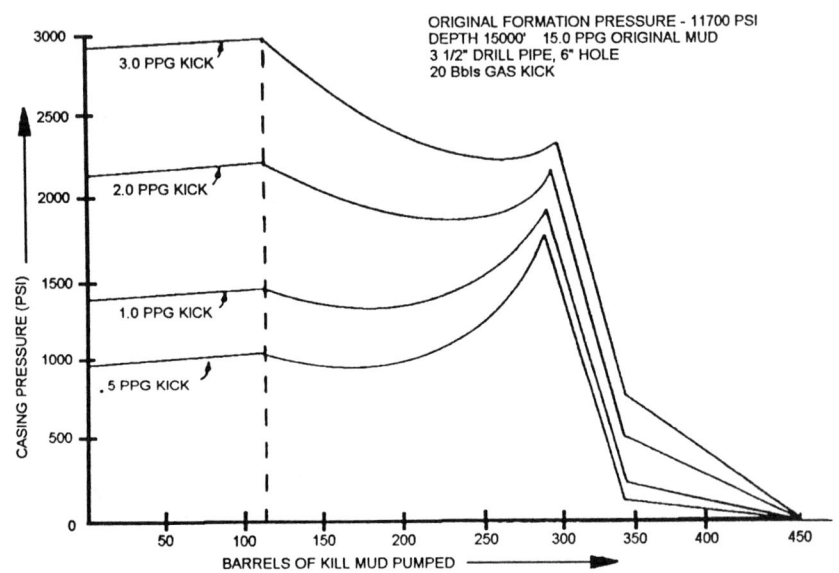

tions are higher than original shut-in pressures for 0.5 ppg (60 kg/m³) and 1.0 ppg (120 kg/m³) kicks. The 2.0 ppg (240 kg/m³) and 3.0 ppg (360 kg/m³) mud weight increases do not show this tendency. The 3.0 ppg (360 kg/m³) kick has a lower gas-to-surface pressure than at initial closure. This is due to suppressed gas expansion, minimizing associated pressures. This is generally observed in kicks requiring greater than a 2.0 ppg (240 kg/m³) incremental increase.

Another important mud weight variation is the difference between kill mud weight necessary to balance bottom-hole pressure and the mud weight actually circulated. If the circulated mud is less than kill mud weight, the casing pressure is higher than if kill mud had been used because of the necessity to maintain a balanced pressure at the hole bottom. (See Figures 2-6 and 2-7.) The equivalent mud weights will be greater, which increases formation fracture possibility.

Circulated mud weights greater than the calculated kill mud weight do not decrease the casing pressure. The situation is synonymous with mud weight safety factors and is termed "overkill." As the extra heavy mud is pumped down the drill pipe, the casing pressure will increase from the U-tube effect. (See Figure 2-16.) This U-tube principle states that the pressures on each side of the tube must be equal. (See Figure 2-17.) These higher casing pressures have associated downhole stresses that increase formation fracture potential.

Well Control Procedures and Principles   131

FIGURE 2-16 • The effect of safety factors (1.0 ppg in this example) causes higher casing pressures than the proper calculated kill mud density

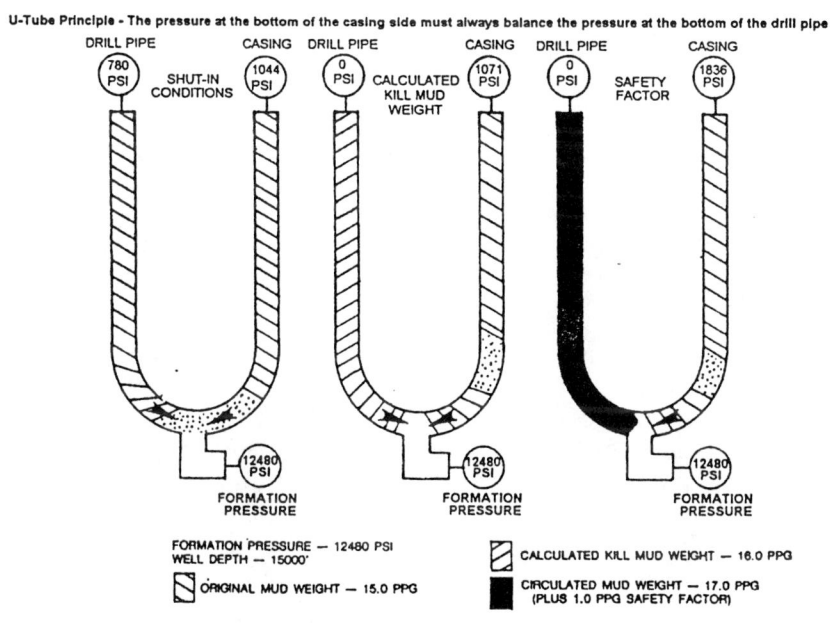

FIGURE 2-17 • Effect of excess mud weight on annulus pressure

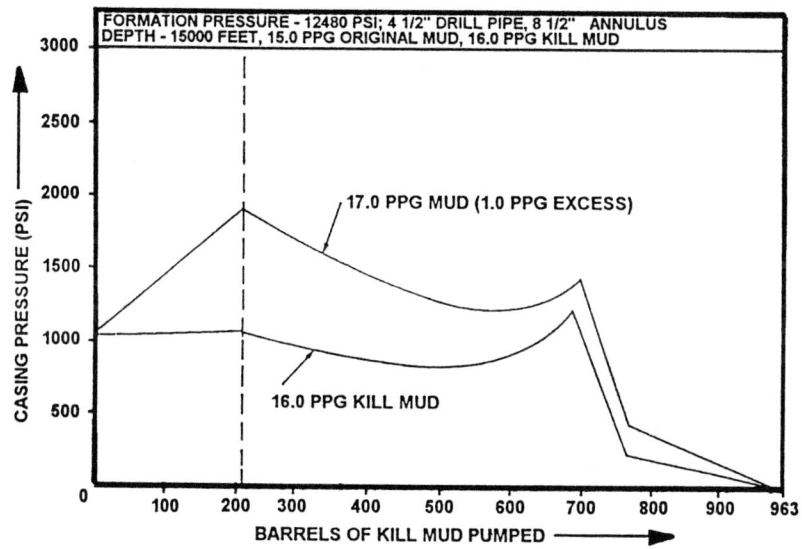

There have been several attempts throughout past years to achieve the benefits of "safety factors" while avoiding the ill effects of high casing pressures caused by the U-tube effect. The most common attempt at this effort is to subtract the hydrostatic pressure supplied by the extra mud weight increment from the final circulating pressure creating a net zero effect from the added mud weight.

In a static situation, the casing pressure is reduced by an amount equal to the safety factor hydrostatic pressure, which results in a zero net effect. From a theoretical standpoint, the approach is based on sound principles. However, field experience has shown this procedure is not practical due to the complexity involved. This procedure is not necessary for proper well control, and it should be used only by experienced well control engineers.

Hole Geometry Variations

In practical kick killing situations, hole and drill string size changes cause the kick fluid geometry to be altered. This is a particular problem in deep tapered holes where several pipe and hole sizes are used. The influx may occupy a large vertical space at the hole bottom creating a high casing pressure. As the fluid is pumped into the larger annular spaces, the vertical height is decreased, thus increasing the mud column height resulting in lower casing pressures. Figures 2–18 (a), (b), and (c) show a typical tapered hole and the associated casing and drill pipe pressure curves.

Implementation of the One Circulation Method

To implement the one circulation method, certain guidelines must be followed to insure a safe kick killing exercise. Although the procedure is relatively simple, its mastery demands a basic knowledge of practical steps taken during the process. Check points indicate any potential problems.

A kill sheet is normally used during conventional operations. It contains certain prerecorded data, formulas for the various calculations, and a graph or other means of determining the required pressures on the drill pipe as kill mud is pumped. Although many operators have complex kill sheets, it is necessary for the kill sheet to contain only the basic required kick killing data. (A kill sheet is shown in an example problem in following sections.)

A summary of the steps involved in proper kick killing is given. The sections not directly applicable to deepwater situations are noted.

1. When a kick occurs, shut in the well using the appropriate shut-in procedures.
2. After pressures have stabilized, read and record the shut-in drill pipe pressure, the shut-in casing pressure, and the pit gain. (If a float valve is in the drill pipe, use the established procedures to obtain the shut-in drill pipe pressure.)

Well Control Procedures and Principles 133

FIGURE 2-18(a) • Tapered hole diagram

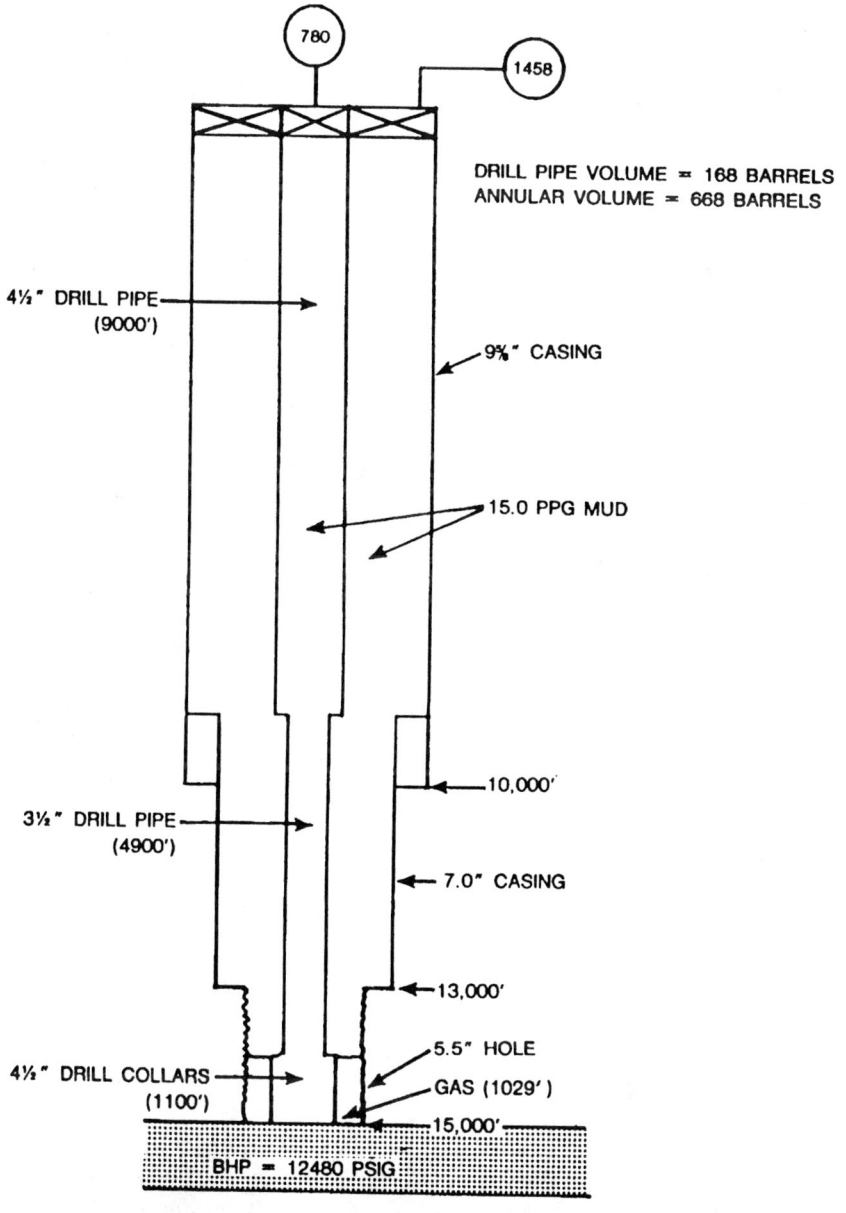

FIGURE 2-18 (b) • Static drill pipe pressure graph for a typical tapered string. Note the pressure underbalance if a linear depth vs. pressure relationship is assumed.

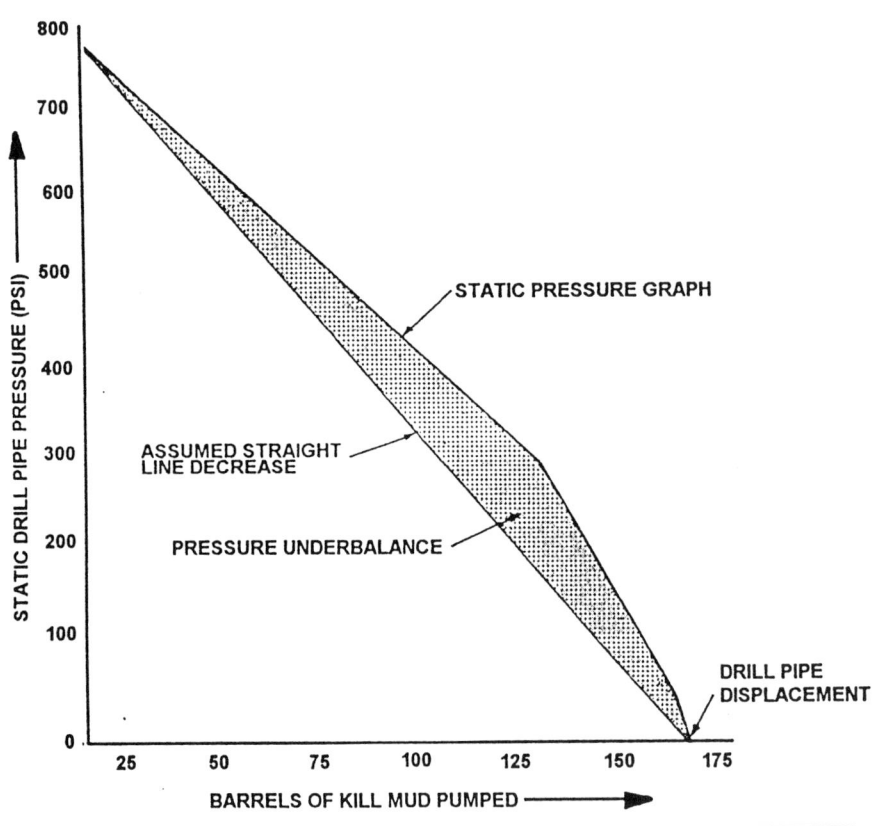

3. Check the drill pipe for trapped pressure.
4. Calculate the exact mud weight necessary to kill the well and prepare a kill sheet.
5. Mix the kill mud in the suction pit. (It is not necessary to weight up the complete surface mud volume initially. First pump some mud to reserve pits.)
6. After the kill mud has been mixed, initiate circulation by adjusting the choke to hold the casing pressure at the shut-in value while the driller starts the mud pumps. (Not applicable in deep water.)
7. As the driller is displacing the drill pipe with the exact kill mud weight at a constant pump rate (kill rate), use the choke to adjust the pumping pressure according to the kill sheet.

FIGURE 2-18 (c) • Effect of hole size changes on casing pressure

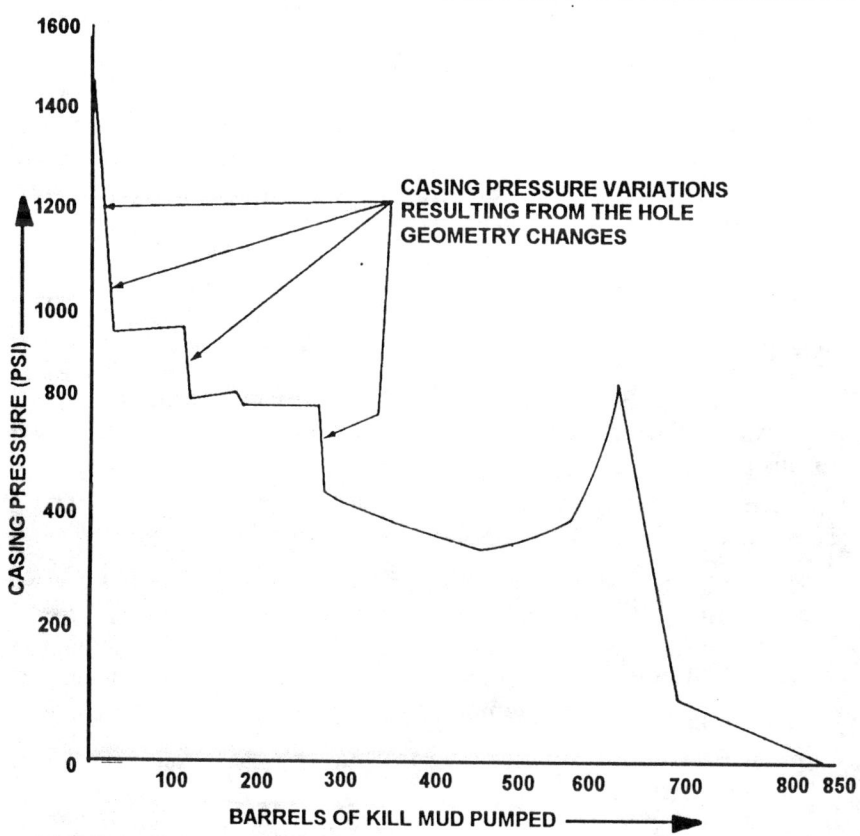

8. When the drill pipe has been displaced with kill mud, consider shutting down the pumps and closing the choke to record pressures. (Note: If the kill mud is highly weighted up, settling and plugging may occur.) The drill pipe pressure should be zero, and the casing should have pressure remaining. If the pressure on the drill pipe is not zero, execute the following steps:

   a. Check for trapped pressure using the established procedures.
   b. If the drill pipe pressure is still not zero, pump 10 – 20 additional bbls (1.5 – 3 m³) to ensure that kill mud has reached the bit. The pump efficiency may be reduced at the low circulation rate.
   c. If pressure remains on the drill pipe, recalculate the kill mud weight, prepare a new kill sheet, and return to the initial steps of this procedure.

9. To displace the annulus with the kill mud, maintain the drill pipe pumping pressure and rate constant by using the choke to adjust the pressures as necessary.
10. After the kill mud has reached the flow line, shut down the pumps and close the choke. The well should be dead. If pressure remains on the casing, continue circulation until the annulus is dead.
11. When the pressures on the drill pipe and casing are zero, open the annular preventers, circulate and condition the mud, and add a trip margin. In subsea applications, the trapped gas under the annular is circulated out by pumping down the kill line and up the choke line with a ram preventer below the annular closed. The riser must then be circulated with kill mud by reverse circulation down the choke line and up the riser before the preventers can be opened.

Well control learning experiences are often best accomplished by observation of an actual kick problem. Example 2.13 has been provided for this purpose.

Example 2.13

1. Pre-kick considerations. While drilling the R.B. Texas No. 1 in the Louisiana Gulf Coast offshore area, the company representative, Mike Smith, carried out his normal drilling responsibilities related to well control in the event that a kick should be taken. Some items that Mike did are listed below:

    a. He read the appropriate MMS orders and complied with the provisions.

    b. The barite supplies were checked to insure a sufficient amount of barite was on board to kill a 1.0 ppg kick, if necessary.

    c. The driller recorded on the driller's book that the kill rate was 21 spm and 800 psi pump pressure.

    d. Mike calculated the drill string volume as follows:

    $$4\ 1/2 \text{ in. drill pipe to } 14{,}000 \text{ ft, and } 6\ 1/2$$

    $$\times\ 2 \text{ in. drill collars to } 15{,}000 \text{ ft}$$

    $$4\ 1/2 \text{ in., } 16.6 \text{ lb/ft pipe capacity}$$

    $$= 0.01422 \text{ bbl/ft} \times 14{,}000 \text{ ft} = 199 \text{ bbl}$$

    $$6\ 1/2 \times 2 \text{ in. collar capacity}$$

    $$= 0.0039 \text{ bbl/ft} \times 1{,}000 \text{ ft} = 3.9 \text{ bbl}$$

    $$\text{Total} = 199 + 3.9 = 202.9 \text{ bbl}$$

2. Shut-in and weight up procedures. The drillers on the rig had just changed tours when a drilling break was observed. The well was checked for flow. A flow was recorded with the pumps off. The following steps were taken:

   a. The kelly was raised until a tool joint cleared the floor. (A jack-up rig was in use.)
   b. The pumps were shut down.
   c. The annular preventer was closed.
   d. Mike was notified the well was shut-in.
   e. The driller told his crew in the mud room to stand by in case the mud weight had to be increased.

   Mike went to the floor and read his pressures as follows:

   $$SIDPP = 240 \text{ psi}$$
   $$SICP = 375 \text{ psi}$$
   $$\text{Pit gain} = 31 \text{ bbl}$$

   After checking for trapped pressures, he recorded the information on his kill sheet. From the kill sheet, he calculated that he needed to raise his mud weight from 13.1 ppg original weight to 13.4 ppg.

   Mike walked to the mud room to tell the derrickman that he needed 13.4 ppg kill mud when he noticed the pits were almost full. He knew the needed barite would raise the mud level, so he instructed the derrickman to pump off a foot of mud, section off the suction pit, and increase the weight to 13.4 ppg. Mike did not particularly want to pump off mud, but he felt it would be better to do so at this time than after the killing operation was started.

3. Pump rates. The pump output was read from the mud engineer's report as 5.2 strokes per bbl for the 6 × 18 in. duplex mud pump. The volumetric output at 21 spm was 0.1916 bbl/stroke × 21 spm = 4.0 bbl/min. Mike knew he could cripple his pumps according to the chart previously provided to him, but he felt that 4.0 bbl/min was not much more than the recommended 1–3 bbl/min as a kill rate.

4. Kill sheet preparation. Mike prepared his kill sheet as shown in Figure 2–19.

5. Working the pipe. While the mud weight was increased and the kill sheet was being prepared, the driller was instructed to work the pipe every 10 minutes by moving it up and down. He was also instructed not to move a tool joint through the annular preventer.

FIGURE 2-19 • Kill sheet

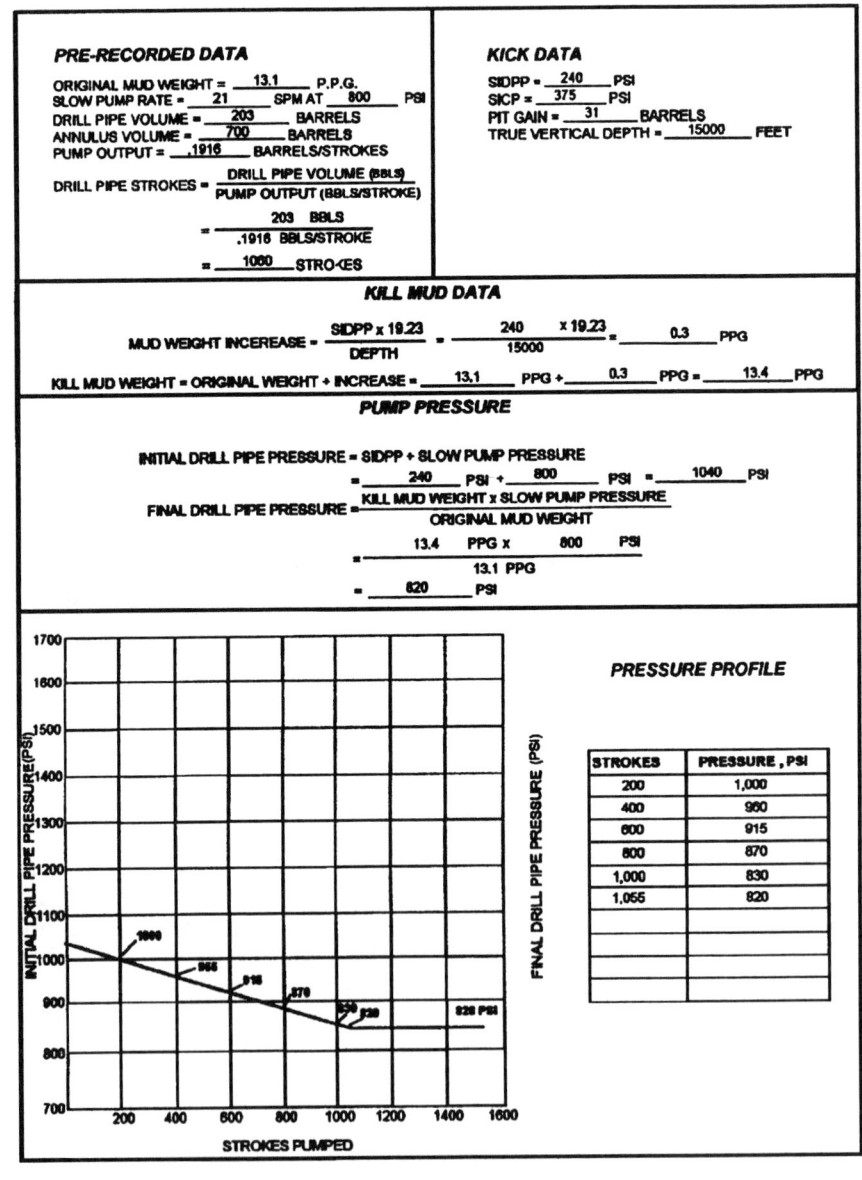

FIGURE 2-19(a) • Kill sheet (metric depths)

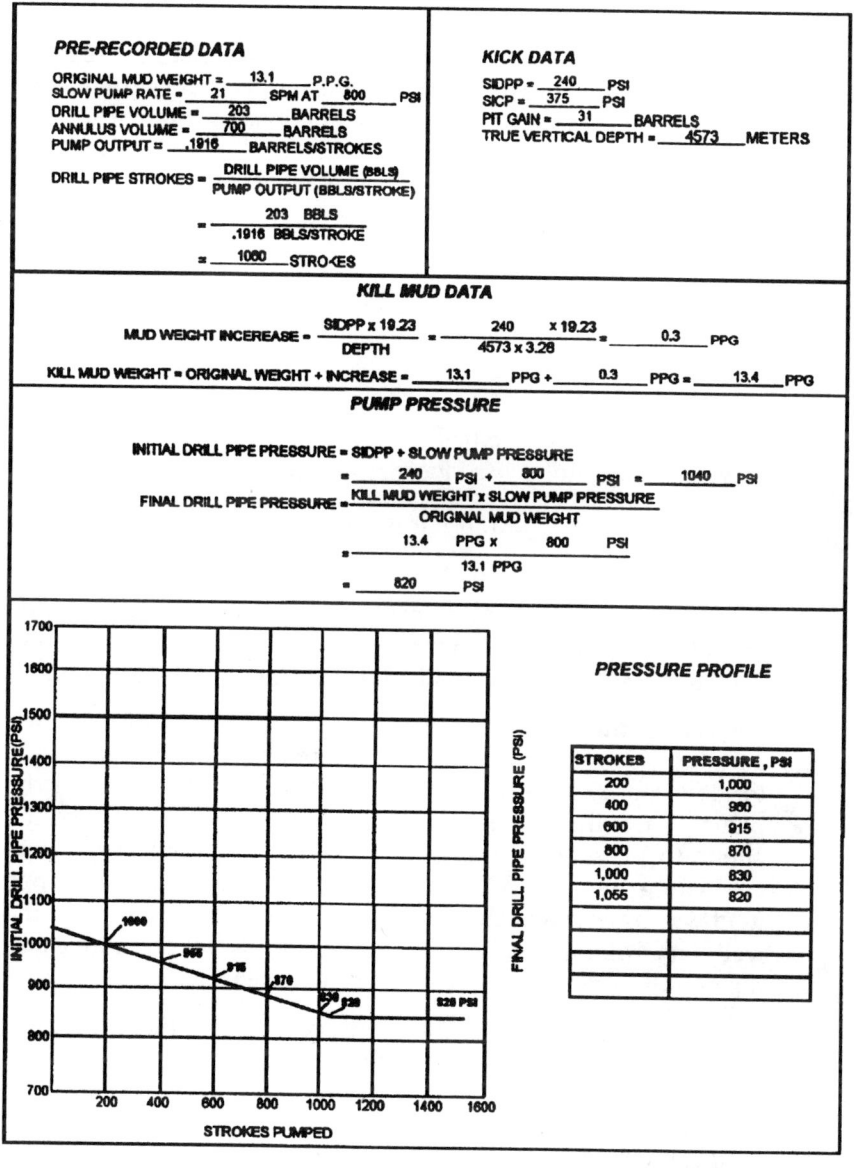

6. Displacing the drill pipe. After the mud was weighted to 13.4 ppg, Mike was ready to displace the drill pipe. He instructed the driller to start his pumps and run them at 21 spm. Mike cracked open the choke slightly and held his casing pressure at 375 psi until the driller had the pumps at the kill rate. The choke was used to control the drill pipe pressure to decrease gradually according to values on his kill sheet. The pressures were maintained as follows:

Strokes	Pressures, psi
200	1,000
400	960
600	915
800	870
1,000	830
1,055	820

When the drill pipe had been displaced, the pump was shut down and the choke was closed. The pressures were then:

$$SIDPP = 0 \text{ psi}$$

$$SICP = 350 \text{ psi}$$

The pressure on the drill pipe told Mike the heavier kill mud weight was sufficient to kill the well. If it had not been at sufficient density, some pressure would have remained on the drill pipe.

7. Displacing the annulus. Mike was now ready to displace the annulus with kill mud. He initiated pumping by adjusting his choke to maintain 350 psi on the casing while the driller started the pumps. After the pumps were running at 21 spm, Mike used the choke to maintain the drill pipe pressure constant at the final circulating pressure of 820 psi. He held this pressure until 13.4 ppg mud was observed at the shaker, at which time he closed in the well. The drill pipe and casing had zero pressure. The choke and the annular preventer were opened. The well was dead.

8. Post-kick considerations. There are several items that Mike considered after the well was dead. He circulated and conditioned the mud in the hole and added a trip margin to the mud weight so he could make a short trip. Additional barite was ordered from the mud company to resupply the bulk tank. Mike also took the time to inspect his equipment to identify any damage sustained from the kick.

NON-CONVENTIONAL WELL CONTROL PROCEDURES

Many attempts have been made to develop well control procedures based on principles other than the constant bottom-hole pressure concept. These procedures may be based on specific problems peculiar to a geological area. An example is low permeability, high-pressured formations contiguous to structurally weak rocks that cannot withstand hydrostatic kill pressures. Oftentimes, non-conventional procedures are used to overcome problem situations that result from poor well design. **These non-conventional procedures, developed for whatever reasons, are not applicable in most situations. Their implications should be understood and their use restricted.**

Low Choke Pressure Method

The low choke pressure method of well control is based on a simple technique. If the casing pressure tends to rise above a predetermined fixed value, the choke will be adjusted, as necessary, to control the pressure at or below the fixed value.

During the initial closure, if the shut-in pressure rises above a fixed value, immediate pumping will begin and the choke adjusted to control the pressure at or below the fixed value. It is intended that the below-necessary, low choke pressure will be sufficient to slow the continuing influx into the wellbore until the hydrostatic pressure needed to control the well can be reached with heavier mud.

The choke pressure limit is commonly called the "maximum allowable casing pressure" (MACP). Industry opinions differ considerably on the logic applied for the MACP and the overall results if it is implemented.

The typical well control situation involving a MACP is as follows:

- Prior to or during a kick, a maximum allowable casing pressure is calculated. A common value is 65–80% of the published internal yield rating of the pipe. There is no standard set for the MACP.
- Kick warning signs are observed and the well is closed in.
- At any time during the closing-in procedure or during kick circulation, the casing pressure will not be allowed to exceed the MACP. The choke is opened to maintain casing pressure at or below the MACP.

Operational procedures after implementation of the MACP vary among operators.

Several considerations are used by operators for implementing a MACP. These include the following:

- Improper casing design to handle the maximum kick situation.
- Uncertainty in well conditions resulting in more severe kick conditions than anticipated.
- Possible mishandling of a kick resulting in larger than anticipated pit gains and casing pressures.
- Casing wear, erosion, or corrosion that is (1) known factually, (2) believed to have occurred, (3) could be possible, or (4) uncertain as to its extent, if any.

Blowout experiences have shown field personnel may occasionally apply safety factors not consistent with logic used on casing design for the well. This point should be known by operators and addressed in their well management.

There is a danger inherent in the low choke pressure method. If the surface pressure needed to maintain a constant bottom-hole pressure is reduced, an underbalanced situation will allow a further influx into the annulus. If this is allowed to continue, the annulus will eventually become filled with influx contaminated mud, which necessitates high surface pressures to kill the well. These pressures will be higher than if the constant bottom-hole pressure method had been followed. (See Figure 2-20.)

FIGURE 2-20 • Casing pressures required to close in various influx volumes

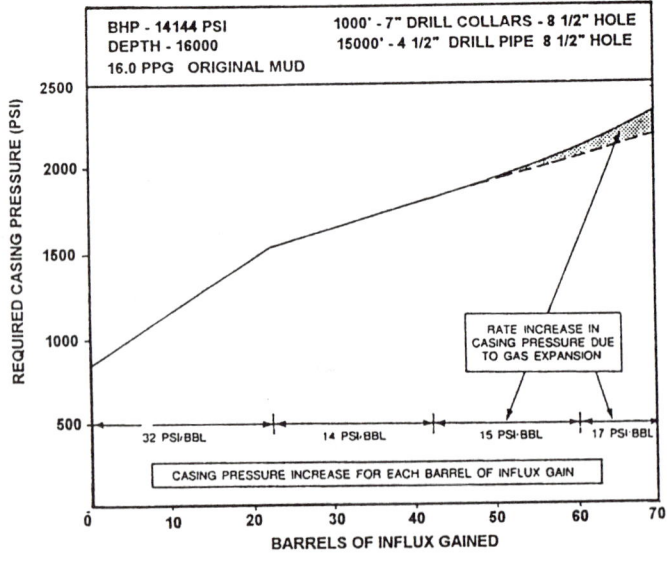

Well Control Procedures and Principles

Many variables affect the low choke pressure method and the possible pressure underbalances resulting from its use. A discussion is presented on each.

Casing Pressure to Be Allowed

The reason for selecting some maximum allowable casing pressure is the basis for this method of well control. Two factors for determining this maximum casing pressure are the casing burst and formation fracture pressures. These pressures are calculated and the lower pressure is designated as the maximum allowable casing pressure (MACP).

The calculation of formation fracture pressures take several variables into account. Among these are casing depth, formation compressive strength, and mud weight. From several authors' works, the formation strength can be approximated for a given set of conditions. The surface pressure that can be safely maintained without fracturing the formation can be approximated according to the following formula:

$$\text{Calculated maximum allowable pressure} = \text{Formation fracture pressure} - \text{Hydrostatic pressure} \quad (2.8)$$

Example 2.14

If the following data is known, calculate the maximum allowable casing pressure.

$$\text{Depth} = 3{,}000 \text{ ft}$$
$$\text{Fracture pressure} = 13.4 \text{ ppg (equivalent)}$$
$$\text{Mud weight} = 10.0 \text{ ppg}$$

Solution (Using Equation 2.8)

$$\text{MACP} = (13.4 \text{ ppg})(0.052 \text{ psi/ft/ppg})(3{,}000 \text{ ft})$$
$$- (10.0 \text{ ppg})(0.052 \text{ psi/ft/ppg})(3{,}000 \text{ ft})$$
$$= 2090 \text{ psi} - 1560 \text{ psi}$$
$$= 530 \text{ psi}$$

If this pressure is lower than the casing burst pressure, it will be labeled as the maximum allowable casing pressure.

Calculations for burst considerations are similar to calculations for formation fracturing. The maximum allowable surface pressure is the difference between the casing burst rating and the resultant hydrostatic pressure. (The resultant hydrostatic pressure at some depth of interest is equal to the mud hydrostatic pressure in the casing minus the back-up fluid outside the casing.)

$$\text{Calculated maximum casing pressure} = \text{Casing burst rating} - \text{Resultant hydrostatic pressure} \quad (2.9)$$

Example 2.15

Using the following data, calculate the maximum allowable casing pressure.

Depth = 3,000 ft

Casing burst pressure = 2,730 psi (13 3/8 in., K-55, 54.5 lb/ft)

Mud weight = 10.0 ppg

Back-up fluid = 9.0 ppg

Solution

1. Resultant hydrostatic pressure

   = (10.0 ppg) (0.052 psi/ft/ppg) (3,000 ft)

   − (9.0 ppg) (0.052 psi/ft/ppg) (3,000 ft) = 156 psi

2. Maximum allowable casing pressure

   = 2730 psi − 156 psi

   = 2574 psi

If the burst pressure is lower than the formation fracture pressure, it is labeled as the maximum allowable casing pressure.

The working pressure rating of the surface preventer equipment may be the criteria for the pressure limitations. When this is the case, the maximum allowable pressure is equal to the preventer working pressure rating. If the preventer selection criteria presented in Chapter 1 is exercised, it is unlikely that the preventer pressure rating will ever be the limiting criteria.

Well Depth

Well depth affects the low choke pressure method by causing changes of the maximum allowable surface pressure. The changes result from deeper and stronger strings of casing. The deeper casing setting depths usually have higher fracture pressures resulting in more allowable surface pressure before fracturing will occur. Also, the burst pressures of deeper strings are normally higher. The variations from well depth changes are most apparent when considering surface casing versus deeper casing strings.

Surface casing in shallow depths increases the danger of placing allowable choke pressures limitations. The detection of a shallow depth kick is often difficult due to the fast penetration rates and fluctuating pit volumes associated with shallow depth drilling. This difficulty may result in a large gain requiring a high casing pressure.

The formation fracture pressure usually associated with surface casing is low. If the casing pressure exceeds the low allowable surface pressure, the well would not be completely closed in, and the influx continues.

The problems with deep wells and long casing strings are not as severe as with shallow kicks. Rig personnel are usually more kick conscious when drilling deeper wells, thus resulting in kicks being closed in with less gain in the pits. Also, the maximum allowable surface pressure limitations are higher due to increased burst pressure ratings of the casing and the increased formation fracture pressures.

There are problems associated with deeper wells. As the depth increases, the hole size generally decreases with each deeper casing string. This gives a larger annular ft/bbl ratio. An influx will have higher surface pressure than the same volume in a large diameter hole.

Regardless of the hole geometry, the maximum allowable surface pressure remains the same, since formation fracture and casing burst pressure remains relatively the same. A small diameter hole is potentially more dangerous because a smaller volume of influx is required to exceed some maximum allowable pressure. (See Figure 2–21.)

Influx Type

The influx type affects the low choke pressure method. Saltwater or oil may tend to promote the effectiveness of this method during a kick. Influxes of gas may increase the danger. Influx type affects the low choke method because of the physical characteristics of the influx and the mud contamination it causes. The physical characteristics include volume expansion capabilities, entry rate into the wellbore, and influx density. The mud contamination effects are friction losses occurring from drilling fluid contamination.

Annular friction pressure acts downward against the fluid in the annulus and against the formation wall. Friction pressure increases total pressures on the formation, which reduces surface pressure necessary to

FIGURE 2-21 • Effect of hole size on casing pressures

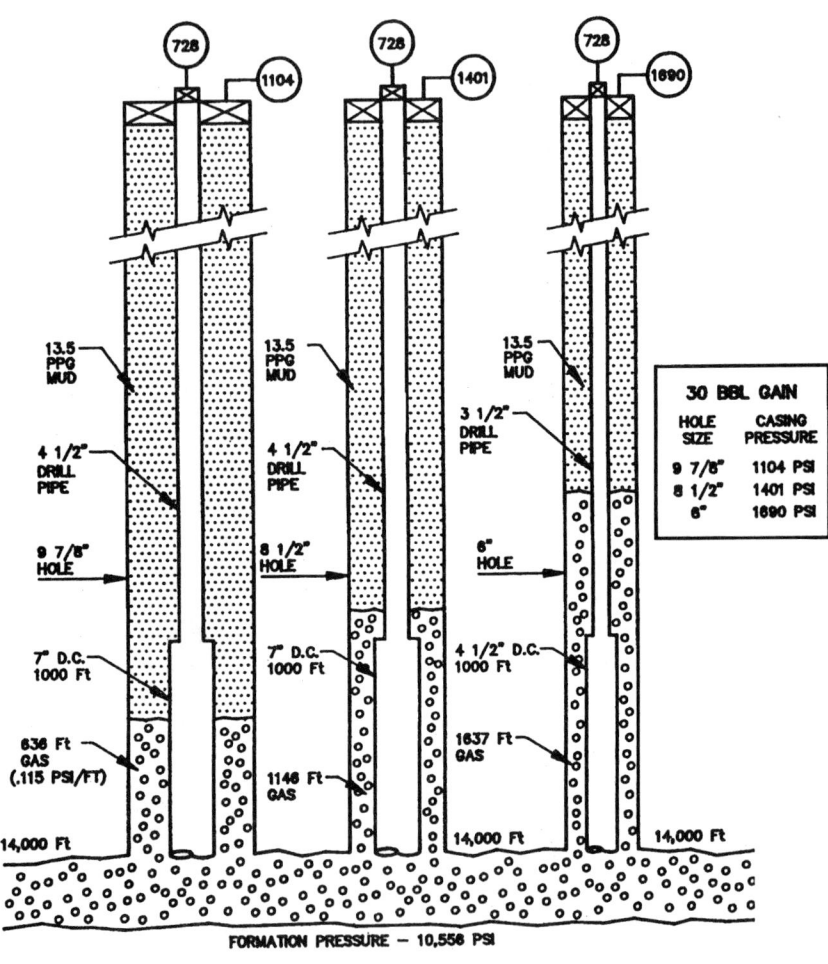

balance formation pressure. Since the annular friction losses are difficult to calculate during a kill operation, this pressure safety factor is seldom taken into account.

The annular friction pressure may aid in the kill operation if the low-choke pressure method is applied. When the casing pressure is reduced below the amounts needed to balance the formation pressure, as is the case when a surface pressure restriction is imposed, an additional influx will enter the borehole and contaminate the drilling fluid. Contaminated fluids generally have higher friction losses. (See Figure 2-22.) If the influx

FIGURE 2-22 • Static and dynamic back pressure curves for saltwater kicks with various degrees of friction loss. The friction loss reduces the necessary casing pressure, in a dynamic system, to zero at some point before the circulation is completed.

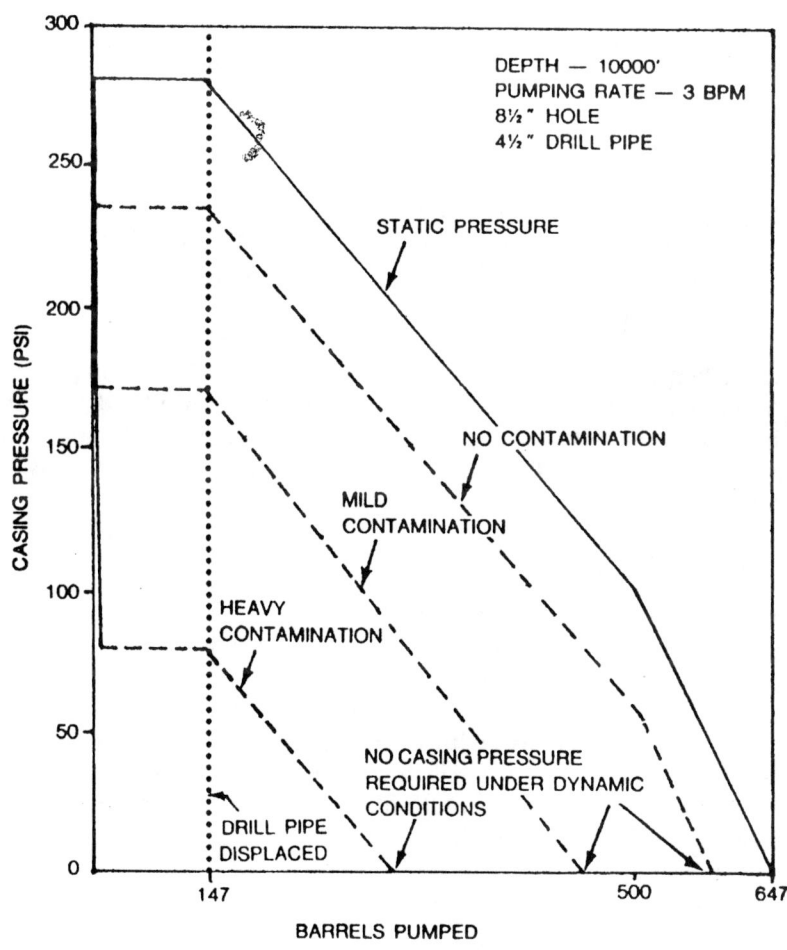

contaminates the mud system, the friction losses overcome the deficit from the low surface pressure.

The entry rate of the initial influx and any succeeding influxes affect the applicability of the low choke pressure method. Under the same pressure and permeability conditions, salt water and oil enter the wellbore at a lower rate than gas. Detection of a salt water or oil kick based on flow rates allow closure with less total pit gain than with a gas kick. Also, secondary or continuing influxes of salt water or oil will be less during the underbalanced pressure situations.

In summation, salt water and oil kicks are less dangerous than a gas kick when the low choke pressure method is applied. This is due to increased friction pressures, higher fluid pressure gradient, nonexpandable volume, and low entry rates. However, any influx is dangerous, regardless of type.

Influx Volume

The concept of minimizing influx volumes is the most important factor in determining the applicability, or practicality, of the low choke pressure method. During a kick killing operation, if the required casing pressure is not allowed to increase as necessary, a new influx occurs. The size of successive gains, or influxes, depends on the influx type, formation permeability, amount of pressure underbalance at the formation face, and duration of pressure underbalance. Since these cannot be determined accurately during the killing operation, it is dangerous to make the assumption that successive gains would be small or that these gains can be controlled.

Original Kill Method Applied

To a degree, the original kill method determines the amount of casing pressure necessary to kill the well. The two workable variations of the constant bottom-hole pressure method are the one and two circulation methods. The one circulation method using the kill mud with no safety factors is the safer method with lower necessary casing pressures and downhole stresses. At any time the low choke pressure method is applied during the use of the two circulation method, the underbalance will be greater than if the one circulation method had been used.

Conclusion

The variables associated with the low choke pressure method are wide ranging and generally not related. Actions of the variables are complicated. It is not safe to assume a variable such as friction pressure loss would nullify any negative actions of variables such as volume expansion, hole geometry, or influx sizes.

Also, the basis of the constant bottom-hole pressure method is that pressure will be maintained on the kicking formation equal to or greater

Well Control Procedures and Principles

than the formation pressure. The low choke pressure method disregards this approach and offers no acceptable substitute. This method *cannot* be considered as a feasible method of well control suitable for general use.

By definition, the low choke pressure method will be applied at any time that the casing pressure exceeds a maximum pressure limitation. This implies that the method will be implemented upon closure if the shut-in casing pressure tends to rise above the fixed value. If this action is taken, several additional complications must be taken into account. First, the two circulation method must be applied originally, since time is not allowed to weight the mud before circulation is begun. Second, no shut-in pressures can be recorded; consequently, the kill weight mud cannot be accurately calculated. Third, the influx is continuous into the wellbore. (See Figure 2-23.) As a result, the low choke pressure method is unpredictable if it is used before complete closure.

FIGURE 2-23 • Continuous influx due to use of low choke pressure method below complete closure

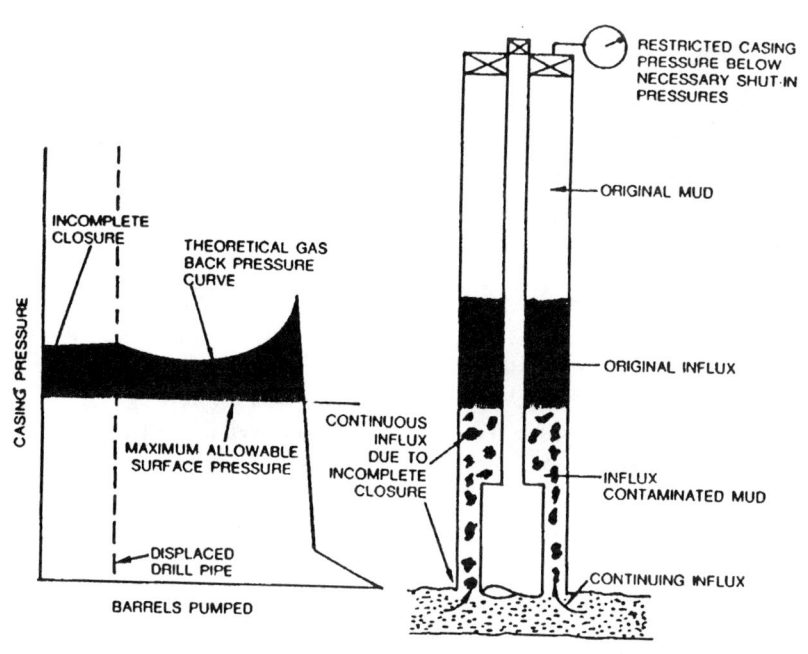

Alternate Options

Options when a MACP situation arises are as follows:

- Continue the circulation using normal procedures and disregard the MACP limit.
- Continue circulating and do not exceed the MACP.
- Pump at high, manageable rates to increase annular friction pressure and reduce the underbalanced circulation time. Use the MACP.

Other options infrequently used in this situation are bullheading and reverse circulation.

The time and pressure when the well is underbalanced are the prime considerations when discussing the MACP implemented as a result of casing strength limitations. Additional influxes will occur when a MACP is implemented. If the "kh" factor (permeability [K]-zone height) is small, the influx may be manageable. But a high "kh" factor will allow a larger influx. The MACP can be implemented with more confidence in areas where formation permeability is known to be small.

Simulations can easily model this situation. However, operational time constraints at the time of the event normally preclude modeling work. If an expert system is being used on a well, the computer capabilities may exist to do the modeling work and give results useful in decision making.

Continue the circulation using normal procedures and disregard the MACP. These procedures do not differ from conventional circulation procedures for removing kicks from a well.

High pressure can cause a casing rupture deep in the well. Well integrity will be lost. Mud losses will occur. An underground blowout may develop.

Signals include rapid and unexpected casing pressure decreases and mud losses. The procedures to fix the problems are the same as any deep mud loss or underground blowout. Also, bullheading can remove influx from the annulus if the kick fluid was above the loss zone.

The casing may rupture near the surface. The well integrity will be lost. Mud losses will occur. If the exterior casing strings fail near the surface, rig integrity could be lost. A surface blowout may occur.

Obtain expert assistance. It may be advisable to shut in the well before abandoning the location. This operation may aggravate the situation so consideration must be given before it is implemented.

Continue circulating and do not exceed MACP. This may create high pressure resulting in casing rupture deep in the well. (Note: Downhole stresses can change even though surface pressure may remain constant.) The procedures to fix the problems are the same as any deep mud loss or underground blowout. Bullheading can remove influx from the annulus if the kick fluid was above the loss zone.

The casing may rupture near the surface. (Note: This situation has a low probability if the casing did not rupture previously from the high casing pressure.) It is possible because downhole stresses may change while surface pressure remains constant. Shut in the well and evaluate the situation. Obtain expert assistance. Consider evacuation.

More influx will be incurred during the time the MACP is implemented. (Discount friction pressure control on the influx.) Procedures to fix the problems:

1. Neglect the MACP and maintain the appropriate amount of casing pressure.
2. Alternatively, pump at high rates to maximize annular friction pressure and reduce the time of underbalance.

Pump at high, manageable rates to increase annular friction pressure and reduce the underbalanced circulating time. Use the MACP. During the time the MACP is implemented, pump as fast as allowed by the rig pumps and still manage to properly maintain control of the drill pipe pressure with the choke. A new drill pipe circulating pressure must be determined if not previously known at the high rate.

This method increases the annular friction pressure to serve as a downhole choke on the influx rate without affecting the MACP. If the additional annular friction pressure can be estimated, the MACP can be decreased by this amount.

When the casing pressure falls below the MACP, reduce the pump rate to the original circulating pressure.

Alternate Solutions
Outrunning a Kick

Many kicks occur when the drill string is not at the bottom of the well. Kill options available to the operator are to strip back to bottom and circulate kill mud, or circulate a heavy mud at the shut-in position to kill the well and then strip back to bottom. This is called a "top kill." An important point to remember is the original mud weight was enough to control the well prior to the kick. The kill mud weight will be equal to the original mud weight with the bit on bottom.

Realizing the drill string must finally be returned to bottom and that stripping is a tedious operation, occasionally an operator will try to "run to bottom" before shut-in procedures are used. This allows a continuous influx necessitating high casing pressures, and it requires handling large volumes of kick fluids when circulated to surface. Even the running of a few stands of pipe is dangerous and should not be attempted.

The success of this method depends on formations with low permeabilities allowing small volumes of influx into the well. This is generally an unknown variable and should not be used as the basis of well control procedures.

Constant Pit Level Method

The censtant pit-level method stems from a logical principle. During a well killing procedure, the choke was adjusted to avoid any changes in the mud pits at the surface. It was believed if the pits were not allowed to gain, no additional influx would occur. Also, since the pit volume would not be allowed to decrease, lost circulation problems would be avoided.

This procedure worked successfully when the influx was saltwater or oil. Since these fluids have minimum expansion capabilities, the pit volumes would not change appreciably. However, gas kicks should have associated gains in the pits. If these gains are not allowed, the casing pressure must be increased to force gas compression, which probably will result in lost circulation.

Pumping a Kick Back into the Formation

Occasionally, an attempt is made to "pump a kick back into the formation" to avoid the necessity of implementing other kick kill procedures. This should not be misconstrued as practical in all situations. It implies the formation must be fractured before pumping can occur. It is not probable the kick fluid will reenter the original zone unless clear water is the circulating fluid. Pore channel plugging with barite and bentonite occurs when mud is in use.

In limited situations such as certain hydrogen sulfide kicks, it may be advisable to pump the fluid into a formation rather than circulate it to the surface. These decisions should be discussed prior to drilling the well, and they should be part of the contingency plan.

PROBLEMS

2.1   Calculate the pressure reduction when 450 ft of 4 1/2 in., 16.6 lb/ft pipe is pulled without filling the hole. The hole diameter is 7 7/8 in. and 15.6 ppg mud is in use.

2.2   Using the data in Problem 2.1, what would be the pressure reduction if 450 ft of 6.0 × 2.0 in. collars are pulled without filling the hole?

WELL CONTROL PROCEDURES AND PRINCIPLES 153

2.3 A well is drilled to 13,000 ft with 13.2 ppg mud. The bottom-hole pressure in a gas sand at that depth is 8,710 psi. The intermediate casing is 43.5 lb/ft, 9 5/8 in. pipe set to 11,000 ft. The drill pipe is 4 1/2 in., 16.6 lb/ft, and the collars are 7 × 2 in. diameter. The operator requires hole filling after 5 stands of drill pipe or collars are pulled. Will the well kick when pulling the drill pipe? Drill collars? (Assume no swabbing effects.)

2.4 Calculate the pump strokes required to fill the hole in Problems 2.1, 2.2, and 2.3 for each of the following pumps. Assume a 90% efficiency in all cases.

a. 6 1/2 × 18 in. duplex

b. 7 × 14 in. duplex

c. 4 × 10 in. duplex

d. 3 × 6 in. triplex

e. 3 × 10 in. triplex

f. 4 × 8 in. triplex

2.5 In an effort to measure the output of a 3 × 10 in. triplex pump under actual conditions, a 20 bbl trip tank was filled with mud. A total of 1,274 strokes were required to empty the tank. What is the pump output? What is the pump efficiency percentage?

2.6 A drill string contains 12,000 ft of 4 1/2 in., 16.6 lb/ft drill pipe and 1,000 ft of 7 × 2.5 in. collars. How many strokes would be required to displace the pipe if a 4 1/2 × 10 in. triplex pump is used at 90% efficiency? At 80% efficiency?

2.7 A kick was taken on a well in which these pressures were recorded:

Shut-in time minutes	SIDPP, psi	SICP, psi	Pit gain, bbl
0	250	375	18
5	290	415	18
10	290	415	18
15	295	415	18

a. What is the true shut-in drill pipe pressure?

b. The drill string contains 10,600 ft (TVD) of 12.1 ppg mud. What is the bottom-hole pressure?

2.8 Using the data given below, what are the true shut-in pressures?

Increment no.	Bleeding volume, bbl	SIDPP, psi	SICP, psi
0	–	760	1,160
1	1	700	1,100
2	1	640	1,040
3	1/2	630	1,020
4	1/2	630	1,020
5	1/2	630	1,030
6	1/2	630	1,045

2.9 Given the solution from Problem 2.8, what is the bottom-hole pressure for the following situations?

Well no.	True vertical depth, ft	Mud weight, ppg
1	10,750	10.4
2	13,500	14.6
3	8,300	9.5
4	15,000	11.7
5	5,500	9.9

2.10* After a kick had been taken, bleeding procedures were implemented to check for trapped pressure. Using the results given below, what are the true shut-in pressures?

Increment no.	Bleeding volume, bbl	SIDPP, psi	SICP, psi
0	–	250	690
1	1	175	615
2	1	100	540
3	1	50	490
4	1	20	470
5	1	0	450
6	1	0	450
7	1	0	460

2.11* Using the answer from Problem 2.10, what is the bottom-hole pressure if the mud weight was 12.1 ppg and the well depth was 13,130 ft?

2.12 A kick was taken on a well in which a float valve was used in the drill string. The SIDPP was read as 0 psi and the SICP was 675 psi. The kill rate and associated pressure were 32 spm at 700 psi. After implementing the procedures to establish the true SIDPP (see Table 2–5), the total pumping pressure was 1,150 psi. What was the SIDPP?

Well Control Procedures and Principles

2.13 Using the data given below, calculate the bottom-hole pressure.

SIDPP = 0 psi (float valve in the drill string)
SICP = 400 psi
Kill rate = 45 spm at 750 psi
Total pumping pressure (initial) = 900 psi
Depth (T.V.D.) = 11,000 ft
Mud weight = 12.9 ppg

2.14 While drilling the Texas Rover No. 1, a kick was shut in with a SIDPP of 400 psi and a SICP of 550 psi. A kill rate had not been established previously. The pumps were started and run at 21 spm while the casing pressure was maintained at 550 psi. The total drill pipe pressure was observed to be 1250 psi. What is the pumping pressure at 21 spm?

2.15 Using the same conditions given in Problem 2.14, the pumps were run at 35 spm, which attained a total pumping pressure of 1,500 psi. What is the pump pressure at 35 spm?

2.16 While killing the kick in Problem 2.14, pump No. 1 washed out a valve while displacing the drill pipe with kill mud. The well was shut in, and the pressures recorded as 175 psi SIDPP and 525 psi SICP. Rather than repair pump No. 1, pump No. 2 was started at 45 spm while the casing pressure was held at 525 psi. The total drill pipe pressure at 45 spm was observed to be 1,475 psi. What was the kill rate for pump No. 2?

2.17 A kick occurred on a well in which the kill rate was not known and a float valve was used in the drill string. The SIDPP was 0 psi and the SICP was 500 psi. A low volume, high pressure pump was connected to the stand pipe and pressure applied. The following results were obtained. What was the SIDPP?

Volume pumped, bbl	SIDPP, psi
0	0
1	0
2	40
3	90
4	140
5	190
6	210
7	215
8	215
9	220

2.18 Using the results from Problem 2.17, the rig pumps were subsequently run at 25 spm with a total pressure of 950 psi. The casing pressure was held constant throughout this procedure. What was the kill rate pressure?

2.19 What is the probable kick influx fluid using the following data? (Assume the kick fluid to be around the drill collars.)

SIDPP = 400 psi

SICP = 600 psi

Pit gain = 25 bbl

Mud weight = 12.0 ppg

Drill collars = 6 in. OD

Hole size = 9 7/8 in.

2.20 Using the data from Problem 2.19, what is the probable influx if the original mud weight was 17.0 ppg?

2.21 If the following data is known, what fluid type entered the well? (Assume no drill collars.)

SIDPP = 800 psi

SICP = 1,400 psi

Pit gain = 20 bbl

Mud weight = 15.0 ppg

Drill string = 3 1/2 in.

Hole size = 6 in.

2.22* Using the same data from Problem 2.21, what fluid entered the well if the SICP was 1,100 psi?

2.23 A kick was taken on a well in which the following data was known. What was the kill mud weight?

SIDPP = 250 psi

SICP = 475 psi

Measured depth = 12,750 ft

True vertical depth = 12,000 ft

Mud weight = 13.4 ppg

2.24 Calculate the kill mud weight for Problem 2.7.

2.25 If the information from Problem 2.8 is known, what must be the mud weight increase if the depth (TVD) is 12,300 ft?

Well Control Procedures and Principles

2.26 Calculate the kill mud weights for the situation developed in Problems 2.8 and 2.9.

2.27* What mud weight increase is necessary for Problem 2.10?

2.28* If the following data is known, what is the mud weight increase necessary to control each situation?

	SIDPP, psi	SICP, psi	Pit gain, bbl	Depth, TVD ft
a.	300	500	30	14,200
b.	300	4,750	108	14,200
c.	450	800	18	21,630
d.	300	500	unknown	14,200
e.	300	unknown	35	14,200
f.	unknown	500	35	14,200
g.	0	500	35	14,200

2.29* Using Figure 2-24, calculate the equivalent mud weights at 1,000; 2,500; 5,000; 7,500; 10,000; 12,500; and 15,000 ft for each pumping interval.

FIGURE 2-24 • Fluid arrangements for various pumping intervals during a typical well control operation

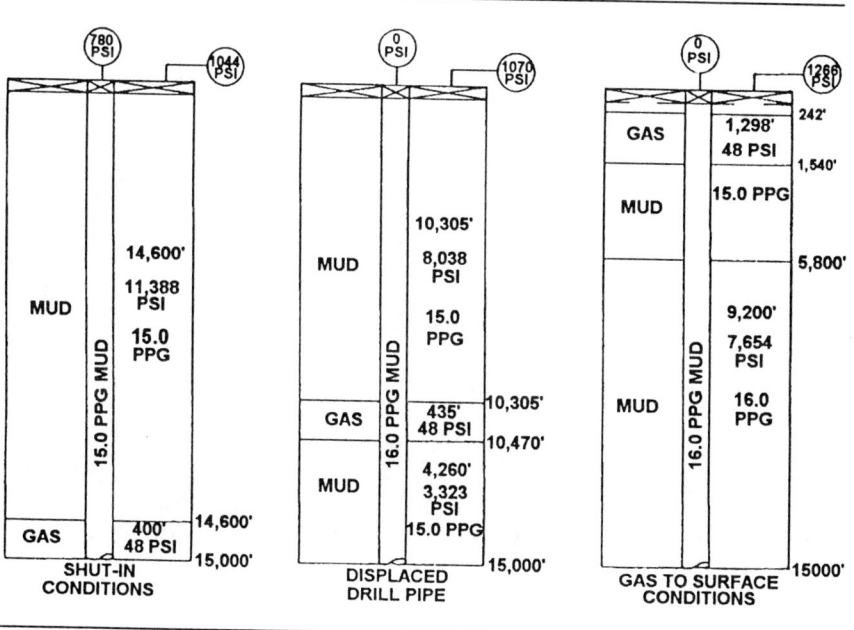

2.30 Prepare a kill sheet using this kick data:

Original mud weight = 13.6 ppg

Slow pump rate = 25 spm at 750 psi

Drill pipe = 11,000 ft, 4 1/2 in., 16.6 lb/ft

Drill collars = 1,000 ft, 6.0 in. OD × 2.0 in. ID

Depth = 12,000 ft (measured depth)

11,500 ft (TVD)

Pump = 6 1/2 in. liner, 16 in. stroke, 2.5 in. rod, 90% efficiency (duplex)

SIDPP = 300 psi

SICP = 600 psi

Pit gain = unknown

2.31 Calculate and plot a drill pipe pressure schedule (kill sheet) for the following kick:

Original mud weight = 10.5 ppg

Slow pump rate = 30 spm at 900 psi

Drill pipe = 8,000 ft, 3 1/2 in., 13.3 lb/ft

Drill collars = 500 ft, 5.5 in. OD × 2.0 in. ID

Depth = 8,500 ft (TVD)

Pump = 3 in. liner, 8 in. stroke, 80% efficiency (triplex)

SIDPP = 250 psi

Pit gain = 15 bbl

2.32* Prepare a static drill pipe pressure graph for the following circumstances.

SIDPP = 800 psi

Depth = 14,000 ft (TVD)

Drill pipe = (1) 12,000 ft, 4 1/2 in., 16.6 lb/ft

(2) 1,500 ft, 3 1/2 in., 13.3 lb/ft

Drill collars = 500 ft, 5.0 in. OD × 1.5 in. ID

Pump = 6 in. liner, 18 in. stroke, 90% efficiency (duplex)

SOLUTIONS

Note: All solutions are approximate depending on user's round off procedures.

2.1 43.1 psi

2.2 389 psi

2.3 Pulling pipe; no kick
Pulling collars; kick

2.4

Pump	Problem/Strokes			
	2.1	2.2	2.3:dp	dc
a.	14	68	15	99
b.	15	73	16	106
c.	67	323	69	469
d.	249	1,197	257	1,709
e.	149	718	154	1,041
f.	104	500	108	725

2.5 0.016 bbl/stroke; 73.4%

2.6 90%; 3,982 strokes
80%; 4,480 strokes

2.7 a. 290 psi
b. 6,960 psi

2.8 SIDPP   630 psi
SICP   1045 psi

2.9

Well	BHP,psi
1	6,443
2	10,879
3	4,730
4	9,756
5	3,461

2.10 SIDPP   0 psi
      SICP   460 psi
2.11 8,261 psi
2.12 450 psi
2.13 SIDPP 150 psi
      BHP   7,528 psi
2.14 850 psi
2.15 1,100 psi
2.16 1,300 psi at 45 spm
2.17 215 psi
2.18 735 psi
2.19 Gas (0.14 psi/ft)
2.20 Probably oil or salt water
2.21 Gas
2.22 Salt water
2.23 13.8 ppg
2.24 12.6–12.7 ppg
2.25 1.0 ppg
2.26

Well	Mud Weight, ppg
1	11.5–11.6
2	15.5
3	11.0
4	12.5
5	12.1

2.27 None
2.28 a. 0.4
      b. 0.4
      c. 0.4
      d. 0.4

e. 0.4

f. unknown (0.7 ppg maximum)

g. 0.0

2.29

Depth, ft	Equivalent Mud Weight		
	Shut-in	Displaced Pipe	Gas to Surface
1,000	35.1	35.6	28.5
2,500	23.0	23.2	17.3
5,000	19.0	19.1	16.2
7,500	17.7	17.7	16.0
10,000	17.0	17.1	16.0
12,500	16.5	16.2	16.0
15,000	16.0	16.0	16.0

2.30 – 2.32   Kill sheets not shown here.

REFERENCES

Adams, N. *Well Control Problems and Solutions*. Tulsa, Oklahoma: The Petroleum Publishing Company, 1980.

Adams, N.J. *Drilling Engineering: A Well Planning Approach*. Tulsa, Oklahoma: PennWell Publishing Company, 1985.

Adams, N.J. *Drilling Problems and Drilling Optimization*, Boston, MA: IHRDC Publishing Company, 1993.

Adams, N.J. *Workover Well Control*. Tulsa, Oklahoma: Petroleum Publishing Company, 1981.

Eaton, B. A. "Fracture Gradient Prediction and Its Application in Oilfield Operation," *Journal of Petroleum Technology*, October 1969, pp. 1353–1360.

Mathews, W.R. and J. Kelly. "How to Predict Formation Pressures and Fracture Gradients," *Oil & Gas Journal*, February 1967.

Prentice, C. M. "Maximum Load Casing Design," SPE 2560, SPE Fall Technical Conference, Denver, Colorado, 1969.

CHAPTER 3

SPECIAL PROBLEMS

DEEPWATER WELL CONTROL

Well control problems occur during deepwater drilling. Although the mechanics of well control are not altered significantly, the implementation may be different due to equipment involved when drilling in deep water and other operational difficulties such as hydrates.

Shallow gas blowouts in shallow or deep water are discussed later in this chapter.

Kick Detection

Kick detection in deepwater environments is becoming an increasingly important issue. Lower fracture gradients than similar land or shallower water cases reduce the kick tolerance margin. Also, the industry's trend toward slim holes makes kick volume reduction a key point.

Two early warning signs of kicks are an increase in flow rate and pit volume. These signs are difficult to detect when drilling from floating vessels due to the nature of the drilling vessel motion. Generally, a floating vessel drills deepwater wells. Wave action moves the drilling vessel and creates pit level fluctuation, even though the total pit volume may remain constant. The same difficulty is observed in detecting flow rate changes.

Equipment compensating for motion-caused fluctuations will minimize these problems in early kick detection. A pit volume totalizing (PVT) system will detect and report overall pit gains by using multiple pit monitors and resolving individual losses and gains reported by each monitor into a single value. (See Figure 3–1.)

FIGURE 3-1 • Pit level fluctuations due to rig heave and associated pit monitor readings (mud volume is the same in all pits). The PVT system accounts for the level changes and indicates true pit volumes.

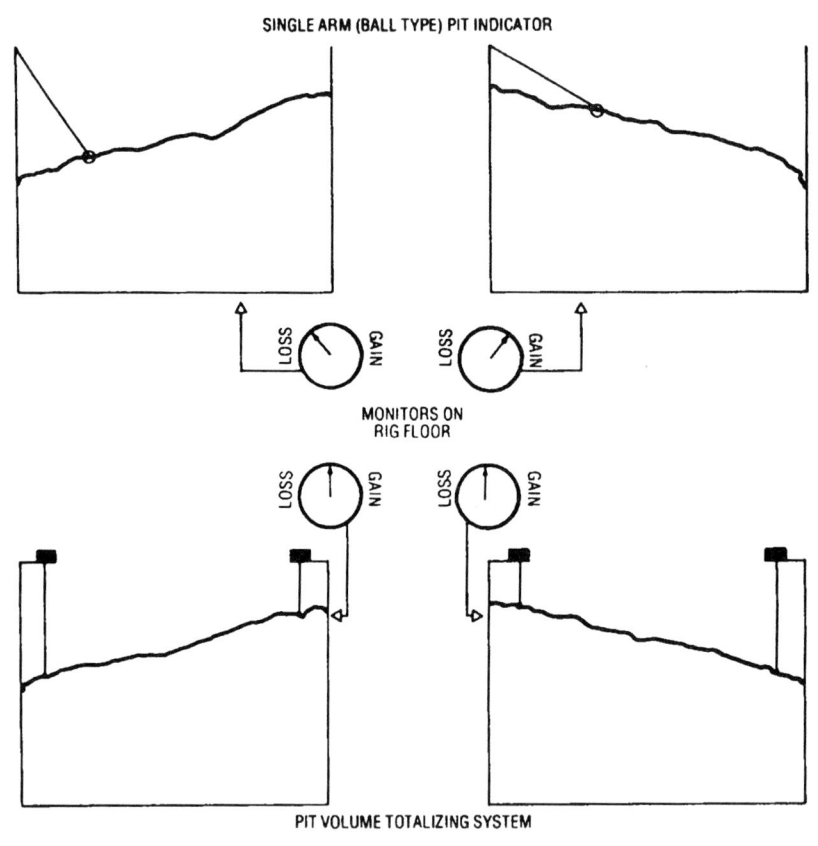

New detection systems are rapidly improving the quick shut-in capability to reduce kick volumes. MWD systems monitor influxes downhole. Also, sensors monitoring mud pulses in the drill pipe and returning annular fluids can quickly detect influxes even with low volumes. These types of devices are essential to improving slim hole drilling and well control.

Fracture Gradients

The open hole formation must exhibit a competency to allow well killing without losing circulation. This implies the formation fracture gradients must be greater than the equivalent mud weights during kill operations.

Special Problems

Fracture gradients in deep water are effectively less than those observed on land or in shallow water at equivalent drilling depths. This reduction is due in part to the lower overburden stresses in the deepwater case because of the water depth and the soil gradation from clay-silt and sand to shale and sandstone.

Figure 3-2 illustrates this concept. The overburden stress on land at 10,000 ft (3,050 m) is 10,000 psi (69 MPa). It is 8,930 psi (61.6 MPa) in the deepwater case at the same depth. Although this example does not encompass all parameters, it illustrates the problem.

Several authors have examined procedures for calculating fracture gradients. Until recently, their work was confined to land operations. Charts and equations are presently available to estimate fracture gradients in various water depths. Figure 3-3 shows a chart prepared by Kendall to estimate fracture gradients in these situations. More research is needed in this area. Additional fracture gradient data collection at deepwater locations is the first step.

Kicks Below Protective Casing

When protective casing is set at a depth sufficient to achieve fracture gradients in excess of kick equivalent mud weights, conventional kill procedures can be implemented. Procedure differences cause deepwater well control to be more complicated than land or shallower water situations.

A procedure has been developed in which a set of pipe rams is closed, and a drill string tool joint is allowed to rest on the rams. These special

FIGURE 3-2 • Offshore overburden

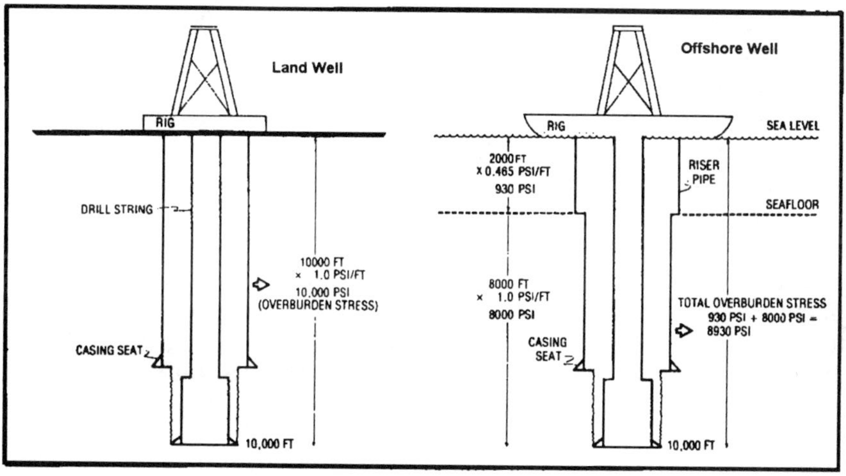

FIGURE 3-3 • Effect of water depth on fracture gradient (Courtesy of Hal Kendall)

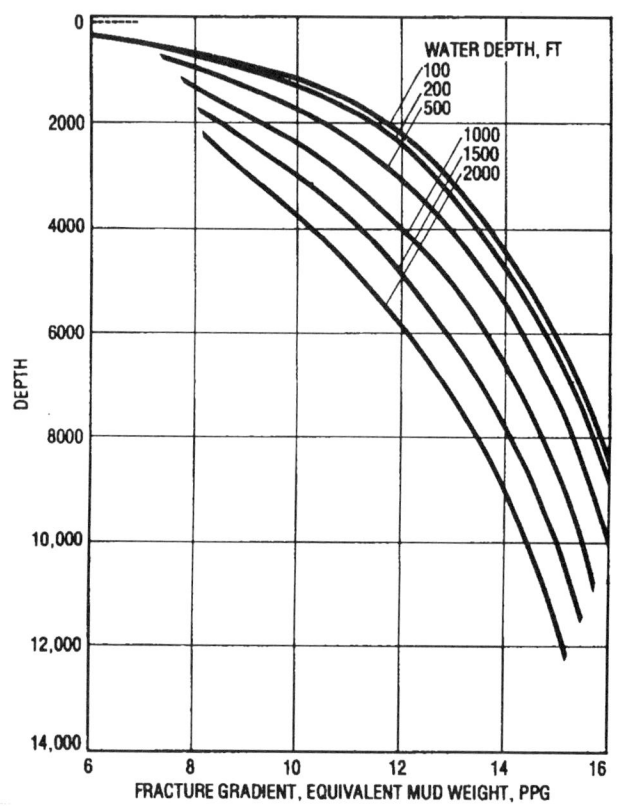

rams will support the drill string weight while holding full test pressures. The drill string remains motionless resulting in no damage to the preventers while the killing operation is in progress.

Basic blowout preventer selection logic, as presented in Chapter 1, has been extended to subsea preventers. Since repair and replacement of worn elements are impractical during a killing operation, several components are added to the stack to provide a back-up system.

The basic stack shown in Chapter 1 includes two annular preventers rather than a solitary unit. Primary and secondary choke lines are used. Shear rams are placed near the top of the stack, in case it becomes necessary to shear the drill string before moving the drilling vessel. Lower pipe rams are used if upper choke lines or flanges fail resulting in the need to close in the well without necessarily shearing the drill string. BOP stack configuration has undergone much discussion during recent years, and sound logic exists for several BOP arrangements.

Special Problems

Choke Line Length

Choke line length has become a critical consideration for several reasons such as associated friction pressures while circulating and small diameters and volumes that can allow complete fluid displacement to occur in a short time. Choke lines used on land and in shallower water often have the same problems but to a minor degree. These problems will necessitate slight alterations in the kill procedure and force the operator to exercise caution.

The friction pressures in long choke lines have been called the "hidden choke effect." As seen from Figure 3–4, it would take 233 psi (1.61

FIGURE 3–4 • Effect of choke line size and flow rate on friction pressure gradients. No allowances are made for internal restrictions.

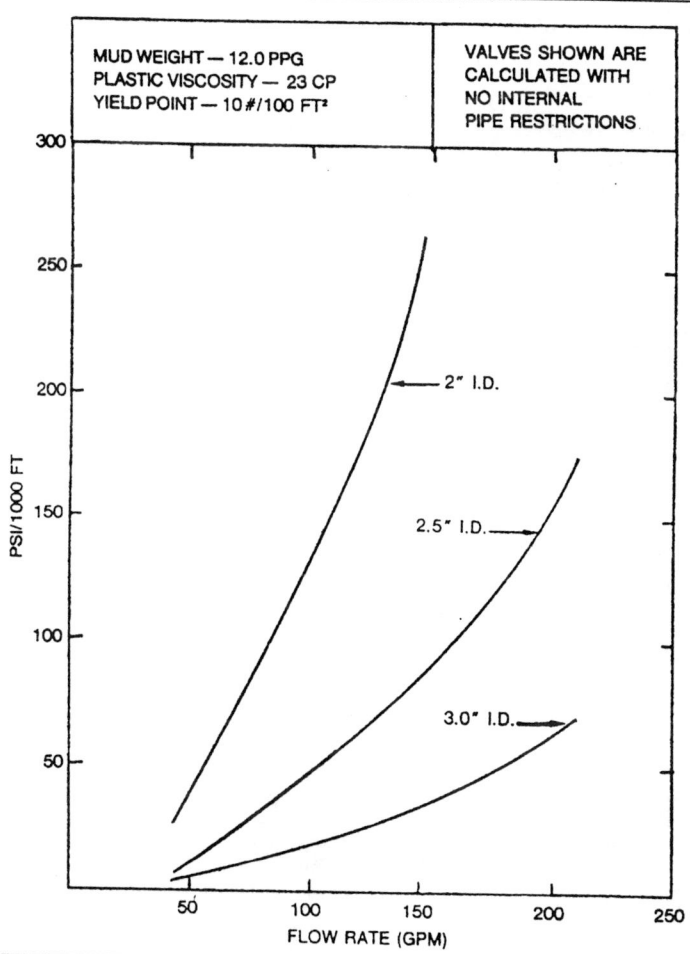

MPa) to pump at a rate of 200 gpm (45 m³/h) through a 2.5 in. (63 mm) ID choke line used in 1,500 ft (460 m) of water. This pressure is required at the base of the choke line, but it is expended within the line and is never recorded at the surface.

Friction pressure is applied at every point below the preventers, as seen in Figure 3-5. This increases the possibility of formation fracture. The friction pressure consideration becomes more severe with higher mud weights or viscosities and increased water depths.

A procedure should be developed to ensure the "hidden choke" friction pressures are not transmitted to the formation during kick killing operations. This is usually done by establishing the kill rate through the open riser and not through the choke line. This procedure effectively subtracts the choke line friction pressures from the total circulating pressures while still maintaining balanced bottom-hole pressure.

FIGURE 3-5 • Equivalent mud weights due to choke line friction

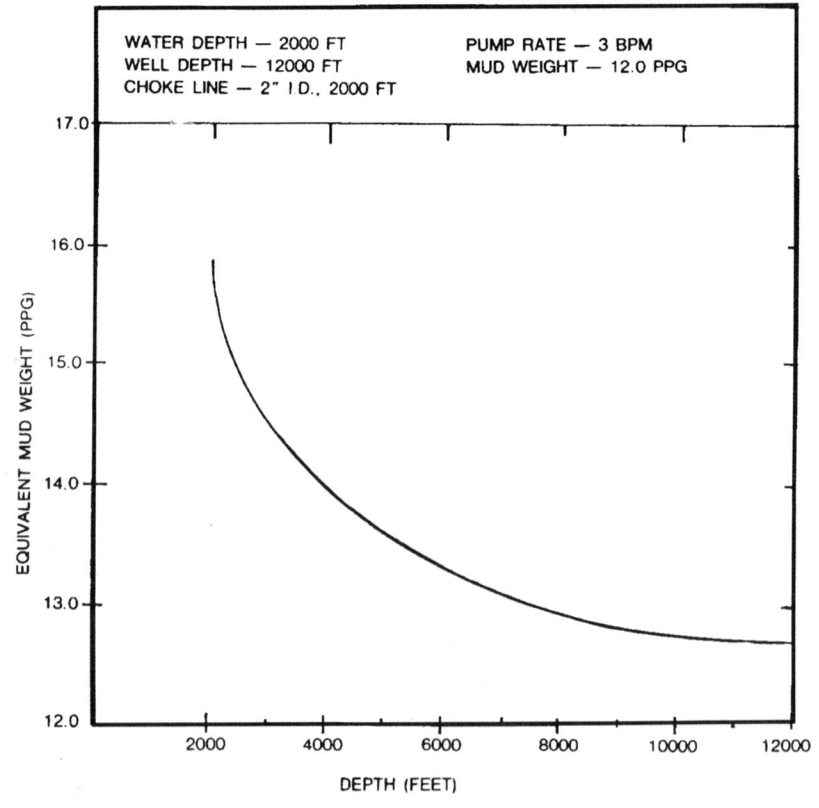

Special Problems

Although the kill rate will be taken through the riser, it is advantageous to know choke line friction pressures. Should a kick occur, these pressures must be known to determine a pumping pressure if one had not been previously established. Procedures presented in Chapter 2 to determine a pumping pressure will give a circulating pressure that includes friction pressures.

Several kill rates should be established daily. The rig pumps should be used to determine pressures at low rates in the range of 1–3 bbl/min (0.2 – 0.5 m³/min) and at approximately 1/2 bbl/min (0.1 m³/min). Some rig mud pumps will not run properly at low rates. It is advisable to determine these rates with another pump (such as a cementing unit). It may be necessary to pump at these low rates. This can result in long pumping times, but it is unavoidable in some cases.

Choke line displacement is perhaps the most severe problem encountered in deepwater well control. As gas displaces mud in the choke line, hydrostatic pressures in the annulus will change rapidly. This necessitates sudden choke adjustments. This is essentially a two part problem based on (1) the frequently large casing pressure adjustment that must be made and (2) the short time allowed to make choke adjustments caused by low line displacement timing.

Some practical solutions to minimize the severity follow:

- Displace the line at low rates using a low volume output pump.
- Use both choke and kill lines to increase displacement time. (This can result in no back up line if damage occurs.)
- Employ large diameter lines that increase the volume to be displaced.
- Use a subsea choke, allow the effluent to enter the riser, and divert at the surface.

Several solutions present associated problems that must be solved.

Figure 3–6 illustrates the displacement problem. The initial displacement occurs when pumping commences and mud enters the choke line. This can be remedied by reverse circulation of the choke line with lower pipe rams closed before initiation of the kill operation. The major problems are noted near the end of the kill procedure when gas enters and is displaced from the line. The short time allowed for the pressure changes can be reduced by procedures previously presented. In most cases, the magnitude of the pressure changes cannot be lessened.

A casing pressure curve of the example from Figure 3–6 is shown in Figure 3–7. The effect of mud initially displacing the saltwater in the choke line is shown. Although the curve shows large pressure changes at various points, the problem is more severe in field cases because the mud and gas do not enter the choke line as fluids with distinct leading and trailing boundaries. Since the invading fluid will probably be mixed with the mud,

FIGURE 3-6 • Choke line displacement with mud and gas. Static pressures are shown for illustration.

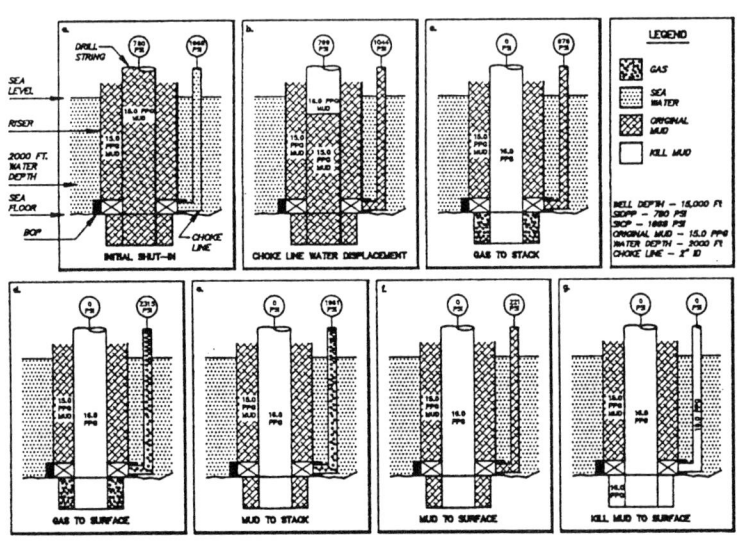

FIGURE 3-7 • Complete casing pressure curve for the situation shown in Figure 3-6

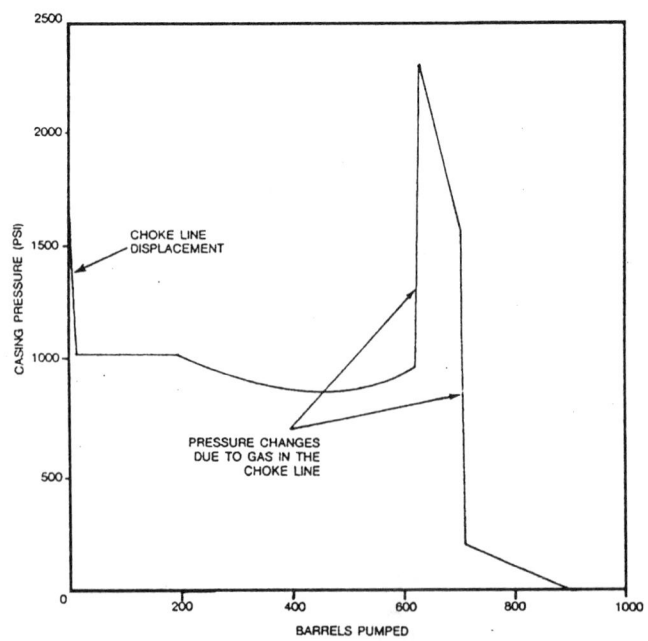

Special Problems

numerous displacements may occur and result in several casing pressure changes. However, the mud and gas mixing has the positive benefit of reducing peak pressure changes.

Casing pressure changes from the displacement may necessitate a variation in initial circulating procedures. In normal operations in shallow water or land operations, the casing pressure is often held constant for a short time while starting the rig pumps. This allows the drill pipe pressure to stabilize before beginning drill pipe pressure control procedures. If this technique is used in deepwater, large overbalances occur which tend to cause formation fracture. For example, in Figure 3-8 the amount of overbalance would be 943 psi (6.50 MPa) if the casing pressure were held constant for two minutes.

FIGURE 3-8 • Effect of holding the casing pressure constant during the initial stages of the circulation shown in Figure 3-7

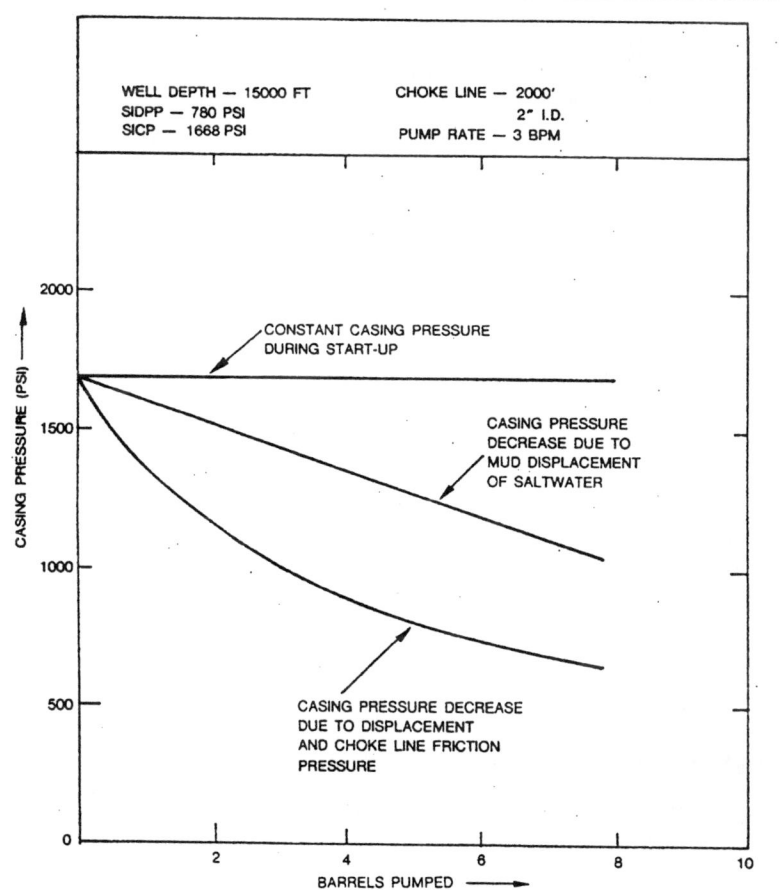

FIGURE 3-9 • Reversing procedures to completely kill the well

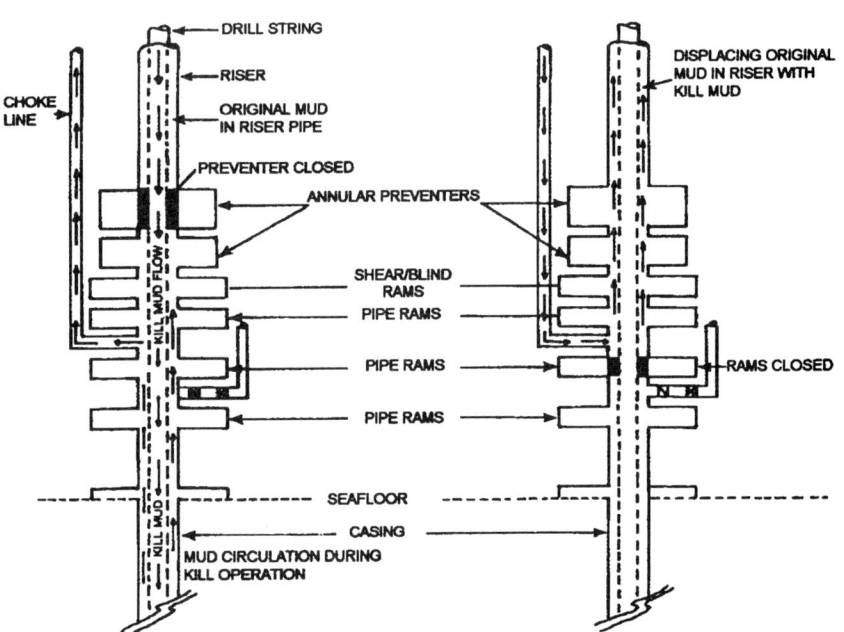

Circulating Gas Out of the Riser

Figure 3-9 shows another problem inherent to deepwater well control. After the well is dead, the annular preventer cannot be opened immediately because the riser contains original density mud. It is necessary to close a lower set of pipe rams and reverse circulate kill mud density through the choke line and riser to kill the system. A secondary gas bubble will be circulated out during this process if the preventer stack trapped gas above the primary choke line during initial kill procedures.

Handling the trapped gas in the BOP stack is an issue that increases in significance with increasing water depth. The increased hydrostatic pressure compresses the gas so the surface expanded volume is larger. If the annular is opened and the gas is reversed out of the riser, the surface volume can be large and unload the riser. This has caused fires and fatalities on some occasions. It has become common to have a diverter at the surface to flow the trapped BOP gas into a safe area for degassing and handling or to circulate across the BOP stack to try to remove as much trapped gas as possible.

Some companies circulate a gel mud across the stack at high rates to sweep some of the gas prior to opening the stack. Another option that works well in some cases is as follows:

Special Problems

- After the well is dead, close the lower rams.
- Circulate the choke and kill lines full of water.
- Open the annular and allow the riser hydrostatic to reverse flow up the choke and kill lines. Fill the riser from the top with mud.

This procedure works *only* if the riser was filled with heavy mud and the trapped BOP gas can be quickly flushed into the choke and kill lines. When the annular is opened, gas migration upward begins quickly and will exceed downward mud movement unless a high differential pressure exists.

KICKS WHILE RUNNING CASING

Causes and warning signs for kicks while running casing are as follows:
Causes:

- An existing gas bubble in the well from the prior trip out of the hole.
- A kick induced by lost circulation while running casing.

Warning Signs:

- A flow rate out of the well greater than the volume of casing run into the well. The incremental flow rate out can be caused by gas migration and expansion or by a continued influx into the well.
- A pit gain volume greater than the volume increase caused by running pipe into the well.
- The well flows without pipe being run.

To control the well, the influx must be removed. It may be necessary to increase mud density sufficiently to control the formation pressure. The required mud density is expected to be the original mud weight prior to the kick or slightly heavier than original mud weight to account for swab pressures when pulling pipe from the hole.

If the kick was induced from lost circulation while running casing, the lost circulation problem must be fixed and the gas bubble removed from the well. Care must be taken to run the casing slower and avoid lost circulation. If the casing can not be run any slower, a decision must be made to run the pipe and fight the lost circulation, to abandon the well, or to cement off the bottom of the well if the kicking zone is near the bottom prior to running casing.

Kill procedure options:

- Strip in the hole with the casing
- Top kill
- Bullhead
- Circulate out the influx
- Strip the casing out of the well and strip/snub in a drill string
- Fix the lost circulation and kill the kick
- Volumetric Kill (discussed in a later section)

Strip in the Hole with Casing

When a kick is observed, close the BOPs. Strip in the hole with casing using standard stripping procedures. Run into the hole to a depth that will allow circulation of the influx or to a depth that will allow a top kill.

If the shut-in pressures create ejection forces that exceed the pipe weight as in the case where only a short section of pipe is in the well, snubbing may be required. A mechanical snubbing unit must be installed.

High shut-in pressures can cause buoyant forces exceeding the casing weight in the well. The pipe may be forced out of the well. Close and lock the rams or chain the pipe to prevent it from moving. The pressure can be reduced with the top kill. A snubbing unit can be installed to snub the pipe into the well.

Top Kill

A top kill involves pumping heavy mud from the bottom of the casing to the top of the well. Ideally, mud should be sufficiently heavy so the hydrostatic pressure increase exceeds the casing pressure caused by the kick.

After the mud is circulated, the BOPs are opened and pipe is run into the well. When the casing enters the gas bubble in the hole, it will elongate the gas bubble and cause the well to start flowing. At this point, the influx can be removed using normal circulation techniques.

Caution must be exercised when pumping heavy mud. Increased hydrostatic pressure may cause lost circulation into some formation.

It may be necessary to use a top kill to lower casing pressure so pipe can be stripped. This procedure would be applied when casing pressure can not be completely killed with heavy mud.

Bullhead

Bullheading is a kill option that handles kicks taken while running casing. Typically, the mud will be bullheaded by pumping down the annulus.

Circulate Out the Influx

Influx fluid can be circulated from the well using conventional techniques if the casing shoe is below the gas bubble. If the gas is near the shoe but below it, gas migration may allow it to move upward past the shoe within a reasonable time.

The circulation techniques for a kick occurring while running casing are the same as with the drill pipe off bottom.

Special Problems

Snub Casing Out of the Well and Snub in a Kill String

A decision may be made to snub casing out of the well and run a kill string. The decision basis is subjective but would usually apply if only a short section of pipe has been run.

If the well has a shut-in casing pressure, snubbing out will be required for some section of the pipe. Mechanical snubbers may be required. It may be possible to perform a top kill to allow removal of the casing without snubbing.

Fix the Lost Circulation and Kill the Kick

If the well can not be stabilized because lost circulation results in an underground blowout, it may be necessary to implement procedures to solve the underground blowout. This may involve fixing the lost circulation and killing the kick. The procedures are the same for running casing as for underground blowouts and lost circulation.

KICKS WHILE CEMENTING

Kicks and blowouts have been experienced on wells during and following apparently successful cementing operations. Some problems are seemingly random unless an investigation is made to determine all contributing factors. Surface blowouts resulting from gas-through-cement kicks can cause problems such as pressure charging other zones affecting offset drilling and the loss of hydrocarbon reserves.

Causes:

- Existing influx in the well
- Water cement spacer causing annulus hydrostatic pressure reduction below formation pressure
- Influx after cementing caused by a hydrostatic pressure reduction in the cement while the cement is curing
- Lost circulation resulting in an annulus hydrostatic pressure reduction

Warning signs:

- Annulus flow
- Abnormal pit gain while cementing
- Pressure build-up on annulus after cementing

Gas-Through-Cement Kicks

Assuming a cement operation was completed without problems such as gas cut mud, lost circulation, or cement channeling, then the circumstances to allow gas migration and resulting kicks to occur have been established by several authors. These circumstances are illustrated in Figure 3–10.

One or more permeable zones should exist above a gas-bearing interval with the upper zone(s) having pressures lower than those in the gas zone. The cement must set on the upper zones and support the drilling fluid hydrostatic pressure. After the cement has set, the pressure in the lower interval must be reduced below formation pressure. Gas flow may occur. If the seal from the upper set cement interval cannot prevent permeation by gas, the flow will continue at an increasing rate due to reduction in annular hydrostatic pressure.

Causes for cement setting on the upper zones prior to setting on the lower zones must be understood in order to establish guidelines for reducing gas flow. Cement setting is a function of many variables, one of

FIGURE 3–10 • Illustration of the circumstances involved in gas-through-cement kicks

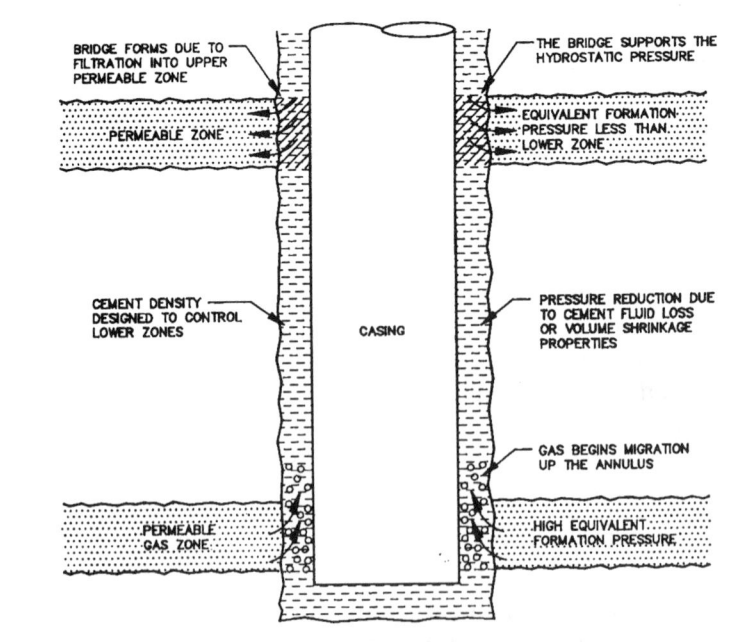

FIGURE 3-11 • High mud weights to control formation pressure may cause large differential pressures at shallow intervals

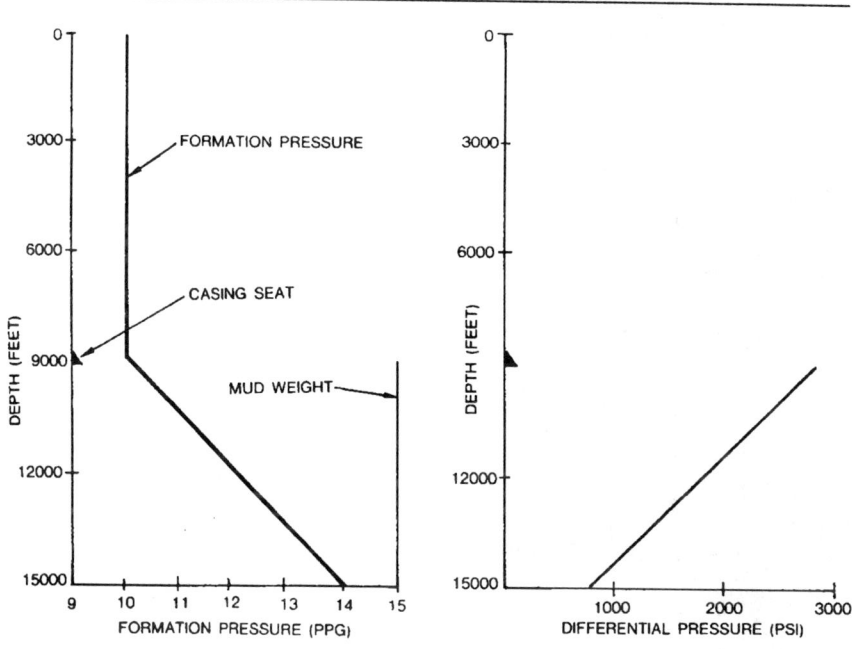

which is temperature. Studies have shown maximum circulating temperatures occur about one-third of the distance from the bottom. This tends to cause the cement to set initially above the hole bottom. If formation pressures increase as shown in Figure 3-11, the larger differential pressures on the upper zone would promote setting of the cement.

Once the cement has set on the upper zone and begins to partially support hydrostatic pressure, pressures within the cement must be reduced to allow a kick. A common cause for this reduction is water lost from the slurry. If the water loss is sufficient to allow fluid pressure to fall below formation pressure, a kick can occur.

Another cause for pressure reduction in the cement is cement expansion-shrinkage characteristics. Cement may have a tendency for initial shrinkage when downhole conditions of temperature and pressure are applied. (See Figure 3-12.) If shrinkage occurs, the pressure will be reduced. To offset shrinkage, commercial cement additives are available that cause cement expansion from initial conditions without any shrinkage. (See Figure 3-13.)

FIGURE 3-12 • Expansion-shrinkage characteristics of a cement sample (Courtesy of SPE, 5701, © 1976)

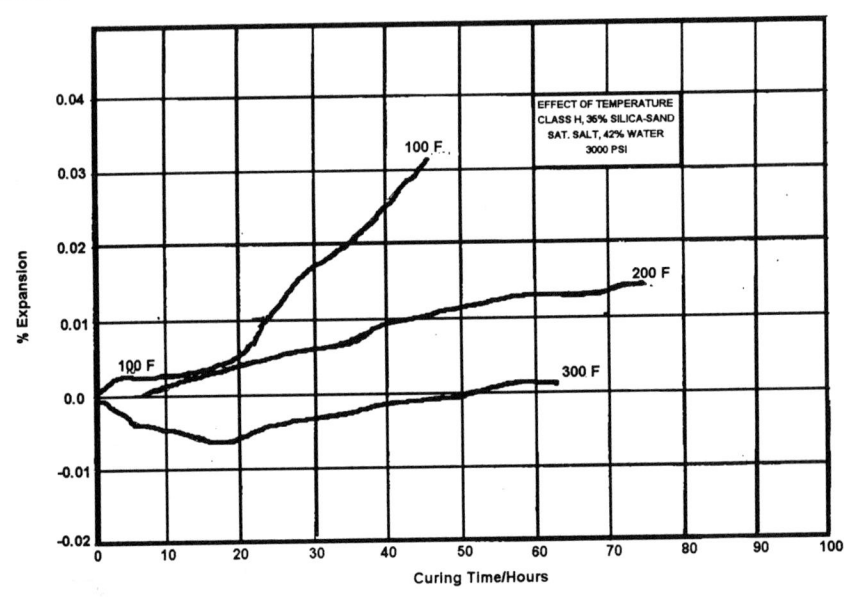

FIGURE 3-13 • Effect of certain commercial additives on expansion properties of cement (Courtesy of SPE, 5701, © 1976)

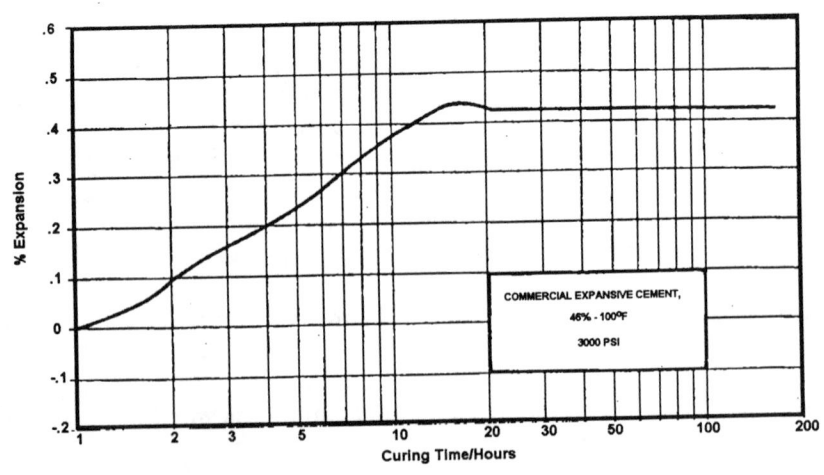

SPECIAL PROBLEMS

Reducing Gas-Through-Cement Kicks

Proper cementing procedures should be exercised during the operation to ensure kicks are not a function of mechanical problems. Procedures include use of cementing aids such as centralizers, proper hole conditioning, and release of surface pressure after cement placement to avoid formation of a microannulus. Cementing operations should be monitored to ensure that lost circulation does not occur.

To maximize preventive measures for gas-through-cement kicks, cement slurries should be tailored with chemical additives to avoid kick causes presented in previous sections. Retarding agents should be blended with cement batches to ensure the slurry will set from bottom to top.

Additives should be used to prevent cement volume shrinkage. The fluid loss from the cement should be reduced as low as reasonable.

Control Procedures

To control the well, the influx must be removed from the well or stabilized in the well. Lost circulation must be handled if it caused the kick or if it causes difficulties while attempting to circulate the influx from the well.

The general procedure is to circulate the influx from the well prior to cementing. If the influx occurs because of the cement spacer, the cement mixing should be stopped. All fluids should be circulated out of the well using conventional kick killing methods.

If an external casing packer is used in the casing, it can be closed to seal the influx below the packer. The annulus above the packer should be circulated and cleaned before a second stage cement job is performed.

If a kick occurs after the cement has been pumped and it has set too much for it to be pumped out of the well, the influx can be controlled with the Volumetric Method. The casing must be perforated or a DV tool opened above the cement, and kill fluids must be circulated to regain control of the well.

The influx can be removed by bullheading. This is not the best method unless lost circulation already exists.

If a kick occurs because of lost circulation, the losses must be fixed before the well can be completely killed.

STRIPPING AND SNUBBING OPERATIONS

When the drill string is pulled from the hole, kicks can occur from swabbing or improper hole fill up. Since the bit is not on bottom, kill mud densities may be high and in some cases will be in excess of practical

FIGURE 3-14 • Required kill mud weights at various bit depths for a kick occuring while tripping

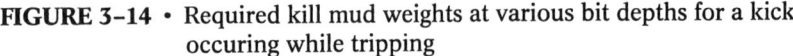

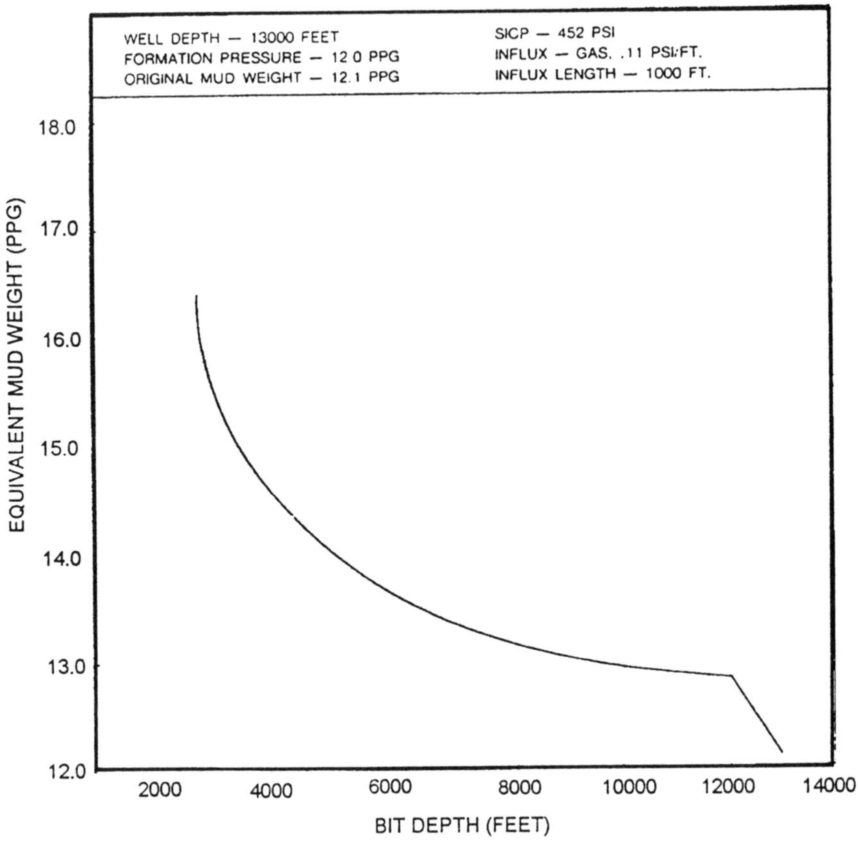

```
WELL DEPTH — 13000 FEET            SICP — 452 PSI
FORMATION PRESSURE — 12.0 PPG      INFLUX — GAS, .11 PSI/FT.
ORIGINAL MUD WEIGHT — 12.1 PPG     INFLUX LENGTH — 1000 FT.
```

ranges. (See Figure 3-14.) When this occurs, it is necessary to lower the drill string into the well while maintaining proper surface pressures to avoid an additional influx. Proper well control procedures under these circumstances are to "strip" or "snub" into the well with the blowout preventer closed. Extreme caution must be exercised if an attempt is made to run pipe to the bottom without closing the preventers.

Stripping and snubbing is the process in which the drill string is moved within the well under pressure. The general case occurs when pipe is lowered to kill an induced kick. Some instances demand the pipe be pulled from the hole to perform some operation. Special equipment and techniques must be used to control the well.

Special Problems

FIGURE 3-15 • Illustration showing the difference between stripping and snubbing. Note that the annular preventer friction forces have been neglected to simplify the calculations.

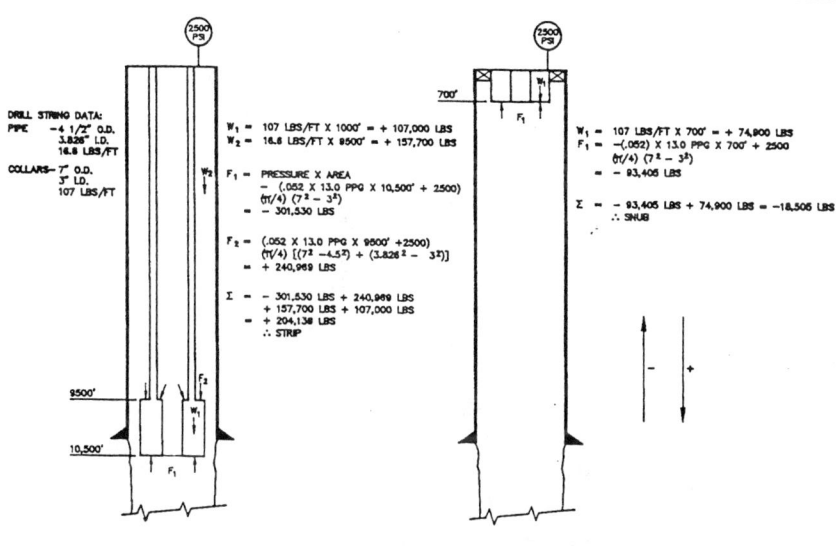

The difference between stripping and snubbing is based on the manner in which kick pressures act on the drill string and the amount of drill string in the well. If the upward force exerted by kick pressures acting on the horizontal surfaces of the drill string exceeds the weight of the drill string, pipe must be snubbed in or out of the well. If the weight is greater than the upward force, the drill pipe must be stripped. (See Figure 3-15.)

Snubbing Equipment

In snubbing, downward forces must be applied and will require special equipment. This equipment may be mechanical or hydraulically operated, and it will often consist of special blowout preventers.

Mechanical Snubbers

This equipment utilizes rig systems to force pipe in or out of the hole. (See Figure 3-16.) Snubbing equipment consists of a set of traveling snubbers to force pipe movement under well pressure and a set of stationary snubbers to prevent pipe movement when the traveling snubbers are released. A wire line (snub line) passes over the center sheave of the

FIGURE 3-16 • Mechanical snubbers

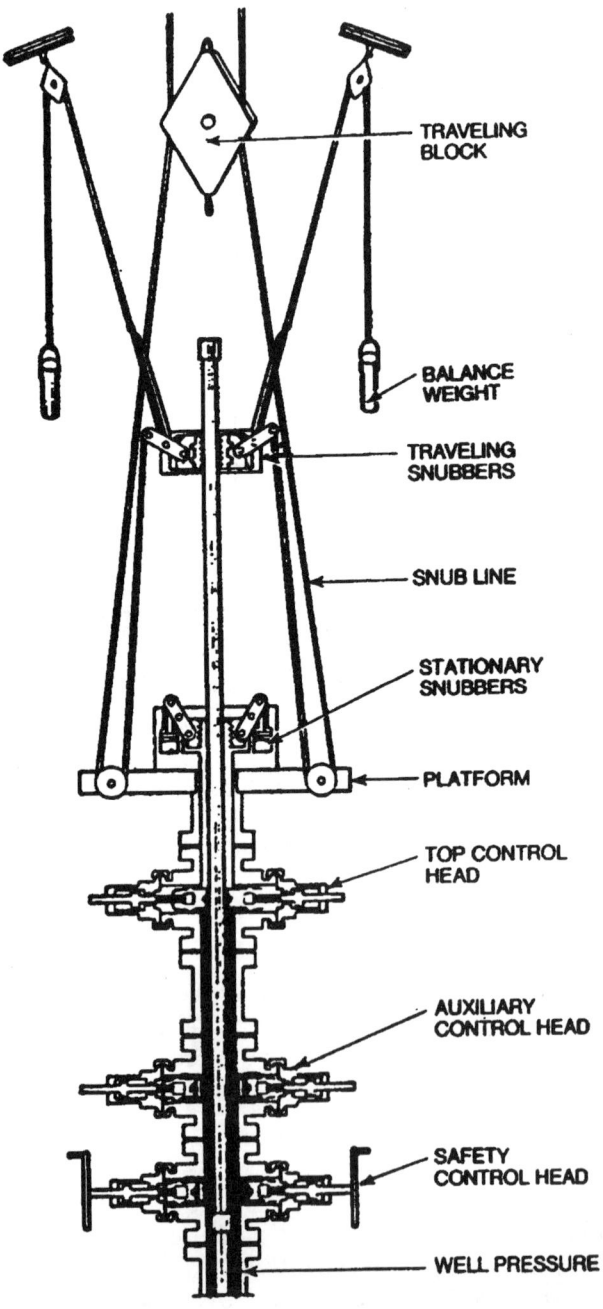

traveling block to clamp to the handle of traveling snubbers. This line is used to transmit a balanced force downward to the traveling snubbers as the block is raised. It eliminates the tendency to bend the pipe by applying equal downward forces. Balancing weights raise the traveling snubbers when the traveling block is lowered.

Stationary snubbers are the lowermost section on the snubbing assembly, and they are bolted to the upper control head. The stationary snubbers are designed, as are most blowout preventers, to take advantage of well pressure. As the well pressure increases the upward force applied to the pipe, the grip of the snubbers increases accordingly.

The control head assembly is used to provide a pressure seal between the casing and pipe. It consists of an upper control head, an auxiliary control head, a safety control head, and a control manifold. The control heads are ram preventers and hydraulically operated for rapid opening and closing. The upper and auxiliary control heads employ special packing. It has less friction drag during pipe movement and is easier to replace than standard rubber rams used in the safety control head.

The control head assembly arrangement depends on the drilling blowout preventers configuration. The control head assembly will usually be added to the preventer stack in place with the safety control head flanged to the annular preventer. An auxiliary control head and a spool are flanged above the safety control head and spool. The spool provides space between upper and auxiliary control heads to allow for lubrication of drill pipe tool joints through the head. It may vary in length to permit lubrication of special tools such as packers.

The control head assembly is designed to prevent the inadvertent opening of all control heads at the same time. The rams in either the auxiliary control head or the safety control head must be closed and pressure released between that head and the upper control head before the upper control head can be opened. The upper control head must be closed and the pressure equalized below it before the lower heads can be opened.

The operation of the mechanical snubbing equipment to allow pipe movement is through the "hand-over-hand" principle. After the equipment has been installed and pressure tested, the first joint of drill pipe with a back pressure valve is lowered into the upper control head and the rams closed. The stationary and traveling snubbers are both engaged on the pipe.

The control assembly is pressurized and the drilling preventers are opened. The traveling snubbers are disengaged from the pipe and moved to a point approximately four feet above the stationary snubbers and engaged. The stationary snubbers are released, and the pipe is forced into the well by raising the block. When the traveling snubbers reach a point immediately above the stationary snubbers, the lower snubbers are engaged and the traveling snubbers are moved up the pipe to grip a new section. This process is repeated to continue pipe movement into the well.

When the upper end of a tool joint is encountered, the traveling snubbers are used to position the tool joint immediately below the stationary snubbers and above the upper control head. The auxiliary rams are closed, and the upper rams are opened before the tool joint is forced into a position between the two rams. The upper rams are closed, and the lower rams are opened to allow for continued stripping until another tool joint is encountered.

The process of snubbing is continued until the weight of the pipe in the hole equals the upward force caused by well pressure acting on the cross sectional area of the pipe. At this point, often called the "snubbing point," the pipe moves easily into the well without having to be snubbed. The snubbing assembly is removed, and the block and elevator are used to proceed with the stripping process.

Hydraulic Snubbers

The hydraulic snubbing unit was developed for application in areas where snubbing was necessary for well control but when a drilling rig was not on the well. The hydraulic unit attains the same result as the mechanical snubber, but it is self-contained and therefore does not require any rig assistance. (See Figure 3–17.)

Pipe movement capabilities of the unit are supplied by a hydraulic jack. Various models generate 150,000 to 600,000 lbs (70–270 metric tons) of lifting force and up to 300,000 lbs (140 metric tons) of snubbing force. The jack may be a single, large cylinder system or a multi-cylinder system, depending on the manufacturer, and it may have a stroke length of 10–15 ft (3.0–4.5 m) for the short stroke jack and approximately 36 ft (11 m) for the long stroke jack. Units are available that can safely handle pipe diameters from 3/4 in. to 7 5/8 in. (19–194 mm).

Two types of sealing devices are commonly used for sealing the OD of the pipe while working under pressure with the hydraulic-snubbing unit. These are ram-type blowout preventers and the solid rubber element strippers. Stripper elements are generally considered adequate for pressure control up to 2,500–3,000 psi (17–21 MPa). These elements are constructed of solid synthetic rubber compounds. They have the ability to stretch as couplings, and some downhole tools are stripped through them. The useful life of the stripper elements will depend on the external condition of the tubing and may range from 10,000–20,000 ft (3,050–6,100 m) at pressures of 3,000 psi (21 MPa) or less.

Stripping preventer rams are solid steel construction, incorporating rubber seals with fiber or teflon replaceable stripper inserts. A second set of rams is employed to seal the well while the other is opened to allow tool

Special Problems

FIGURE 3-17 • Hydraulic snubbing unit

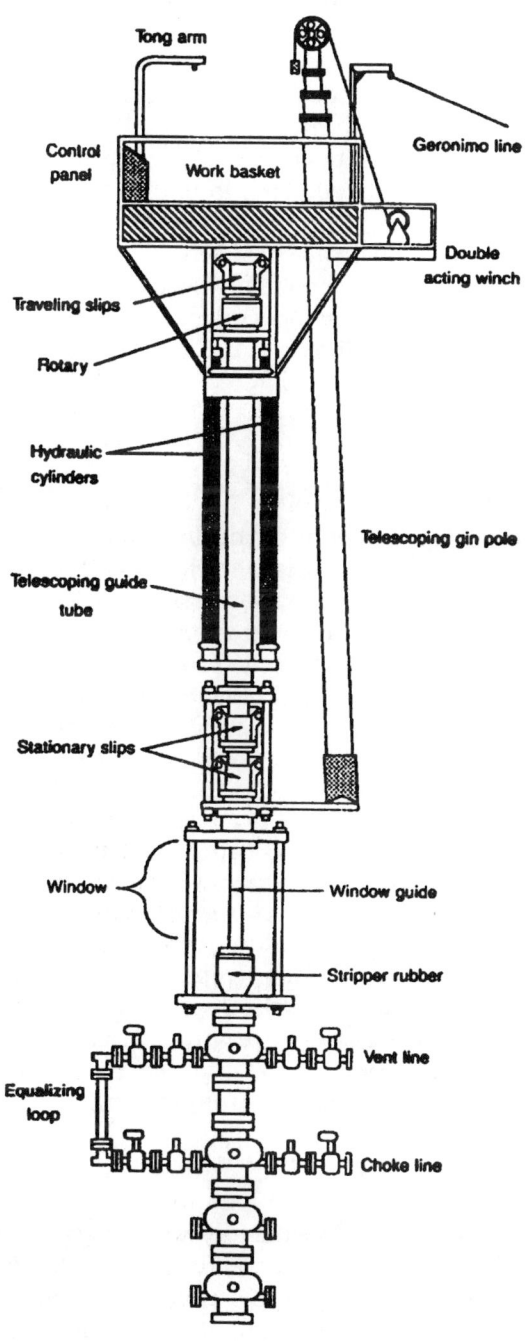

joint passage. This process of alternately opening and closing the working blowout preventers requires deliberate stop and start motions by the operator.

In a typical operation, the lower working rams are closed on the pipe, and the upper rams are opened. Well pressure is held by the closed rams while the joint is lowered by the hydraulic unit. When the tool joint enters the equalizing chamber, the upper ram is closed. The equalizing chamber has two valves hydraulically operated from the operator's console. One valve vents the chamber to the casing pressure below the lowermost standby preventer. The other valve vents the chamber to the atmosphere through a bleed-off line. Before opening the lower working rams to allow the tool joint to pass, the proper valve on the chamber is operated, and casing pressure is vented to the chamber. The pipe can again be moved downhole.

Snubbing and Stripping Control Procedures

Regardless of the equipment used during pipe movement, pressure control procedures must be exercised to ensure that formation fluids are not allowed into the well and that excessive pressures to fracture the formation or burst the casing are not maintained. The two processes most often used are the volumetric and the pressure methods. Although the volumetric method is easier to comprehend initially, the pressure method may be more accurate due to ease of implementation.

Volumetric Method

As the pipe is moved into the well, the pressures observed at the surface will tend to increase from fluid compression in the wellbore. If the compression is allowed to continue, the pressures will eventually build to a value that will fracture the formation. To compensate for compression, a volume of mud equal to the volume of pipe forced into the well is allowed to escape. The pipe volume will be the displacement of the string plus the capacity because a back pressure valve is used (Example 3.1). As a section of pipe is lowered into the well, the choke allows a mud volume equal to total pipe displacements to bleed off from the well.

Example 3.1

A kick developed on a well while the pipe was being pulled for a new bit. If the pipe is to be stripped into the well using the volume method, what amount of mud should be allowed to escape for each 93 ft stand of pipe; each 31 ft joint? Use the data given on the following page.

Special Problems

Pipe size = 3 1/2 in.

Pipe weight = 13.3 lb/ft

Pipe displacement = 0.0054 bbl/ft

Pipe capacity = 0.007421 bbl/ft

Solution

1. Total displacement = displacement + capacity

    = (0.0054 + 0.007421) bbl/ft

    = 0.012821 bbl/ft

2. For each 93 ft stand

    = 93 ft × 0.012821 bbl/ft

    = 1.192 bbl/93 ft stand

3. For each 31 ft stand

    = 31 ft × 0.012821 bbl/ft

    = 0.397 bbl/31 ft joint

Example 3.1 shows problems associated with the volumetric method. Running the 93 ft (28 m) stand required 1.192 barrels (0.19 m^3) of mud leave the well. The difficulty arises in maintaining this value and not some slightly larger amount from improper choke operation. In some reported field cases, a volume 50% larger than calculated volume was allowed to escape from the well. This allows additional influxes into the well resulting in larger surface pressures.

Pressure Method

Surface pressures are those needed to balance bottom-hole formation pressure and prevent further influxes. The pressure method for stripping and snubbing uses the same concept with the exception that dynamic pressures are substituted for static pressures imposed by the blowout preventers. This method provides more accurate fluid control as well as being applicable for pipe movement into and out of the hole. The volume method is more easily applied when going into the hole.

Equipment necessary for the pressure method is shown in Figure 3–18. The equipment should be available on most drilling rigs and requires only a small amount of special preparation. The pump output is directed through the choke and is designed to supply the pressures necessary for control purposes. The downstream side of the choke returns

FIGURE 3-18 • Diagram of equipment used in the pressure method for stripping

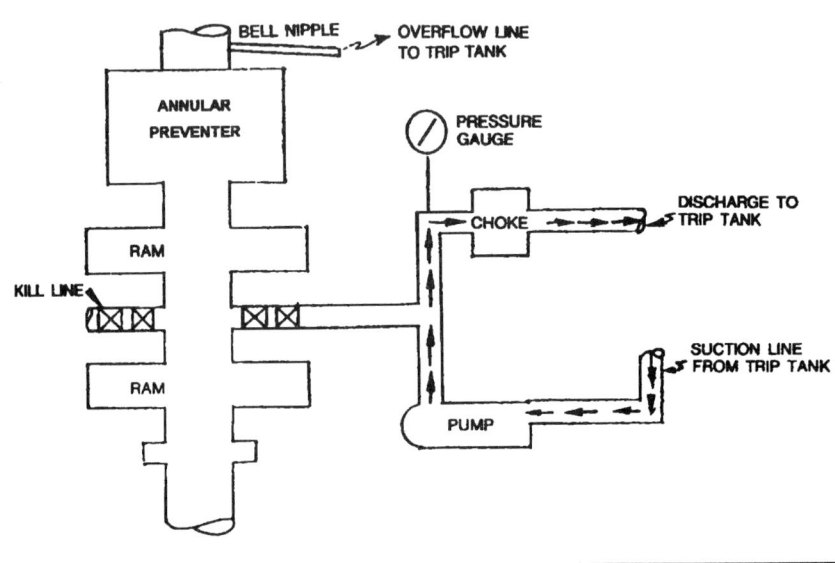

the mud to a trip tank where it is picked up by the pump. A small return line may be connected between the bell nipple and the trip tank to retrieve any mud escaping through the preventers during tool joint passage on stripping operations.

The procedure is implemented by starting the pump and using the choke to control pressures at a value slightly greater than the well pressure. The control valve is opened and pipe movement commences. Since the confining pressures are greater or equal to those necessary to control the well, no additional influx occurs. The pressure method can be monitored by recording volume increases in the trip tank. These increases should be equal to the calculated total pipe displacement as shown in Example 3.1.

The advantage of the pressure method is the manner in which the fluid is allowed to escape. In the volume method, the procedure alternates between a static and dynamic state. The pressure method is in a dynamic state throughout the process.

The pressure method is applicable for pipe movement in either vertical direction. The volume method is most easily applied to pipe movement into the well.

Fluid migration up the hole must be taken into account. Gas migration results in volume expansion causing fluid expulsion at the surface and larger confining pressures. To compensate with the pressure method, the

Special Problems

choke pressure should be increased by small increments (± 50 psi) (0.3 MPa) when it is observed that the original casing pressure is not sufficient to control the well. If the volume method were used, the same pressure increases would be required.

Entry into the influx column will cause confining pressures to increase from elongation of the fluid column. (See Figure 3-19.) When this entry is noted at the surface, pipe movement into the well should continue until the pressures approach a safe maximum level or until the pipe is moved through the influx. The kick fluid should be circulated from the well.

When stripping has been completed, the drilling fluid should be circulated to clean the hole. The density necessary to control the well

FIGURE 3-19 • Entry of the drill string into the influx column

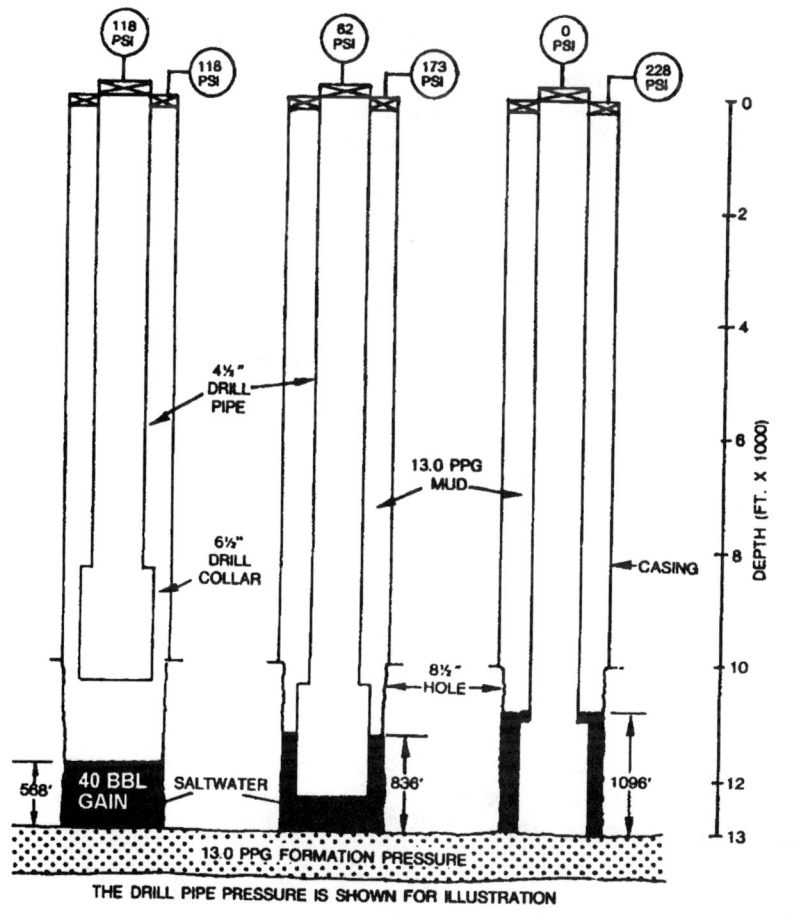

should be the same as used prior to the kick. Although a trip margin of mud weight is necessary to control swab pressures, additional mud weight for pressure control is not needed. This point is often misunderstood, and it has resulted in losing wells from fractured formation caused by circulation of excessive mud weights after the stripping process was successful.

KICKS WHILE PIPE IS OUT OF THE HOLE

The drill string or work string can be out of the hole for various operational reasons. Some are as follows:

- Pull out to change the bit, BHA, down hole tools, etc.
- Logging
- Run casing
- Nipple down/nipple up BOPs, Christmas tree, etc.
- Wellhead installation
- Cementing, waiting on cement
- Wirelining

Kicks while pipe is out of the hole are generally caused by the following:

- Improper hole fillup while pulling pipe
- Swabbing
- Gas existing in the well prior to the trip (The gas could have been taken as gas cut mud, core volume cut mud, gas below a test tool or packer, gas in the well from a drill stem test, or other possible sources.)
- Mud losses

Warning signs for these kicks are improper hole fillup while coming out of the hole, a flow from the well, and/or a pit gain.

If a gas bubble, oil, or condensate has been swabbed into the well while pulling pipe, but an underbalance does not exist while the pipe is stationary, the flow from the well while the pipe is out of the hole will be due to bubble expansion during its upward migration. The initial rate of expansion will be small, and the pit gain will be small. After a sufficient amount of wellbore mud has been displaced to allow the hydrostatic pressure to fall below formation pressure, the reservoir will begin to flow, and the fluid exit rate at the surface will increase.

The flow rate from the well could be as follows:

None: If the influx during swabbing was salt water or oil.
Abnormal-low: Very small if gas expansion is occurring without further influxes from the formation.

Special Problems

Abnormal-high: High if the pressure conditions are allowing further influxes or the gas bubble is approaching the surface, 0–2,000 ft (0–610 m).

To control the well, the influx must be removed from the well. It may be necessary to increase the mud density sufficiently to control the formation pressure. The required mud density to control the well is expected to be the original mud weight prior to the kick or slightly heavier than the original mud weight to account for swab pressures when pulling pipe from the hole.

Mud can be replaced by (1) running, snubbing, or stripping pipe into the well and circulating, (2) bullheading, or (3) volumetric kill techniques.

Running Into the Well

For low flow rates from the well, the BHA and pipe string can be run into the well until the flow rate becomes abnormal-high. An "abnormal-high flow rate" is defined as a flow rate caused by formation fluid influxes in addition to fluid expansion in the well. This approach is not recommended unless a computer model is running real time to monitor all activities.

It is not generally desirable to run pipe with a flow from the well. However, kicks while pipe is out of the hole are unusual situations that require special consideration. Extreme caution must be given to the situation.

The selection of an allowable flow rate from the well is an important decision. The flow rate from the well will be small if only gas expansion is occurring unless the gas is near the top of the well. Mud can be circulated with the pipe in the well.

Snub Into the Well

For abnormal-high flow rates, the blind ram BOPs should be closed. Provisions should be made to snub/strip with rig snubbers or install a coiled tubing unit or a snubbing unit. Mud is circulated with pipe in the well. The well must be controlled with the Volumetric Method until snubbing or coiled tubing equipment arrives at the rig.

The difference between low and high flow rates need not be subjective. A flow rate based on a worst case assumption of expanding gas near the surface, as an example within 1,000–2,000 ft (300–600 m) of the surface, can be calculated and set as the allowable flow rate from the well. If the fluid exit rate exceeds the allowable value, close the BOPs.

Bullheading

Bullheading the well to displace influx fluids back down the well with mud is an optional kill technique.

Volumetric Kill

A volumetric kill is an optional technique if the influx is gas. Saltwater or oil influxes will not migrate quickly enough to allow the implementation of the volumetric kill method.

Run Into the Hole with the BHA and Pipe String

Pick up the BHA and pipe string and run into the hole using normal running procedures. Run the BHA as it is sitting on the drill floor (i.e., if the bit has been removed, do not take time to make up the bit). A back pressure valve (float-type) must be run in the string.

If the pipe string is easily accessible without picking up the drill collars, run the pipe string into the well. It can be run quicker than the collars.

If the flow rate from the well exceeds the allowable rate while running pipe into the well, close the annular BOP and continue stripping pipe into the well.

If the well is closed in and the shut-in pressure exceeds an allowable limit when the BOPs are closed, the pipe must be chained down or the pipe rams must be closed and locked to prevent the pipe from being ejected out of the well. A snubbing unit must be installed on the well.

The allowable pressure under shut-in conditions to prevent the pipe from being ejected out of the well is a function of the well pressure, BHA/pipe geometry, and weight. If the pressure acting upward on the bottom of the string exceeds the actual weight of the pipe in the well, this ejection force will tend to push the pipe out of the well. If the pipe weight exceeds the ejection force, it will not be pushed out of the well.

Mud is circulated after pipe is in the well.

Snub Into the Well

Pipe must be snubbed into the well with the BOPs closed if the flow rate out of the well is abnormal-high with open BOPs. Pipe stripping operations can be accomplished without special equipment if the forces caused

Special Problems

by the well pressures are less than the string weight. Snubbing is required if the pressure-induced forces exceed the string weight. Snubbing will be required initially to get the pipe in the well.

Conventional snubbing can be accomplished with mechanical snubbers, a hydraulic snubbing unit, or a coiled tubing unit.

High shut-in pressures can cause ejection forces exceeding the weight of the pipe in the well. The pipe will begin to move upward through the stack and the well. If not secured mechanically, it will be ejected from the well. Close and lock the rams. Rig up a snubbing unit or coiled tubing unit; or bullhead.

The BOP stack may not contain an annular or proper rams for stripping, the annular preventer rubber is worn out, or some other BOP/control system problem exists that prevents stripping into the well. A design limit or fault with the BOP stack can prevent stripping or snubbing with the stack. The BOP stack may not shut in the well, or it may not restrain the pipe if it starts moving upward. Install new elements in the stack if blind rams are positioned near the bottom of the stack to allow upper rams to be changed. Otherwise, close the blind rams and install a new stack on top of the existing stack; or bullhead.

Bullheading

Bullheading involves pumping mud down the well into an exposed formation. High pressure on the casing may cause problems. Well pressure may approach the casing allowable working pressure. The original casing integrity may be reduced because of damage or wear.

The casing can burst if the internal yield pressure is exceeded. This situation exists for the entire section(s) of the innermost string(s) of casing. The well can blow out at the surface or underground.

If the pressure is too high to bullhead, although the casing strength limitation has not been exceeded, pipe can be snubbed into the well with a coiled tubing unit or snubbing unit.

The bullheading pressure and mud hydrostatic pressure may exceed the formation fracture pressure at the casing seat or some other exposed formation. The formation will be fractured, and mud may be lost into the formation. This may be undesirable depending on the rock type and other well conditions. Cratering may occur if the fluids enter the formation and flow to the surface. Mud may be pumped into the well with no mud or only part of the mud returning. It is not possible to differentiate if the mud is going into a formation at the casing seat, into an exposed formation, or through a hole in the casing. Do not bullhead if it is not desirable to fracture the formation.

Abnormal-High Flow

Abnormal-high flow rate is defined as a flow out of a well caused by (1) fluid expansion as it moves up the well and (2) additional influx entering the well from the formation. The expected flow rate from gas expansion at any time is calculated as follows:

$$V_f = \frac{V_2 - V_1}{t_2 - t_1} \tag{3.1}$$

Where:
V_f = volume flow rate, gal/min
V_2 = volume of migrated gas at time t_2, gal
V_1 = volume of gas at initial conditions, time t_1, gal
t_2 = new time step, min
t_1 = initial time of gas influx, min

Observed rates significantly greater than V_f could mean an additional influx entering the well.

The new volume, V_2, is calculated with Boyles Law.

Gas expansion within 2,000 ft (610 m) of the surface can be rapid. If the gas influx is calculated to have migrated to within this depth range of the well, the well should be shut in and not allowed to flow regardless of the calculations determining the type of flow (i.e., abnormal-low or abnormal-high).

Ejection Equation

The ejection force is the sum of pressures acting on all cross sectional, horizontally exposed surfaces of the pipe in the well. The pressures to be considered are the casing pressure and the hydrostatic pressure of the mud and gas down to the horizontal surfaces. If the ejection force exceeds the pipe weight, the pipe will be forced out of the well unless physically restrained.

The ejection force can be calculated as follows:

$$F_e = [\sum (P_h + P_{sicp})_n A_n] - W_p \tag{3.2}$$

Where:
F_e = ejection force, lbs
W_p = pipe weight, lbs
P_h = hydrostatic pressure, psi
P_{sicp} = shut-in casing pressure, psi
A_n = area, horizontal exposed surface, in^2

Case History—Gulf of Mexico

A Gulf of Mexico well was being reworked when it blew out. The well was being recompleted after the lower sand had watered out.

A drill stem test had been conducted on the well with 2 7/8" (73 mm) tubing inside of 9 5/8" (244 mm) casing and a 7" (178 mm) liner. (See Figure 3–20.) The 14.0 ppg (1,680 kg/m³) fluid in the well was displaced

Figure 3–20 • Well B23 blowout configuration

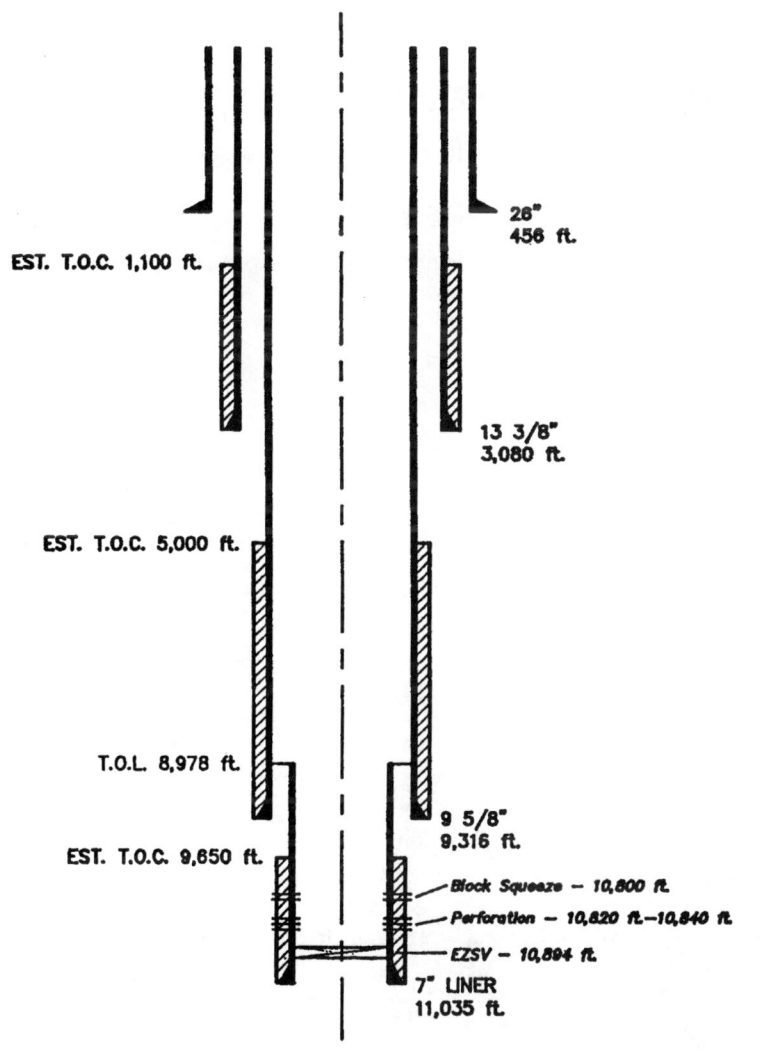

with a filtered 12.0 ppg (1,440 kg/m³) brine. The tubing was pulled from the well in the early morning hours of 9 September 1990. The tubing string was out of the well at approximately 0700–0730 hrs when bubbling was noticed in the annular fluid. The pipe was started back in the hole, but the well kicked up the pipe I.D. before 2 stands were run. After several efforts to stab a TIW valve, the well blew out. An unsuccessful attempt was made to close the BOPs from the remote closing unit. All personnel were safely evacuated.

The Neal Adams team was directed to plan all details of the relief well to the point that a rig would be mobilized. At that time, a decision would be made by the operator and its partners concerning the disposition of operations and determine if the well would be drilled.

The well was planned by the team. After the blowout bridged at 1304 hrs on 12 September 1990, the intensity of the planning was curtailed so a relief well overview was completed instead of detailed drilling and well killing operational plans.

Some areas addressed during the work were as follows:

- Determination of kill requirements for pumping
- Relief well site selection
- Directional planning
- Drilling planning
- Killing plans
- Casing design
- Ranging
- Ellipse of uncertainty calculations
- Advising on recompletion attempts after killing the well
- Review of snubbing plans
- Determination of blowout causes

A report was prepared and submitted.

Analysis of Blowout Causes

An analysis was made into blowout causes. The analysis was developed with the facts known at the time of the blowout.

Reference is made to the strip chart shown in Figure 3-21. The annotations are provided by Neal Adams Firefighters. The strip chart is from the PVT monitoring system on the rig used for the workover operations.

The general operations were as follows:

- The 14.0 ppg (1,680 kg/m³) fluid in the well was displaced with a 12.0 ppg (1,440 kg/m³) filtered brine. (Steps 1–7)
- The gas at the bottom of the well was removed by reversing the tubing string. (Steps 9–10)

FIGURE 3-21 • Strip Chart (1 of 4)

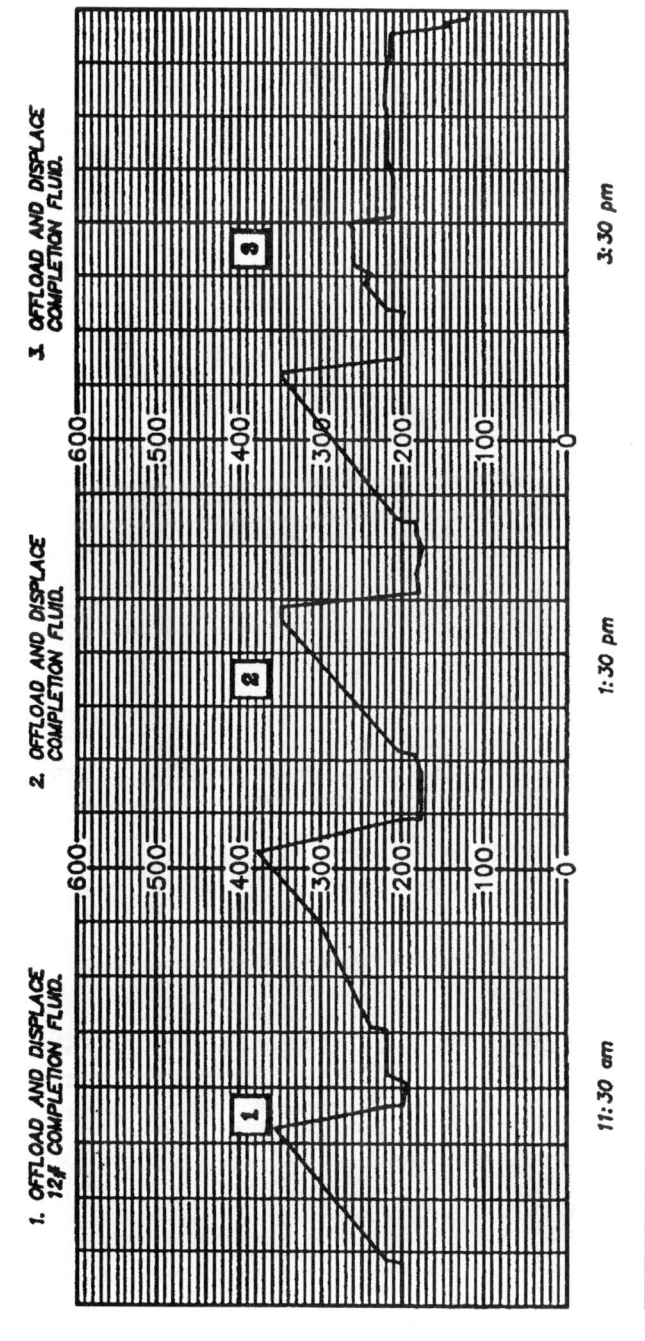

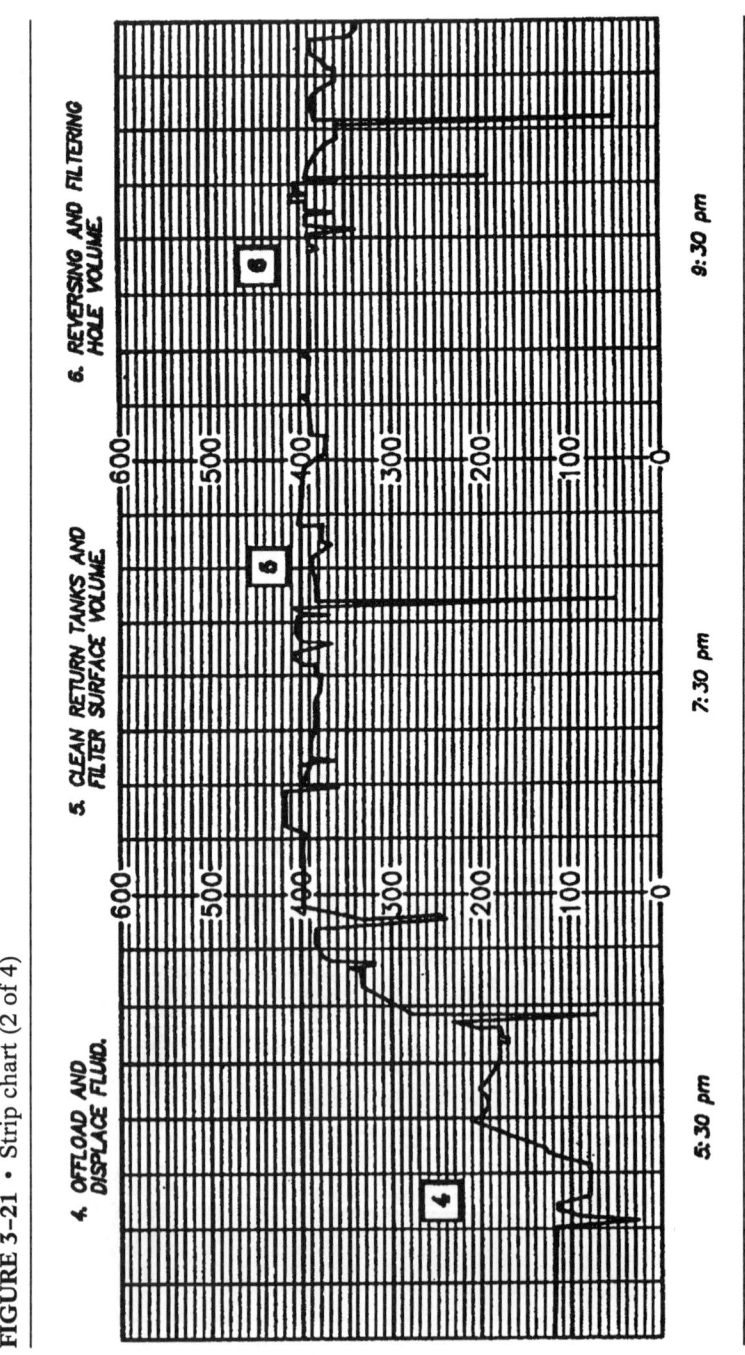

FIGURE 3-21 · Strip chart (2 of 4)

FIGURE 3-21 • Strip chart (3 of 4)

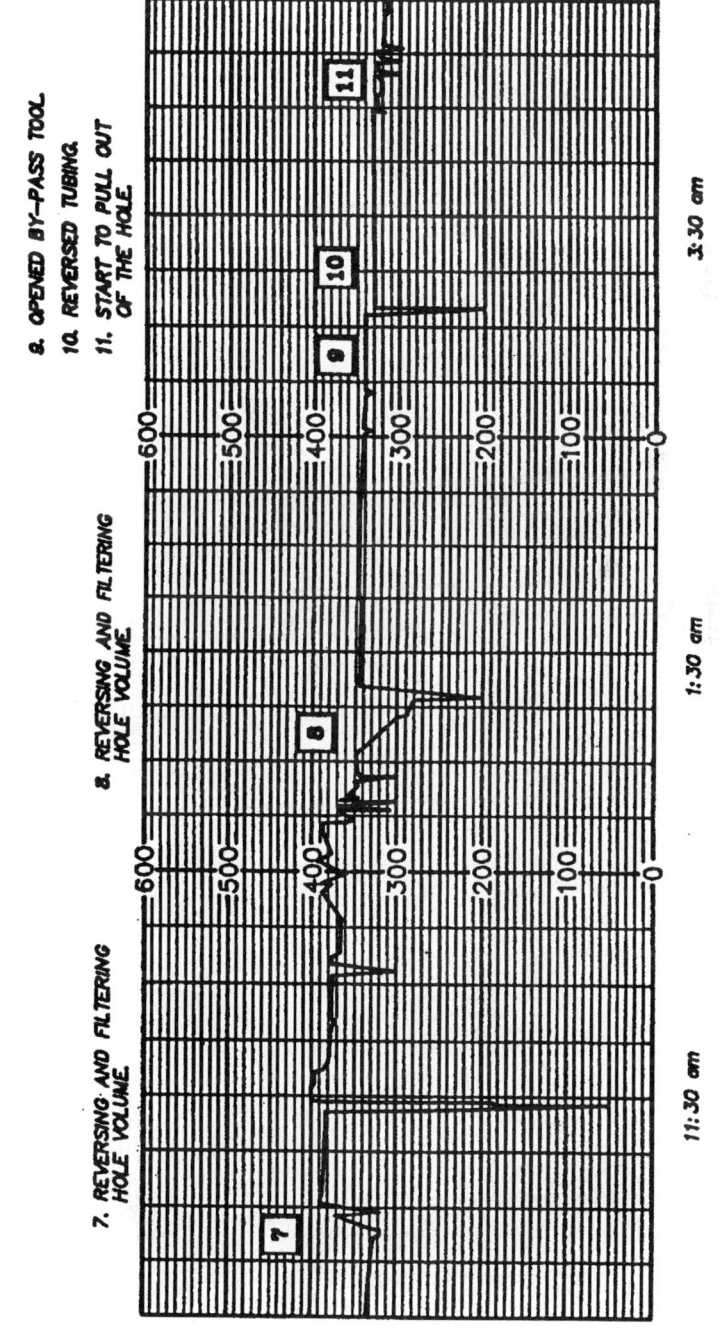

FIGURE 3-21 • Strip chart (4 of 4)

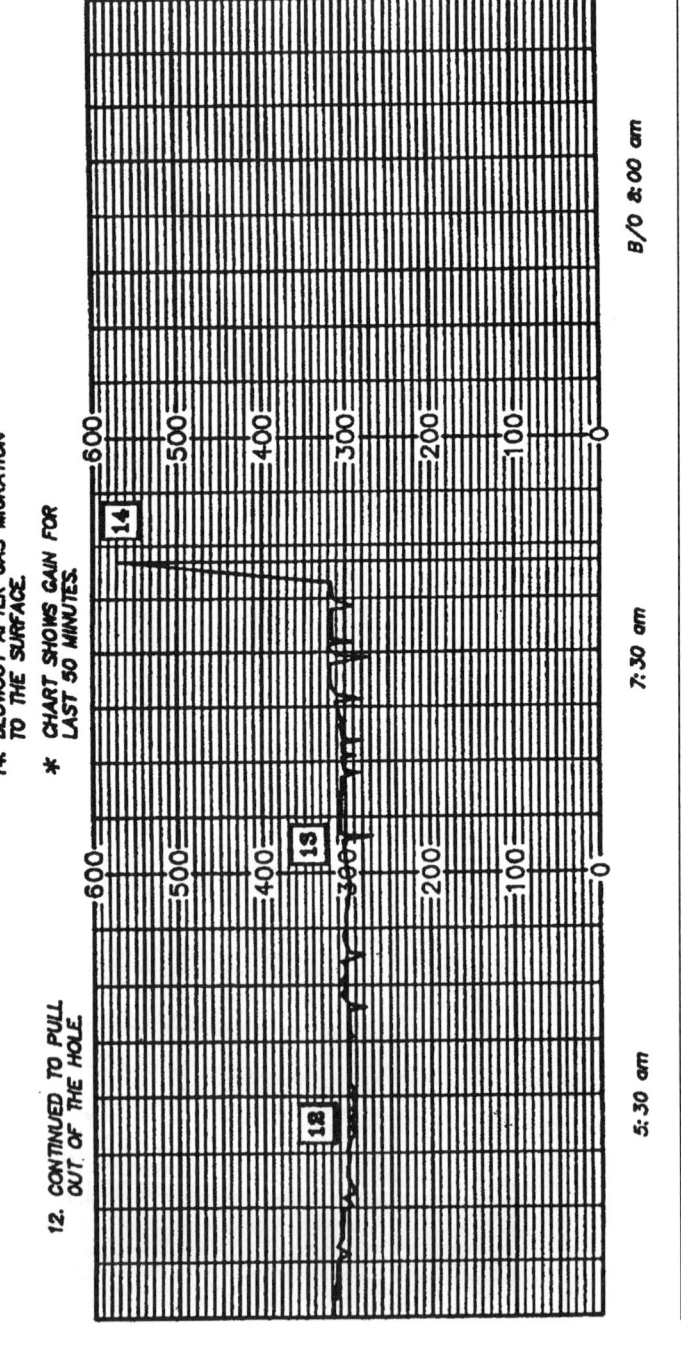

200

Special Problems

- The 2 7/8" (73 mm) tubing was pulled from the well.
- According to the morning reports, tripping operations were halted approximately 1/2 of the distance out of the well to observe the well for 30 minutes. No flow was observed. (Step 13)
- The remaining pipe was pulled from the well. After being out of the well for a short time, the well was observed to be bubbling. Unsuccessful efforts were made to run pipe back into the well. The well blew out. (Step 14)

According to the strip chart, all operations appeared normal from the start of the trip. From steps 11 to 12, it appears that the displacement was correct, or perhaps some slight lost circulation may have been occurring. The suggestion of lost circulation is that the hole fill from step 11 to 12 was approximately 30 bbl (5 m^3), which should have been the total fill for the trip. It is possible that an underground transfer was occurring at this point, but it is not believed to be likely.

The last part of the trip, particularly after the 1/2 hr flow check, showed that the well was not taking the proper amount of fluid to equal the pipe displacement. The total pipe displacement for the trip was approximately 30.2 bbls (5 m^3).

The driller should have noticed that the hole was not taking the proper amount of fill during the 0730 hrs period. It was not noticed and the well subsequently blew out. If the warning signs of the problem had been observed by the driller, the pipe could have been run into the well without incident, and the influx could have been circulated from the well. The incident became critical immediately after a crew change so a new driller finished the trip out of the hole when most of the obvious warning signs were occurring.

Refer to Figure 3-20. It is believed that 2-3 bbls (0.3-0.5 m^3) was in the well at the time of the start of the trip. The volume below the packer above the perforations was calculated to be 4.85 bbl (0.8 m^3). Invariably, some of this amount was removed during the circulation and reverse circulation process. However, not all of it is believed to have been removed.

The fact that some gas remained in the well would not have been apparent to the supervisors. The circulation process should have been satisfactory to remove the gas.

Figure 3-22 shows the volume expansion of the gas as it moves up the well due to gravity segregation. The illustrations show most of the expansion occurs near the surface.

The figure also shows the migration rate and position of the gas in the well, assuming that it starts migration at about 0130 hrs. It would reach the surface from 0730-0800 hrs, which coincided with the observations on the rig. The migration rates are from confidential calculations used by Neal Adams Firefighters in their blowout control work.

FIGURE 3-22 • Volume expansion

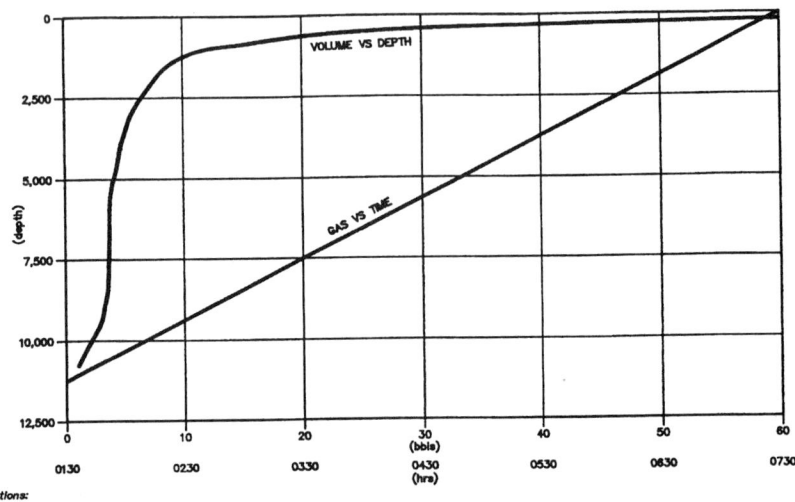

As an example, assume that the well was observed for flow at about 0530 hrs. The gas top would be at 3,100 ft (945 m). It would have expanded to 5 barrels (0.8 m³) at this depth from the original 3 bbl (0.5 m³) at the 10,800 ft (3,290 m) starting point. The differential expansion was 2 bbls (0.3 m³). In practical terms, it would not be possible for this small level of expansion to have been observed with the rig sensors at this point. The well would not be seen to be flowing at this time.

According to Figure 3-22, noticeable expansion would occur with the gas above the 1,500 ft (460 m) depth, which occurred from 0630 hrs. This was observed on the rig at this time.

Swabbing

An evaluation was made to determine if the gas influx was originally swabbed into the well at the start of the trip. The fluid in the well was 12.0 ppg (1,440 kg/m³) filtered brine with a hydrostatic pressure of approximately 6,676 psi (46.03 MPa). The formation pressure was 6,460 psi (44.54 MPa). The differential was 216 psi (1.49 MPa).

The swab pressure for 2 7/8" (73 mm) tubing inside 9 5/8" (244 mm) casing with a filtered brine was less than 100 psi (0.70 MPa). The differential pressure of 216 psi (1.49 MPa) was more than adequate to control the swab pressure.

Special Problems

The 216 psi (1.49 MPa) differential is equivalent to 0.39 ppg (47 kg/m^3) EMW. A standard swab margin for muds is often considered to be 0.3 ppg (36 kg/m^3). Therefore, the margin on the B23 well exceeded the average industry rule of thumb.

Furthermore, if swabbing had started while pulling off the bottom of the well, it would have been probable that it would have continued even with the annular geometry changes from a 7" (177 mm) liner to 9 5/8" (244 mm) casing. At some point, the well would have begun to flow before the pipe was out of the well. This did not occur. Also, the well appeared to have taken the correct fill during most of the early part of the trip.

The conclusion is that swabbing probably did not occur.

HANDLING DRILL PIPE UNDER PRESSURE

Occasionally, pressure will be trapped in the drill string. Problem situations are (1) removal of the kelly under pressure to install a valve or additional drill pipe, (2) below a fish when pulling a joint of pipe with possible pressure trapped below the tools, or (3) removal or repair of malfunctioning equipment under pressure. Drill string handling under these conditions is dangerous and must be approached with caution. Processes used to solve these problems are valve drilling, hot tapping, and freezing.

Valve Drilling and Hot Tapping

When valves or sections of the drill string have pressure trapped beneath or within, some means must be used to bleed the pressure before safe handling techniques can be used. If conventional equipment cannot be made operable under these conditions, special tools must be employed. Valve drilling and hot tapping are designed to meet this requirement by drilling entry ports into the pressured equipment. The term "hot tap" means entry under pressure.

The tool (see Figure 3–23) consists of a bit drive shaft adaptable to hand operation or power tools; a ratchet assembly to apply pressure to the bit by transmitting a downward push on the drill shaft; a rod clamp acting as a pressure point on the rod shaft; a stuffing box to pack off the drive shaft; a bleed-off valve through which pressure can be equalized, bled off, or for circulation; a quick union for ease in make-up and disassembly; a full opening plug valve that can be used to close upon removal of bit and drive shaft; and a saddle clamp to adapt to concentric objects.

An alternate clamp can be used with the tool to drill gate valves, plug valves, or ball valves. Standard tapping equipment as shown is rated to 10,000 psi (69 MPa) and has experienced 9,000 psi (62 MPa) in field service. The valve drilling equipment has a 15,000 psi (103 MPa) rating and has been exposed to 12,500 psi (86 MPa) in the field.

FIGURE 3-23 • Schematic of valve drilling equipment

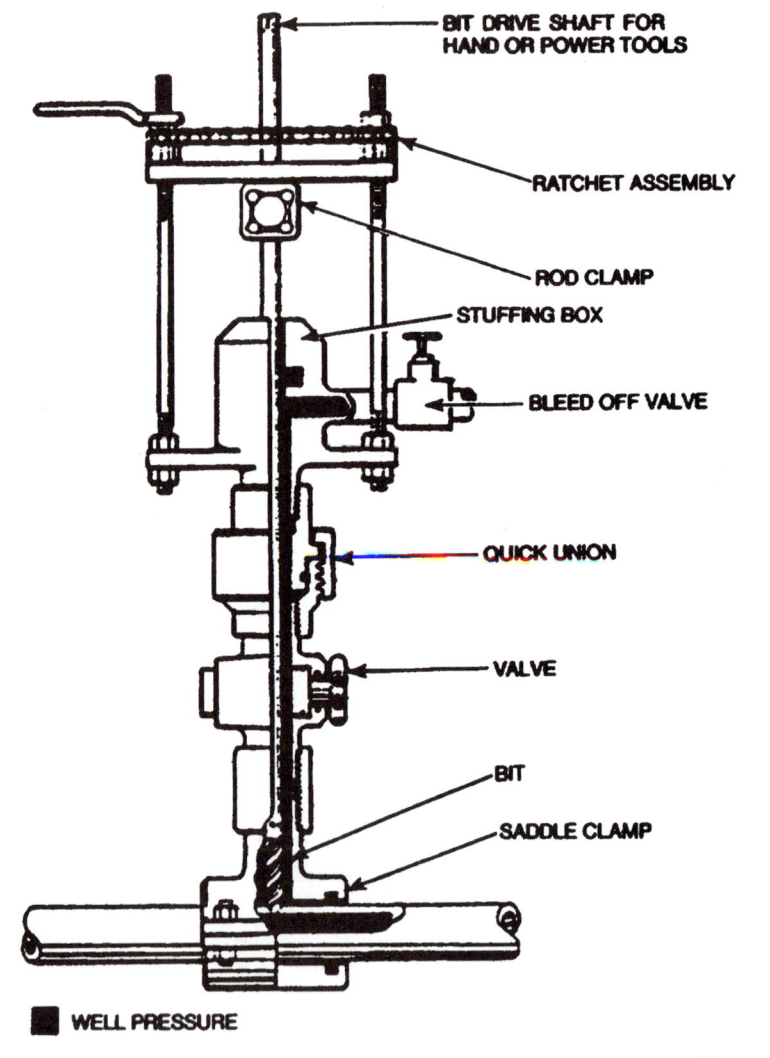

Freeze Process

In some cases, simple entry into the pressured equipment does not offer the complete solution. An example is the case of drill pipe under extreme kick pressures, since it would be impossible to bleed off the pressures through the hot tapping process. The freeze process has been developed to offer additional solutions. The process has been used successfully in such cases as (1) the need to remove the kelly to install a valve, (2) below a pressured fish or (3) below blowout preventers that have failed or have been damaged.

Special Problems

The process involves the use of specially prepared dry ice in a sufficient quantity to freeze a solid plug or a bridge of ice to allow for the safe removal of equipment above the plug. The procedure involves wrapping the pipe with a container of dry ice and allowing the fluid within the pipe to freeze. A test sample is often obtained and frozen in a pup joint on the rig floor to determine the proper setting time. The dry ice causes the fluid in the pipe to reach approximately −142°F (−96.7°C). This develops a plug that can withstand as much as 15,000 psi (103 MPa) differential pressure. Liquid nitrogen systems are also available.

Some primary requirements and recommendations for a successful process are that the pipe must contain a static water-based fluid, the pipe should not be frozen in tension unless necessary, and a plastic coated pipe should not be frozen. Specialists should be consulted before attempting the process.

DRILL STEM TESTING

When drilling a well, it is often desirable to test various formations for potential productivity to determine the future course of the well. This testing, generally known as "drill stem testing," can aid in the selection of casing setting depths and also can be used to avoid setting expensive and unnecessary strings of pipe. Drill stem testing may be used to collect formation fluid samples or to perform pressure testing on a zone to determine its extent and flow capabilities. Although the testing is beneficial in decision making, it poses special well control problems.

The equipment usually consists of the drill string, a packer, and surface control systems such as chokes and special manifolds. The packer is designed to isolate the formation from upper open hole sections, and it should have the capability of allowing flow from (1) the formation up the drill pipe or (2) between the drill pipe and annulus with the formation isolated. The manifold is used to change flow direction as necessary to the drill pipe, flare area, or pumping units.

Drill stem testing principles rely on a hydrostatic pressure reduction below formation pressure of the test zone. Constant pressure is maintained on the other hole sections. This is generally accomplished by placing a packer immediately above the test zone and reducing the hydrostatic pressure in the drill pipe. The pressure reduction allows formation fluids to enter the wellbore and flow through the drill pipe to the surface. After the test is completed, the packer is manipulated to isolate the formation and allow formation fluids in the drill pipe to be displaced with drilling fluid. (See Figure 3–24.)

Flow testing through the drill pipe presents several problems. As low density formation fluids displace the pipe, the static surface pressures must increase to maintain control of the well. (Example 3.2) During flow testing, the equipment may begin to leak and expose formation fluids at

FIGURE 3-24 • Diagram of a drill stem test tool in position

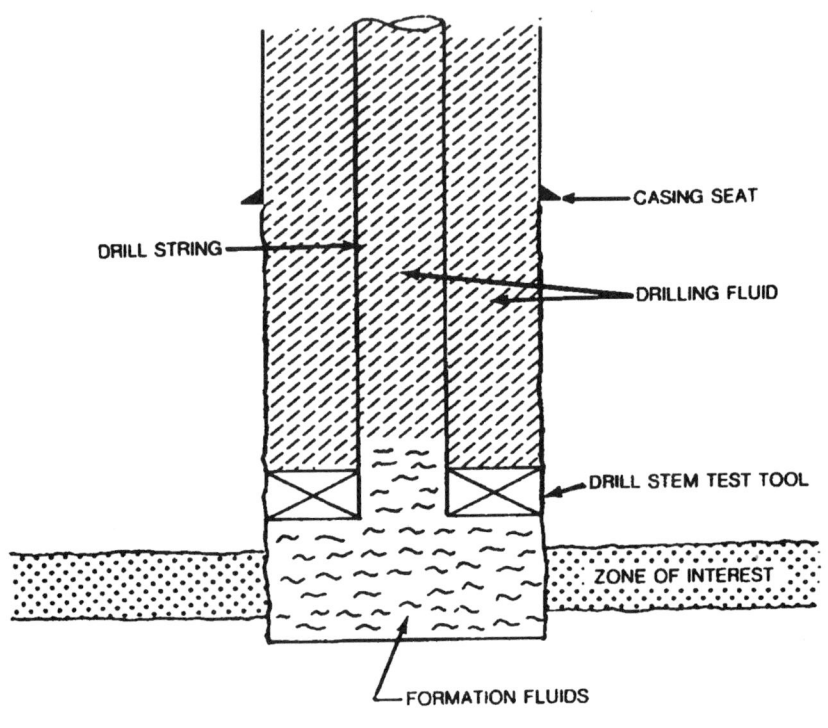

the surface. The problem of handling hydrocarbons at the surface must be taken into account.

Example 3.2

Assume an oil-bearing zone exists at 13,000 ft with a formation pressure equivalent to 14.0 ppg mud. If the oil has specific gravity of 0.5, what is the static surface pressure during a drill stem test?

Solution

1. Formation pressure $= 0.052 \times 14.0 \text{ ppg} \times 13,000 \text{ ft}$

   $= 9,464 \text{ psi}$

2. Oil hydrostatic pressure

   $= 0.052 \times 0.5 \times 8.33 \text{ ppg} \times 13,000 \text{ ft}$

   $= 2,815 \text{ psi}$

3. Surface pressure = Formation pressure − oil hydrostatic pressure
= 9,464 psi − 2,815 psi
= 6,649 psi

After testing has been completed, the well must be killed. The kill procedure involves displacing low density fluids from the drill pipe as well as below the packer. The difficult part of the procedure is displacing fluids below the packer because of the inability to circulate drilling fluid in this interval. If this fluid is not completely removed from the well while the packer is in the hole, it may cause a kick while pulling the packer from the hole or when drilling resumes.

DRILL STRING IMPAIRMENT

During the course of normal kick killing operations, drilling fluid is circulated down the drill string and up the annulus. This flow path uses high strength drill pipe to withstand the circulating pressures from pumping kill weight mud. Problems may impair the drill string and require remedial procedures. These may be a hole, or washout, in the drill pipe or bit and an obstruction or plug that prevents circulation.

Hole in the Drill String

A hole in the drill string is commonly called a "washout." The hole is generally the result of drilling fluid erosion, and it may occur at tool joint connections that were pitted or not properly made up; jet ports in the bit; an isolated metal impurity within the string; or a fracture or crack not previously detected. Proper equipment inspection will locate most causes for washouts before they become a serious problem.

The most common indicator of a washout while drilling is not used in well control. When a washout occurs during normal drilling, a pump pressure decrease will be noted. This indicator is not as applicable in well control because of the fluctuating pressures used during the control procedures. For example, an increase in casing pressure necessitated by a washout may be erroneously considered to be caused by gas expansion or by a change in hole geometry.

In well control, continued growth of a washout may eventually sever the drill string. This is a serious problem, particularly if the pipe should part near the surface.

Another problem is reduction of the circulation path for the kill mud. If the hole occurs near the surface early in the kill procedure, the heavy mud will exit the drill string at this point and flow up the annulus. This exit leaves the annular volume below the hole filled with original mud and kick fluids, with no procedure to displace either. If the hole occurs near the bottom of the well, the seriousness is diminished.

If the washout continues to grow in size, control procedures will be difficult and dangerous. The increased size will decrease the amount of required circulation pressure for a given rate. A new kill rate and pressure must be established periodically to account for hole size change.

If this is not done, excess drill pipe pressure will be used and will tend to promote formation fracture.

Remedial Procedures

An important step in solving this situation is locating the washout. If the point of interest is near or at the bit, the primary concern is using the correct kill pumping pressure as the washout erodes. On the other hand, if the washout is near the surface, the operator must execute procedures to avoid a severed drill string as well as the circulation path reduction.

One common method used to determine the depth of the hole is by observing the number of strokes when kill mud reaches the surface. Although some cases will indicate depths below the actual washout, it is the most accurate method without logging tools. Example 3.3 illustrates the procedure.

Example 3.3

During the course of kick killing operations, the kill mud reached the surface after 1,200 strokes had been pumped. Using the following data, what is the deepest possible depth of the washout?

Pump = 16 in. × 5 in. duplex (90% efficiency)

Drill pipe = 4 1/2 in., 16.6 lb/ft, 12,000 ft

Drill collars = 7 × 2 in., 1,000 ft

Casing = 9 5/8 in., 40.5 lb/ft, 10,000 ft

Open hole = 8 1/2 in., 3,000 ft

Solution

1. −Pump output is 0.1020 bbl/stroke.

    −Drill pipe capacity is 0.0142 bbl/ft.

    −Annular capacity is 0.0562 bbl/ft

2. The volume pumped when the mud reached the surface was

   1,200 strokes × 0.1020 bbl/stroke = 122.4 bbl

3. The drill pipe capacity plus the annular capacity above the washout will equal the volume pumped.

   (0.0142 + 0.0562) bbl/ft × depth = 122.4 bbl

   Depth = 122.4 bbl/0.0704 bbl/ft = 1,738 ft

A field-developed procedure is to note the number of strokes required to pump a plugging material to the washout. The plug material is usually a rope or the equivalent that can be entangled on the sharp edges of the washed out hole. When the plug reaches the hole, it should cause a pump pressure increase that can be recorded at the surface.

Solving the problem of a washout is difficult. The solution depends on the location of the hole, the severity of the problem, the time at which it occurs in the overall kill process, and the kick influx type. Control procedure decisions must be made upon detection of the washout to minimize further complications.

A drill string hole near the surface may be remedied by stripping the pipe and replacing the bad joint(s). If the drill pipe does not contain a back pressure valve, it will be necessary to plug the pipe with a mechanical plug or a pill of viscous, high density mud before stripping operations are initiated. If the hole is located some distance below the surface, the time required to strip out and replace the joint must be weighed against the gas migration rate.

Many operators have attempted to solve the problem by plugging the washout with rope or a similar material. It is pumped down the drill pipe. The material may snag on the hole and plug the washout, which allows fluid movement down the drill pipe. If the plug is successful, use low pump rates through the remainder of the kill procedure to avoid pumping the plugging material out of hole.

Plugged Drill Pipe

Occasionally, the drill string will become completely plugged. The most common cause is barite plugging caused by adding weight material without sufficient suspension agents such as gel (bentonite). It is necessary to remove the plug or provide a circulation path other than through the bit.

When plugging occurs, it may be possible to reverse flow up the drill pipe and free the plug. If this is not successful, the plug can perhaps be jarred loose by surging the pipe with pressure from the pumps. Although this procedure may form a tighter plug, it is occasionally successful.

If plug removal attempts fail, an alternate fluid circulation path must be provided. The drill string may be perforated or severed with a cutter. The primary concern is to perforate or cut at the deepest possible interval. This is usually at the top of the drill collars. Caution must be given to the size of the perforation charge when perforating opposite a casing string.

Some operators have exercised initial preventive measures relative to jet plugging by using a primer cord charge on wireline to blow the jets out of the bit. This procedure is useful when coarse materials are to be pumped and time permits a logging unit to rig up and shoot the drill string. It may not always be possible to implement this technique. Probability for success decreases with the distance the charge is placed above the bit.

WELL CONTROL WITH MWD IN DRILL STRING

MWD systems monitor mud properties, formation parameters, and drill string parameters during circulation operations. The system is not only widely used for drilling, but also has application for well control including the following:

- Drilling efficiency data such as downhole weight on bit and torque can be used to differentiate between ROP changes due to drag and those caused by formation strength. Monitoring bottom-hole pressure, temperature, and flow through the MWD tool is not only useful for early kick detection but can also be of value during a well control kill operation.
- Formation evaluation capabilities such as gamma ray and resistivity measurements can be used to detect influxes into the wellbore, identify rock lithologies, and predict pore pressure trends.
- The MWD tool allows monitoring of the acoustic properties of the annulus for early gas influx detection. Pressure pulses generated by the MWD pulser are recorded and compared at the standpipe and at the top of the annulus. Full-scale testing has shown that the presence of free gas in the annulus is detected by amplitude attenuation and phase delay between the two signals. For water-based mud systems, this technique has demonstrated the ability to detect gas influxes consistently within minutes, before significant expansion has occurred. Further development is currently underway to improve the system's capability to detect gas influxes in oil-based mud.

SPECIAL PROBLEMS

Some MWD tools feature kick detection by means of ultrasonic sensors. In these systems an ultrasonic transducer emits a signal that is reflected off the formation back to the sensor. Small quantities of free gas significantly alter the acoustic impedance of the mud. Automatic monitoring of these signals permits detection of gas in the annulus. It should be noted that these devices will only detect the presence of gas at or below the MWD tool.

Table 3-1 shows some manufacturers' information on features of MWD tools applicable to well control.

The MWD tool offers kick detection benefits if the response time is less than the time for surface indicators to be observed. The tool can provide early detection of kicks and potential influxes as well as monitor the kick killing process. Tool response time is a function of the complexity of the MWD tool and the mode of operation. The sequence of data transmission (e.g., gamma, resistivity, tool face orientation, etc.) determines the update times of each type of measurement. Many MWD tools allow for the reprogramming of the update sequence while the tool is in the hole. This feature can enable the operator to increase the update frequency of critical information to meet the expected needs of the section being drilled. If the tool response time is longer than the amount required for surface indicators to the observed, the MWD can only serve as a confirmation source.

WELL CONTROL WITH DOWNHOLE MOTOR IN DRILL STRING

A downhole motor in the drill string uses drilling fluid circulation to rotate the bit. The most commonly used motors are positive displacement motors (PDM), which utilize a metal rotor inside a nitrile rubber stator. Other motors utilize multiple turbine stages (usually between 100 to 300) to transmit torque to the bit. Each turbine stage consists of a fixed metal stator directing fluid flow on the movable turbine blades. In both cases, the bit is connected via subs to the rotor, and the bit speed is a linear function of the circulation rate.

Both types of motors have recommended minimum and maximum flow rates. Excessive flow rates through positive displacement motors may cause failure of mechanical components due to vibrations from the eccentric motion of the rotor as well as hysteretic failure of the rubber stator from high temperature. The resulting destruction of the stator often causes partial or complete obstruction of the bit nozzles. Maximum flow rates can be significantly increased by the addition of a rotor nozzle that

TABLE 3-1 • MWD comparison tables

	ANADRILL SCHLUMBERGER		COMPUTALOG	EASTMAN CHRISTENSEN		EXLOG		HALLIBURTON GEODATA	
	MWD	LWD		ACCUTRAK	RG	DLWD	RG	AGD/BGD	RGD
Tool OD available, in.	7, 8, 9	6 1/2, 8, 9 1/2	4.75, 6.25, 6.50, 6.75, 8, 9, 9.50	6.75, 7.75, 8, 9, 9.50	6.75, 8	6.25, 8, 9.50	6.75, 8	4.75, 5, 6.25, 7, 7.75, 8, 9.50	6.75, 7.75, 8, 9.50
Maximum operating temperature	150C 302F	150C 302F	150C 302F	125C 257F	150C 302F	125C 257F	150C 302F	150C (battery) 150C (turbine)	150C 302F
Maximum working pressure, psi	20,000	18,000	20,000	20,000	20,000	20,000	20,000	20,000	20,000
Mud flow rate range, gal/min (tool OD)	210-750 (7) 210-1,200 (8) 350-1,200 (9)	600 (6.50 CDR) 800 (6.75 CDR) 850 (8 CDR) 1,200 (8.25 CDR) 450 (6.50 CDN) 850 (8 CDN)	250 (4.75) 900 (6.25, 6.50, 6.75) 1,400 (8, 9, 9.5) maximums	No restrictions, clear 2" ID	No restriction	200-900 (8.25) 200-1,750 (8)	No restriction	Battery: 50-250 (4.75-5), 50-700 (6.25), 50-1,500 (6.75-9.50); Turbine: 50-300 (4.75-5), 50-1,500 (6.25-9.50)	250-1,500
Lost circulation material, max. size and concentration	Medium nut plug 40 lb/bbl	Medium nut plug No limit	Medium nut plug 40 lb/bbl	No restriction	No restriction	Fine to medium nut plug, no limit	No restriction	Fine-medium nut plug 30 lb/bbl, consult field engineer	Fine-medium nut plug 30 lb/bbl, consult field engineer
Surface mud screen required?	Recommended	Recommended	Preferred, not required	Not required	No	Recommended	No	Not required (battery except 4.75), Recommended (turbine)	Recommended
Maximum bit pressure, psi	No limitation	No limitation	2,000	No limitation	No limitation	300 min, 3,000 max	No limitation	500 min, 2,500 max	500 min, 2,500 max

Nomenclature: CDR - compensated dual resistivity
CDN - compensated density neutron

Special Problems

allows a portion of the fluid flow to bypass the power section and travel through the hollow rotor directly to the bit and into the drill pipe annulus. Maximum flow rates for turbine drilling motors are calculated to prevent damage from hydraulic thrust forces on the bearing section and possible contact between the stator and turbine blades.

The minimum flow rate for a positive displacement motor is the rate necessary to close the bypass, which is a hydraulically actuated directional valve placed in the drill string immediately above the motor. This minimum rate must be attained to direct fluid flow into the power section, thereby transmitting torque to the bit.

Most PDMs are designed to allow 3% to 5% of the total fluid flow to bypass into the bearing section for lubrication. In the absence of a bypass valve, the minimum flow rate is that necessary to create adequate dynamic forces that will cause pressure to build in the rotor/stator cavity and cause rotation.

A turbine drilling motor must have sufficient fluid velocity to create the tangential force on the turbine blades to overcome the rotational inertia of the bit plus any applied torque. Flow rates below this minimum will pass directly through the turbine and into the annulus.

Table 3–2 gives a description of common motors and recommended minimum and maximum flow rates.

Consequences of a Motor in a Well Control Event

Motors do not have any serious adverse effects during conventional well control. The normal circulation rate of 1–3 bpm, 42–126 gpm (0.16–0.48 m^3/min) is lower than the manufacturers' maximum recommended flow rate. (See Table 3–2.)

If the circulation rate is maintained below this level, the flow path will be predominantly through the bypass valve. This will not alter the kill procedure, since the valve is located relatively close to the bit.

If the circulation rate is above the minimum to close the bypass valve, the bit will rotate during the circulation, well killing process. This does not create any problems for well control.

Nonconventional Well Control

Bullheading does not create any problem with the motor. The typical maximum rate of 5 bpm (210 gpm) (0.79 m^3/min) is less than the manufacturers' maximum recommended flow rate.

The volumetric method has no impact relative to a motor because fluid circulation does not occur.

TABLE 3-2 • Positive displacement motor comparison tables

	OUTSIDE DIAMETER	FLOW RATE	SPEED	MAXIMUM MOTOR PRESSURE DROP
COMPUTALOG	4.75 in. (120.7 mm)	100/250 gpm (375/950 lpm)	225/565 rpm	530 psi (3650 kpa)
	6.31 in (160.3 mm)	200/400 gpm (750/1500 lpm)	210/425 rpm	500 psi (3250 kpa)
	6.75 in (171.4 mm)	200/450 gpm (750/1700 lpm)	190/430 rpm	500 psi (3250 kpa)
	8.06 in (204.7 mm)	220/550 gpm (825/2100 lpm)	140/385 rpm	500 psi (3250 kpa)

	OUTSIDE DIAMETER	FLOW RATE *	SPEED	MAXIMUM MOTOR PRESSURE DROP
DAILEY	4.75 (120.7 mm)	100/250 gpm (375/950 lpm)	56/212 rpm	450 psi (3100 kpa)
	6.5 in (165.1 mm)	200/600 gpm (750/2270 lpm)	50/170 rpm	450 psi (3100 kpa)
	7.75 in (196.9 mm)	300/900 gpm (1135/3400 lpm)	48/140 rpm	350 psi (2400 kpa)

	OUTSIDE DIAMETER	FLOW RATE *	SPEED	MAXIMUM MOTOR PRESSURE DROP
DRILEX	4.75 in (120.7 mm)	100/250 gpm (375/950 lpm)	140/350 rpm	750 psi (5100 kpa)
	6.75 in (171.4 mm)	200/650 gpm (750/2460 lpm)	55/185 rpm	500 psi (3250 kpa)
	8.25 in (210 mm)	350/750 gpm (1325/2840 lpm)	110/235 rpm	625 psi (4300 kpa)
	9.5 in (241.3 mm)	700/1100 gpm (2650/4170 lpm)	115/180 rpm	500 psi (3250 kpa)

	OUTSIDE DIAMETER	FLOW RATE *	SPEED	MAXIMUM MOTOR PRESSURE DROP
EASTMAN CHRISTENSEN	4.75 in (120.7 mm)	80/240 gpm (300/900 lpm)	100/300 rpm	725 psi (5000 kpa)
	6.75 in (171.4 mm)	185/475 gpm (700/1800 lpm)	100/260 rpm	725 psi (5000 kpa)
	8.0 in (203.2 mm)	315/685 gpm (1200/2600 lpm)	85/190 rpm	580 psi (4000 kpa)
	9.5 in (241.3 mm)	395/740 gpm (1500/2800 lpm)	100/190 rpm	800 psi (5500 kpa)

* - Indicated maximum can be increased via installation of rotor nozzle

Special Problems

The dynamic kill, or dynamic-heavy slug method, used for a shallow gas kick involves flow rates usually exceeding the manufacturers' maximum recommended flow rate. The motor may be damaged or destroyed. It is not likely, however, that it will impact the kill operations. Since killing the kick is more important than damage to the motor, the kill operation should proceed in a normal method.

DRILL STRING SHEARING

Shear rams are commonly considered as having the capability to shear certain sizes and weights of pipe (See Table 3-3.) and seal on the open hole as blind rams after shearing. These rams are called "shear/blind rams" or simply "SBRs." Shear rams that do not have blind sealing capability require special planning.

Preplanning is required for the blowout preventer equipment to effectively shear pipe. The key aspect of planning is related to hydraulic requirements. The control system must develop sufficient pressure to activate the shear rams and cut the pipe. Table 3-3 gives shear pressure vs. drill pipe information (e.g., size, grade, density) for a common make and model of shear/blind rams. (Manufacturers should be consulted for confirmation and details.) Remaining system pressure after various combinations of component operations should be determined with the accumulator pump off. This will help simulate shut-in and hang-off activities that may have to be done under rig power shut-down conditions due to safety concerns. Various manufacturers' and A.P.I. recommended practices are available as guidelines for accumulator sizing. The operator can follow these guidelines or develop their own directives regarding accumulator sizing and performance.

Various BOP modifications are recommended to ensure adequate shear force. These alterations typically involve changes to the hydraulic system that convert standard BOP bonnets to "shear bonnets." Another option is the addition of hydraulic boosters to increase the shear force directly. Both modifications may be included in the BOP stack design as an additional safety measure.

Shear rams will not cut most drill pipe tool joints. BOP stack configuration should allow for adequate vertical distance between the pipe ram and the shear ram to ensure the shear blades are adjacent to the tube rather than a tool joint. Most manufacturers recommend a minimum of 30 inches (762 mm). In addition to clearance, the stack should also be configured to allow well control measures to be initiated after shearing. This should include access to the wellbore below the shear-blind rams for pumping, venting, and pressure monitoring.

TABLE 3-3 • Shearing capabilities of Cooper shear rams (Type U BOPs) (Coursty of Cooper Oil Tool)

PIPE SIZE	GRADE	11" 3-10M SBR				11" 15M SBR				13-5/8" 3-10M SBR				13-5/8" 15M SBR				16-3/4" 5-10M SBR				18-3/4" 10M SBR	20-3/4" 3M / 21-1/4" 2M SBR				21-1/4"* 10M SBR
		SB	SBT	LB	LBT	B	BT	LB	LBT	SB	SBT	LB	LBT	B	BT	LB	LBT	SB	SBT	LB	LBT	B	SB	SBT	LB	LBT	SB
3-1/2" 13.3 #/FT	E	X	X	X	X	X	X	X	X	X	X	X	X	X	X	X	X	X	X	X	X	X	X	X	X	X	X
	G	X	X	X	X	X	X	X	X	X	X	X	X	X	X	X	X	X	X	X	X	X	X	X	X	X	X
	S	X	X	X	X	X	X	X	X	X	X	X	X	X	X	X	X	X	X	X	X	X	X	X	X	X	X
3-1/2" 15.5 #/FT	E	PL	X	X	X	PL	X	X	X	PL	X	X	X	X	X	X	X	X	X	X	X	X	PL	X	X	X	X
	G	PL	X	PL	X	PL	X	PL	X	PL	X	X	X	X	X	X	X	X	X	X	X	X	PL	X	X	X	X
	S	PL	X	PL	X	PL	X	PL	X	PL	X	X	X	X	X	X	X	X	X	X	X	X	PL	X	X	X	X
4-1/2 16.6 #/FT	E	PL	X	PL	X	PL	X	PL	X	PL	X	PL	X	X	X	X	X	PL	X	X	X	X	PL	X	X	X	X
	G	PL	PL	PL	X	PL	PL	PL	X	PL	PL	PL	X	X	X	X	X	PL	X	X	X	X	PL	X	PL	X	X
	S	PL	PL	PL	X	PL	PL	PL	X	PL	PL	PL	X	PL	X	X	X	PL	X	X	X	X	PL	X	PL	X	X
5" 19.5 #/FT	E	PL	PL	PL	X	PL	PL	PL	X	PL	PL	PL	X	PL	X	X	X	PL	X	X	X	X	PL	X	PL	X	X
	G	PL	PL	PL	X	PL	PL	PL	X	PL	PL	PL	X	PL	X	X	X	PL	PL	X	X	X	PL	PL	PL	X	X
	S	PL	PL	PL	PL	PL	PL	PL	PL	PL	PL	PL	PL	PL	X	X	X	PL	PL	X	X	X	PL	PL	PL	X	X
5" 25.6 #/FT	X	PL	PL	PL	PL	PL	PL	PL	PL	PL	PL	PL	PL	PL	PL	X	X	PL	PL	X	X	X	PL	PL	PL	X	X
	G	GL	GL	GL	GL	GL	GL	GL	GL	GL	GL	PL	PL	GL	PL	X	X	PL	PL	X	X	X	PL	PL	PL	X	X
	S	GL	GL	GL	GL	GL	GL	GL	GL	GL	GL	PL	PL	GL	GL	X	X	PL	PL	PL	X	X	PL	PL	PL	PL	X
5-1/2" 24.7 #/FT	G	GL	GL	GL	GL	GL	GL	GL	GL	GL	GL	PL	PL	GL	GL	X	X	PL	PL	PL	X	X	PL	PL	PL	PL	X
	S	GL	GL	GL	GL	GL	GL	GL	GL	GL	GL	GL	PL	GL	GL	X	X	PL	PL	PL	X	X	PL	PL	PL	PL	X
6-5/8" 25.2 #/FT	G	GL	GL	GL	GL	GL	GL	GL	GL	GL	GL	GL	PL	GL	GL	X	X	PL	PL	PL	X	X	PL	PL	PL	PL	X
	S	GL	GL	GL	GL	GL	GL	GL	GL	GL	GL	GL	GL	GL	GL	PL	X	PL	PL	PL	X	X	PL	PL	PL	PL	X
6-5/8" 27.6 #/FT	G	GL	GL	GL	GL	GL	GL	GL	GL	GL	GL	GL	GL	GL	GL	PL	X	PL	PL	PL	X	X	PL	PL	PL	PL	X
	S	GL	GL	GL	GL	GL	GL	GL	GL	GL	GL	GL	GL	GL	GL	PL	PL	PL	PL	PL	X	X	PL	PL	PL	PL	X

X = capable of shearing at 2800 psi closing pressure
PL = Incapable of shearing due to operating pressure limitation
GL = Incapable of shearing due to bore size/ram geometry limitation

B = Standard Bonnents
BT = Standard Bonnets with Tandem Boosters
SB = Shear Bonnets
SBT = Shear Bonnets with Tandem Boosters
LB = Large Bore Shear Bonnets
LBT = Large Bore Shear Bonnets with Tandem Boosters
SBR = Shearing Blind Rams

* = Will not seal effectively if pipe is hung-off

Note - Shearing capability information requested from other manufacturers was not supplied.

Special Problems

The limits and capabilities of a particular BOP should be compared to the requirements of each operational phase of the drilling program. For example the Cooper 20 3/4–3,000 psi (21 MPa) and the 21 1/4–2,000 psi (14 MPa) type U BOPs do not have the capability to seal after shearing if the pipe is hung off. Some Shaffer shear rams have adequate clearance to allow shearing/sealing of up to 2 in. (51 mm) OD bar or wireline tools.

It is important to note that significant differences in required shear pressures have been observed for similar size, density, and grade drill pipe having different manufacturers. These differences become particularly obvious between pipe manufactured in different countries.

Causes for Shearing

The drill string may require shearing for several reasons. Some are as follows:

- Flow up the drill string which cannot be stopped with a valve.
- Leaks on the surface which endanger the rig or crew. The leaks may result from malfunctioning BOP equipment. The leaks include gas escaping from a subsea riser and a subsea BOP stack.
- An emergency drive-off of a floating rig is required and the pipe cannot be hung-off in the rams and backed-off.

Normal Procedures—Shearing

Normal preferred procedures for shearing pipe are as follows:

- After a cause for shearing is observed, the drill string is hung-off on a set of pipe rams below the shear rams.
- The shear rams are activated after the hydraulic control system is appropriately regulated for shearing.
- Well control procedures must be implemented that account for possible pressure below the shear/blind rams.

The same procedure is followed for an emergency drive-off of a floating vessel.

If the drill pipe is hung-off prior to shearing, all attempts should be made to ensure that any pressure trapped between the pipe rams and any upper seal (i.e., annular BOP or rotating head) is bled off. This will be of benefit for two reasons:

- Pressure will act against the rod that forces the shear ram into the well bore. This will require more closing force from the accumulator.
- Pressure between the pipe rams and an upper seal creates an upward thrust force which, if greater than weight of the remaining pipe, will cause the pipe to jump or be blown out of the hole.

An alternate approach is to shear the pipe without hanging it on a lower set of pipe rams. Contrary to popular belief, studies indicate that drill pipe tension has little effect on required shear pressure.

WELL CONTROL WITH HYDRATE BUILDUPS

Hydrates are a solid substance with ice characteristics. It is a compound of water and hydrocarbon gas. Several classes dependent on the gas have been described in the literature.

Hydrates can block flow lines, BOP stacks, drill stem test strings, or other tubulars and cause wellhead connector release difficulties. The blockage is usually complete and prevents any fluid circulation. The hydrates must be removed before the previous well control or flow testing operations can be resumed. In-situ hydrate formations are not uncommon and can cause drilling difficulties.

The conditions surrounding the formation of hydrates can be described generally as follows:

- Water and gas must be present
- Temperatures near 0°C
- High pressures
- Motionless time

Hydrates have been produced in laboratory conditions. However, the exact formation conditions are not repeatable so it is difficult to predict precise formation environments.

Chemistry Review

Natural gas hydrates are an ice-like crystalline structure. A spherical lattice of water molecules is stabilized by an encapsulated "guest molecule." The specific gravity of the hydrates, or clathrates, is 0.88–0.90 sp.gr.

Figures 3–25 and 3–26 show the temperature-pressure relationship for hydrate formation. As water depth increases, the water hydrostatic pressure increases upward along the vertical axis. The low seabed temperatures of 34–36°F (0–2°C) fall in the hydrate formation region.

Gas hydrates crystallize in one of two cubic structures, known as "Structure I or II." The type of structure depends primarily on the size of the hydrate-forming molecule. Structure II hydrates are formed only by

Special Problems

FIGURE 3-25 • Effect of natural gas specific gravity on hydrate forming conditions

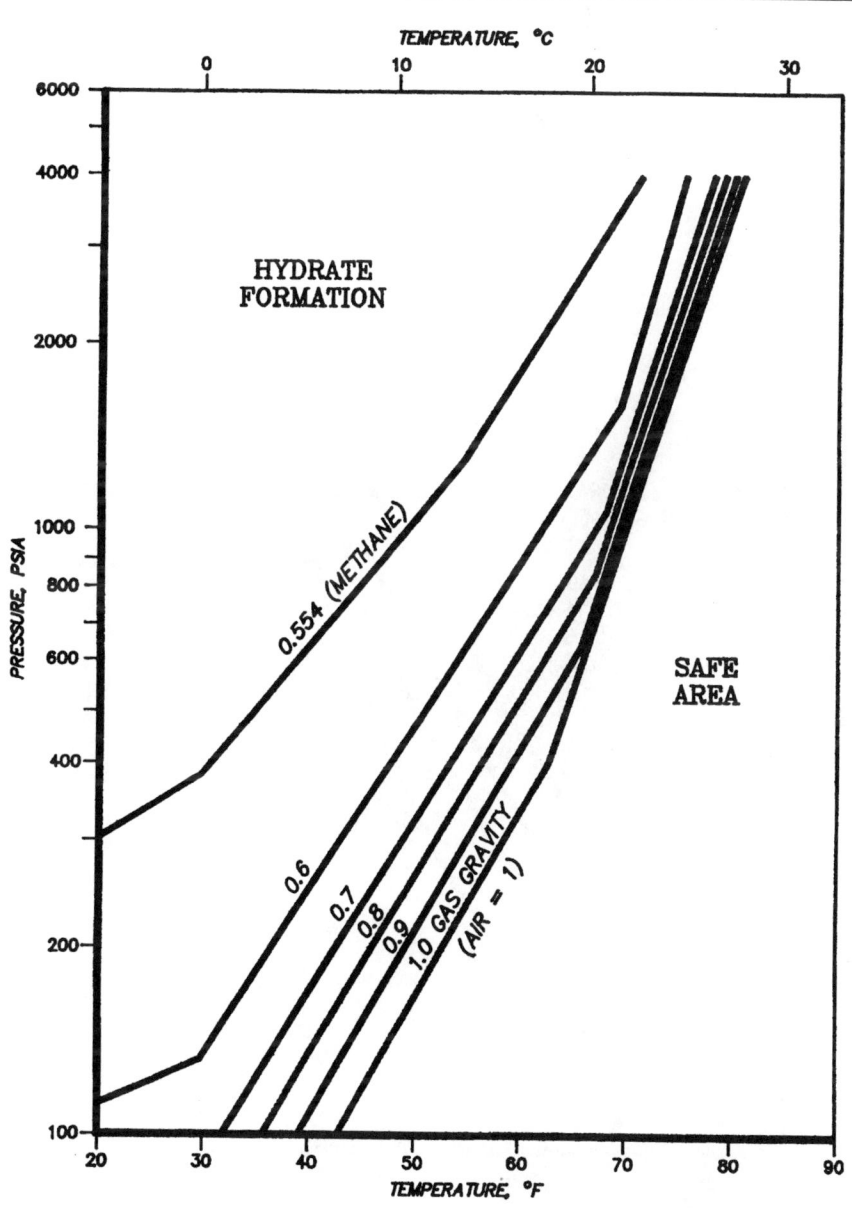

FIGURE 3-26 • Depression of hydrate formation temperature by inhibitors

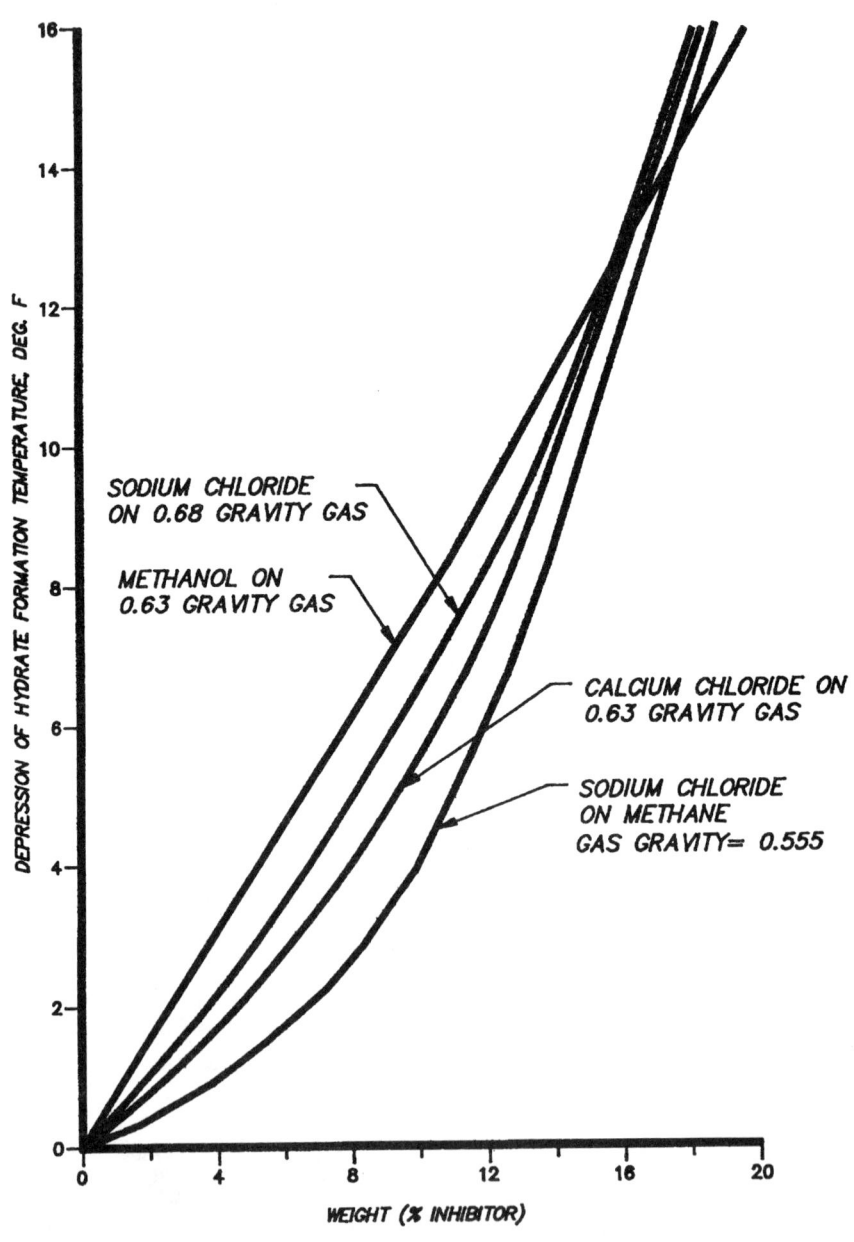

Special Problems

TABLE 3-4 • Hydrate lattice properties

	Structure I	Structure II
Water molecules per unit cell	46	136
Cavities per unit cell		
Small	3	16
Large	6	8
Cavity radius, Angstroms		
Small	3.97	3.9
Large	4.30	4.73
Typical gases that form in each cavity of this structure		
	methane(S)	propane (L)
	ethane (L)	i-butane (L)
S = small cell		n-butane (L)
L = large cell		n-pentane (L)

those "guest molecules" that are too large to fit in the cavities of Structure I hydrates. Each structure type has a small and large type of cavity. (See Table 3-4.)

- Composition
- Temperature
- Pressure

Four methods or conditions exist for preventing the formation of hydrates. If these conditions are allowed to coexist and natural gas is available in sufficient quantities, then hydrates will not form.

Hydrate occurrence in deepwater drilling is a complex problem. However, a basic observation can be used to develop conceptual approaches to resolution or mitigation. The hydrate control parameters are as follows:

- Reduce the pressure below that of the hydrate formation for the given temperature and composition.
- Maintain the temperature of the gas above the hydrate formation temperature for the given pressure and composition.
- Reduce the dew point of water vapor in the gas below the generating temperature by drying with agents such as glycol.
- Introduce substances such as electrolytes or alcohols into the gas reservoir to lower the hydrate formation temperature.

Prediction of hydrate formation conditions is based on these parameters.

A key parameter associated with hydrate prediction is time. Laboratory studies do not sufficiently document this variable. Some field case histories suggest that long dormant times in the range of 24 hours may be required. Other cases show that only several minutes are required. Needless to say, the results are confusing.

Hydrate Problems

Problems related to hydrates are numerous and many do not have simple solutions. Some are as follows:

- Difficult to predict hydrate formation
- Possible BOP operations impairment
- Inability to circulate
- Difficulty in hydrate removal
- Wellhead connector release difficulty

Example situations shown in this section give insight into the difficult and unpredictable nature of hydrates.

Hydrate Prevention

Several procedures have been developed to assist in preventing hydrates. Most are based on altering fluid composition or temperature environments. None appear to have guaranteed results.

In most cases, experience has shown that some amount of motionless time is required for hydrates to form. If phase envelope conditions exist for hydrate formation, a preventive aide is to maintain fluid movement. Most hydrate examples show formation after circulation has been shut down for a period of time. An exception to the motionless time guide is drill-stem testing. Examples show hydrate formation while flowing the well. An explanation is hydrates formed along pipe walls and then congested at one location to block flow suddenly.

Fluid (Hydrate) Composition

Composition of the hydrate is a factor affecting its formation. As shown in Figure 3–25, the heavier hydrocarbons will form hydrates at lower pressures and higher temperatures than lighter fractions such as methane. Most produced gas in drilling operations is methane rich, which favors the more difficult hydrate formation conditions.

Impurities in the water affect hydrate generation. Salt inhibits hydration by reducing the activity of the water. H_2S and CO_2 tend to depress the temperatures and pressures required for hydration.

Special Problems

As noted by Wooley et al., the composition of the liquid phase is a factor in hydrate equilibrium. The effect of the following substances on hydrate formation were investigated:

- Methanol
- Seawater (4% NaCl)
- Chemical additives
- Diesel oil

These substances behave as inhibitors. Their presence requires that at a given pressure that a lower temperature is required to form hydrates. Figure 3-27 shows the impact of these inhibitors on hydrate formation conditions and the effect of inhibitor concentration.

For drilling mud solids and chemical additives, only qualitative results are available. The presence of impurities in the water phase has an inhibiting effect on equilibrium conditions. Furthermore, if the impurity is a polar compound, it could link to water molecules so that its behavior is very similar to alcohols, and therefore inhibit hydrate formation like alcohols. The addition of diesel to the liquid introduces a fourth phase due to its immiscibility with water.

Comparing the effectiveness of various contaminants as inhibitors, 10% methanol in water is the most effective inhibiting solution investigated, although it is not practical. A more feasible approach is seawater, which provides the same inhibition as approximately 2% methanol.

Salt Water Drilling Fluids

Due to the recent experiences of several operators with hydrates, a trend towards salt mud usage is developing. The saturated salt system alters the water's activity level and lowers the freezing point. A sodium chloride (NaCl) system has been used on several occasions after selecting it from a field of brine options. The selection criteria included cost vs. performance and drilling requirements.

Drilling fluid additives are currently being studied in the laboratory by some operators. The purpose is to identify the additives that promote and inhibit hydration. The chemicals that promote hydration can be restricted to wells without potential hydrate problems.

Methanol Applications

Methanol is often injected into a production gas stream to prevent hydrate formation. It is frequently used for intermittent or continuous injection in natural gasfield-gathering systems and transmission lines to protect against hydrate formation when the gas is cooled by the environment or

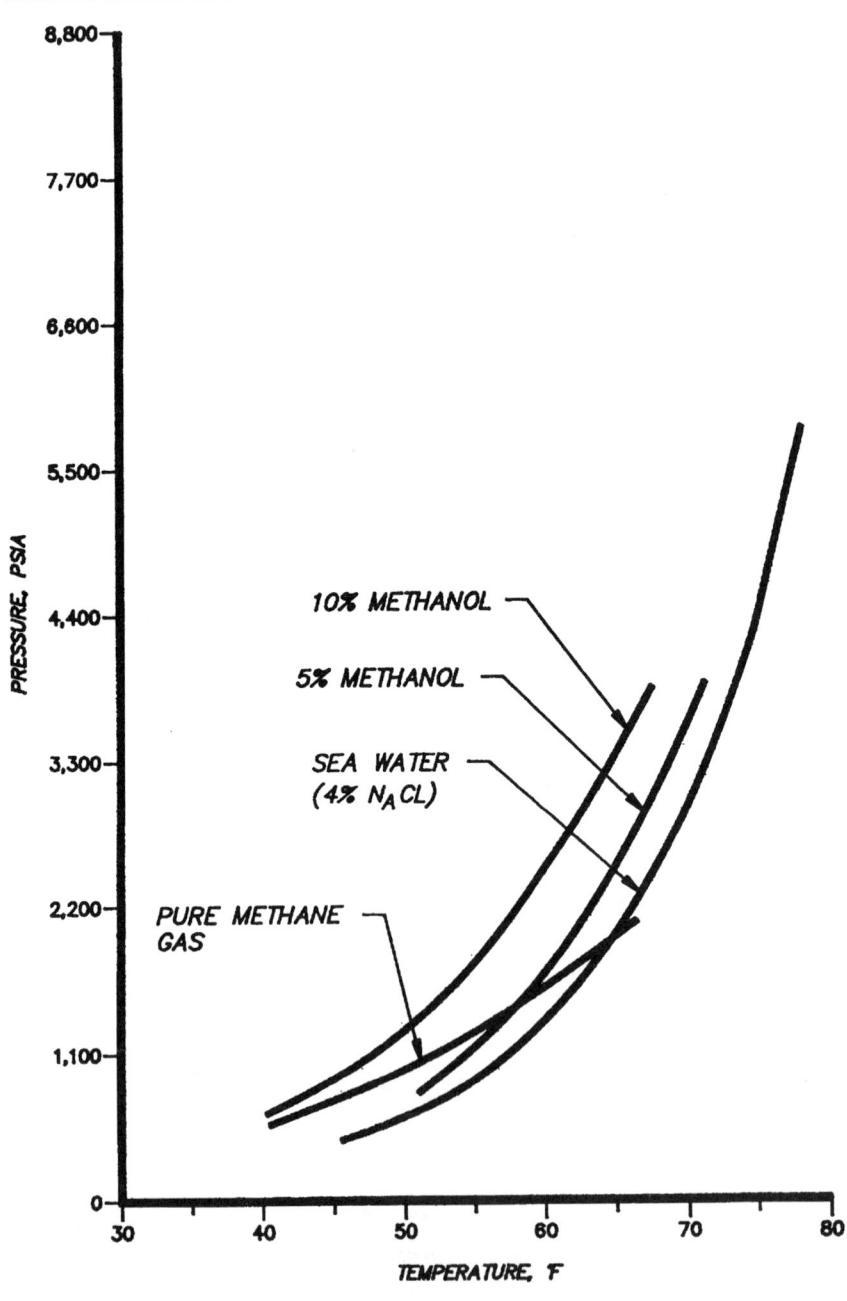

FIGURE 3-27 • Effect of sea water and methanol on hydrate formation

pressure drops. In gas processing plant operations, intermittent injection of methanol is frequently used when there is a slow build up of hydrates.

Due to its widespread applications in related fields of hydrate prevention, methanol (or similar substances) should be considered for deepwater applications. Two approaches are (1) prevent hydrate formation and (2) thaw hydrates after they have formed. Key questions are as follows:

- What volume of methanol is required?
- Is intermittent injection satisfactory or will continuous injection be required?
- After hydrates have formed, is methanol effective to some degree?
- If a post-formation methanol injection is effective, how can the fluid be injected in a frozen BOP stack or choke/kill lines?

Answers to these questions are not currently available.

The required volume of methanol is uncertain but can be estimated. The difficulty involves the associated kick parameters. Is the gas in a slug or spread through the mud system? Did the gas slowly trickle into the mud stream so that it is uniformly distributed in small bubbles? What is the mud volume? How do the mud chemical additives affect hydration?

The probable response for the volume question is that significant quantities will be required. As an order or magnitude qualification, it is more likely that 10–50 bbl (2–8 m^3) will be required than 10–50 gallons (0.04–0.19 m^3). The questions that arise from this estimate relate to the negative aspects of injecting large volumes of methanol into the mud system (i.e., compatibility, environmental safety, and personnel hazards).

It is not considered likely that intermittent injections to prevent hydrate formation will resolve the problem. It may be effective if it can be injected at the time that gas is near the BOP stack. However, it is impractical to consider that the location of the gas in the well can be tracked with sufficient accuracy to allow for effective intermittent injection. Also, gas distribution throughout the well mitigates intermittent usage success. As a result, continuous injection appears to be the most appropriate alternative.

For methanol to be effective at the time of the kick, it must be injected at the BOP. Mud-pit mixing at the surface is not satisfactory because the gas would reach the BOP before the methanol would be in the hole.

Large volume fluid injections to prevent hydrate formation must be done with the kill line. The choke line will be reserved for circulation. An attached accumulator bottle-reservoir concept was discarded because of the relatively high expected fluid volume requirements. If subsequent calculations show that small methanol volumes are required, an attached pressurized system may be satisfactory. The attractiveness of the kill line usage is that it is controlled from the surface and any volumes can be used.

If necessary, it can be used in an emergency as a choke line if the primary line fails. If the kill line-methanol injection concept is pursued, it can be implemented quickly and without any major equipment modifications.

Methanol effectiveness on a post-formed hydrate is questionable. At best, its effectiveness is limited to the contact surface. Additional research is appropriate for this topic.

Hydrate Removal

Hydrate removal is difficult. The ice structures have a high resistance to differential pressure so pumping out with high pump pressures is not practical.

Current experiences have shown the only reliable means to remove hydrates is by increasing temperature. Hydrates will dissolve slowly. Although the method appears to be reliable, it is not easily implemented for several reasons:

- A means is not directly available to increase temperature on part or all of the mud systems.
- High heat losses occur down the choke line or riser so these paths do not provide a hot fluid to melt hydrates.
- Circulation is more difficult if the choke lines are plugged.
- Coiled tubing or snubbing units are required for circulation if the hydrate is well below the BOP stack in the drill pipe-casing annulus.

Pressure reductions to remove hydrates have not proven to be practical.

Normal Procedures for Hydrate Prevention and Removal

The planning process involves several points. Some precautions conflict, so an optimum selection must be made.

Determine if the planned operating conditions fall in the hydrate phase behavior envelope. If the operating conditions are within the hydrate formation region, planning to prevent their occurrence should be done. Worst conditions occur at the seafloor where the coldest well temperatures will be found. Fluids below the seafloor typically have higher temperatures due to geothermal gradients and fluid circulation.

Oil muds should be a first priority. The oil mud minimizes the available free water to mix with any gas for hydrate formation. Also, small quantities of gas will dissolve in the oil and remove some of the hydrate-forming constituent.

Salt muds should be used when oil muds are not practical. The salt reduces the hydrate-forming temperatures by several degrees. This action

Special Problems

can move the operational conditions outside the hydrate phase behavior envelope.

Motionless time should be avoided when gas is in the well. Situations for hydrate formation are during kicks and well testing. Shutting down kick operations should be avoided when gas is in the depth range of 100–200 ft (30–60 m) below the BOP stack up to expulsion from the well. These depths are cold because of seawater. Shut-downs when gas is deep in the well is not a problem.

Motionless time during well testing (i.e., DSTs) should be avoided but cannot be completely prevented for operational reasons. An interesting discussion point is that several DST situations have experienced hydrate formation while flowing the well.

Hydrates have blocked BOP functions. If possible, BOP components should be functioned occasionally if operations must be shut down while gas is in the cold region of the well.

The principal technique for hydrate removal is increasing temperatures around the ice plug. It will melt slowly. A consideration is that pressure may exist below the plug, so pressure control measures must be maintained.

Specific removal techniques depend on the situation. General requirements are circulating a warm fluid around or near the plug. As an example, circulating down the kill line and up the riser may be possible.

Small diameter tubing can be run down the riser to the plug. Warm fluids are circulated. Coiled tubing is an ideal choice. Snubbing units have application if allowances can be made for the drill string or flow string in the well. Small pipe such as 2 3/8"–2 7/8" (60–73 mm) tubing can be run by the rig under some conditions.

WELL CONTROL IN HORIZONTAL WELLS

Well control theory for vertical wells has been defined for many years. The theory addresses areas such as kick causes, warning signs, and kill procedures. Directional wells not considered in the horizontal category can easily apply well control theory for vertical wells.

The theory has a basic principle relative to pressure transmission. The wellbore can be used as a pressure conduit from the formation to the surface if the pressure transient can move vertically. A practical application is using surface pressure readings to infer differences between bottom-hole formation pressure and mud hydrostatic pressure. Drill pipe pressure is a better indicator than casing pressure because it is not contaminated with influx.

Horizontal wells have a nonvertical hole section. Pressure differences can not be directly measured if the influx is in this section of the well because pressures are transmitted vertically to the surface. The best kick indicator after the well is shut-in will be an increased pit volume. The shut-in pressures on the drill pipe and casing will be zero if the influx remains in the horizontal section and the kick was due to swabbing.

Well Control Differences Between a Horizontal and Vertical Well

Basic well control procedures are the same for horizontal and vertical wells. These include kick causes, warning signs, shut-in procedures, and kill methods.

Well control differences exist between the two well configurations. These differences include the following:

- Influx volumes can be greater for horizontal wells than vertical wells under the same conditions of differential pressure and time of underbalance.
- Shut-in pressures for the drill pipe and annulus will be zero for a swabbed kick if the gas remains in the horizontal section at shut-in.
- Shut-in pressures for the drill pipe and annulus will be the same and greater than zero for a kick taken while drilling underbalanced assuming the influx is in the horizontal section of the well.
- The drill pipe pressure schedule used to displace kill mud down the drill pipe is different than for a vertical well.
- Horizontal drilling through several faults with different pressure regimes can result in an underground blowout.

The differences will be discussed.

Greater Influx Volumes

A horizontal well is designed to expose more productive formation to the wellbore. A pressure underbalance between the formation and mud column will result in a larger influx for the same conditions of pressure differential and time of underbalance. For horizontal wells drilled through long sections of productive formation, the influx rate into the wellbore could be high.

The result is a large volume gained that must be handled when the influx enters the vertical hole section. This causes high pressures that can rupture casing or cause an underground blowout.

Shut-In Pressures

Refer to Figure 3–28(a). The gas in the horizontal section was swabbed in the well. The shut-in pressures will be zero for the drill pipe and annulus

SPECIAL PROBLEMS

FIGURE 3-28(a) • Swabbed gas kick response in a horizontal well

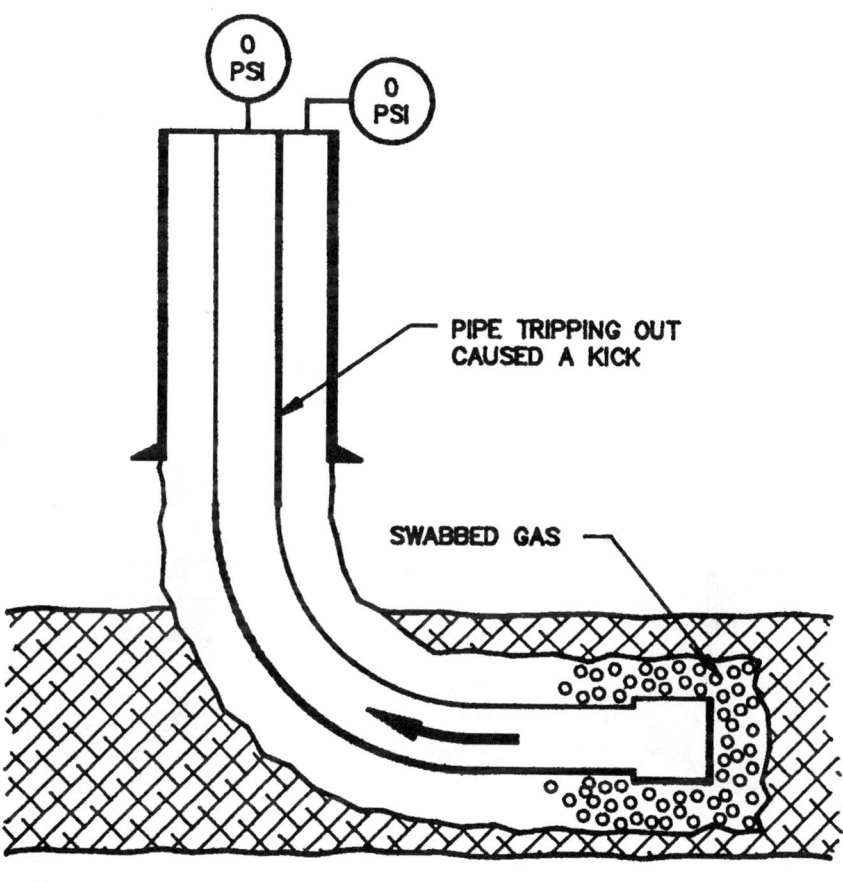

because the mud hydrostatic in the vertical section is sufficient to control the well in a static condition.

Figure 3-28(b) shows a horizontal well drilled into a higher pressured zone across a fault. The shut-in pressures are the same for the drill pipe and casing if the influx is in the horizontal section.

This situation is different than a swabbed kick in a vertical well with the bit above the gas. The shut-in pressures will be the same. However, the horizontal well shown in Figure 3-28(b) will require a mud weight increase for proper control, whereas the swabbed kick in a vertical well can be controlled with the original mud weight with the bit at the bottom of the hole.

FIGURE 3-28(b) • Gas kick from higher pressured zone

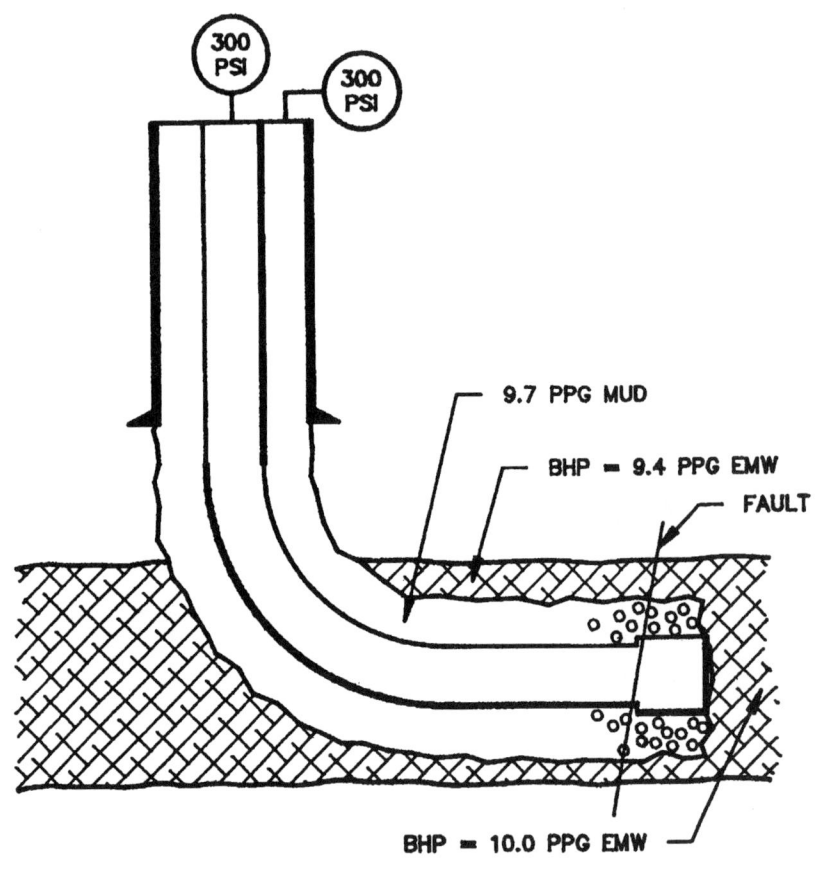

Drill Pipe Pressure Schedule

Displacement of kill mud down the drill string is usually done with the aid of a pressure schedule. It shows the amount of drill pipe pressure required to balance bottom hole pressure. The pressure schedule will be different in a horizontal well. (See Figure 3-29.) The build angle and horizontal sections will not have linear pressure decreases as in a vertical well. An overbalanced situation will occur if a schedule is developed using techniques for a vertical well. Lost circulation may occur.

Drilling Through Faults

Some geological areas such as the Austin Chalk, Texas, have parallel fractures in the producing zone. Intersecting a fracture is similar to the

Special Problems

FIGURE 3-29 • General relation for drill pipe pressure schedule for horizontal wells (same measured depth)

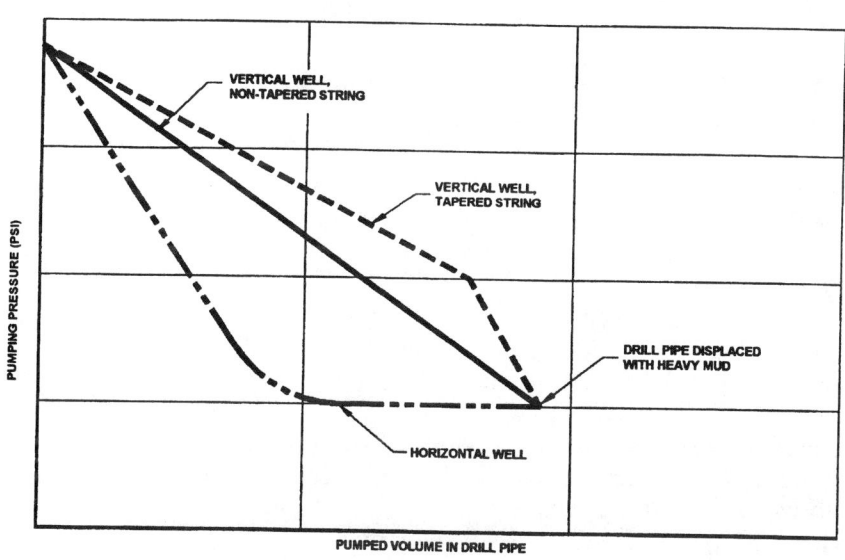

desired results of a fracture job because the flow potential is increased. Drilling horizontal wells to a perpendicular intersect with several faults increases flow potential and overall well production.

These naturally occurring faults may not have the same pressure regimes. A connection of two or more faults via an intersecting wellbore creates a situation conducive to an underground blowout. Also, lost circulation often occurs in faults with pressure less than the wellbore hydrostatic pressure.

Kick Killing

Well control concepts do not change for horizontal wells. However, differences exist in data interpretation and calculation methods for well control operations. The normal procedures are as follows:

- Observe kick warning signs.
- Pick up the kelly.
- Shut in the well.
- Read shut-in pressures.
- Calculate kill mud weight and drill pipe pressure pumping schedule.
- Circulate out the kick from the well.

Procedural and other differences in horizontal wells are discussed in the following problems.

Potential Problems

Drill pipe and casing pressure are zero. A zero shut-in pressure recorded for a vertical well indicates that a kick has not occurred. For horizontal wells, this indicates that the kick influx volume must be in the horizontal section. Typical kick causes are improper hole fillup on trips or swabbing.

Drill pipe and casing pressure will be zero, but the pits show a volume gain. Also, returning flow rate prior to shut-in has increased.

To fix the problem, evaluate the situation and understand the well control theory. The kick influx as evidenced by the pit gain is in the horizontal hole section. Circulate out the kick using the Driller's Method.

Drill pipe and casing pressure are equal and greater than zero. This situation indicates a swabbed kick with the bit above the influx for a vertical well. It represents a kick from a higher pressured zone in a horizontal well. The influx is in the horizontal hole section. Drill pipe and casing pressures are greater than zero and equal. A pit gain has occurred. Calculate the kill mud weight and circulate out the kick. Use a drill pipe pressure schedule for horizontal wells.

The well will be overbalanced if a vertical well drill pipe pressure schedule is used for a horizontal well. Lost circulation could occur. The pressure schedule will show a straight line decrease from the point mud starts in the drill pipe until displaced at the bit. Prepare a correct pressure schedule. For constant drill pipe sizes, the pressure should decrease linearly in vertical sections. It should decrease nonlinearly according to the vertical component of the measured depth in the build section. It will not decrease in the horizontal hole section. A computer program should be used to prepare the drill pipe pressure schedule.

Underground blowouts may occur while drilling through faults. A formation will flow from one fault into another. Mud can not be circulated effectively. Gas may migrate if the flow is near the vertical section of the hole. Underground blowout signs will be apparent. Mud can not be circulated effectively. Gas may come up the annulus. Implement techniques to solve the underground problem by fighting lost circulation or killing the kick. Solving the loss problem may be easier in horizontal holes than resolving the kick.

Friction pressure may cause lost circulation. Friction pressure required to pump through a long horizontal hole section may exceed the fracture gradient and cause lost circulation. Lost returns while drilling a long horizontal section are an indicator of this problem. Condition the mud to reduce viscosity, decrease the mud density, or add lost circulation material.

Special Problems

Gas removal is impaired in the horizontal section. Gas will segregate to the top side of the hole. Mud moving at low velocities during kick circulation may not sweep the gas effectively. More circulation time may be required to remove all gas. Gas will be circulated from the well longer than anticipated according to bottoms-up calculations. Circulate for a longer time. An alternative is to improve gas removal efficiency by circulating at a higher rate, although this is not necessary and may cause lost circulation due to increased friction pressure.

Case History—Horizontal Well Control

The prognosis for this "horizontal well" called for a true vertical depth of 8,196 ft (2,498 m), a measured depth of 11,541 ft (3,518 m), a horizontal length of 3,600 ft (1,100 m), and an inclination of 88°. The pay objective was the Austin Chalk Formation, a highly fractured oil reservoir. (See Figure 3–30.)

A rig contracted on a daywork basis to drill the well and moved onto the location. The well was spudded and drilled to 870 ft (265 m) where 10 ¾ in. (273 mm) surface casing was set and cemented in the hole.

The blowout preventers were installed and tested, and drilling continued to 7,616 ft (2,321 m). The BOP stack consisted of a rotating head, annular preventer, blind rams, and pipe rams. It appears after the fact that the rotating head did not have the rubber element installed, so it was not functional as a pressure control device. (It was a common, though not necessarily safe, practice to wait on rubber element installation until a kick was taken from one of the fractures existing in the Austin Chalk.)

The well was logged and 7,614 ft (2,321 m) of 7 in. (178 mm) intermediate/production casing set and cemented in the well. The blowout preventers were again installed and tested, and the float equipment drilled out. The casing seat was then tested to a 12 ppg (1,438 kg/m^3) equivalent and drilling resumed.

At that point, directional drilling equipment, specifically Measurement-While-Drilling (MWD) tools, was incorporated into the drill string as the well was kicked off to begin building angle to drill the horizontal section. The angle was built to 88° by May 6, 1990, with a true vertical depth of 8,112 ft (2,472 m), a measured depth of 8,466 ft (2,580 m), and a horizontal distance of 532 ft (162 m). The well was being drilled with a water-copolymer mud system.

During the 1800 hrs to 0600 hrs tour, in which the event occurred (about 0430 hrs), the crew was working on mud pumps. They changed two swabs and one liner on the hole pump. After this pump was put on-line, they began working on the other pump. (It is not certain if this pump work had any relationship to kick detection problems to occur later.)

FIGURE 3-30 • Geology and hole configuration for a blowout in a horizontal well

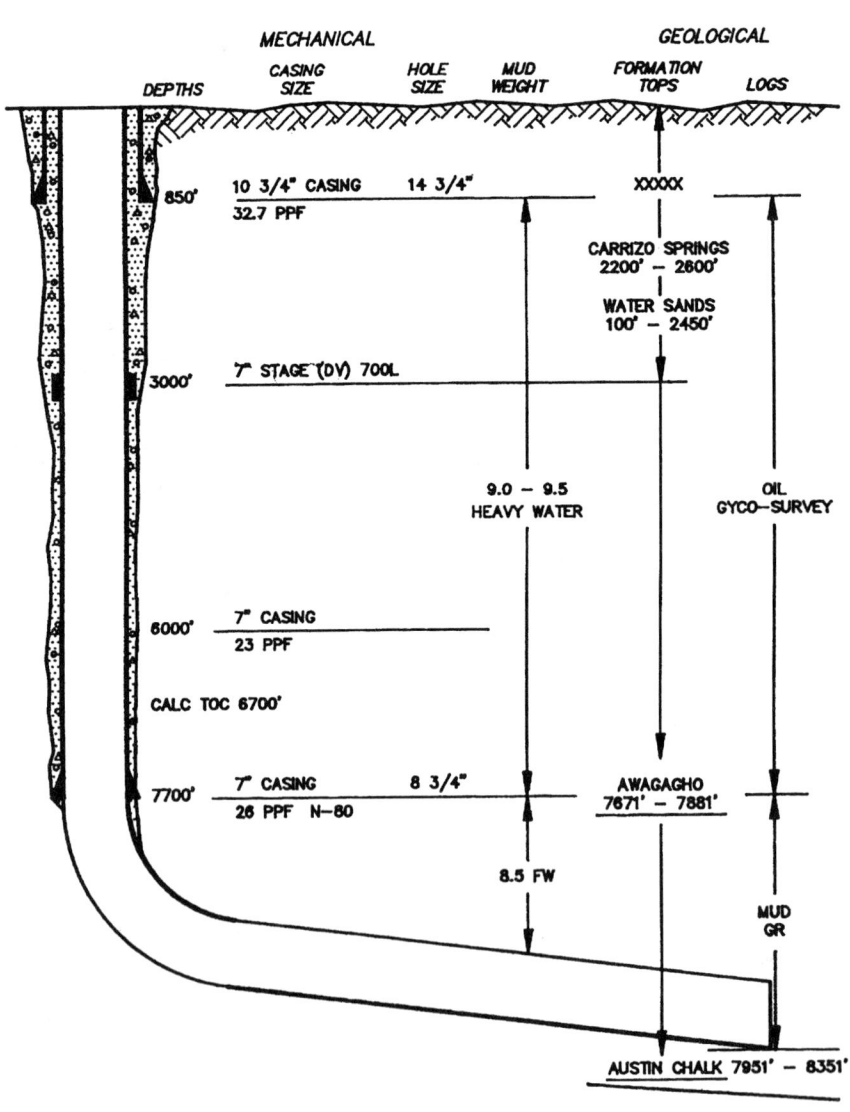

At 0415 hrs on May 8, 1990, horizontal drilling of the 6 in. (152 mm) hole was proceeding at 8,116 ft (2,474 m) true vertical depth, 9,143 ft (2,787 m) measured depth, and 1,207 ft (367.9 m) horizontal distance. A connection had been made about 15 minutes earlier and the mud logger reported to the driller that there were 30 units of connection gas and 150 units of background gas. At this time, it is not known if the pumps were shut down to check for flow at that point or if the rotary was turning.

The derrickman would normally work on the pits during the drilling situation. However, the driller ordered him this evening to inventory pump parts, presumably connected to pump difficulties earlier in the evening. When the derrickman went back to the pits, he noticed they were overflowing. He cut off the water and dumped the sand trap. The sand trap immediately filled up again. The derrickman has reported that there did not appear to be any bubbling in the possum belly at this time, and he thought there was no mud coming over the bell nipple at this time.

Instead of immediately shutting in the well, the driller called the mud logger to determine if there was an increase in gas. The mud logger reported about 30 units. During this conversation, mud began to come over the bell nipple.

When drilling fluid began to flow from the well about 40–50 ft (12–15 m) above the rig floor, the driller attempted to pick up the drill string and shut the well in. For reasons unknown at this time, it appears the driller attempted to close the blowout preventer pipe rams on the six sided (hex) kelly. These rams, which are designed to close only on 3 1/2 in. (89 mm) drill pipe or tubing, did not effect a seal and the well continued to flow uncontrolled.

Immediately thereafter, a spark of unknown origin caused an explosion believed to have started beneath the rotary table. The flame instantaneously engulfed the drilling floor. After the explosion, fire was seen from the shaker and across the pits. The rig was destroyed and several fatalities occurred.

BULLHEADING

Operators cannot always solve critical well control problems with conventional circulation down the drill string and up the annulus. Kick or blowout control under unusual situations sometimes requires different techniques.

One technique is bullheading, or pumping into the annulus (or drill string) of a closed well, so mud and formation fluids are displaced back downhole into the weakest exposed open hole interval.

Bullheading is a valuable tool for fighting kicks under many such circumstances. The following are some examples:

- The influx contains more H_2S than the operation can tolerate.
- Plugged or parted drill pipe cannot get kill mud to bottom.
- Too big a kick foreshadows excessive surface pressure.
- A weak zone below the kick takes mud too fast for a kill.
- To gain time when short of material, skilled personnel, or equipment.

But bullheading also has several crucial disadvantages:

- Crews do not fully understand when the technique should be used.
- Fluid will go to the weakest interval and may not follow the preferred path.
- Potential is created for an underground blowout and/or a surface eruption.
- Even a successful bullhead may not kill the well.

Figure 3–31 illustrates a typical bullheading situation. All imposed pressures opposite corresponding burst resistances must be evaluated before pumping. In this instance, the casing shoe and all open hole sections have potential to fracture before the original kick zone. And if severely worn casing exists uphole, pumping may rupture the pipe, charge the casing-casing annulus, and leave the gas kick in the hole. While this is not an example of a particular well, it does show what has gone wrong during some kicks.

FIGURE 3–31 • Typical bullheading situation (Courtesy of *World Oil*, March 1988, pp. 46–48.)

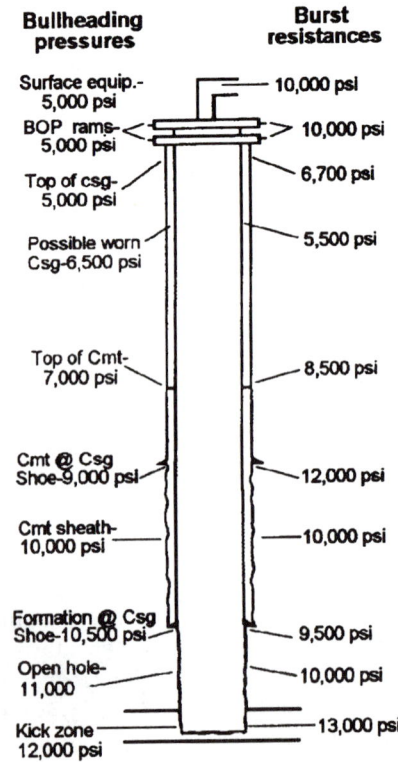

Special Problems

Bullheading is common and effective in workovers, but drillers seldom employ it. A typical workover consists of a cased hole with perforations and workover fluid consisting of solids-free salt water that is easily pumped into the perforations when bullheading. A normal workover might take several bullhead kills.

Bullheading during drilling is different from a workover, because a long section of formation usually is exposed to mud that contains solids with wall building qualities. When bullheading occurs, mud will follow the least resistant path, and fracturing usually occurs at the casing seat. Since mud builds filter cake on a sand body, it is unlikely that kick fluids will re-enter their original formation. This, in turn, means lost circulation occurs elsewhere in the hole.

When to Bullhead

Bullheading is not routine and the previously outlined conditions mean the well already has big trouble. Competent authorities should be consulted before attempting high pressure pumping. Most blowout specialists and firefighters will provide free telephone advice, and some will prepare a summary kill plan that can be helpful in determining appropriate methods to use and whether outside services are advisable.

Key considerations for bullheading include the following.

Escaping Hydrogen Sulfide

If hydrogen sulfide escapes at the surface, even small concentrations can be very dangerous. Thus, kick handling requires a thorough contingency plan. If routine surface control of an H_2S kick is not possible, it is better to bullhead the gas into the formation. This is particularly true in populated areas.

Plugged Drill String

Plugged bit nozzles or drill string may occur after poor mixing of barite or lost circulation material. Perforating the drill pipe above the plug gives up valuable depth for a hydrostatic kill. If perforating seems impractical, the alternate control method is to bullhead the kick back down the hole.

A well can require more than one bullhead kill. If kick fluids contain gas, gas can migrate up the hole, past a lost returns zone, and build pressure at the surface casing. It may be necessary to flush the annulus again.

Drill String Twistoff or Shallow Washouts

Parted drill pipe prevents conventional bottom-hole circulation. Also, if a washout occurs early in the kill effort, heavy mud cannot be pumped to the problem zone without bullheading.

Excessive Casing Pressure

Allowing a kick to get too big or flow too long can be a reason for bullheading. A large gas expansion can generate excessive pressure on surface casing and equipment.

Pressure limits depend on more than just initial casing and BOP stack ratings. The weakest member will fail first, so lowest rated components control bullhead pressure. Casing wear reduces burst rating while depth increases burst resistance.

Some companies define excessive pressure (or operating limit) as 80% of API burst rating with adjustments for tension. Other companies use the 100% value with the belief that API ratings are conservative. One major operator uses 50% as the cutoff. Personal and company experiences influence the selection of this upper pressure limit.

No substitute exists for an exhaustive comparison of burst versus imposed pressure for each portion of the pressure vessel. (See Figure 3–31.)

One recent major H_2S blowout occurred because the operator incorrectly evaluated bullhead loads. On pumping, the well ruptured uphole, far higher than expected. The well cratered and ultimately had to be killed with a relief hole.

Lost Circulation

Losing mud during a kick can create the need for bullheading. One can pump gas, migrating above a thief zone, back down the hole into the first weak interval.

Research and field data show gas migrates at up to 1.0 fps (0.3 m/s) in clear, low density fluids, and several fpm in viscous muds. Downward annulus bullhead rates must exceed upward gas velocities to clean the annulus.

Tubing and casing size affects rise rate. Davies and Taylor, et al., published data relating to bubble rise in tubing. Figure 3–32, derived from a compilation of research data, can help determine minimum bullhead rates. The example uses a casing volume of 140 bbl (22 m^3).

Underground Blowouts

Underground blowouts are an important consideration during bullheading because they can often develop into surface blowouts.

An underground blowout can occur if bottom-hole pressure in the kicking interval, less hydrostatic pressure, is still greater than the fracture propagation pressure of another formation exposed in the open hole section of the well. However, uncertainties exist in all cases. For example, the hole can bridge or cave in, when an underground blowout would otherwise occur. In some instances, bullheading should be considered as a well control technique even though an underground blowout is prob-

FIGURE 3-32 • Minimum bullheading pump rates (Courtesy of *World Oil*, March 1988, pp. 46–48.)

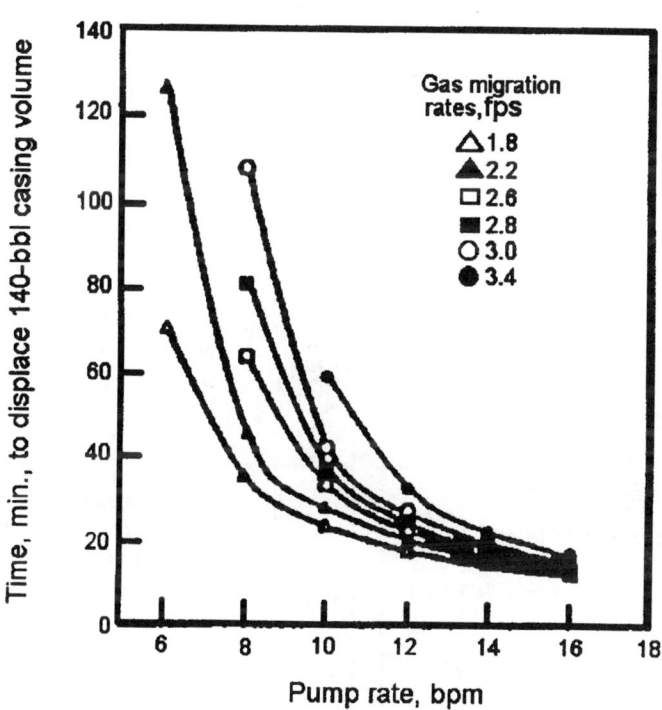

able. The well should always be monitored closely for early signs of such an event so appropriate action can be taken.

How to Bullhead

Guidelines for bullheading emphasize safety and are simple:

- The possibility of required bullheading must be addressed early in the operation so the situation does not get out of hand to where bullheading is not an option.
- If pump-in pressures approach the rig pump limit, use a cementing unit for better control and adequate pressure rating.
- Have large mud volumes available. Bullheading creates lost returns and possible continued pumping of mud that will not return.
- Select safe tie-in points at the BOP stack for pump lines. Connect lines above BOP rams that can be closed if necessary.
- Install check valves in the pump-in lines at a point convenient for repair and hydraulic-actuated valves near the BOP.

Remember that extra pressure is required to fracture the well for bullheading. For example, a South Louisiana well was partially full of gas, but shut-in as in Figure 3-33. The operator elected to bullhead because the drill string was plugged. Maximum surface pressure was established at 5,000 psi (34 MPa). The calculation shows that 5,689 ft (1,734 m) of mud had to be pumped, or lubricated, into the well before it fractured at 10,600 ft (3,231 m).

In Figure 3-34, the pressure limitation at surface is 10,000 psi (69 MPa). This well will fracture before reaching 10,000 psi.

FIGURE 3-33 • Bullheading example (Courtesy of *World Oil*, March 1988, pp. 46–48.)

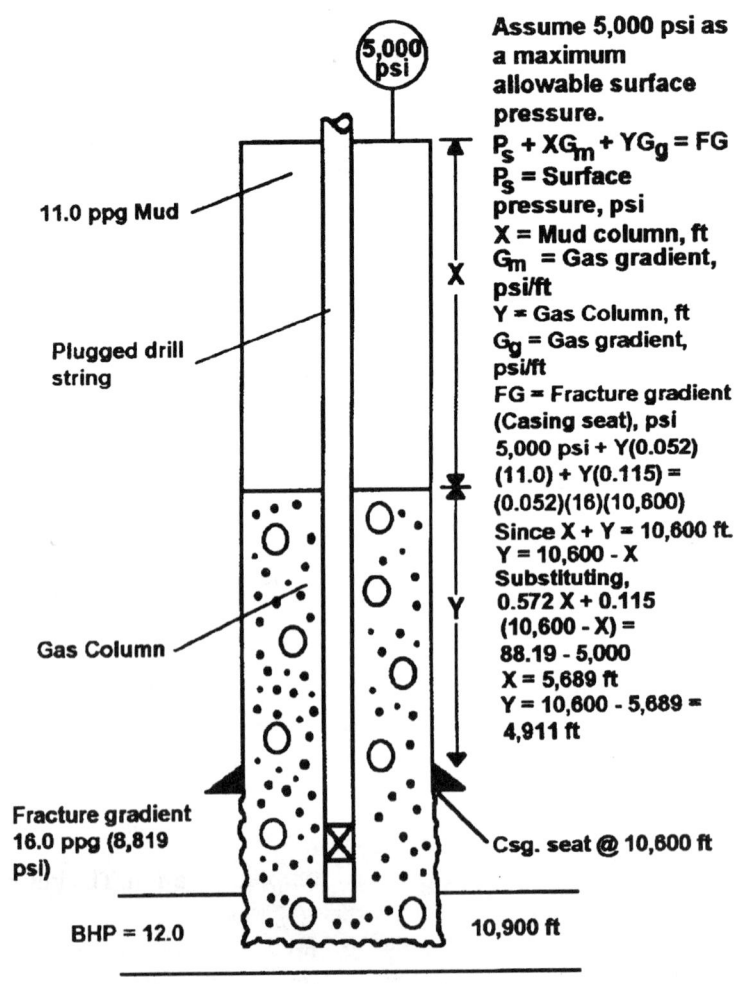

SPECIAL PROBLEMS

FIGURE 3-34 • Bullheading example using a 10,000 psi surface pressure limit (Courtesy of *World Oil*, March 1988, pp. 46-48.)

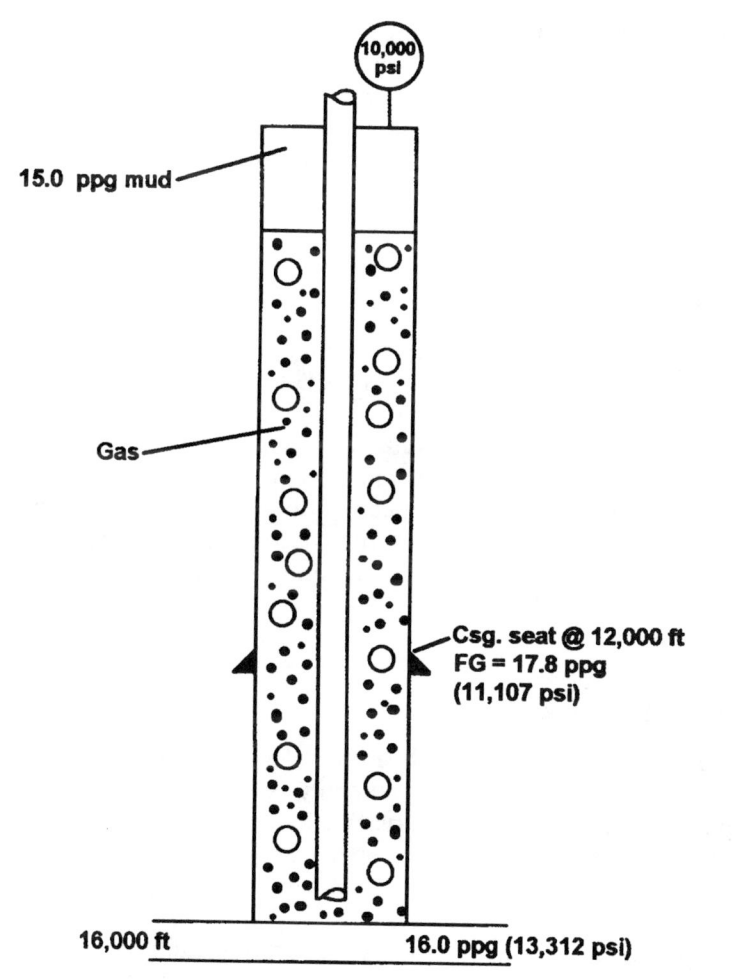

In both previous cases, a high pressure pumping unit is required.

If excessive pressures develop at the surface, lubricate and bleed procedures may help reduce surface pressure before bullheading. Lubricating has important guidelines for correct implementation. The key rule, however, is to remain patient long enough for the mud to fall in the well after pump-in. This takes several minutes to one-half hour. Many bad, but controllable, situations have turned into blowouts after poor lubricating attempts.

HYDROGEN SULFIDE

Drilling in a potential H_2S (hydrogen sulfide-bearing) zone requires precautionary steps to ensure the safety of the rig crew and the continued integrity of the control equipment. H_2S can cause death and mental function failures in a relatively short time, and it can complicate an already serious problem like a kick or blowout. As a result, a working knowledge of the effects of H_2S and proper control procedures is necessary to safe planning and drilling techniques.

Rules and Regulations

National, state, and local regulations may govern drilling and production practices in potential H_2S areas. In the U.S.A, these agencies may include federal bodies such as the Occupational Safety and Health Administration (O.S.H.A.), National Institute for Occupational Safety and Health (N.I.O.S.H.), Minerals Management Service (M.M.S.), or others. U.S.A. state agencies and some European nations require adherence to guidelines that may be more stringent than U.S.A. federal rules. Regardless of the body that has jurisdiction, criteria common to most rules and regulations must be followed.

Radius of Exposure

A procedure for determining the radius of a certain level of concentration of H_2S gas to public exposure is generally used to establish which rules and regulations, if any, apply to an area. Various concentration levels and radii may be used, and different plans must be implemented depending on the results of each radius of exposure calculation and the surrounding surface conditions.

One common scheme used to determine radii of exposure is the Pasquill-Gifford equation. The concentration levels generally employed are 100 ppm and 500 ppm. The Pasquill-Gifford equation for each of these levels is as follows:

1. For determining the 100 ppm exposure radius

$$x = [(1.589)(H_2S)(Q)]^{(0.6258)} \qquad (3.3)$$

2. For determining the 500 ppm exposure radius

$$x = [(0.4546)(H_2S)(Q)]^{(0.6258)} \qquad (3.4)$$

Where: x = radius of exposure, ft

Q = maximum escape (blowout) volume in cfd (at standard temperature and pressure)

Special Problems

H_2S = mole fraction of hydrogen sulfide in the gaseous mixture available for escape

The volume used as the escape rate in determining the radius of exposure is generally based on the following guidelines:

1. The maximum daily rate of gas containing H_2S handled by that system element for which the radius of exposure is calculated.
2. For existing gas wells, the current adjusted open-flow rate, or operator's estimate of the well's capacity to flow against zero backpressure at the wellhead.
3. For new wells drilled in development areas, the escape rate shall be determined by using the current adjusted open-flow rate of offset wells, or the field average current adjusted open-flow rate, whichever is larger.
4. For drilling a well in an area where insufficient data exist to calculate a radius of exposure, but where H_2S may be expected, a 100 ppm radius of exposure equal to 3,000 ft (910 m) is usually assumed. A lesser assumed radius is often considered when a written request with adequate justification is given.

After the calculations for radii of exposure have been made, efforts must be made to determine if the well in question is subject to the rules and regulations. The guidelines most commonly used for this purpose are:

1. The 100 ppm radius of exposure is in excess of 50 ft (15 m) and includes any part of a city, town, village, park, dwelling, school bus stop, or similar area that is expected to be populated.
2. The 500 ppm radius of exposure is greater than 50 ft (15 m) and includes any part of a road owned by and maintained for public access or use.
3. The 100 ppm radius of exposure is greater than 3,000 ft (910 m).

Example 5.6

Assume that a proposed drilling location for a development well in an area in which the maximum open hole rate of the best offset well or the rate carried by the production system was as follows:

$Q = 2$ MMcfd

$H_2S = 10\%$ or 0.1 mole fraction

For 100 ppm radius:

$x = [(1.589)(0.1)(2,000,000)]^{(0.6258)} = 2,775$ ft

For 500 ppm radius:

$x = [(0.4546)(0.1)(2,000,000)]^{(0.6258)} = 1,268$ ft

In this example, both the 100 and 500 ppm radii exceed 50 feet (15 m), and the requirements would apply if a public area is within 2,775 ft (846 m), or if a public road is within 1,268 ft (386 m). If neither was the case, the operation would not be subject to the requirements of the regulations because the 100 ppm radius was calculated to be less than 3,000 ft (910 m).

If the escape rate (Q) of this same gas (10% H_2S) had been 2.5 MMcfd (71,000 m²/d), the 100 ppm radius would be 3,190 ft (972 m). In this case, the operation would be subject to the regulations because the exposure exceeded 3,000 ft even though no public road or dwellings were within the radius. However, a 10 MMcfd (280,000 m³/d) well or system with an H_2S concentration of 2% in the same geographical location (public not included) would not be under the rules because the radius of exposure is less than 3,000 ft, [i.e., 2,775 ft (846 m)].

Characteristics and Effects of Hydrogen Sulfide

Characteristics of H_2S include its explosive nature, its toxicity, and its ability to cause sudden metal failure. Characteristics of hydrogen sulfide taken from N.I.O.S.H. standards show the gas is colorless and exhibits a rotten egg odor only when inhaled in small concentrations. Its specific gravity of 1.192 is higher than air, which has a specific gravity of 1.0 at 60°F (16°C). H_2S forms an explosive mixture with air or oxygen. For comparison, hydrogen sulfide will burn when mixed in 4.3–45% air, while methane will burn only when mixed with 5–15% air. The gas ignites at 500°F (260°C). For comparison, methane autoignites at 1,000°F (538°C). It is soluble in water.

One primary characteristic relative to detection that is most often misunderstood is the ability to smell hydrogen sulfide gas as a rotten egg odor. In low and safe concentrations, the gas does smell similar to rotten eggs. However, when the concentration reaches high, toxic levels, olfactory nerves are quickly deadened and can no longer be used as a detection device.

Since H_2S is heavier than air, the gas will settle and collect in low places near the rig such as the cellar and around the mud pits. Proper consideration must be given to the location of monitors to ensure that the gas will be detected before it reaches dangerous levels.

Effects on Personnel

Sour gas, or hydrogen sulfide, has serious effects on the health and safety of rig personnel (see Table 3–5). Death is not the only serious consideration. Since unconsciousness may cause rig crewmen to fall from heights or into mud pits, serious injury may also result.

TABLE 3-5 • Toxicity of hydrogen sulfide

PPM (Parts per Million)	0-2 Minutes	2-15 Minutes	15-30 Minutes	30 Minutes To One Hour	1-4 Hours	4-8 Hours	8-48 Hours
5-100				Mild conjunctivitis: respiratory tract irritation			
100-150		Coughing; Irritation of eyes; loss of sense of smell	Disturbed respiration; pain in eyes; sleepiness	Throat irritation	Salivation and mucous discharge; sharp pain in eyes; coughing	Increased symptoms*	Hemorrhage and death*
150-200		Loss of sense of smell	Throat and eye irritation	Throat and eye irritation	Difficult breathing; blurred vision; light shy	Serious irritating effect*	Hemorrhage and death*
250-350	Irritation of eyes, loss of sense of smell	Irritation of eyes	Painful secretion of tears, weariness	Light shy, nasal catarrh; pain in eyes, difficult breathing	Hemorrhage and death*		
350-450		Irritation of eyes; loss of sense of smell	Difficult respiration; coughing irritation of eyes	Increased irritation of eyes and nasal tract; dull pain in head; weariness; light shy	Dizziness; weakness; increased irritation; death	Death*	
500-600	Coughing; collapse and unconsciousness*	Respiratory disturbances; Irritation of eyes; collapse*	Serious eye irritation; light shy palpitation of heart; a few cases of death	Severe pain in eyes and head; dizziness; trembling of extremities; great weakness and death*			
600 or greater	Collapse;* unconsciousness;* death						

* Data secured from experiments on dogs which have a susceptibility similar to men.
Source: National Safety Council data sheet D-chem. 16

The long-term effects of hydrogen sulfide are not well understood. For example, personnel that have been exposed to hydrogen sulfide tend to have a reduced resistance to its effects when again exposed to the gas.

Effects of Hydrogen Sulfide on Metal

It is generally known that hydrogen sulfide affects metals. Some terms applied to the hydrogen sulfide-metal reaction are "hydrogen blistering," "hydrogen embrittlement," "stress cracking," and "sulfide stress cracking." Each is similar in its effect and, in fact, is caused by the same phenomenon.

When hydrogen atoms are formed on a metal surface by a corrosion reaction, they often combine to form gaseous molecular hydrogen, which is released into the environment. Some hydrogen atoms are absorbed by the metal. The atomic hydrogen migrates to the grain boundaries of the metal and recombines to form molecular hydrogen, which occupies a greater volume than the hydrogen atoms. The formation of the molecular hydrogen causes internal stresses to increase, which in turn causes hydrogen blistering or embrittlement. The blistering will occur with metals that have an average yield strength of less than 90,000 psi (620 MPa), while embrittlement occurs with metals having a higher yield strength.

When hydrogen sulfide is present in the electrolyte, the sulfide ion reduces the rate at which the hydrogen atoms combine outside of the metal. This creates a larger concentration of atomic hydrogen on the metal surface. A greater portion of the hydrogen atoms entering the metal increases the tendency for blistering or embrittlement.

Failures due to hydrogen embrittlement often do not occur immediately after application of the load or exposure to the hydrogen-producing environment. This is referred to as "delayed failure." The time before failure is referred to as the "incubation period" during which hydrogen is diffusing to points of high stress. The time to failure decreases as the amount of hydrogen absorbed, applied stress, and strength level increases.

Spontaneous brittle failure that occurs in steel and other high strength alloys when exposed to moist hydrogen sulfide and other sulfuric environments is frequently referred to as "sulfide stress cracking." It is generally thought to be a form of hydrogen embrittlement. Although the mechanism of sulfide cracking is not completely understood, it is generally accepted that four conditions must be present before cracking can occur: hydrogen sulfide, water, high strength steel, and applied or residual stress.

Stress level, either applied or residual, affects sulfide cracking tendencies. The time to failure decreases as the stress level increases (see Figure 3–35). In most cases, stress results from a tensile or bending load or from the application of pressure. However, residual stresses and hard spots can be created by welding or cold working the material.

FIGURE 3-35 • Effect of stress level on time to failure (Courtesy of *OGJ*)

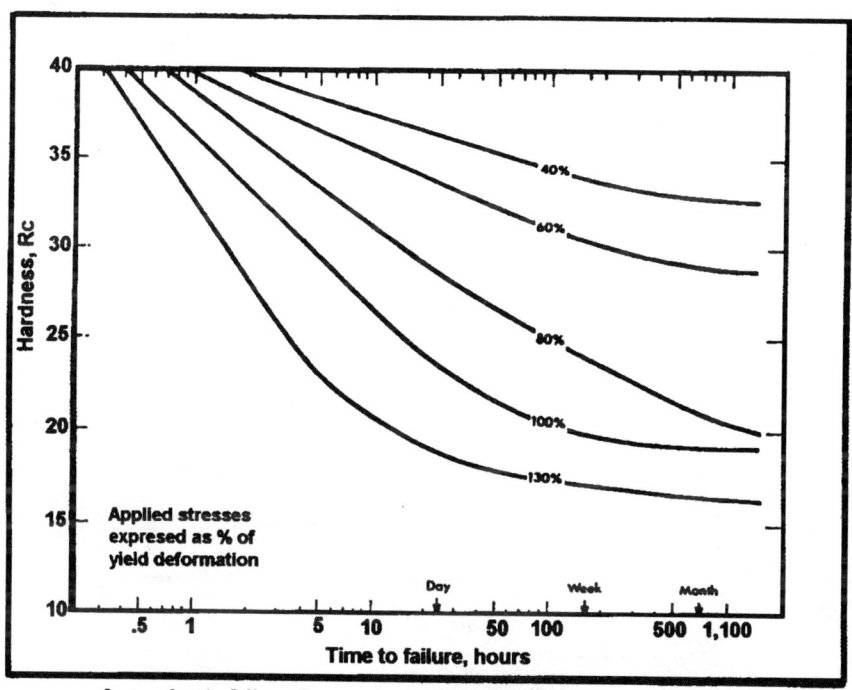

Approximate failure time vs. hardness and applied stress for carbon steel, 3,000 ppm H_2S in a 5% solution of NaCl

The time to failure decreases as the hydrogen sulfide concentration increases. (See Figure 3-36.) Although delayed failure can occur at very low concentrations, the time to failure becomes great. There is evidence that cracking susceptibility decreases above 175-200°F (79-93°C) regardless of the sulfide concentration. Although the exact temperatures are not yet defined, this principle can be used in well planning for tubular goods.

Hydrogen Sulfide, H_2S, Detection

Equipment used to detect hydrogen sulfide may include fixed location monitors, personal detectors, mud monitors with electronic probes, or chemicals for analysis of the drilling fluid. The monitors may be qualitative or quantitative and may function with chemical or electronic sensors.

The most important concern with any H_2S detector is proper placement of the sensor units. Since H_2S is heavier than air, it will settle in low

FIGURE 3-36 • Effect of hydrogen sulfide concentration on time to failure (Courtesy of *OGJ*)

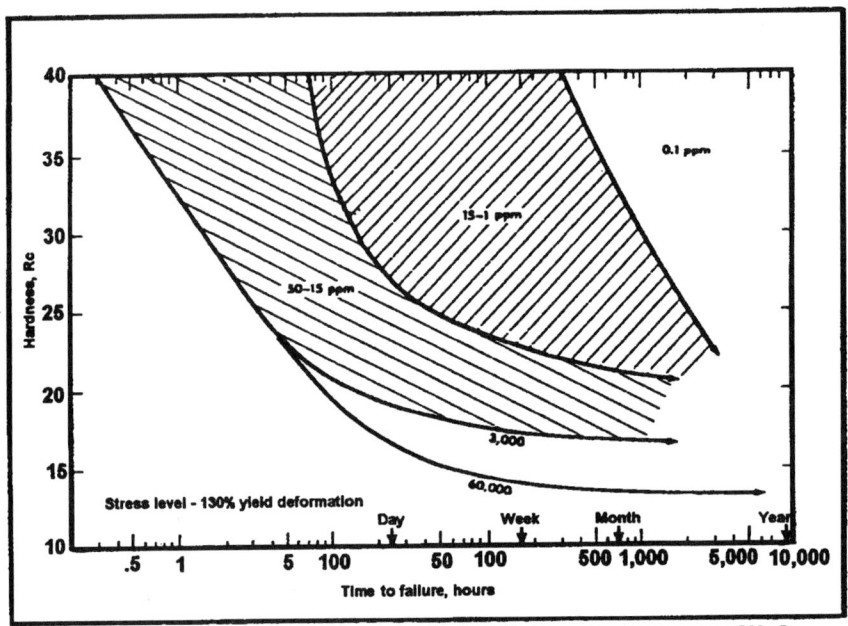

Approximate failure time of carbon steel 5% NaCl and various ppm of H_2S

areas. The personal units should be attached to the clothing or carried level with the waist. The electronic rig monitors have portable sensor heads that should also be placed near the shale shaker, since it is the first location where the mud will receive exposure to the air.

Lead Acetate Paper Detectors

Several reasonably semi-quantitative detectors for hydrogen sulfide are based on lead acetate paper. As the gas contacts the paper, the lead acetate impregnated in the paper reacts with the gas to form lead sulfide, which causes the paper to change color from white to various shades of brown or black. The degree of color change depends on the H_2S concentration, which can be roughly estimated by comparing the observed color to a control chart or table.

The primary advantage of these detectors is that they are carried by each crewman, enabling him to detect the gas wherever he may be. This provides an additional measure of safety to each crewman as well as an atmosphere of security. When necessary, the paper can be changed to provide a new chemical surface for gas detection.

Special Problems

The reaction time required for the detector to function is a disadvantage of the tool. The 3–5 minutes necessary can be excessive and dangerous when large concentrations of hydrogen sulfide are encountered. Also, it is advisable to consider the lead acetate paper as a qualitative indicator rather than as a precise gauge of the concentration.

The two paper detectors most often employed are the badge-type and the spot check. The badge-type is clipped to the clothing, while the spot check can be carried in the hand or pocket. Color codes are used to determine H_2S concentrations.

Capsule Detector

The capsule detector resembles an ammonia-type capsule and is filled with chemical granules. The capsule is broken and attached to clothing. If hydrogen sulfide contacts the granules, a brown discoloration occurs.

This detector should be used only as an indicator of H_2S because of the limitations of the capsule. The tube life is approximately six days after it is broken. The maximum concentration of the gas that can be measured accurately is 20 ppm.

Draeger Detector

The Draeger unit is one of the most widely used tools for quantitative gas detection. It can be altered to measure almost any type of gas and, as a result, is used extensively in hydrogen sulfide detection.

The tool consists of a calibrated glass tube filled with lead acetate granules. A pump is used to draw gas samples into the tube, and the level of color change denotes the H_2S concentration. Several scales are generally presented on the glass tube to denote high and low concentrations. The pump is usually the bellows type.

The simple operating procedure increases the utility of the tool. The tips of the detector tube are broken and inserted into the suction outlet of the Draeger unit. Ten compressions of the bellows are required to ensure an accurate reading in low concentrations of hydrogen sulfide. As the gas is drawn into the tube by the bulb, the lead acetate granules become discolored denoting the quantitative gas measurement.

Accuracy depends on training and practice of the personnel using the unit. As varying amounts of air are drawn into the unit, the measurements will be different than if 10 compressions were used. In high concentrations of H_2S, only one compression is required to activate the high scale on the unit.

The measurements obtained with the Draeger unit are usually reliable. Since there are no electronic parts, the unit is not subject to electronic malfunction. The shelf life of an unbroken tube is approximately two years, and the tube can be used after the tips are broken as long as no indication of H_2S is present.

Belt Detectors

The belt-type hydrogen sulfide detector is an electronic unit usually attached to the crewman's belt. The unit is operated by rechargeable and/or replaceable batteries. The detector has a sensor head that will monitor hydrogen sulfide gas and report in a visible readout for concentrations of 5–10 ppm. An audible alarm can be used and is usually pre-set to respond at 20 ppm. The response time for the unit is approximately 35 seconds.

Fixed Location Monitors

The rig monitor is a fixed-location, quantitative, electronic device designed for permanent, full-time operation. Sensor heads are placed at various locations on the rig and attached to the detection unit, which is housed in a hard plastic or metal case. On the monitor, a readout in ppm concentration will be shown on a needle-type indicator.

A rotating beacon or strobe light attached to the unit will activate automatically when a specified amount of gas has been detected. An audible alarm can be used to denote a higher level of gas concentration. The response time for the monitor is approximately 35 seconds for concentrations of 0–10 ppm.

The detection unit, depending on brand and model, can have from one to twelve sensor channels. The most common units have four to six channels.

The rig monitor or the belt-type detector must be calibrated and tested periodically to insure that it is functioning properly. A calibration instrument is used. A known concentration sample is placed in the machine and fed into the monitor. If the monitor does not respond accurately, it is adjusted accordingly. The calibration unit is fully self-contained with rechargeable batteries.

Mud Analysis for Hydrogen Sulfide

Several testing procedures are available for evaluating hydrogen sulfide, sulfide, and sulfide scavenger concentrations in the drilling fluid. Some of these include the Hach test, Garrett gas train, iodine test for the determination of hydrogen sulfide in water, quantitative copper carbonate concentration, quantitive Ironite Sponge concentration, and electronic probes such as the Mud Duck (Delphian Corp.). The most commonly used methods during drilling are the Garrett gas train and the electronic Mud Duck.

Chemical means to analyze sulfides in the drilling muds and filtrate should be quick, accurate, reliable, and simple. The Garrett gas train meets these requirements. The equipment and procedures do not respond falsely to other fluid components, primarily because this method eliminates most

Special Problems

interference problems by using a gas train to separate H_2S gas from the liquid phase. Self-contained Garrett gas train analysis kits are available from major mud companies.

Electronic sulfides analysis in the mud system can be accomplished with the Delphian Corp. patented Mud Duck. The tool uses special electrodes and processes their signals with electronic circuiting developed specifically for this purpose. The equipment also provides a constant monitoring of the drilling fluid pH, since the form of dissolved hydrogen sulfide in the mud system depends on the hydrogen ion concentration. It has proved accurate and reliable in field service.

Breathing Apparatus

When drilling is conducted in an environment containing harmful concentrations of hydrogen sulfide gas, protective breathing apparatus must be supplied to and worn by the crew members. The apparatus must be a pressure demand, supplied air respirator; and it can be either a self-contained breathing apparatus (SCBA) or a hose line-supplied air respirator. Strict rules and regulations govern the use of breathing apparatus and ensure that:

- Proper, certified equipment is used.
- The equipment is worn and maintained properly.
- Each individual receives personal fitting and use instructions.
- The individual meets all medical and physical requirements for respirator use.

The self-contained breathing apparatus (SCBA) has a limited supply of air that may range from 5 to 20 minutes depending on the size of the bottle, the amount of air in the bottle, and the user's physical activity while wearing the unit. The SCBA has an audible alarm to notify the user when the air supply is almost depleted. A pressure regulator maintains a slight, positive pressure within the facepiece to ensure that the user has a constant supply of air upon demand.

In drilling situations where the worker must perform his routine duties in a contaminated atmosphere, the SCBA is not ideal because it must be refilled at frequent intervals and is cumbersome. When this is the case, a supplied air respirator is usually employed. The respirator uses the same facepieces and regulator as on the SCBA, but draws its air supply from a hose line connected either to an air compressor or to a series of large volume, compressed-air bottles. The hose line unit, or "work unit," will usually have a 5-minute capacity bottle to be worn by the worker so that he may escape from the area if the hose line unit fails for any reason.

Proper fitting of the facepiece is important when wearing breathing apparatus. If the facepiece does not fit the user's face pressure-tight, it will reduce effectiveness of the entire breathing apparatus. Some physical conditions that prohibit a proper face seal are growth of beard, sideburns, a skullcap that projects under the facepiece, temple pieces on glasses, and the absence of one or more dentures. Wearing contact lenses is prohibited because they may "float" on the eye when the breathing apparatus pressurizes the facepiece.

Mud System and H_2S Corrosion

The corrosivity of any mud system will depend to a large degree on the conductivity of the electrolyte. Corrosion rates increase with higher mud or electrolyte conductivity. If a mud has essentially no conductivity, corrosion will be very low. This is the case when using an oil mud as a drilling fluid, since the oil mud has a very low conductivity. Table 3–6 shows the effect of H_2S corrosion rates in various common drilling fluids.

Corrosion control is a greater concern in water-based fluids than oil-based fluids. As seen in Table 3–6, the primary corrosion effects were in the water-based muds with the lignite/lignosulfonate additives while significant levels of corrosion were noted in the nondispersed system. Since these two systems are being used extensively in the industry for reasons not related to corrosion, special efforts must be made to make the systems more resistant to corrosion.

TABLE 3-6 • Effect of mud type on corrosion*

Mud Types	H_2S Presence on Coupons	Hydrogen Embrittlement	Corrosion Rates, MPY
Invermul (3 lb/bbl lime)	No	No	5.30
Invermul (8 lb/bbl lime)	No	No	3.99
Low lime	No	No	3.23
High lime	No	No	3.42
Nondispersed-low lime with saturated salt, polymer, starch	Yes	Yes	26.60
Lignite/lignosulfonate (starting pH 9–11)	Yes	Yes	107.47
Lignite/lignosulfonate (starting pH 11)	Yes	Yes	70.02

*Series of tests using mild steel coupons and prestressed bearings contaminated with 2,400 ppm H_2S rolled 16 hr at 150°F. MPY = mils/year.

Special Problems

TABLE 3-7 • Corrosion fatigue of steel in brine (Courtesy of *OGJ*)

Dissolved Gas	% Decrease from Air-Endurance Limit
H_2S	20
CO_2	41
CO_2 + Air	41
H_2S + Air	48
H_2S + CO_2	62
Air	65

Corrosion rates are influenced by the amount of dissolved gases in the electrolyte. As seen in Table 3-7, the corrosion of steel in brine shows a significant increase as a gas is dissolved in the fluid. The increase is greater when several gases are dissolved in the electrolyte.

Reduction in dissolved gases can be attained in several ways. The most effective methods are to minimize the entrance of gases into the mud (1) by proper utilization of surface equipment such as mud hoppers and guns to reduce the mud and air mixing and (2) by prevention of formation fluid entries into the well that may contain dissolved gases as CO_2 or H_2S. Another method is with chemical agents that remove gases.

pH Control

The pH is a measure of the hydrogen ion concentration. The pH values are presented on a scale of 1 to 14 with 1-7 considered to be acidic and 7-14 considered as basic. A value of 7 is neutral (Figure 3-37).

The pH of the mud systems must be monitored and controlled since most mud additives function more effectively in a pH range of 9-11. Also, corrosion rates are affected by the hydrogen ion concentration, and personnel safety problems are increased in certain pH ranges. Although most mud systems function more effectively in higher pH ranges, the

FIGURE 3-37 • pH ranges

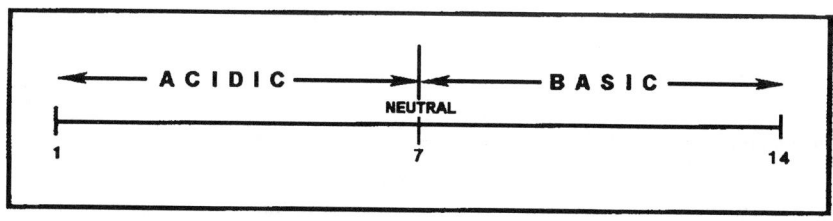

individual additive pH values are often low. For example, chrome lignosulfonate solution has a pH of about 3.6. A pH adjuster, such as sodium hydroxide (caustic soda), can raise the level to the desired range.

The corrosion rate is affected significantly by the pH or hydrogen ion concentration of the electrolyte. As the pH decreases, the embrittlement tendencies increase due to the larger concentration of hydrogen ions. Figure 3-38 shows the effect of pH on the time to failure of a sample metal ring. Note that the time to failure is high at a pH value of 9.5.

Another effect of pH is the gas solubility at higher pH levels. (See Figure 3-39.) The total concentration of dissolved gas exists as hydrogen sulfide at pH ranges of 3-6. It begins to convert hydrogen sulfide and sulfide ions at pH ranges of 6-14. In the range of 6-9, a mixture of hydrogen sulfide and sulfide (monovalent and divalent) ions are present.

The presence of the sulfide ion has only a small corrosion effect itself, but it does increase the tendencies for hydrogen embrittlement and sulfide stress cracking by retarding the rate at which atomic hydrogen is allowed to escape from any corrosion point on the metal surface.

Caution must be exercised when using pH control as a corrosion preventive measure. As noted in Figure 3-39, high pH converts the

FIGURE 3-38 • Effect of pH on time to failure (Courtesy of *OGJ*)

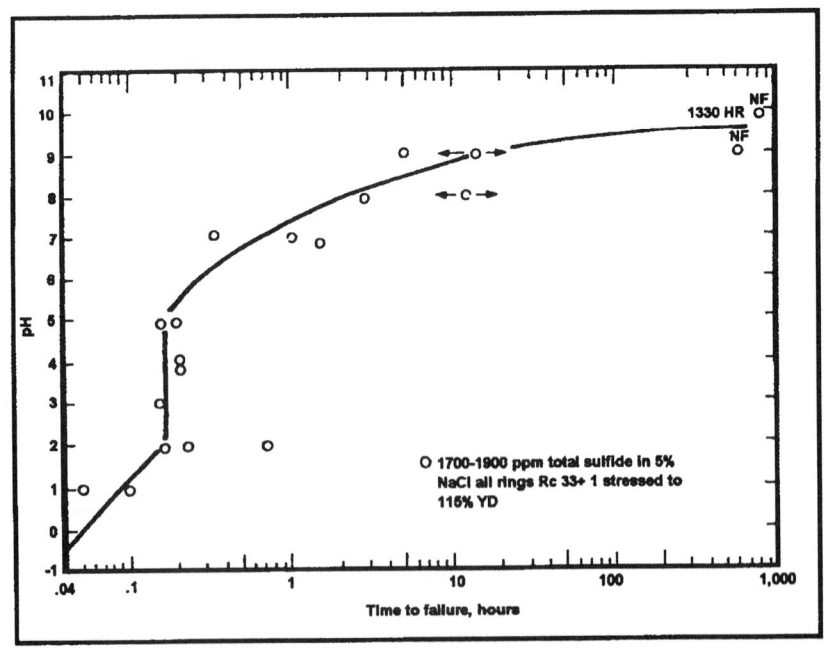

FIGURE 3-39 • Effect of pH on hydrogen sulfide and sulfide ions (Courtesy of *OGJ*)

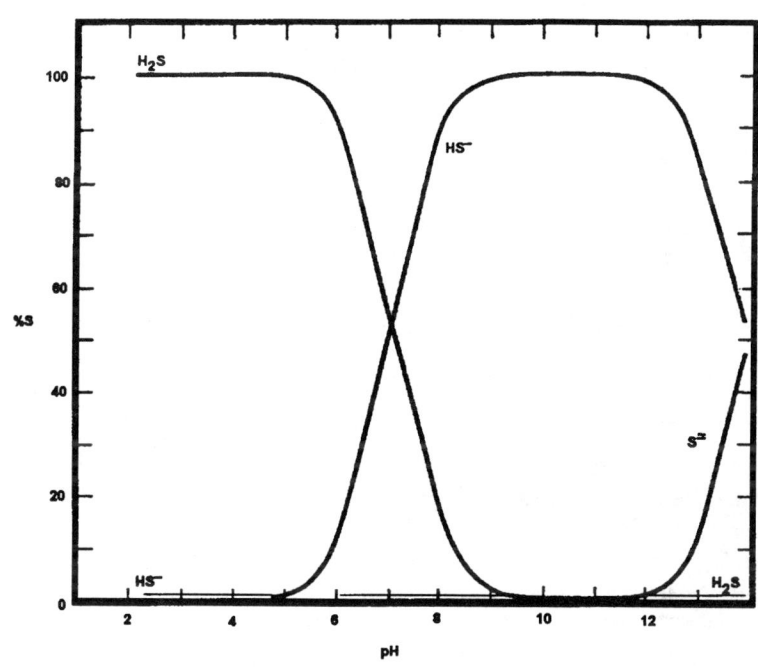

hydrogen sulfide molecule to ionic sulfides, which are not as directly dangerous to human life. This tends to mask the problems by hiding the hydrogen sulfide. If the pH is reduced for any reason, the sulfides may revert to H_2S and cause severe problems.

Another problem area is the use of scavengers to remove the hydrogen sulfide while the pH of the system is not conducive for the use of the particular additive employed. Certain agents require high pH values to function, while others need lower ranges. A high pH additive generally will not function in a low pH system, while the converse is true for the low pH additive. Therefore, additives and pH ranges must be properly chosen for effective hydrogen sulfide control.

Corrosion Inhibiting Fluids

One widely used method of protecting downhole equipment is chemical inhibition. Since corrosion is a reaction on metal surfaces, any modification of the steel-electrolyte interface will affect the corrosion rate. Inhibitors are chemicals that when added in small quantities to a corrosive system will alter the steel-electrolyte interface and reduce the corrosion rate. The action of these materials may be described as oil wetting the steel surface.

Most inhibitors are amine-fatty acid salts formulated into either oil soluble or oil soluble water dispersible materials. The inhibitor molecule can be imagined as having a polar end and an oil soluble end. The polar end can adhere to solid surfaces such as the steel tubular goods. The oil soluble end absorbs a film of oil that protects the surface. The dual nature of the inhibitor molecule also gives detergent properties to the chemical.

Inhibitors are used for continuous or periodic treatments. The main requirement for treatment is to contact the metal surface with sufficient inhibitor for a long time to obtain a film coating. New treatments should be made at regular intervals to maintain the film.

Hydrogen Sulfide Scavengers

Scavengers are additives designed to remove a contaminant from the mud system. The additive generally does not prevent corrosion such as hydrogen embrittlement. It reduces the severity by sequestering the hydrogen sulfide or sulfide ion that would have increased the embrittlement tendencies. There are many types of additives with different properties.

Most scavengers function in either a surface adsorption manner or through ionic precipitation. If the scavenger is based on the surface adsorption technique, the mud must be thoroughly mixed to insure that a sufficient number of collisions occur between the hydrogen sulfide and the scavenger for a completion of the process. In the ionic reactions, the solution characteristics of the scavengers must be studied to insure that pH and salinity are conducive to the additive.

The primary hydrogen sulfide scavengers used in the industry are metallic compounds based on copper, zinc, or iron.

Copper Products

An effective copper derivative used in hydrogen sulfide scavengers is copper carbonate. This product was the first material to be widely used because of its efficiency as a scavenger. Even though basic copper carbonate is essentially insoluble in the drilling fluid, it has been found to be sufficiently reactive to precipitate hydrogen sulfide as copper sulfide, which *is* insoluble. Copper carbonate does not adversely affect the drilling fluid properties.

Copper derivatives have side effects that limit their application. Copper compounds may react with iron (the drill string) causing the steel to corrode or copper plating to occur. Therefore, when copper products are used as hydrogen sulfide scavengers, one corrosion problem is simply replaced with another. Copper products are rarely used worldwide in the current industry.

Zinc Products

In an effort to avoid the corrosion problems posed by copper, metallic compounds with oxidation potentials closer to steel have been used. The most common metal that will meet this requirement and still readily form a sulfide is zinc.

Zinc carbonate is one of the most widely used sulfide scavengers in the industry. It utilizes an ionic reaction as well as a surface fraction for scavenging, and it functions most effectively at pH values above 9. The compound is relatively safe and poses no immediate hazards.

Although zinc carbonate is an effective scavenger, there are problems. Concentrations above 3 lb/bbl in weighted, high solids systems cause high gel strength. The zinc ion may cause a reaction similar to calcium contamination in some muds. The compound does not prevent hydrogen embrittlement. A high pH is required to solubilize the product; and since its specific gravity is roughly the same as barite, it may settle in muds and brines that have a low carrying capacity.

Organically chelated zinc compounds may be used in brines where the zinc carbonate would settle. These compounds are efficient but contain only one fourth as much zinc as conventional zinc carbonate. The product may also function as a thinner in nondispersed muds.

Zinc chromate is an efficient scavenger that also minimizes hydrogen embrittlement and maintains low corrosion rates. The compound uses an oxidation reaction to form a sulfate when the pH is greater than 9. The problem with zinc chromate is the environmental restriction on the use of any chromate product.

Iron Products

Ironite Sponge, a trade name for iron oxide, is a hydrogen sulfide scavenger. The surface reaction is not restricted by temperature or time, and the product does not degrade mud properties. The iron oxide, or Sponge, reacts with the hydrogen sulfide to form the stable iron sulfide, pyrite.

Sponge has a specific gravity of 4.4, which allows it to replace barite on an even basis as a density additive. The use of a high specific gravity material is advantageous because it limits the buildup of low gravity solids. The particle size averages 6–8 microns and ranges between 1.5–50 microns, which is comparable to barite. The uniform spherical shape and size of the Sponge particle creates a low abrasion level.

Ironite Sponge is ferromagnetic. It is strongly attracted to a magnet but will not retain magnetism. High saturation magnetism is the basis for the simple field test to determine Sponge concentration. The low remnant magnetism prevents attraction to the drill pipe or casing.

The reaction between the hydrogen sulfide and iron oxide appears to be stable and irreversible under most conditions. One pound of Ironite Sponge will react with 0.7 pounds of hydrogen sulfide. The reactions occur most effectively in a pH range of 6–9. Values above 9 restrict the scavenging of H_2S because it is in the sulfide form (see Figure 3–39). This fact must be approached with caution because it may lead rig personnel to assume that an excess of Sponge is present when actually the hydrogen sulfide is in a form that cannot readily be scavenged by the additive. If the pH drops, the gas or liquid may revert to hydrogen sulfide and react out the existing Sponge, while leaving an excess of unreacted hydrogen sulfide.

SHALLOW GAS HANDLING

The drilling industry and well control specialists are rapidly acknowledging that shallow gas handling is a difficult challenge. It is perhaps the most complex of all well control problems and poses more technical challenges than blowout capping, snubbing, hot tapping, freezing, or other tasks routinely addressed by blowout specialists.

Primary and remedial control techniques for handling shallow gas are separate and distinct. Primary control techniques are implemented with existing equipment and systems generally found on drilling rigs. These are usually performed by the rig crew at the time of the blowout.

"Remedial techniques" are defined as those procedures that are generally performed by blowout specialists with special techniques and equipment.

New remedial control techniques, such as direct vertical intervention and offset kills, and improved primary control diverter systems have been developed in recent years. They have been successfully field tested. These techniques are based on new technological developments and would have been considered impossible during the early 1980s.

Shallow gas blowouts have caused more offshore drilling rigs to be lost than any other type of well control problem. There have been many other cases where shallow blowouts resulted in severe rig damage, although the rig was not totally lost.

A shallow gas blowout on a platform can cause large financial losses. For example, a recent 1987–1988 shallow gas blowout severely damaged a platform. The losses were reported to be US$200,000,000. Other similar events have resulted in the total loss of platforms.

The data in Tables 3–8 through 3–11 are a partial list of rigs and platforms that have been damaged from shallow gas blowouts. An industry-wide comprehensive list is not available for assessing rig losses and damage from blowouts because a common source does not exist to compile the data (i.e., government body, insurance company, etc). The data in Tables 3–8 through 3–11 are derived from an in-house proprietary data base and several public domain databases.

A shallow gas flow is a critical issue because field experience and mathematical modeling have shown that it is difficult, and almost impossible, to control or stop a flow with existing rig equipment once it begins. Formation bridging is responsible for stopping many shallow blowouts. Often shallow gas blowouts that do not bridge have to be controlled by some remedial means such as a direct intervention tech-

TABLE 3–8 • Platforms damaged by shallow gas blowouts

Year	Platform	Damage	Location
1957	South Pass 27	Light	Gulf of Mexico
1962	Grand Isle 9	Extensive	Gulf of Mexico
1962	Middle Ground Shoals	Extensive	Cook Inlet, Alaska
1965	S. Marsh Island 48	Extensive	Gulf of Mexico
1967	S. Timbalier 67	Extensive	Gulf of Mexico
1974	E. Cameron 338	Light	Gulf of Mexico
1974	High Island A-563	Total Loss	Gulf of Mexico
1976	Fateh L.	Total Loss	Arabian Gulf
1976	High Island A-511	Extensive	Gulf of Mexico
1976	Eugene Island 380	Moderate	Gulf of Mexico
1977	S. Marsh Island 96	Moderate	Gulf of Mexico
1977	S. Marsh Island 146	Light	Gulf of Mexico
1978	West Cameron 180	Total Loss	Gulf of Mexico
1978	West Delta 79	Light	Gulf of Mexico
1978	Vermilion 23	Light	Gulf of Mexico
1980	High Island 368	Total Loss	Gulf of Mexico
1981	Khafji 156	Extensive	Arabian Gulf
1982	Eugene Island 361	Extensive	Gulf of Mexico
1982	Campeche	Moderate	Bay of Campeche
1983	Forties Delta	Extensive	North Sea, UK
1983	East Breaks	Extensive	Gulf of Mexico
1985	Grayling	Moderate	Cook Inlet, Alaska
1987	Steelhead	Extensive	Cook Inlet, Alaska

TABLE 3-9 • Bottom-supported rigs (jack-ups and submersibles) damaged by shallow gas blowouts

Year	Contractor	Rig	Damage	Location
1958	Odeco	N/A	N/A	Gulf of Mexico
1968	Fluor	Little Bob	Total loss	Gulf of Mexico
1972	Reading & Bates	M.G. Hulme	Total loss	Java Sea
1972	Marine	J. Storm II	Total loss	Gulf of Mexico
1974	Offshore	Meteorite	Total loss	Nigeria
1975	Zapata	Topper III	Total loss	Gulf of Mexico
1978	Penrod	Penrod 61	Light	Gulf of Mexico
1979	Odeco	Ocean Patriot	N/A	Gulf of Mexico
1980	Reading & Bates	Ron Tappmeyer	Extensive	Arabian Gulf
1981	Sedco	Sedco 250	Total loss	Angola
1983	Penrod	Penrod 52	Total loss	Gulf of Mexico
1983	Santa Fe	Santa Fe 134	Light	Kalimantan
1985	Beaudril	Molikpaq	Moderate	Beaufort Sea
1988	Sedco	Sedco 251	Total loss	Java Sea
1989	Sedco	Sedco 252	Total loss	India
1989	Teledyne	Teledyne 16	Total loss	Gulf of Mexico
1989	Beaudril	Molikpaq	Light	Beaufort Sea

TABLE 3-10 • Semisubmersibles damaged by shallow gas blowouts

Year	Contractor	Rig	Damage	Location
1971	Odeco	Ocean Driller	Light	Gulf of Mexico
1973	Santa Fe	Mariner 1	Extensive	Trinidad
1973	Santa Fe	Bluewater 2	Light	Gulf of Mexico
1975	Santa Fe	Mariner II	Light	Gulf of Mexico
1978	Sedneth	Sedneth 1	Moderate	Gulf of Mexico
1980	Sedco	Sedco 135C	Total loss	Nigeria
1981	Wilhelmsen	Treasure Saga	Moderate	North Sea, Nor.
1981	Odeco	Ocean Scout	Light	Gulf of Mexico
1984	Wilhelmsen	Treasure Seeker	Moderate	North Sea, Nor.
1985	Smedvig	West Vanguard	Extensive	North Sea, Nor.

nique or a relief well. Occasionally, a well will blow for years before depletion will cause it to cease flowing.

A key parameter in shallow blowouts is the small tolerances between formation pressure overbalance with the drilling mud and the rock integrity (fracture gradient). The low fracture gradient is the focal point of the issue. The pressure overbalance is small as compared to situations in deeper drilling environments. Due to the low overbalance margins, relatively moderate amounts of swabbing or core volume gas cutting can initiate gas flows that result in blowouts.

Special Problems

TABLE 3-11 • Drillships/barges damaged by shallow gas blowouts

Year	Contractor	Rig	Damage	Location
1964	Reading & Bates	C.P. Baker	Total loss	Gulf of Mexico
1969	Reading & Bates	E.W. Thornton	Moderate	Malaysia
1970	Offshore	Discoverer II	Light	Malaysia
1970	Offshore	Discoverer III	Moderate	Java Sea
1971	Fluor	Wodeco II	Total loss	Peru
1971	Atwood Oceanics	Big John	Total loss	Brunei
1975	Offshore	Discoverer I	Light	Nigeria
1981	Petromarine	Petromar V	Total loss	S. China Sea
1982	Global Marine	Conception	Moderate	Kalimantan
1988	Viking Offshore	Viking Explorer	Total loss	Balikpapan

Typical Shallow Gas Scenarios

A discussion on typical shallow gas scenarios is useful to provide an insight into the phenomena that can be expected in a shallow event. Case histories have been used to develop these typical scenarios.

Definition of Shallow Gas

Two useful definitions can be given for shallow gas. The first is gas encountered at shallow depths for which the fracture gradients are low and do not allow kicks to be controlled with conventional shut-in techniques. These depths range from initial soil penetration down to the casing setting point for either conductor or surface casing. Recent cases showed large blowouts occurring from a zone at 300 ft (90 m) subsurface on a land blowout and 400 ft (120 m) below the mud line, or 700 ft (210 m) RKB, on an offshore event.

Another common type of shallow gas blowout includes those blowouts occurring in open hole below conductor or surface casing or behind surface casing caused by phenomena such as gas migration through cements. The flow cannot be closed in and killed with conventional techniques. A common example is blowouts through the annulus of surface-conductor casing or intermediate-surface casing resulting from gas migration.

Causes of Shallow Kicks

The following list contains causes of shallow gas flows. The list was developed from the case history database. The causes are common for shallow and deep kicks.

- Swabbing
- Core volume cutting

- Improper hole fill-up on trips
- Abnormal pressures—charging or structural crest of gas zone
- Lost circulation during drilling or cementing
- Gas migration through cement

Special emphasis must be given to several causes of kicks. Flows during tripping are a frequent cause of blowouts. The data does not indicate whether the cause of the flow is swabbing or improper hole fill-up.

Common causes of shallow gas blowouts are related to the cementing process. The following list describes blowouts during cementing:

- Gas-through-cement channeling allowing a hydrostatic pressure reduction and a blowout. This situation often occurs when the BOP stack is nippled down, thus creating a situation where the well cannot be shut in.
- Lost circulation during the casing running operation or cementing. Many wells are cemented with slurry densities in excess of the fracture gradient.
- Cement spacers of water or diesel that reduce the hydrostatic pressure of the column. Also, one recent blowout occurred because the water spacer filtered into the permeable formation opposite the spacer. The mud level in the annulus dropped, which allowed the oil and gas formations to flow. Annular casing valves were inadvertently left open.

One observation is particularly significant. Gas blocking agents were not used on any of the blowouts where sufficient data was available to assess cement chemical additives. However, data is not available to determine the ratio of wells using the chemical blocking agents to stop gas flow versus wells that do not use the agents.

Reaction Time Following The Initial Flow

Two aspects of time are important relative to shallow gas blowouts. Most shallow events due to cementing problems are observed within two to six hours after the cement slurry is in place. Often the situation starts while the BOP stack is being nippled down. Methods to prevent the flow used by some operators are (1) maintain a small amount of annular BOP pressure during this time to allow the cement to set and minimize the flow possibility or (2) leave the BOP stack nippled up until the cement has set. These methods seem to be effective, but do not guarantee that the flow will not start.

The reaction time after the flow starts will usually be short, and the hole can unload before key signs are observed and any action can be taken. Two recent shallow gas flows on a well in Alaska unloaded the hole and riser from 700 ft (210 m) RKB without any apparent warning signs from the detection equipment. This fact is interesting because a full MWD system was being used in conjunction with a complete computerized mud logging system to monitor downhole and surface indicators of a kick. Other operators have indicated similar experiences with reaction times from shallow blowouts.

The short reaction time warrants a diverter system that closes quickly. This issue will be discussed later. However, it is important to note that accidents have not been recorded with respect to a slow operating diverter system where a faster operating system would have prevented the damage. This fact does not mitigate the need for a quick response system.

Typical Results From A Shallow Gas Blowout

Shallow gas blowouts have predictable results that can be categorized into bridging, diverter system failures, flow outside the casing, and cratering. Once a flow has started, it is almost inevitable that a blowout will occur and fall into one of these categories. This situation will be explained further in the following section, Kill Techniques.

Bridging. Bridging occurs when the wellbore caves into the open hole and stops the flow. In the case of a casing annulus blowout, the formation caves in on the casing. Bridging can occur because of two phenomena. As the well flows, the pressure in the formation is drawn down. If the rock becomes unstable due to the lower pressure, it can cave or bridge.

Another common cause for bridging is rock fragments being dragged into the wellbore and up the well because of the high fluid flow rates. This action can horizontally erode the blowing zone until the strata immediately above the zone collapses from structural instability. During one shallow blowout in 1988, the well depth was increased by 15 ft (4.6 m) from debris being blown out of the well. It is believed that the wellbore size was increased spherically with a horizontal radius approximately equal to the depth increase of 15 ft (4.6 m). A blowout in the South China Sea in 1985 resulted in an average wellbore diameter enlargement from 17.5 in. (0.4 m) to 17 ft (5.2 m). These observations have an impact on the viability of the dynamic kill technique, which will be discussed in a later section.

Historical records show that if a shallow gas blowout does not bridge within the first one to two days, then the well will probably continue to blow for an extended period of time (i.e., weeks or months). Some have continued for years.

Diverter system failures. Diverter system failures occur at such an alarming rate during shallow gas blowouts that contingency plans should perhaps be based on their anticipated failure rather than an expectation that they will function effectively. Previously published studies show that failure rates range from 50–70% of all applications.

Many operators and contractors now design diverter systems for the primary purpose of providing time to evacuate the rig. They do not plan to remain on the rig and attempt to control shallow gas blowouts.

Specific causes of the failures are addressed in the next section. Recently developed diverter technology is also discussed with examples of its successful applications.

Flow outside casing. Flow outside casing usually results in severe situations such as a damaged well and rig or platform loss. The flow often starts as a result of gas-through-cement migration, hydrostatic reductions from cement spacers, or lost circulation during cementing. The flow can be up casing string annuli if a surface flow path exists, such as open annulus valves or nippled down BOPs. Also, the flow can exit to the surface through fault planes or around poorly cemented casing.

Cratering. Cratering occurs when flow outside the casing displaces large volumes of surface sediment. The eruptive force of blowouts can be dramatic, and it has been documented to lift large boulders weighing several hundred pounds into the air and drop them as much as 150 ft (45 m) from the well site.

The areal extent can be large. One well had a crater with dimensions of 1,300 ft × 250 ft × 300 ft deep (400 m × 75 m × 90 m). The actual depths of craters are not easily determined. Large rigs and platforms have been lost in craters without any evidence of the rig remaining at the surface.

A 1987–1988 event illustrates the severity of the cratering forces. A well had been drilled with the following casing strings:

30 in. (762 mm)	479 ft (146.0 m)
20 in. (508 mm)	765 ft (233.2 m)
13.375 in. (340 mm)	2,240 ft (682.8 m)

All depths are RKB and nearly vertical. A flow started in the 17.5 in. × 13.375 in. (444 mm × 340 mm) hole-casing annulus while cementing. A blowout resulted. After the well was finally controlled months later, it was determined that the 13.375 in. pipe was gone. Probably, it had been cut by the erosive action of the blowout and fallen down along the side of the crater. The 20 in. (508 mm) was cut off at the base of the 30 in. (762 mm) pipe. The 30 in. and the 20 in. casing strings were hanging from the platform unsupported by the cratered hole. Also, the leg through which the well had been drilled was standing unsupported, which caused the platform to deflect 7 in. (178 mm) downward on that corner. The crater had a 125 ft (38.1 m) diameter at the mudline and was at least 765 ft (233 m) deep.

Frequently, diverter systems may fail and then a crater can be formed later. Bridging is often the cause for stoppage of the flow. Some wells have blown out underwater for years, so bridging of a blowing well after a short period of time is not necessarily a valid assumption.

Special Problems

The depth of an underwater crater is an important issue relative to new remedial control techniques. The depth is believed to extend downward to near the top of the blowing zone. The inverted cone shape probably narrows considerably near the bottom. This observation is made from recent experiences with shallow blowouts, and it has resulted in the development of new proprietary techniques for controlling underwater blowouts with a direct intervention approach.

Diverter System Failure Modes

A variety of failures have been experienced on diverter systems. Failure analyses have revealed other potential failure modes. Diverter system failures are frequently due to more than one factor. The failure modes can be grouped into categories as follows: erosion, blockage, excessive back pressure, diverter unit limitations, control system failure, improper materials, poor component selection, support failure, and slip joint packing leaks.

Erosion

Erosion primarily depends on fluid velocity, abrasiveness of the entrained solids, and the angle of impact of the solids against system components. Erosion can be mitigated by reducing one or more of these factors. Erosion failures have generally been due to undersized lines and flow path upsets that cause turbulence. Diverter systems with diameters of 10 in. (254 mm) and smaller are common. Shallow gas flow rates have generally been grossly underestimated. Bends, bore size changes, and flow path discontinuities produce high particle impact angles and local increases in velocity.

Blockage

Many shallow gas flows contain large quantities of debris (e.g., rocks, sand, etc.). There have been several documented cases where this debris packed off at a bend in the diverter line or other obstruction. The pressure surge due to the blockage resulted in failure of other components.

Excessive Back Pressure

The back pressure from the diverter system during a shallow gas flow can result in broaching in the open hole section and failure of components. Line sizing and flow restrictions are the key factors that affect the back pressure for a given flow. Figure 3–40 shows steady-state back pressure curves for various diverter line sizes on a typical shallow gas flow.

FIGURE 3-40 • Diverter system steady-state backpressure (dry gas) (Courtesy of SPE/IADC 16129, © 1987)

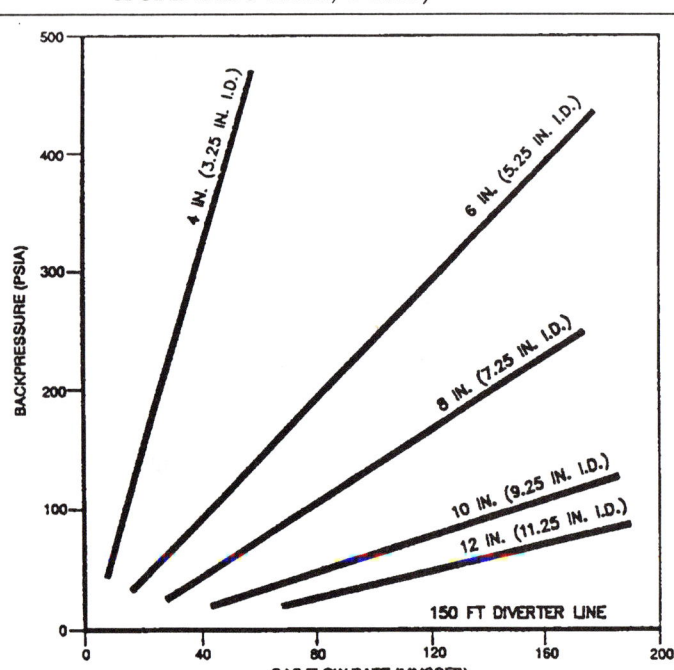

Diverter Unit Limitations

The insert-type diverter cannot divert the flow if the pipe is out of the hole or if the insert bushing is not in place and latched. This type of unit cannot close on open hole. Also, it may not be possible to strip back into the well with the bladder sealing element used in the insert-type diverter. These limitations have caused problems in several documented cases.

Control System Failures

Control failures have typically occurred on complex systems with several optional modes of operation. Portions of the controls were locked out for testing or maintenance and not returned to full operational readiness. When an attempt was made to divert, the system either totally shut in the well or the wrong valve operating sequence resulted. Some control failures have been due to poor component selection or improper materials.

Improper Materials

Material failures have generally occurred where the material being used was not suitable for the abrasion and pressure surges seen in most shallow

SPECIAL PROBLEMS

FIGURE 3-41 • General pressure vs. time for shallow gas kick (unloading phase) (Courtesy of SPE/IADC 16129, © 1987)

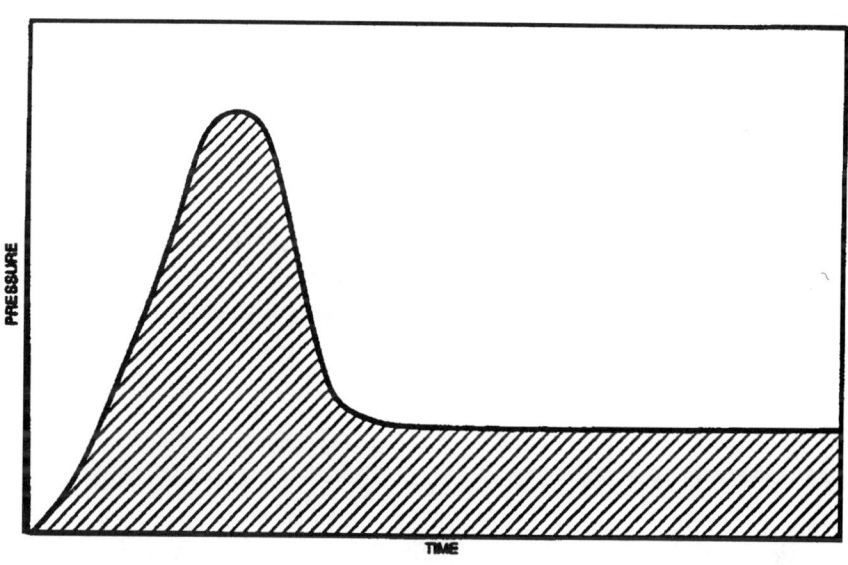

gas flows. Again, line sizing and flow restrictions have a significant effect on pressure build-up in the system. The failures have included rubber hoses used for connecting diverter lines and thin wall pipe in the vent lines.

A typical pressure-time plot is shown in Figure 3-41. The peak pressures occur during the well unloading phase. Full-scale tests have shown that these pressure peaks can be as high as 900 to 1,000 psi (6.2 to 6.9 MPa).

Poor Component Selection

In some cases, valves and other key elements have been installed that could not handle the dynamic conditions, pressure surges, and debris-laden flow. Low pressure, butterfly and guillotine valves have failed to operate as required or caused line blockage. Gate valves, in general, have had problems with trash in the guides. Valve operators have been installed that could not actuate the valves under the existing conditions.

Support Failures

Diverter unit and pipe support failures have been recorded. On at least two occasions, surge pressures during the unloading phase destroyed the diverter supports on a floating operation. The diverter and riser scoped up, which dislocated the rotary table and separated the diverter vent lines. Also, inadequately supported diverter piping has been damaged from vibrations and whipping caused by high flow rates.

Slip Joint Packing Leaks

The slip joint packing element has historically been a source of problems during shallow gas flows on floating drilling operations. The typical bladder element is designed for containing low pressure fluids while accommodating the heave motion of the floating vessel. When a flow occurs, the inflation pressure on the bladder is increased to seal off the higher pressure in the riser. The packing elements have typically failed due to the flowing pressure exceeding the sealing capabilities of the bladder or due to the bladder seizing to the riser and rupturing at the increased inflation pressure.

Disconnect Difficulties

When drilling with a riser on floating operations, it has been a standard practice with most companies to release the riser and move off if a shallow gas blowout occurs. However, it has been reported that the riser connector failed to release on a significant number of these drive-off attempts.

Damage Due to Disconnect Failures

Failure of the riser to disconnect has generally forced the operator to drive off and physically separate the riser. In most cases, the casing and/or wellhead were either bent or fractured. Resulting inclination angles of 2° to 45° have been recorded. Some cases of angles in excess of 45° were reported, but not fully documented. The larger inclination angles would severely restrict any direct remedial intervention on the well.

The bend usually occurs immediately below the mudline. The bending casing and wellhead are resisted by passive soil pressures as they move laterally. It is not uncommon that the first casing connector below the mudline will snap or part. The BOP stack (if installed) and wellhead have a higher section modulus and have demonstrated a greater bending resistance than the casing.

Mode of Failure

Certain types of hydraulic connectors have had a much higher failure rate than others. These connectors swallow the mating surface on the wellhead or BOP stack in an arrangement that is a relatively tight tolerance cylinder within a cylinder. The riser must be aligned almost exactly

Special Problems

vertical above the BOP stack before an unrestricted release can be made. The connector must be able to move vertically several inches before the connector can tolerate any significant misalignment. If the riser is not aligned vertically, interferences exist between the two cylindrical surfaces.

The other type of connector releases freely when actuated. It can reportedly accommodate release angles to 30°.

Vessel and Platform Losses

Damage due to shallow gas blowouts has extended to platforms and all classes of mobile offshore drilling units. In many cases, the blowout resulted in severe damage or total loss of the unit. Tables 3–8 to 3–11 are a partial list of rigs and platforms that have been damaged by shallow gas blowouts. There are probably other undocumented cases in the industry.

Platforms

The investment in a platform's structure, facilities, rig, and existing producing wells can be large. Therefore, the cost of potential blowout property damage to a platform can be much greater than that of a MODU in a similarly sized blowout. For example, damage and well control costs exceeded US$200 million on one platform shallow gas blowout. Also, entire platforms have been destroyed.

Fixed platforms are particularly susceptible to structural foundation damage from cratering. If cratering is extensive, the platform may tilt or, in the worst case, collapse completely.

Production vessels with hydrocarbon inventories, tie-ins to pipelines, other hydrocarbon handling facilities, and live wells increase the risk of major damage if the blowout results in an explosion and fire. It is difficult to isolate the production facilities with 100 percent assurance unless they are located on a separate platform.

Jack-ups and Other Bottom-Supported Rigs

Bottom-supported rigs are also vulnerable to foundation failure. The units have either independent legs or rest on a barge or mat-type base. Cratering can lead to capsizing. Table 3–9 shows that a large number of the jack-ups were total losses from blowout damage.

In some circumstances, the unit can be moved off of location. However, this is generally not possible due to the risk of fire.

The diverter system arrangements on jack-ups present a special set of problems. Jack-ups have multi-position substructures on cantilever extensions or slot "fingers" at one end of the rig. The arrangement of the diverter vent lines must direct the flow from the diverter to at least two points overboard. The lines need to be straight, but also must accommodate the various positions of the substructure and the movement of the diverter unit.

Semisubmersibles

Semisubmersibles, as a class, have performed better in blowouts than the other rig types. (See Table 3–10.) Semis can drill riserless in water depths of 250–300 ft (75–90 m) or greater. Most cases of damage to semis occurred because a riser was being used and it provided a direct, large bore conduit for the gas to reach the rig. Failure of the riser connector to unlatch prevented some of the rigs from driving off of the well quickly.

Loss of buoyancy in a gas boil has not been a problem. Stability and adequate draft have been maintained. The reduction in specific gravity from the rising gas is the greatest near the sea surface. Semisubmersible pontoons and portions of their columns are located below this maximum aerated zone at the surface. (See Figure 3–42.) Also, the large air gap between the water surface and the rig facilities allows for dispersion of gas before it reaches the rig's decks. The only potential problem would be if the wind is calm and the volume of gas from the blowout is extremely large.

FIGURE 3–42 • Semisubmersible in a blowout boil typical specific gravity variations

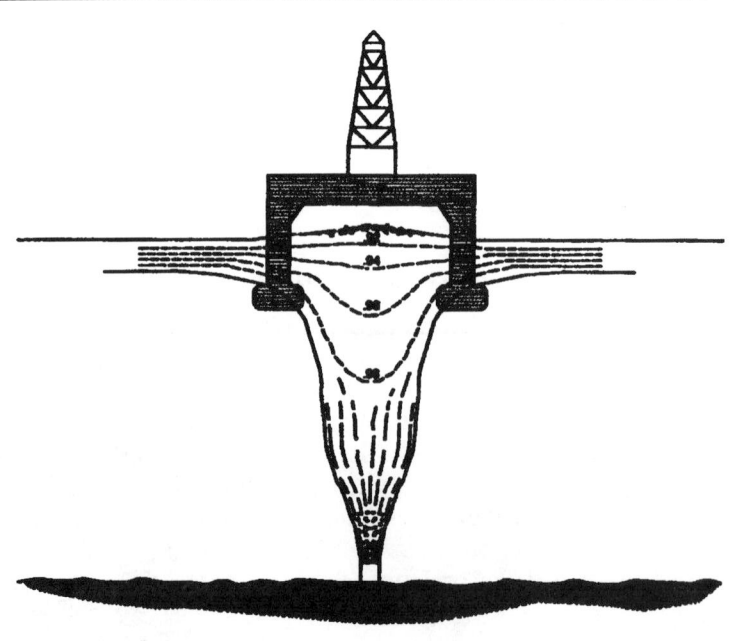

Special Problems

It has been observed that gas from a subsea blowout is dissipated by wind and sea currents.

Drillships and Barges

Moored drillships and barges operating in relatively shallow water have had problems with shallow gas blowouts. (See Table 3–11.) Gas is more likely to envelope the rig and shut down the power due to the low freeboard and lack of an air gap. The rise in the sea surface in a blowout boil has flooded several vessels that were operating with hatches open or that suffered hull damage during the blowout.

Surface currents from a blowout acting on the hull tend to set the rig off to one side of the boil. This, in conjunction with the mooring lines, applies an overturning moment to the vessel and causes a list that further reduces the freeboard on the boil side of the vessel. The effects of the blowout and evolved gas are reduced as water depth increases.

Dynamically positioned drillships have the benefit of being able to readily drive off in an emergency. There are no reported cases of significant shallow gas blowout damage to a dynamically-positioned drillship.

Handling Techniques

Several primary kill techniques have been developed for handling a shallow gas blowout. Unfortunately, the techniques are either not completely reliable, have a low success ratio, or are not suitable for implementation on all rigs and situations. Shallow gas blowouts have no reliable primary control methods when using existing diverter technology. Recently developed proprietary diverter systems should prove more effective. However, some industry designs that have been proposed are not believed to be able to control worst case situations for extended periods.

Prevent Kicks

It is desirable to prevent kicks in most drilling situations. It is even more critical to prevent kicks in shallow gas environments. Field experiences and mathematical modeling have shown that a kick from the typical high permeability shallow formation is difficult to control with equipment readily available on most rigs. This is a disturbing situation, and it becomes more serious when coupled with the previous discussions on cratering and rig losses.

Shut-in

Some operators, when drilling offshore from a floating vessel, have adopted the practice of shutting in on shallow flows. The practice has been developed as a last resort effort. It is based on the conclusion that gas diverted at the surface cannot be controlled and that diverter system failure potential is high.

The shut-in well has the potential to fracture in the open hole and cause a crater. This has not had a high occurrence based on reported operators' experiences. If cratering does occur, the rig moves off location. The only loss is the shallow well, the subsea diverter, and the guide base/casing left on the well. Although this is not an insignificant equipment loss, some operators consider it to be a viable alternative to the damage that could be caused if the well is diverted at the surface with a resultant diverter system failure. If the well continues to flow and does not bridge, it can be controlled with a proprietary direct intervention kill technique.

Unfortunately, a shut-in procedure is not acceptable for bottom-supported rigs or platforms. Cratering could result in loss of the rig or platform.

Pilot Holes

Many operators advocate the use of "pilot holes" during the drilling of the shallow hole sections. A pilot hole is a small diameter hole drilled below a previously set casing string. The hole section is drilled at a controlled rate to minimize core volume gas build-ups in the mud. Circulation rates are usually high to flush gas out of the hole.

Pilot holes usually range from 8.50–12.25 in. (216–311 mm) diameter. Logic applied with this hole size selection is that the well can be dynamically killed if a flow starts. Unfortunately, it is usually not possible to kill the well dynamically, particularly with hole sizes larger than 9.875 in. (250 mm). An 8.50 in. diameter hole is recommended.

Pilot holes are a viable technique for evaluating the hole sections below the casing. However, even if pilot holes are drilled with maximum precautions, they do not guarantee that a blowout will be prevented. Since the hole diameter is small, minimal gas cutting or swabbing can initiate a flow. After the flow starts, the small diameter hole is evacuated more quickly than a large diameter hole.

Pilot holes are a recommended procedure for shallow gas evaluation. However, they provide no certainties and caution should be maintained.

Dynamic Kill

The dynamic kill technique has been proposed as a kill method for shallow gas blowouts. It was originally developed for deep well kills where the well bores have smaller diameters than shallow hole sections. The technique relies on annular friction pressures to assist in a dynamic well kill.

The dynamic kill is not effective in most shallow gas situations for several reasons. The hole sizes are large and do not allow the build-up of high friction pressures to control the flow. Mathematical modeling has shown that dynamic kills are remotely feasible in hole sizes of 9.875 in. (250 mm) or smaller if the well is drilled riserless and the hole section below the previous casing string is of some reasonable length. If the flow starts soon after drilling out of a casing string, sufficient annular friction pressures cannot be generated to control the flow. The same situation is true for riser drilling.

Hole sizes of 12.25 in. (311 mm) are impossible, in a practical environment, to kill dynamically if the flow rate is moderate or severe. Various studies have reached the same conclusions. Field experiences confirm the results.

Furthermore, hole sizes of 9.875 in. (250 mm) or smaller are still difficult to kill dynamically. The shallow sediments are usually soft and quickly erode to large diameters during the drilling or the initial flow.

The most conceivable situation for a successful dynamic kill on a shallow gas blowout is with low productivity formations. Again, this is not the common situation in shallow environments.

Quick response to the flow does not significantly improve the feasibility of a dynamic kill. As discussed previously, the response time based on field experiences is low and virtually nonexistent in many cases. Sufficient time does not exist in most cases to recognize the situation, close in the diverter, and begin the kill operation before the flow becomes uncontrollable.

A detailed analysis of factors in the dynamic kill method in conjunction with mathematical modeling is required to give a full appreciation of the difficulties with using the method as a viable shallow gas-handling technique.

Heavy Slug

Heavy slugs of mud have been considered as a means to build hydrostatic pressure and stop the well flow. Although the concept seems logical and reasonable, practical considerations prevent it from being effective for most situations. The response time to recognize a flow, close the diverter unit, line up on the mud, and pump it far exceeds the allowable time before the well has blown out. Rig mud pumps can not operate at rates required to cut the gas flow and build hydrostatic pressure.

The issue becomes more complicated when considering the appropriate mud weight to use as a kill fluid. It should be sufficiently high to stop the flow, yet it should not fracture the last casing seat. The mud weight is sensitive to well depth, and it would require density adjustments as the hole is being drilled to meet the two criteria of pressure control and fracture gradients.

Dynamic-Heavy Slug

The technique that holds the most promise is a combination of the dynamic kill and the heavy slug (i.e., pumping a heavy slug at high rates). Mathematical modeling shows that the technique can be effective under rigid constraints. However, it still suffers from the same problems as the two prior separate techniques.

MWD Usage

Shallow gas sections are sometimes drilled with MWD systems as a means to prevent or detect shallow flows. The MWD system is supposed to detect hydrocarbon zones; and with certain tools, it can be used to detect gas flow in the annulus.

From a practical operational view, MWD systems have little value as a shallow gas detection and blowout prevention tool. The tool is located above the bit, usually about 40 ft (12 m), which means that the zone has been penetrated before the MWD system can detect it. Also, the down hole computer cycle time and signal coding/decoding time is much greater than the allowable response time to control a shallow gas blowout even when the MWD computer functioning times are relatively small.

These observations are based on field applications of an MWD for shallow blowout detection and comments received from operators with similar experiences.

Riser vs Riserless Drilling

On floating drilling operations, the shallow hole sections can be drilled without a riser if the water depth is sufficient. Semisubmersibles and dynamically positioned drillships can operate safely in most shallow gas situations in 250–300 ft (75–90 m) water depths or deeper. Potential buoyancy reduction, gas concentrations around the rig, and other factors become increasingly significant when considering operations in shallower water depths. As previously mentioned, moored drillships can be hazardous in shallow water if shallow gas is encountered, with or without a riser.

Drilling with a riser has the following advantages and disadvantages:

- The riser provides a direct conduit for gas to the rig floor.
- The large bore conduit allows for a higher blowout flow rate than the riserless case.
- It is more difficult to kill the well with a riser in place (large diameter).
- A disconnect is required before the rig can move off.

Special Problems

- Historically, riser disconnect has been difficult.
- For deepwater operations, there is the potential for riser collapse if the well unloads.
- A higher risk of lost circulation occurs from drilling fluid returns to the flowline.
- The use of a riser involves the added time and associated costs of running a riser.
- Structural casing may have to be set deeper.
- The operation is more complex than riserless drilling.

Riserless drilling considerations include:

- It is only applicable to floaters.
- Gas is not allowed to be conducted up to the rig.
- The constant sea water hydrostatic helps to reduce the flow if a blowout occurs.
- Drilling with sea water does not offer the added hydrostatic to control flows and the hole stabilization provided by drilling mud.
- If mud is used, it is dumped and lost at the sea floor.

Recommended Handling Program

The historical data, theoretical investigations, and actual field experiences provide an insight as to the procedures and equipment that offer the best odds for successfully handling shallow gas. They are grouped into predrilling investigations, equipment, and operational practices.

Predrilling Investigation

Shallow hazard surveys should be run for the proposed location(s). These should be interpreted with the assistance of professional analysts.

Shallow hazard surveys and drilling records for other wells in the area should be reviewed. This includes discussions with operator and drilling contractor personnel knowledgeable about those operations.

If practical, sites should be avoided that have a high probability of encountering shallow gas. In some areas, the drilling location can be moved a short distance to miss a shallow gas anomaly without compromising the drilling program.

Casing setting depths and mud weights must be selected that maximize the ability to maintain control of the well.

Equipment

Riserless drilling is recommended for floating drilling in water depths of 400–500 ft (120–150 m) or greater. With special precautions, riserless drilling can be extended into shallower water depths of 250–300 ft (75–90 m). The benefits, such as the sea water hydrostatic effect and dissipation of gas by the sea currents, outweigh any slight reductions in buoyancy and other factors in water depths over 250 ft.

It is sometimes necessary to use a riser for drilling shallow hole sections to circulate mud for hole stability or other requirements. If a riser must be used, a riser connector should be selected that can dependably release at high angles. A subsea diverter or shut-in at the seabed with a subsea stack should be considered as an option.

For bottom-supported units (e.g., jack-ups, submersibles) and platforms, a subsea diverter may also be an option. However, further development is needed in this area.

The diverter system must be upgraded to meet a design criteria that has a reasonable chance at handling a shallow gas kick and provides adequate time to evacuate, if necessary. Most systems in operation today are not capable of meeting these standards.

Recently introduced MWD and advanced flow indicator systems may be successful in providing a warning for some shallow gas kicks. Their use is warranted as a precautionary measure.

Operating Procedures

The following procedures are recommended for drilling shallow hole sections.

- Drill pilot holes no larger than 8 1/2 in. (216 mm) diameter.
- Control the rate of penetration—no more than 100 ft/hr (30 m/hr).
- Clean the hole thoroughly.
- For drilling from moored vessels:
    - Mooring winches should be kept on the brakes, not dogged off, while drilling shallow hole sections, weather permitting.
    - Secure all hatches.
    - Evaluate means of releasing the mooring lines that do not require power.

If a shallow gas flow starts, these steps should be taken.

- Activate the diverter system.
- Immediately pump as fast as possible and switch to kill mud.
- On bottom-supported units and platforms, alert all personnel and commence evacuation of nonessential personnel. If kill attempts are unsuccessful, evacuate all remaining personnel.
- If kill procedures are unsuccessful for floaters drilling with a riser, release the riser.

Special Problems   277

- Drillships and drill barges should move off immediately. Depending on the magnitude of the flow, semisubmersibles may remain on location and continue kill operations, wait for the well to deplete or bridge, or move off.
- If a riser will not release, physically pull it apart using the mooring system or dynamic positioning system as the vessel is being moved off of location.

Conclusions

Shallow gas blowouts are some of the most difficult well control situations to handle. Operators and drilling contractors should place a high priority on providing the equipment and procedures to deal safely with potential shallow gas blowouts. The records show that failure to do so can result in loss of lives and major property damage.

KICKS WITH OIL-BASED MUDS

Gas kicks in oil muds pose problems not encountered in other situations. The gas is soluble to certain degrees in the oil mud. This disguises key kick warning indicators such as pit and flow monitoring. This situation does not occur significantly in gas kicks with water muds or oil/water kicks in either oil or water muds.

As gas enters the well, some dissolves in the oil mud, since both are hydrocarbons. The quantity remaining in a separate phase at the hole bottom is a function of pressure, temperature, pump rate, influx rate, and influx composition. If most of the influx dissolves in the oil mud, only a small pit gain may be observed. Flow rate out of the well will not be increased to the same level as a gas kick in a water-based mud.

The result is that a large gas influx can be taken without the same surface warning signs as seen with a water mud. A 20 bbl (3 m^3) influx will not usually result in a 20 bbl gain. More typical results might be a 4–6 bbl (0.6–1.0 m^3) gain.

As the gas and mud are circulated to the surface, the pressure is reduced. When the bubble point pressure is reached, the gas evolves as a separate phase and the pits begin to gain. This can occur near the surface and surprise the driller or supervisor.

The net result is that gas entering the hole bottom will reach its full expanded volume at the surface. However, it may not be seen as clearly as in a water-based mud.

The handling procedures do not change. The well should be shut in quickly and with the understanding that oil muds can disguise kick warning signs. Kick circulation procedures do not change. The crew should be advised the surface expansion may occur dramatically near the end of operations.

ULTRA-SLIM HOLE WELL CONTROL

Well control in small diameter, or "slim," holes is a topic receiving much interest by the oil industry in the early 1990s. Slim holes offer some advantages over larger holes, such as significant cost savings and the ability to use small rigs in remote locations. Slim hole drilling will always occupy a segment of oil and gas drilling.

The oil industry has given a broad classification to smaller diameter holes. These are called "slim holes." The exact hole diameter has not been defined. One industry expert defines hole sizes as slim hole if they are less than 7 in. OD (178 mm). Although this classification is satisfactory, "slim hole" might be more appropriate for hole sizes less than 5 in. OD (127 mm) and the term "small hole" drilling for hole sizes of 5–7 in. OD (127–178 mm). This difference might be considered as insignificant. However, an investigation into the well control differences between these hole sizes illustrates the need for a separate term for the smaller hole size. Also, the smaller hole size may often be drilled with special small rigs or coiled tubing units, whereas the size of 5–7 in. can be drilled with conventional rigs.

Drilling slim holes, although somewhat new to the oil industry, is an old and proven concept to the mining industry. Slim holes have been drilled to 18,000–20,000 ft (5,500–6,100 m) for many years. Small rigs, purpose-built for this application, have been used. Also, special pipe and tools have been developed and used. Fortunately, most of the applications have not experienced well control problems.

Well Control Parameters

Well control in small or slim holes does not change if the constant bottom-hole pressure method is to be used. An alternative method will be discussed later. Chapter 2 gives a detailed explanation of well control with examples for small holes.

Slim holes utilize the same basic well control technology as larger holes. An invading fluid displaces drilling fluid. This is seen at the surface by increases in flow rate and pit volume. The driller should respond by

Special Problems

closing the well. Mud weight calculations are made. The new drilling fluid is pumped into the well, and the invading influx is removed.

Two recognized parameters that cause concern in slim hole well control are (1) the annular geometry and volume displacement and (2) the annular friction losses. The annular geometry is reduced in slim holes. A small influx of fluid causes a large vertical mud column removal from the annulus into the rig pits. Since this mud provides the hydrostatic pressure controlling the well, the flow rate into the well can increase substantially quicker than in larger holes. Also, the shut-in pressures will be greater than for an equivalent influx volume in a larger hole. Surface response time is reduced before the well unloads the mud. Lack of attention by the drill crew for a short time can allow complete hole evacuation.

The annular volumetric capacity per unit of vertical height results in higher casing pressures. This also results in more pressure on the casing seat. If the equivalent mud weight caused by the kick exceeds the fracture limit, lost circulation occurs.

Annular friction loss is a key element in the discussion of slim hole well control. The pressure losses are a function of several variables including pump rate, mud properties, and hole geometry. As the hole size is reduced, pressure losses increase dramatically for equivalent pump rates. Mathematical examples are provided in Chapter 2. The annular pressure loss is a function of hole geometry raised to a power of approximately 2. If the hole geometry is reduced, the square root of the typical large hole pump rate must be used for drilling and well control if the pressure loss is to be the same as seen in larger holes. This is not done as a practical matter, so the industry must learn to work with higher annular friction losses.

Pressure losses are manifested in several ways in well control. High pressure losses during drilling can mask a hydrostatic pressure underbalance. The well can flow when the pumps are shut off if a permeable zone is exposed. During kick circulation, the pressure losses are added to the casing pressure to give a high equivalent mud weight as in the case of certain pump startup procedures for well control kick circulation.

Well control in slim holes does not change to account for these two issues. However, more attention must be given to the well control and shut-in procedures to avoid gaining large fluid volumes during the initial kick or causing lost circulation with annular friction pressures. Accurate rig site instrumentation capable of detecting small changes is a requirement to adequately handle slim hole well control matters. However, some small rigs may not have this instrumentation.

Most operators recognize these differences between well control in large and small diameter holes. They do not utilize any special or new procedures. They have shown that proper implementation of existing technology (CBHP) is satisfactory.

Alternate Well Control Procedures

Amoco was the first operator in the late 1980s to promote an alternate form of well control for slim holes. The techniques were developed from the SHADS research that investigated slim holes in the 3–6 in. OD (76–152 mm) range. The ultimate objective was to find a low cost means to investigate stratigraphic plays with smaller rigs. They started their research by studying mining (coring) rigs. They used a small rig at an Oklahoma test site to do actual well testing with nitrogen kicks.

They recognized the impact of the parameters discussed in the prior section. Amoco developed a method using the dynamic kill procedure for handling slim hole kicks. Their system was based on computer monitoring and modeling of well activities and implementation of kick control procedures. Also, flow monitoring was done with special magnetic flow meters. The proposed final form of the system seems quite capable of handling the problems. However, it requires proven computer hardware and software and highly reliable flow sensors not typically available on most rigs.

The industry has not yet universally accepted the dynamic kill as a means to control slim hole kicks. When computer monitoring and handling of rig site operations advances to a more refined stage, the dynamic kill method may receive much wider usage.

PROBLEMS

3.1   What amount of mud should be allowed to escape from a well during a stripping operation for each 93 ft stand of 3 1/2 in. OD, 15.5 lb/ft drill pipe?

3.2   If 7 in. OD × 2.5 in. ID drill collars are to be snubbed into a well, what amount of mud should escape for each foot of collars forced into the well?

3.3   A kick was shut in after the complete drill string was removed from the well. Using the following data, what would be the total volume of mud displaced by the drill string when it reaches the bottom of the hole?

Drill collars = 6 in. OD × 2 in. ID, 1,200 ft

Drill pipe = 4 1/2 in. OD, 20.0 lb/ft

Well depth = 13,600 ft

Special Problems

3.4* A kick occurs when the bit is at 4,500 ft. In order to return to bottom, will it be necessary to strip or snub? (Assume no annular preventer frictional forces in the following problem.)

SICP = 4,500 psi

Drill pipe = 4 in. OD, 14.0 lb/ft, 3,600 ft

Drill collars = 6 in. OD × 2 in. ID

Mud weight = 13.0 ppg

3.5* In Problem 3.4, what would be the "Snub point," or the depth at which the summation of vertical string forces is zero?

3.6* Using the following data, is stripping or snubbing required if the SICP is 1,500 psi? 3,000 psi? 6,000 psi?

Drill pipe = 4 1/2 in. OD, 16.6 lb/ft, 6,400 ft

Drill collars = 6.5 in. OD × 2 in. ID, 900 ft

Mud weight = 14.5 ppg

3.7 What is the maximum surface pressure during a drill stem test under the following conditions?

Formation pressure = 11.5 ppg (equivalent)

Depth = 8,600 ft

Formation fluid = Oil, 0.6 specific gravity

3.8 Using the conditions given in Problem 3.7, what would be the surface pressure if the formation fluid had a gradient of 0.2 psi/ft?

3.9 In many North Sea areas, regulatory agencies require a gas gradient of 0.10 psi/ft in all well planning. Using this design criterion, what API pressure rating equipment must be used in a drill stem test under the following conditions?

Formation pressure = 16.6 ppg (equivalent)

Depth = 15,800 ft

3.10 During a kick killing operation, kill mud reached the surface after 800 strokes had been pumped. Using the following data, what is the deepest possible depth of the washout?

Pump = 5 in. × 18 in. duplex (95% efficiency)

Drill pipe = 4 1/2 in., 16.6 lb/ft, 11,600 ft

Drill collars = 6 in. × 2 in., 1,000 ft

Casing = 10 3/4 in., 54.0 lb/ft, 9,500 ft

Open hole = 8 3/8 in., 3,100 ft

3.11 Using the data from Problem 3.10, what is the deepest possible depth if 5,750 strokes had been pumped?

3.12 Rework Problem 3.11 using a 6 in. × 14 in. duplex pump at 95% efficiency.

3.13* Warning signs of a washout were noticed during a kill operation after the drill pipe had been displaced with mud. A piece of rope was pumped down the drill pipe in an effort to determine the depth of the hole. A pump pressure increase was noted after 260 strokes were pumped. Using the following data, what was the approximate depth of the washout?

Pump = 5.5 in. × 18 in. duplex (95% efficiency)

Drill pipe = 4 in., 14.0 lb/ft, 13,500 ft

SOLUTIONS

Note: All solutions are approximate depending on user's round off procedures.

3.1 1.182 bbl (~1.2 bbl)

3.2 0.048 bbl/ft

3.3 302.4 bbl

3.4 Strip

3.5 Snub until collars enter the hole then strip.

3.6 Strip, strip, strip

3.7 2,908 psi

3.8 3,423 psi

3.9 15,000 psi stack

3.10 ~1,100 ft

3.11 ~7,900 ft

3.12 ~9,550 ft

3.13 ~3,600 ft

REFERENCES

"Accidents Connected with Federal Oil and Gas Operations on the Outer Continental Shelf Gulf of Mexico," Vol. 1, 1956–1979, U.S. Geological Survey Conservation Division, December 1979.

"Accidents Connected with Federal Oil and Gas Operations on the Outer Continental Shelf Gulf of Mexico OCS Region," Vol. 2, January, 1980–December, 1984, U.S. Department of the Interior/Minerals Management Service Gulf of Mexico OCS Region, June 1985.

Adams, N.J. "Deepwater Poses Unique Well Kick Problems," *Petroleum Engineer*, May 1977.

Adams, N.J. *Drilling Engineering: A Well Planning Approach*, Tulsa: PennWell Publishing Co., 1985.

Adams, N. *Well Control Problems and Solutions*, Tulsa: Petroleum Publishing Company, 1980.

Adams N.J. and L.G. Kuhlman. "A Discussion on Casing Settling During Shallow Gas Blowouts," SPE/IADC 27502, SPE/IADC Drilling Conference, Houston, Texas, 15–18 February 1994.

Adams, N.J. and L.G. Kuhlman. "Shallow Gas Blowout Kill Operations," SPE/IADC 21455, SPE Middle East Oil Show, Bahrain, 16–19 November 1991.

Adams, N.J. and L.G. Kuhlman. "How to prevent or minimize shallow gas blowouts–Part 1," *World Oil*, May 1991, pp. 51–58.

Adams, N.J. and L.G. Kuhlman. "How to prevent or minimize shallow gas blowouts–Part 2," *World Oil*, June 1991, pp. 66–71.

Adams, N.J. "What to Remember About Bullheading," *World Oil*, March 1988, pp. 46–48.

Adams, N.J. *Workover Well Control*, Tulsa: PennWell Publishing Co., 1981.

Adams, N.J. and D. Cater. "Hydrogen Sulfide in the Drilling Industry," 1979 Deep Drilling Symposium, Amarillo, TX.

Adams, N.J. and L.G. Kuhlman. "Case History Analyses of Shallow Gas Blowouts," SPE/IADC 19917, SPE/IADC Drilling Conference, 27 February–2 March, 1990.

Adams, N.J., B. Hansen, A.D. Stone, J. Voisin, and S. Clement. "A Case History of Underwater Wild Well Capping-Successful Implementation of New Technology on the SLB-5-4X Blowout in Lake Maracaibo, Venezuela," SPE 16673, 62nd Annual Technical Conference, Dallas, 27–30 September 1987.

Amoco Chemical Corp. Bulletin WT-1A, "Sour Corrosion and its Prevention."

Beall, Joe, "Riserless Shallow Blowout-Control Method Is Safe and Effective," *Oil & Gas Journal*, 2 August 1976, p. 125.

Beck, F.E., J.P. Langlinais, and A.T. Bourgoyne, Jr. "An Analysis of the Design Loads Placed on a Well by a Diverter System," SPE/IADC 16129, SPE/IADC Drilling Conference, 15–18 March 1987.

Beirute, R. and A. Tragesser. "Expansive-Shrinkage Characteristics of Cements Under Actual Well Conditions," SPE 4091, *Journal of Petroleum Technology*, August 1973.

"Blowout Database Offshore Blowouts," Part 1, 1955–1974, Trondheim, Norway: Marintek Sintef-Gruppen.

"Blowout Database Offshore Blowouts," Part 2, 1975–1985, Trondheim, Norway: Marintek Sintef-Gruppen.

Bourgoyne, A.T.; W.R. Holden, R.R. Desbrandes, J.P. Langlinais, and W.R. Whitehead. "Final Report—A Study of Improved Blowout Prevention Systems for Offshore Drilling Operations," U.S. Department of the Interior/ Minerals Management Service, February 1986.

Carter, G. and K. Slagle. "A Study of Completion Practices to Minimize Gas Communication," *Journal of Petroleum Technology*, September 1972.

Cook, C. and L.G. Carter. "Gas Leakage Associated With Static Cement," Drilling-DCW, March 1976.

Davies, R.M. and G.I. Taylor. "The Mechanics of Large Bubbles Rising Through Extended Liquids and Through Liquids in Tubes," Proc. Royal Soc., 200, Series A, London, 1950, pp. 375–390.

Eaton, B.A. "Fracture Gradient Prediction and Its Application in Oilfield Operations," *SPE Journal*, October 1969, p. 1353.

Fagerjord, O. "Worldwide Offshore Accident Databank," *Veritec Marine Technology Journal*, Noroil Publishing House, Ltd. A/S, 1986.

Fanneløp, T.K. and K. Sjøen. "Hydrodynamics of Underwater Blowouts," New York: American Institute of Aeronautics and Astronautics, 14–16 January 1980.

Fosdick, M.R. "Compilation of Blowout Data from Southeast U.S. Gulf of Mexico Area Wells," The University of Texas at Austin, College of Engineering, August 1980.

Garcia, J.A. and C.R. Clark. "An Investigation of Annular Gas Flow Following Cementing Operations," SPE 5701, January 1976.

Garrett, R.L. "A New Field Method for the Quantitative Determination of Sulfides in Water-Base Drilling Fluids," *Journal of Petroleum Technology*, September 1977, pp. 1195–1202.

Grinrød, M., O. Haaland, and B. Ellingsen. "A Shallow Gas Research Program," IADC/SPE 17256, IADC/SPE Drilling Conference, 28 February–2 March 1988.

Hadden, D.M. "A System for Continuous On-Site Measurement of Sulfides in Water-Base Drilling Muds," SPE 6664, 1977 Sour Gas Symposium, Tyler, Texas.

Higdon, A., E.H. Ohlson, W.B. Steles, J.A. Wesse, and W.F. Riley. *Mechanics of Materials*, New York: John Wiley and Sons, Inc., 1976.

Hughes, Virginia M. "Reducing Blowout Incidents Through a Computer Assisted Analysis of Trends Among Gulf Coast Blowouts," Report No. UT 86-3, Texas Petroleum Research Committee, University Division, August 1986.

Ilfrey, Alexander, Neath, Tannich, and Eckel. "Circulating Out Gas Kicks in Deepwater Floating Drilling Operations," SPE 6834, October 1977.

"Joint Industry Program for Subsea Blowout Control Capability," Phase I Report, managed by Adams Engineering Inc., Houston, 1986.

Kelly, O.A., A.T. Bourgoyne, Jr., and W.R. Holden. "A Computer Assisted Well Control Safety System for Deep Ocean Well Control," International Well

Control Symposium/Workshop, Baton Rouge, Louisiana, 27–29 November 1989.

Kendall, H.A. "Why You Lost Your Offshore Rig," *Ocean Engineering*, February 1977.

Koederitz, W.L., F.E. Beck, J.P. Langlinais, and A.T. Bourgoyne, Jr. "Method for Determining the Feasibility of Dynamic Kill of Shallow Gas Flows," SPE 16691, SPE Fall Technical Conference, 27–30 September 1987.

Løes, M. and T.K. Fanneløp. "Concentration Measurements Above an Underwater Release of Natural Gas," *SPE Drilling Engineering*, 1989.

Matthews, W.R., and J. Kelly. "How to Predict Formation Pressure and Fracture Gradient," *Oil & Gas Journal*, 20 February 1967.

Milgram, J.H. and R.J. Van Houten. "Plumes from Blowouts and Broken Gas Pipelines," MIT Report No. 82-7, 1982.

Milgram, J.H. "The Response of Floating Platforms," Massachusetts Institute of Technology, Department of Ocean Engineering Report No. 82-8, July 1982.

NIOSH Publication No. 77-158, "Occupation Exposure to Hydrogen Sulfide," May 1977.

NL Baroid BTC Laboratory Report # STB-461.

Patton, C.C. "Corrosion Control in Drilling," *Oil & Gas Journal*, 29 July 1974.

Podio, A.L., R. Fosdick, and J. Mills. "Analysis Gives Blowout Causes, Trends, Cost," *Oil & Gas Journal*, 7 November 1983.

Raymond, L.R. "Temperature Distribution in a Circulating Drilling Fluid," *Journal of Petroleum Technology*, March 1969.

"Recommended Practices for Well Control Operations," API RP 59, 1st ed., 20 August 1987.

Redmann, K.P., Jr. "Understanding Kick Tolerance and Its Significance in Drilling Planning and Execution," *SPE Drilling Engineering*, December 1991.

Rehm, B.: "Deepwater Drilling Poses Special Pressure-Control Problems," *Oil & Gas Journal*, 3 May 1976.

Reynolds, D.: "How to Cope with Sulfide Corrosion," *Drilling DCW*, June 1976.

Sandlin, C.W. "Drilling Safely Offshore in Shallow Gas Areas," SPE 15897, 1986.

Santos, O.L.A., H.R. de Paula Lima, and A.T. Bourgoyne, Jr. "An Analysis of Gas Kick Removal from the Marine Riser," SPE/IADC 21968, 1991 SPE/IADC Drilling Conference, Amsterdam, 11–14 March 1991.

Tarvin, J.A., I. Walton, P. Wand, and D.B. White. "Analysis of a Gas Kick Taken in a Deep Well Drilled With Oil-Based Mud," SPE paper 22560, presented at the SPE 66th Annual Technical Conference, Dallas, TX, 6–9 October 1991.

Tannich, J.D., C.W. Sandlin, and A.N. Gist. "Well Killing: Possibilities and Limitations," Shallow Gas Seminar, Norwegian Petroleum Directorate, 27–28 August 1987.

Tucker, J., L. Nunenmacher, and H. Williamson. "Shallow Gas Events," United States Department of the Interior/Minerals Management Service, 1985.

"Veritas Claims Diverters Fail in 50% of Blowouts," *Offshore*, November 1986, p. 54.

Westergaard, R.H. "All About Blowouts," *Norwegian Oil Review*, 1985.

Wilder, O. "Handling Drill Pipe Under Pressure," *Gulf Coast School of Drilling Practices*, October 1977.

CHAPTER 4

• • • • • • • • • • • •

BLOWOUTS

A "blowout" is an uncontrolled flow of formation fluids. The flow may be to an exposed formation and termed an underground blowout, or it may flow uncontrolled at the surface. Regardless of the type, remedial control procedures can be expensive, often difficult to implement, and not always successful. This chapter discusses the following:

- Contingency planning for blowouts
- Surface control (capping) operations
- Relief well drilling
- Underground blowouts

Numerous blowout problems are not considered in conventional well control. A primary concern is environmental protection (considering governmental supervision and possible intervention). A surface blowout can release large volumes of potentially dangerous formation fluids. This includes oil, gas, and salt water, but it may also include hydrogen sulfide. In the case of toxic gases, the safety of human life becomes a serious and potentially paramount consideration.

Loss of hydrocarbon reserves is another problem associated with blowouts. In a field case in Sumatra, the reservoir was reported to have lost over 450 MMcfd (13×10^6 m^3/D) for approximately three months before control could be regained. Although this example is abnormal, blowouts can cause reservoir depletion and/or productivity impairment to a point where the zones are no longer commercial.

The drilling rig equipment can be destroyed during a blowout. Equipment may be lost by fire and/or cratering around the rig. Figure 4–1 shows a large rig after a blowout.

Blowout cost factors can be large, although seldom given primary consideration during the actual event. Cost variables include losses of equipment and hydrocarbons, personnel and equipment costs incurred in

FIGURE 4–1 • Damaged rig after blowout

regaining control of the well, and the expenses involved in protection of the surrounding areas. A recent blowout in the North Sea is reported to have cost over $250 million before control could be regained.

BLOWOUT CONTINGENCY PLANNING

Overview of the Blowout Contingency Plan

The blowout contingency plan should be part of an overall emergency plan, which is a guide giving directions and procedures required for handling any type of emergency, regardless of the nature. The emergency plan could be used in conjunction with blowout plans, hurricane or typhoon contingency planning, fire control operations, or other similar disaster situations.

A blowout contingency plan should contain directives for handling most aspects of blowout management, personnel assignments and responsibilities, and selection of specialists for specific assignments. It may contain specialized equipment requirements.

Blowout contingency plans depend on the operator's personnel and management structures. Since these differ among companies, it is not common to find identical plans for several companies. It should be customized to utilize specific personnel and management, and it must be designed to handle the most likely disasters and the surrounding conditions.

The contingency planning described herein should be used as a guide. Each company must weigh alternatives and make selections to form their unique contingency plan.

This discussion identifies topics found in contingency plans including the following:

- Management structure
- Stage 1–Disaster Early Response
- Stage 2–Disaster Containment
- Stage 3–Disaster Control
- Blowout data and information requirements
- Blowout capping equipment (basic list)
- Specialist third party personnel
- Kill technique selection

The format for the contingency plan will vary among operators. Also, it may be prepared as a written document and/or computerized.

Relief well contingency planning is a subset of blowout control contingency planning (Stage 3). Relief well plans can be large and are often prepared as a separate document. This is due to the fact that most information is available on a preliminary basis to prepare the relief well plan. Capping operations cannot be preplanned to the same level of detail as relief wells, since capping is dependent on blowout conditions and the resulting damage. Relief well contingency planning will be described briefly.

Blowout management can be separated into several stages. A possible division is as follows:

- Stage 1–Early response
- Stage 2–Blowout containment
- Stage 3–Blowout control

The "early response" phase should be predetermined operations implemented without significant influence by characteristics of the blowout. These operations are executed by the operator without assistance of specialist companies.

"Blowout containment" can be described as operations designed to mitigate or reduce maximum possible damage from the blowout. The time period for initiating containment operations is short. The blowout will often do most damage within one to two hours. Operations must be assessed and implemented within this time frame to mitigate substantial damage. This phase of operations is influenced by the type and severity of blowout. On-site decisions must be made. Preplanning and drills can assist the operator. The responsibility for this phase rests on the operator because third-party blowout specialists will not normally be on-site at this time. Some interim advice may be gained by telephone conversations with blowout specialists.

"Blowout control" is usually implemented with the assistance of specialists. An on-site inspection is required before planning and operations can start. The specialist will require some information about the well before making recommendations.

A training program for company personnel should be defined and presented in the contingency plan. The training may be restricted to developing an understanding of the contingency plan, or it can be extended to cover advanced well control topics designed to develop a knowledgeable group of internal company specialists. A periodic review of job functions will refresh personnel with their roles and task requirements.

Some companies organize their management so a team of in-house well control specialists can take an active part in operations. In some cases, this includes capping specialists and relief well engineers. An advanced well control school designed for these personnel is appropriate.

Management Structure

The company management structure defines the organizational chain of command for the operator in charge of the well. Groups reporting to the operations manager may include the following:

- Capping operations
- Relief well operations
- Operations support/contractor personnel
- News media interface
- Regulatory interface
- Logistics
- Insurance adjusters
- Pollution control

Figure 4-2 shows the relationship. Several organizational structures will usually be required to handle various tasks. Sublevels of management structures should be identified.

Each position must be identified in terms of its responsibility, personnel filling the position, and back-up (relief) personnel.

The operations manager is a key position. Several philosophies exist about the individual assigned to this post.

FIGURE 4-2 • Organization chart with blowout advisor to the operator

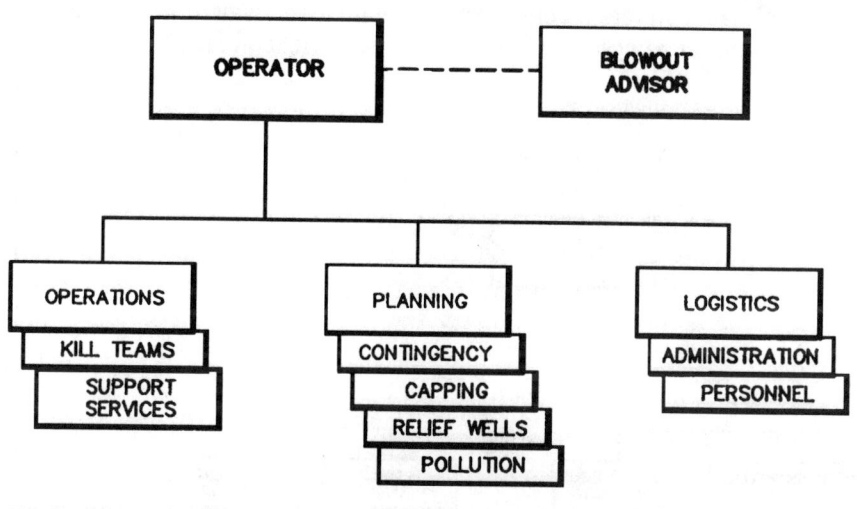

- The operations manager should be from outside the district. The existing district manager can continue his normal duties. The (blowout) operations manager can perform his duties without the burden of blame and fault judgment that often accompanies disasters.
- The operations manager should be from within the district. He is knowledgeable of previous conditions, current operations structure, and suppliers.

A good option is an operations manager currently assigned outside the district but who formerly served in a top role within the district.

Contractor Interface

The recommended contractor-operator structure is shown in Figure 4–2. Other options for relating to blowout specialists are shown in Figures 4–3 and 4–4. Some blowout specialists historically have taken charge of all operations and given few options to the operator. This structure is changing as operating personnel become more knowledgeable about blowout control and disaster management.

Command Center

A command center is the focal point for operations on the well, third-party relations, and contact with parties outside immediate involvement with the well. The command center should be located away from the well site so it will not be affected by the well's noise or heat and traffic to the well.

FIGURE 4–3 • Typical capping job

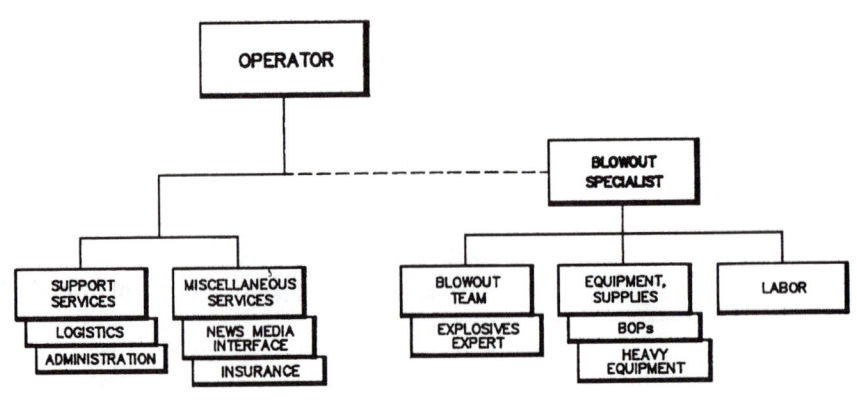

FIGURE 4-4 • Typical relief well

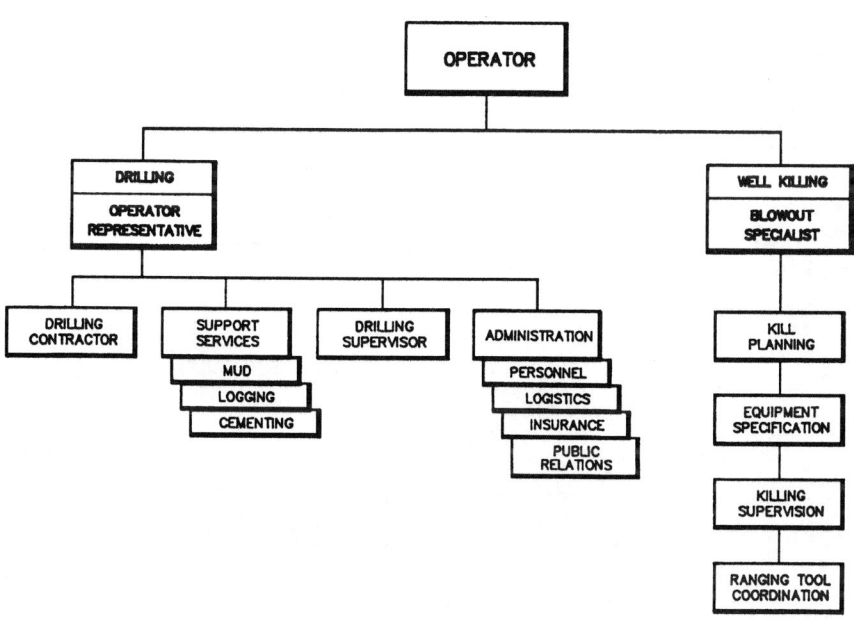

The center should be located within a reasonable distance to the well so operations personnel can easily commute to the well as necessary.

The facility should contain the following:

- Telecommunications including telephone, radios, fax, and telex links. Multiple telephone lines should be available.
- Meeting rooms including a conference room.
- A computer system with all appropriate software for management, word processing, contingency planning, relief well drilling, and well killing.
- Resting/sleeping accommodations for two to four persons.
- Audio-visual equipment including TV, VCR, overhead vugraph projector, video recorder, camera, and marking boards.
- Multiple copies of the site plan, local maps, telephone lists, blowout contingency plan, and well records.
- Reference material such as applicable government regulations, emergency equipment lists, photographs and videos of the site before the blowout, etc.
- Rig layout drawings (particularly for offshore installations), emergency equipment layouts, firefighting equipment layout, power generation system information, tank configuration and operating

manual for semisubmersibles, jacking system and operating manual information for jack-ups, and details of major systems and structures for platforms.

Drills (Exercises)

Regular drills should be conducted to maintain operational readiness of personnel. The drills should be scheduled so crews can plan and organize their actions before the event. Also, unannounced drills are necessary to improve sudden action responses. Field and office drills should be conducted.

It is beneficial to plan the exercises in conjunction with a blowout specialist. He can offer real situations from his experiences and prompt company personnel on recommended actions. Drills simulating actual conditions are valuable.

Stage 1—Disaster Early Response

Early response involves a set of preplanned operations that are almost independent of blowout characteristics. Some operations include the following:

- Issue an SOS if offshore.
- Implement an ESD (emergency shut-down) if on an offshore rig, or shut-down all engines on a land rig.
- Move the rig off location if rig is floater, water depth is less than 200 ft (60 m), flow is violent, flow has ignited, and/or flow has high percentage of H_2S.
- Evacuate rig personnel.
- Evacuate the surrounding area if on land.
- Consider igniting an H_2S well.
- Initiate the notification process.
- Take appropriate blowout control actions: shut in the BOPs, etc.

Offshore rig evacuation is a serious event. Fatalities have occurred during evacuation where none were recorded by the actual blowout. The evacuation decision can be based on a preestablished set of events that would trigger the decision. The on-site decision is often complicated by noise, confusion, and stress. Realistic drills are valuable aids to the decision makers.

An H_2S well generally must be ignited if concentrations are lethal. Flares are usually used. Ignition can not be done until personnel are evacuated. Assigned crew members should practice occasionally with the guns. Flares should be replaced on a periodic basis.

BLOWOUTS

Well ignition can be difficult in some situations. The criteria for ignition and practical ignition methods should be established. This should be recognized by crew members. This information on hydrogen sulfide ignition should be defined in the contingency plan.

Ignition is a terminal event—the rig will likely be a total loss. In case of offshore H_2S blowouts away from populated areas and remote from any other rigs or platforms, it will probably be prudent not to ignite the flow immediately, if at all. Conversely, an H_2S land well blowout near a town should probably be ignited as soon as the crew has safely evacuated the rig. In any case, the prescribed action can be thoroughly considered and preplanned during blowout contingency planning and training drills. The final decision should be left to the on-site person-in-charge after he has evaluated existing conditions.

Notification of the blowout to certain parties should begin as soon as possible. This can include the following:

- Onshore personnel
- Company management
- Evacuation support and medical treatment services
- Local fire brigade
- Regulatory authorities
- Coast Guard personnel (if offshore blowout)
- Local law enforcement agency

The list is defined by the well's location, type of well, and company structure.

Maintenance of the communications section of the blowout contingency plan requires significant effort. Personnel in key positions will change on a relatively frequent basis. Also, phone, fax and telex numbers will change. This task is well suited for computerization where a single change can be accessed immediately by all concerned parties.

Stage 2—Blowout Containment

Blowout containment is an effort to mitigate damage from the blowout. Typical containment results might include the following:

- Prevent blowout ignition.
- Confine the damage, fire, and so forth, to one section of the platform or rig.
- Channel or pool oil spills, or deploy containment devices.
- Safely remove rig equipment, if practical.

A list of operations designed to contain damage is extensive and depends on well and rig conditions. Typical operations may include some of the following. Sequence of operations must be specified for each blowout scenario.

- Spray water if the well has not ignited.
- Shut down all power sources.
- Shear the pipe and shut in the well.
- Inject water into the wellhead to reduce erosion on the BOP or lines and reduce chances of ignition.
- Flare to an overboard line.
- Deploy booms and skimmers for offshore events, or build containment dikes and establish flow channels for oil to avoid location flooding on land sites.
- Implement "active bridging" techniques.
- Position a tool joint immediately above the rig floor.
- Close and lock pipe/casing rams to prevent pipe from being blown out of the well.
- "Drive off" the location if the well is blowing out through the riser on a floater.
- Ignite an oil spill if possible or practical.

Stage 3—Blowout Control

Blowout control efforts may require a few hours to many months. The options are usually separated into surface and subsurface operations, commonly called "capping" and "relief well drilling," respectively.

The starting point to develop a contingency plan for blowout control is a definition of worst case scenarios that might reasonably be encountered. These might be defined as follows:

- High control difficulty to include a multiwell platform fire or a shallow gas blowout with possible rig or platform fire
- High environmental impact such as an oil production platform blowing out and collapsing below the water line or a subsea oil well blowout

A blowout database is beneficial to assist in developing worst case scenarios that will coincide with the type of production the operator currently has offshore.

Blowout control services can be classified as follows:

- Direct services related to capping hands-on work or relief well planning and supervision.
- Support services for the direct operations including pumping, equipment fabrication, well drilling, engineering, pollution control, transport/logistics, etc.

- Associated services are necessary but not linked to direct services. These can include news media interfacing, working with regulatory authorities, preparing insurance claims, etc.
- Required support services are dependent on kill technique selection.

Blowout Data and Information Requirements

Data and information should be provided for blowout specialists to select and implement the kill techniques. A sample list is included here. Some situations may require other data.

1. Well name/number/identification, drilling contractor and rig names
2. Well location (onshore/offshore), directions and arrangements to travel to the well
3. Maps of the area, data on distances to available transport centers (e.g., airports, supply bases, boat docks, etc.)
4. Type of blowout—shallow gas, hydrocarbons (moderate to deep), geothermal, etc.
5. Well type—exploration, producing, workover, etc.
6. Probable cause(s) of the blowout
7. Status at the time of the blowout:
   - Date and time of the blowout
   - Total depth
   - Hole configuration
   - Bit location
   - Casing/liner sizes, setting depths, liner top
   - Drill string, bottom-hole assembly, tubing, and other downhole equipment sizes, and configuration
   - Mud type/weight/properties
   - Operation in progress (activities, timing, pressures, etc.)
   - Type, description, rating, and drawings (if available) for rig/snubbing unit/workover unit, BOP, wellhead equipment, pumps, and other surface equipment
   - Water depth
8. Status at the time blowout specialists are called:
   - On fire or only blowing?
   - Approximate flow rate (if known).
   - Fluid type—oil, water, natural gas, H_2S, CO_2, etc.
   - Fire/blowing stream description and configuration
     - Cratered?
     - Single point?
     - Vertical flow/flame? Height?
     - Horizontal flow/flame? Length/direction?

- Smoke color
- Description of probable leak locations
– Ground fires (land well)
– Pollution situation
– Wellhead/BOP/xmas tree damage
– Description and configuration of equipment or structures around the blowing well
– Firefighting boats or other intervention equipment in the area
– Weather conditions–current and forecast
9. Operator/drilling contractor/service personnel–names, assignments, authority, etc., under any blowout contingency plan in place
10. Pertinent reservoir/geological data
11. Well records–daily drilling/production reports, logs, surveys, etc.

Blowout Capping Equipment

A blowout contingency plan may include basic capping and firefighting equipment. The list should include equipment that will handle most capping and firefighting jobs on land. Some will be applicable offshore. Operator's personnel should have a basic knowledge of its design and purpose.

Specialists

Various specialists will be required to handle a blowout job. A few or many may be required depending on the blowout. Some are as follows:

- Firefighting and capping
- Relief well planning, drilling, and supervision
- Blowout engineering and technology development
- Pollution control
- Insurance adjusters
- Media interface
- Regulatory authorities interface

A description of each is given below.

Each commercial blowout company offers a range of services. Some are extensive while others are more limited. As an example, several companies offer well capping and firefighting services exclusively, while others may offer capping, firefighting, relief well operations, and blowout engineering services within the same company. It is wise to visit with each company and become familiar with its services.

Contract terms should be discussed with the specialists prior to a blowout. Fees and terms of work should be defined. The operator is at a commercial disadvantage if required to call a specialist after the blowout has occurred.

BLOWOUTS

A list of approved specialists should be given in the blowout contingency plan. It should contain telephone, telex, and fax numbers. Preferred contact personnel should be included.

Firefighting and Capping Specialists

Firefighting and capping specialists provide surface kill capability. They usually can provide equipment and experienced personnel on a call-out basis. The equipment stocked by most blowout specialists is designed for land wells.

Relief Well Planning, Drilling, and Supervision

Contrary to some industry opinions, relief well drilling is substantially more complicated than routine directional drilling. The specialists in this area should be experienced, have a working knowledge of ranging tool theory, understand kill dynamics, and have access to specialized kill software. The optimum approach is to have the relief well supervised by the same group that plans the relief well.

Blowout Engineering and Technology

Some jobs require unique solutions not previously implemented. Examples include subsea snubbing and operations of a semisubmersible rig over a live blowout. Few specialists provide these services. Their capabilities should be explored and identified in the contingency plan. An ideal situation occurs if the developers of the advanced technology can also supervise the operations.

Pollution Control

Control of spilled oil requires a separate contingency plan. Specialists should be identified.

Subsea blowout pollution control requires unique and novel solutions. This approach is particularly applicable in a rough sea environment where surface containment is difficult.

Insurance Adjusters

Most operating companies maintain insurance coverage for blowouts. The insurance underwriters contract the services of third-party adjusters to settle claims on a blowout event. Some adjusting companies will provide an on-site representative to monitor operations during control efforts.

Effective interfacing with the adjusters usually expedites claims payments. The adjusters are ultimately responsible to the underwriters to provide an accurate claims assessment. An "open door" policy between the operator and adjuster usually assists in a quicker settlement.

News Media Interface

Although it may seem initially counterproductive, it is important to provide immediate and direct contact with news media representatives. The news staff have a perceived obligation to report news events to the public. Some situations have occurred where the reported events were not completely correct concerning the actual facts. Experiences have shown that formal news releases with factual content is the best approach to containing adverse publicity.

The operator's news media contact needs several qualifications. He should be knowledgeable of news media requirements, be technically competent, and be an articulate speaker. It is recommended to provide a written statement daily.

Regulatory Authorities Interface

The operator should provide a contact person to interface with governing groups. Regulatory groups are required by law to monitor blowouts. The level of monitoring varies with each agency. Most groups have some degree of legislated power to interact with the operator. Some have the ultimate power to shut down operations if they believe prudent actions are not being carried out.

Kill Technique Selection

It is a fairly obvious observation that a blowout should be controlled with the optimum approach. However, history of blowout control efforts shows that optimum approaches are not always used. In some cases, inappropriate techniques have been used that result in loss of the well or platform or that cost huge sums of money without yielding success. As a result, some operators are now taking control of the decision-making process away from firefighters and blowout specialists. Unfortunately, this seems the most appropriate action in many situations.

Factors constituting an "optimum approach" for kill technique selection include the following:

- Probability that the technique will work under the blowout conditions
- Time, cost, and logistical requirements for the technique
- Terminal nature of the technique
- Safety of personnel
- Comparison to other techniques

These factors warrant discussions.

Success Probability

An important question relates to the probability that the proposed technique will be successful under reasonable conditions. It is important that a strong differentiation be made between "probability of success under reasonable conditions" versus "technical possibility."

One suggestion involves using the "decision tree" approach to determine the best kill option. (See Figure 4-12.) This could result in a kill procedure that takes into account most variables. Advance preplanning is necessary for this approach. Unanticipated conditions and circumstances at the site must be considered in the decision tree process.

Time, Cost, and Logistical Requirements

For each possible kill technique, an evaluation must be made for time to complete the kill, cost, and logistical requirements. The time aspect relates to the point at which the well is safely killed or controlled.

Logistical requirements can be extensive in some situations. Remote locations can pose transportation problems. Movement of explosives can cause significant "red-tape."

If only one kill option exists for a blowout, the time and cost evaluation has little significance.

Cost is an important topic that should be discussed. The typical approach to cost considerations for most drilling wells is to get "best value for the money." In dealing with most aspects of blowout control, the recommended approach is to prioritize the best service available and then compare costs if the services are nearly equal. Real savings does not mean accepting the lowest bidder, but rather using the best service available that can safely do the required task effectively and efficiently.

Terminal Nature of the Technique

A proposed kill approach must be evaluated to determine if it could eliminate other options if unsuccessful (i.e., if it does not work, it terminates other options).

For example, suppose that a well is blowing out from 10,000 ft (3,000 m) in which only 2,000 ft (600 m) of surface casing is set. One kill option is to cap the well, close the BOPs, and bullhead into the well. If the formation fractures at 2,000 ft (600 m), will the well crater and eliminate other capping options? Is it a more prudent decision to take a more time consuming approach and cap the well, divert it through a flare line, and snub into the well?

If the casing is set to 7,500 ft (2,300 m) in this example, the decisions become much easier.

Safety of Personnel and Well/Facilities

Without understating the issue, personnel safety must always be the highest priority. During the final stages of an intense kill operation, it is easy to become "tunnel-visioned" on well control objectives and lose sight of personnel safety. For the well control specialist, the safety issue must always remain in the forefront of his operations.

Firefighters and blowout specialists are often involved in operations containing some risks. They are supposed to know how to handle these risks. However, other personnel involved with the killing operations often want to provide assistance, sometimes in a very eager manner. They usually do not understand the risks and related safety procedures. They can expose themselves to the danger of an accident. It is incumbent on the well control leader to be cognizant of this potential problem area.

Some kill methods are more hazardous than others. Personnel approaching a nonburning sour gas blowout are at significantly greater risk than drilling an offshore relief well for a deep intersect in 500 ft of water. Fire presents a different set of problems. Safety of the well and facilities are also key considerations, although not as high a priority as personnel safety. Actions that result in the destruction of a platform or a deep, expensive well can result in massive financial losses for the operator.

The safest kill technique is bridging, since the blowout is contained downhole. If bridging occurs near the surface, however, broaching around the surface casing can occur resulting in a crater.

Safety to personnel off location should also be considered. High volumes of sour gas, accumulations of combustible hydrocarbons, and large fires can pose a hazard to people working and living near the blowout. Panic and flight from the area during evacuation can also result in injury. The selection of a "quick" kill technique may be warranted in such a situation, even though it may have a lower probability of success than other techniques. This, of course, presumes that personnel directly involved in the kill are adequately protected.

Comparison to Other Techniques

Consideration must be given to all kill options prior to making a final decision on one approach. Recent situations have occurred in which one approach was followed against recommendations of other groups for alternative approaches that had significantly more technical merit and a definite safety advantage. The alternative approaches were not given due consideration. Ultimately, the initial approach resulted in failure and tremendous financial losses. The alternative solutions were finally used efficiently and effectively but only after major efforts were expended on a "brute force" initial approach. In summary, all options should be evaluated on an equal basis and then a decision made for a kill technique.

BLOWOUTS

The operator's staff must participate in these evaluations. They should be the experts with respect to drilling and reservoir conditions for the blowing well. Without their input, an inappropriate or less-than-optimum technique could be used. A decision tree, prepared by the blowout specialists and/or other team members, is suggested to allow the operator to conduct an informed comparison of the various kill techniques.

Description of Available Techniques

A variety of blowout kill techniques are available. Some are applicable only in certain situations, while others are more universally applicable. An example is capping and snubbing into a land well. This technique does not have easy applicability on underwater offshore blowouts. Relief wells can be used almost universally.

Kill techniques can be separated in two broad categories:

- Top kill techniques involve surface control methods such as well capping and subsequent bullheading or lubrication of mud.
- Bottom kills require that mud be circulated from the bottom to the top of the well.

Some require a combination of surface control and a bottom kill. An example is capping and diverting a well followed by snubbing pipe for a bottom kill.

Common kill techniques are as follows:

- Bridging
- Capping/shut-in
- Capping/diverting
- Surface stinger
- Vertical intervention
- Offset kill
- Relief wells

Each is briefly described in the following subsections.

Bridging

Many blowouts have been killed by well bridging. The formation around the wellbore collapses and seals the flow path. (See Figure 4–5.)

Typically, bridging occurs within 24 hours after the well blows out. This observation is confirmed by a computerized database of almost 1,000 blowouts. If the well does not bridge within 24 hours, it is likely to blow for an extended time or until it is killed. Bridging does occur, however, on wells after the 24-hour period in some situations. Technical reasons exist for the 24-hour bridging phenomenon. These involve near-wellbore

FIGURE 4-5 • Wellbore bridging

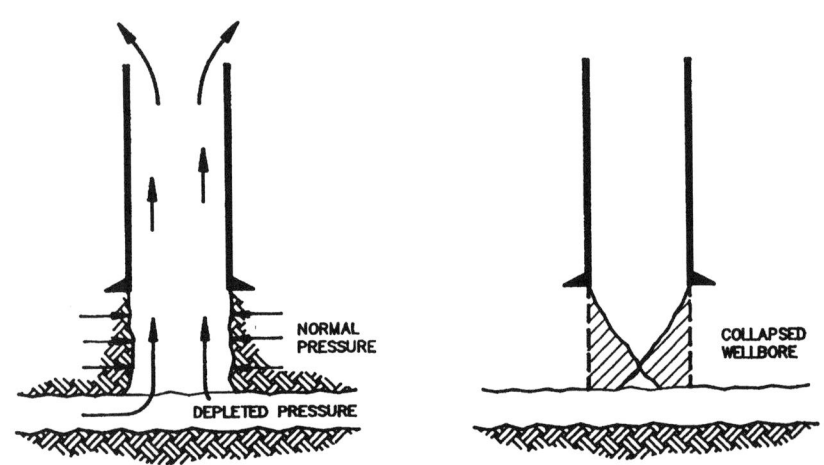

PRESSURE DRAWN DOWN IN THE BLOWING ZONE ALLOWS EXPOSED NORMAL ENVIRONMENTS TO COLLAPSE

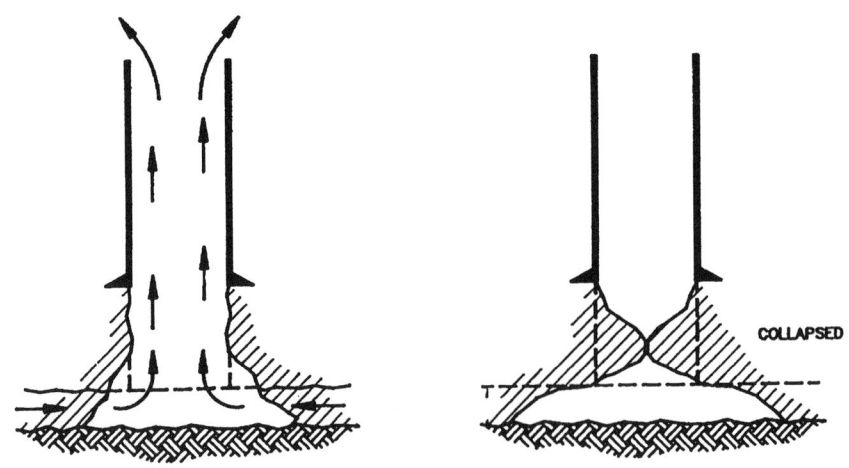

EROSION ALLOWS WELLBORE DESTABILIZING AND BRIDGING

pressure drawdown, erosion of wellhead and BOP components, and formation integrity under open flow conditions.

Bridging is typically considered a passive technique. The term "passive" means that it is subject to formation properties and generally is not influenced by kill attempts. In simple terms, the well bridges or it does not bridge, but no one has much control over it.

However, techniques are available for active bridging. Some firefighters and blowout specialists can implement techniques to accelerate the bridging. An active bridging technique involves opening the BOP/diverter stack or removing damaged, leaking wellhead component(s) to allow accelerated entry of reservoir fluids resulting in high annulus velocities and subsequent bridging.

Factors generally found in bridging situations include:

- Shallow casing strings
- Formation instability under drawdown situations
- Gas blowout fluids
- High flow rates
- Land or shallow water depths
- Long open hole sections

Also, salt water flows in deeper wells can cause the formation to become unstable and bridge after some time has elapsed.

Capping/Shut-in

"Capping" means, in simple terms, to put a cap on a blowing well. (See Figure 4–6.) Typically, this involves clearing debris, removing the old BOP stack and wellhead, installing a new wellhead and stack, then closing the BOPs.

If the well is shut in, access to a competent casing string is required. The casing string must have integrity and must be sufficiently deep to have a fracture gradient that will withstand shut-in conditions. Reservoir drawdown pressures should be evaluated and compared with the fracture gradient at the casing seat before the decision is made to shut in the well.

Considerations for capping and shutting in the blowout include the following:

- Access to a casing string with the necessary pressure rating.
- Fracture gradients sufficient to withstand shut-in pressures. Initial or drawdown pressures must be considered.
- Sufficient blowout flow rates for the fluids to extend some distance above the top of the casing or BOPs or the fire must be extinguished.
- If H_2S is present, the well generally must be capped on fire. All equipment must be H_2S serviceable.

Typically, a casing string is set deep to achieve the desired fracture gradient.

Several kill methods are commonly used for a capped well. Bullheading is probably most common. However, it requires initial fracture pressures to break down the formation. Historically, many firefighters have bullheaded with 18.0 ppg (2,160 kg/m³) mud or some mud weights sufficiently in excess of the level required to control the well. Many operators are

FIGURE 4–6 • Capping and shut-in on a blowout with a competent casing string

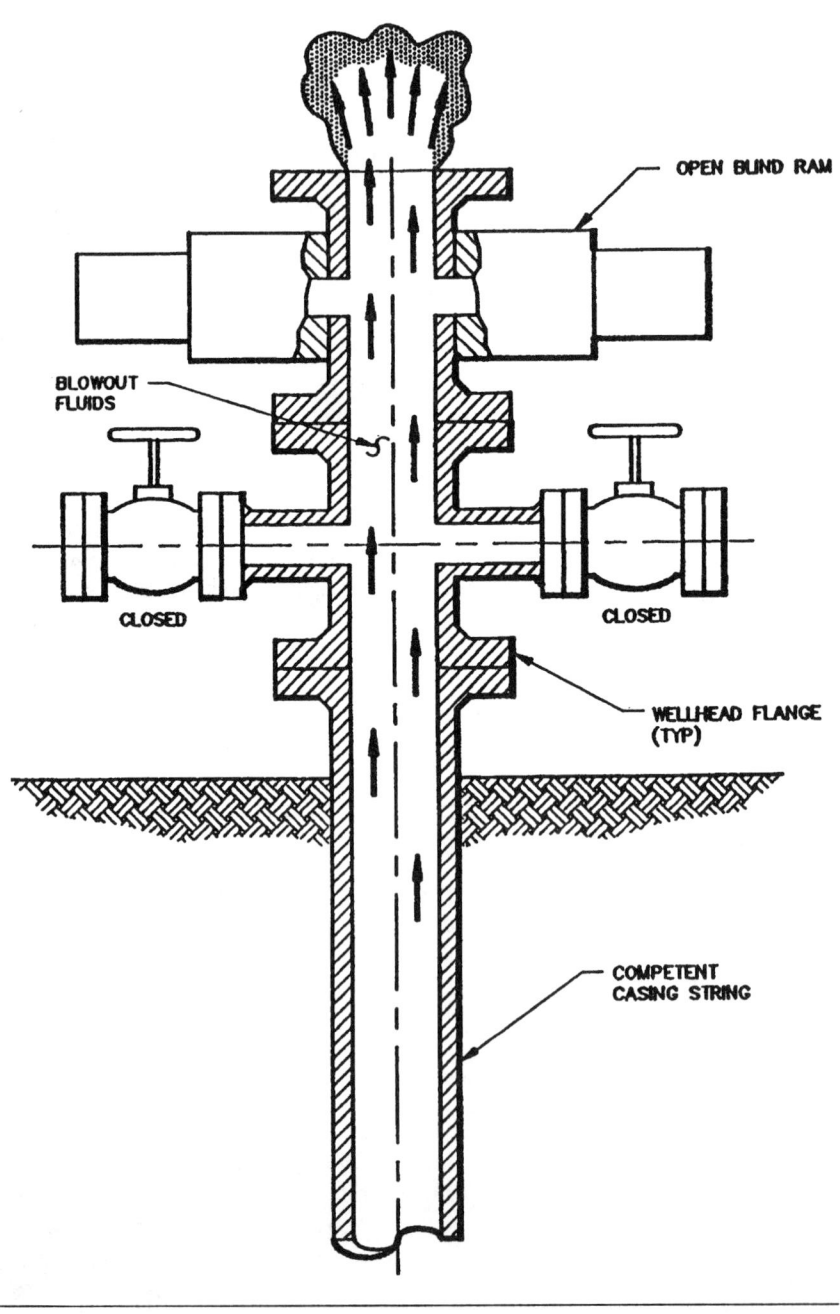

currently changing this practice and using engineering to determine mud weights needed to control the well.

Bullheading can also be performed below a packer stung into the blowing well. This has the advantage of isolating an eroded or damaged BOP/wellhead component and any casing near the mouth of the blowout that lacks structural integrity.

Bullheading applies considerable stress to the wellbore. Pressure from the formation is trapped inside the wellbore by the slug of descending kill fluid. This pressure can compromise casing shoes, break down exposed formations in the open hole by exceeding their parting pressure, and burst casing. This increases the possibility of the blowout being altered to an underground blowout with a different set of consequences.

Another kill method for a shut-in well is to lubricate mud into the well. The procedure is effective with gas wells, but it does not work with oil or saltwater wells. It is a time-consuming task, but it generally applies less wellbore stress than bullheading.

Pipe can be run into the well with a snubbing unit. Mud can be circulated from the bottom in a common kick circulation technique.

Capping/Diverting

A capped well must be diverted when the shut-in pressures would exceed the casing integrity or the formation fracture gradient. The capping assembly normally has a blind ram and 1 or 2, 4–6 in. (102–152 mm) diverter lines. (See Figure 4–7.)

Pipe is snubbed to bottom and mud or water is circulated. The pipe can be ran with a snubbing unit or coiled tubing unit. The coiled tubing is easier and faster to rig up and run, but it has certain strength limitations, notably little resistance to collapse. Some coiled tubing units have a 5,000 psi (34 MPa) burst limitation.

The pipe size is important because of hydraulic constraints. If the well has not depleted or drawndown to a lower level, the kill may require high mud weights or flow rates. Usually, larger pipe sizes are desirable to avoid excessive fluid friction. They also require larger snubbing or coiled tubing units.

Access to the inner casing string is required for this technique to be effective. Also, if the well is flowing H_2S gas, the well must be capped on fire, and all flow lines and BOPs must be H_2S serviceable.

Surface Stinger

A quick and effective approach to handling certain blowouts is with a surface stinger. The stinger may be some type of packer forced into drill pipe, tubing, or casing and hydraulically set. Metal cones with a through-

FIGURE 4-7 • Capping/diverting a blowout

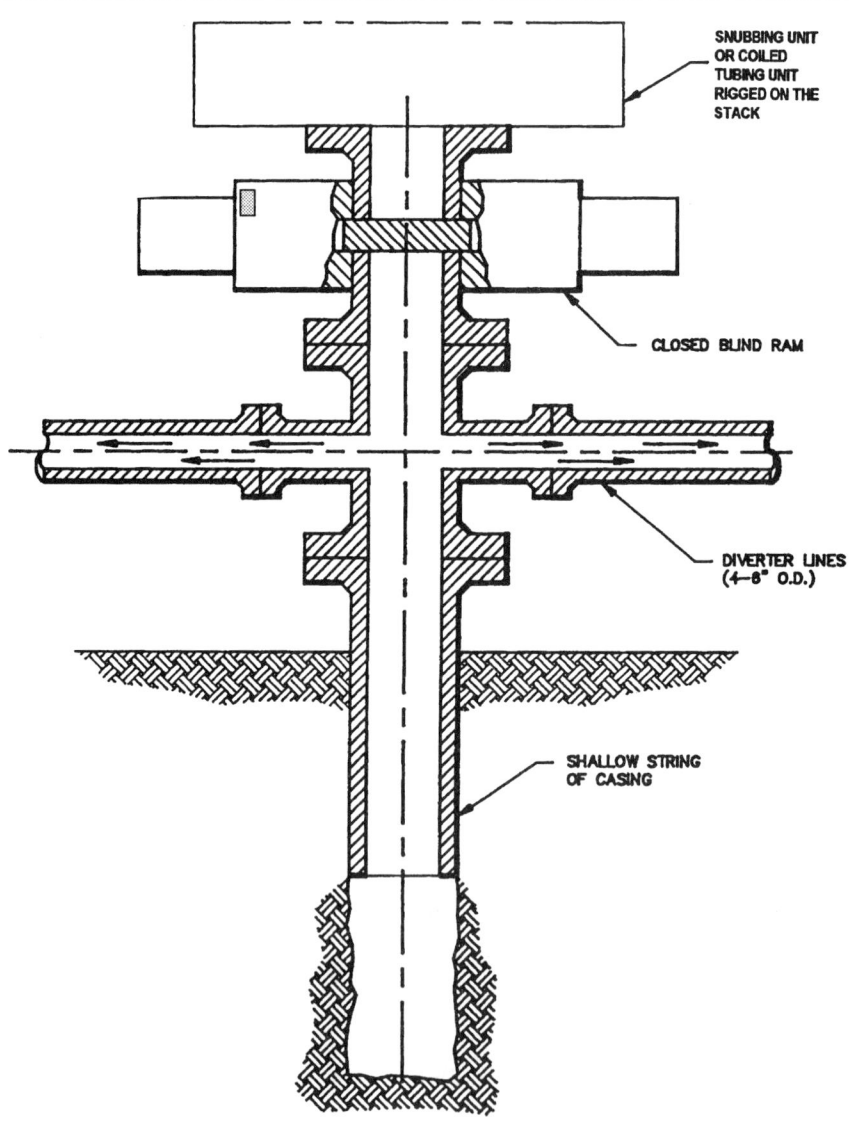

bore fluid path may be used as an alternative to a packer. Fluid is pumped into the well through the stinger.

The most frequent application of the stinger is with blowouts where access to the drill pipe, tubing, or casing is available. Methods have been developed in certain situations to stab a small packer into the pipe and set it hydraulically. Kill fluid is pumped into the pipe. Stingers are limited generally to wells with low flow rates and pressures. Many wells on Piper Alpha and in Kuwait were killed with stingers.

Fire does not prohibit the use of a stinger. Water monitors are arranged to keep the packer and pump lines as cool as possible. Also, the fire does not generally damage the top part of the drill pipe or tubing to the extent that it fails upon the introduction of cooler kill fluids.

It is not considered feasible in blowouts with moderate-to-high flow rates to stab a packer into a casing string. The flow out of the well prevents stabbing. The US-DOE salt dome blowout in Hackberry, Louisiana, was killed in the mid-1970s with a packer shoved into the casing. The oil was not flowing at a high rate.

Vertical Intervention

The term "vertical intervention" was coined by Adams in 1986–1987. It has received widespread industry acceptance since that time.

The operations are restricted to offshore blowouts. A semisubmersible is moved directly (vertically) over a live subsea blowout. (See Figure 4–8.) Work is done on the blowout from the vertical position. The work can include killing a shallow gas blowout, entering a blowout through the casing string, explosively removing a wellhead or BOP stack, or other similar operations.

The range of capabilities for the approach has not been completely explored at this time. New technology and field experiences continue to add capabilities to vertical intervention.

Vertical intervention techniques are complicated, and they should be employed only by groups with knowledge and experience in this type of operation. Many safety systems must be employed. Also, only semisubmersibles are applicable for most situations. Current technology expressly prohibits the use of jack-ups. Drill ships can only be used in deep water.

Offset Kill

The offset kill technique was developed simultaneously with the vertical intervention method. It also is applicable offshore. With the offset approach, the service vessel works near to the blowout at the surface but slightly offset of the center. The service vessel can be a semisubmersible, derrick barge, or some other type of floating vessel. It is possible that a jack-up could be considered for use with an offset kill but with several restrictions. (See Figure 4–9.)

One advantage to the offset kill is that it can be implemented if the well is on fire. The heat normally precludes the implementation of the vertical intervention approach.

FIGURE 4-8 • Vertical intervention involves working a semisubmersible over a live blowout under strict safety/operating conditions

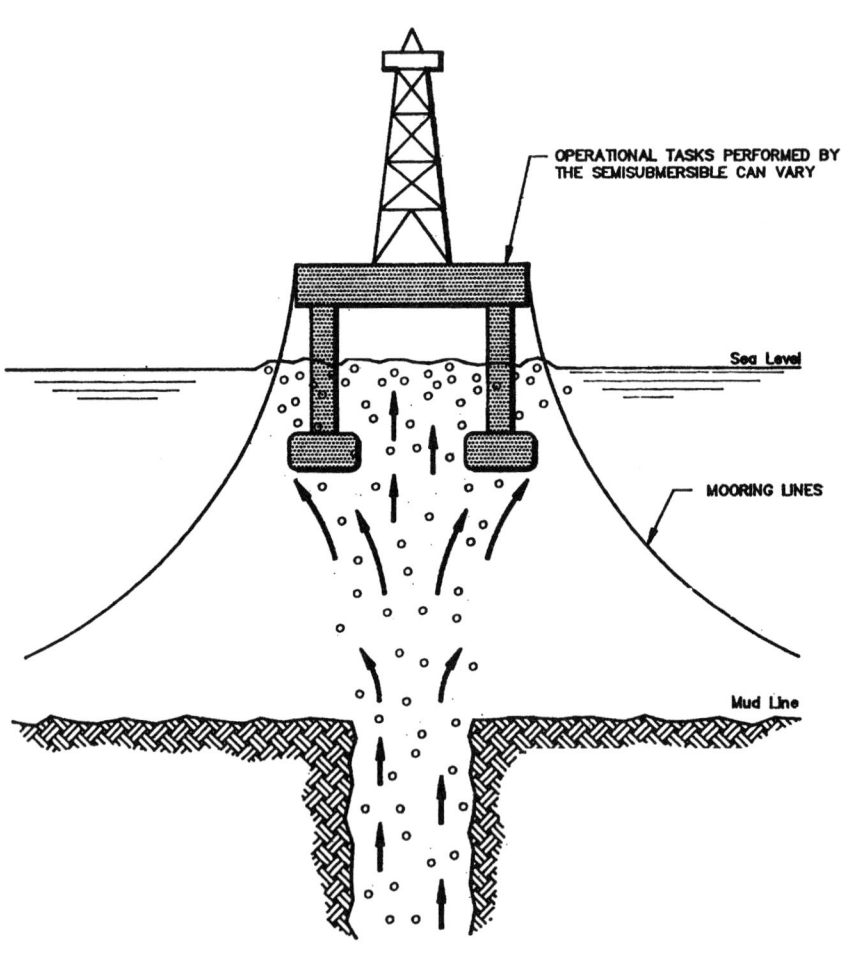

Relief Wells

One of the most well-known blowout control methods is the relief well. (See Figure 4-10.) It uses the bottom kill approach by intersecting the blowout well with a directionally controlled well. Contrary to the opinion of many operators, the relief well is not just another directional hole. It

FIGURE 4-9 • Offset kill

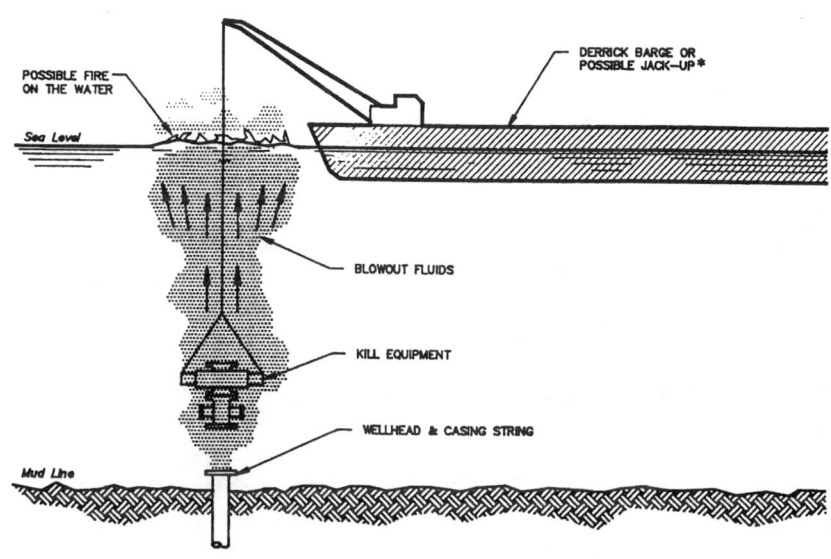

involves complex operations and requires a skilled technical engineering approach combined with experience in relief well drilling. Kill techniques used in relief wells include the dynamic kill or reservoir flood. Reservoir depletion is an important factor that has been seldom considered.

Factors required for a successful relief well are:

- Casing or drill pipe in the well to at least as deep as the minimum intercept point.
- Reasonable quality surveys indicating the general bottom-hole location.
- Ability to locate the surface site of the blowout well. This presents difficulties if the blowout is in a deepwater environment.

The well must be blowing out for a relief well to be successful. If the well is shut in under high pressure and surface intervention is not a safe option for any reason, a relief well can be highly effective if the problem well can be flowed from the top in a controlled manner.

FIGURE 4–10 • Relief well

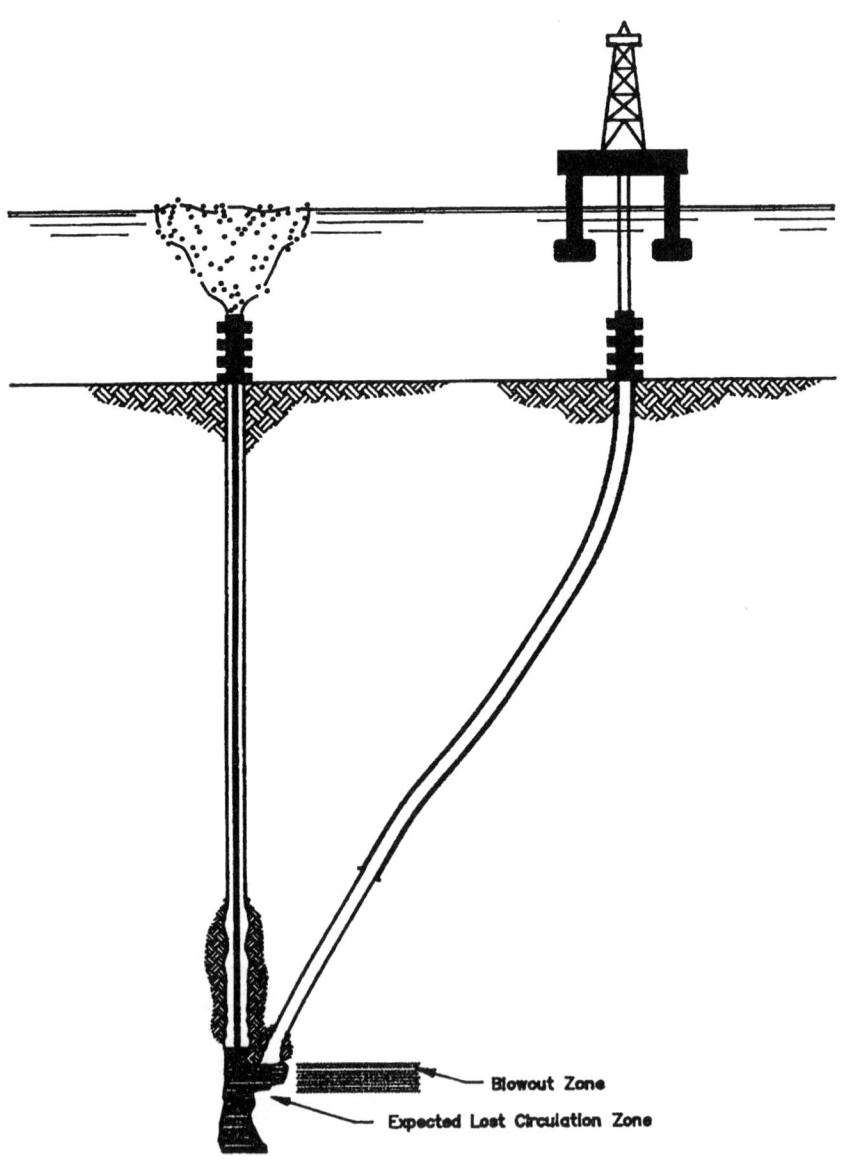

Simultaneous Implementation of Several Kill Techniques on a Blowout

Considerations should be given to the simultaneous implementation of several kill techniques on a blowout.

Some reasons for such considerations include the following:

- The primary kill technique selection has a high factor for risk or uncertainty. An example is capping a sour gas well when the integrity of the casing is uncertain.
- Public pressure or media response is intense and negative.
- The blowout fluid is oil.
- The initial kill selection has a high degree of complexity and/or will require a long time to implement.

Oil Blowouts

Simultaneous kill operations are strongly recommended when the blowout fluid is oil. Key issues are pollution, its associated clean up cost, and the public's perception of the incident. The situation is more severe if the blowout is offshore.

Offshore oil blowouts should be ignited if possible. Every effort must be made to maintain the ignition. This includes working to maintain structural stability of the platform so it does not sink below water level. Release of some pollutants into the air will result. Rapid dilution of these compounds in the atmosphere will reduce the average pollutant concentrations to acceptable levels. This is a more attractive alternative than allowing a large visible oil slick to form.

Complex Operations or Long Implementation Times

A simultaneous operation should be planned if the primary approach is complex or will require a long time to implement. A complex operation has the potential for failure because of the uncertainties associated with blowouts. Also, long operational times, such as those required for deep relief wells, support the need for considering a simultaneous approach.

Single Approaches

Many blowouts do not warrant the time and expense associated with a simultaneous kill operation. These situations can include the following:

- Capping operations that are reasonably quick and "routine."
- The blowout fluid is sweet gas rather than oil or sour gas.

- A second approach is not technically, financially, or realistically feasible.

Other reasons probably exist for using a single kill method.

Most wells have been killed by using a single approach. However, some wells have caused extended problems because the initial approach failed and a second approach had not been implemented.

Requirements for a Simultaneous Implementation of Kill Options

The question arises as to the requirements for the simultaneous implementations of kill operations. The answer is simple. Simultaneous operations can include any of the kill options that are technically possible for the given situation yet do not interfere with each other.

As an example, capping a well as a control method is not consistent with implementing techniques to let it bridge. Also, capping and shutting in a well is not consistent with capping and diverting.

Ranking of Kill Technique Viability for Different Blowout Scenarios

An effort has been made to rank the kill options for common scenarios. Figure 4–11 contains the results. When several kill options are available for a scenario, some ranking is given on a 1,2,3 basis where 1 represents the preferred approaches and 2 or 3 represent secondary approaches.

Figure 4–11 should be used as a guide. Circumstances for each blowout should be evaluated to determine the best kill approach.

Fire has a great effect on the kill technique selection. For the purposes of Figure 4–11, it is assumed that blowouts in water depths greater than 300 ft (90 m) will likely not have a fire, or the fire will extinguish itself. The previously mentioned blowout database supports this assumption. Also, the dynamics of deepwater, subsea blowouts make it difficult for a fire to sustain itself.

The impact of water depth on a subsea H_2S well is interesting to note. It appears that the water may strip the H_2S and create sulfuric acid. The escaping gas is sweet. The key variables are the gas concentration and water depth. It is believed that in depths of 500–600 ft (150–180 m), the gas will be sweetened. Lesser water depths have appeared to have sweetened the gas in the few field cases that are available. However, it is clearly recommended to conduct ongoing tests to evaluate this situation if plans involve working near a sour gas blowout.

Figure 4–12 (a–d) is a flow chart used for kill technique selection.

FIGURE 4-11 • Kill scenarios for various blowouts

LAND/OFFSHORE WITH A RIG OR PLATFORM

Access to a competent casing string with sufficient fracture gradient for a shut in

    1-Cap/shut in
    2-Cap/divert
    3-Relief well

Access to an incompetent casing string or a casing string with insufficient fracture gradient for shut in.

    1-Cap/divert
    2-Relief well

Shallow gas/no crater

    1-Bridging
    2-Cap/divert
    3-Relief well

Access to deep string of drill pipe, tubing, or casing

    1-Stinger
    2-Cap
    3-Relief well

OFFSHORE/UNDERWATER/0-300 FT (0 -91 m)OF WATER

Shallow gas/no crater

    1-Bridge
    2-Vertical intervention or offset kill
    3-Relief well

Shallow gas/crater/no fire

    1-Vertical intervention
    2-Offset kill
    3-Relief well

Shallow gas/crater/fire

    1-Offset kill
    2-Vertical intervention
    3-Relief well

Deep blowout/no fire

    1-Relief well
    1-Vertical intervention
    2-Offset kill

Deep blowout/fire

    1-Relief well
    2-Offset kill

Access to deep string of drill pipe or tubing casing

    1-Vertical intervention with stinger
    2-Offset kill with stinger

OFFSHORE/UNDERWATER/ > 300 FT OF WATER

Shallow gas/with or without crater

    1-Vertical intervention
    2-Relief well

Access to a competent casing string with sufficient fracture gradient to shut in the well

    1-Vertical intervention
    2-Relief well

FIGURE 4–12 (a) • Kill technique selection

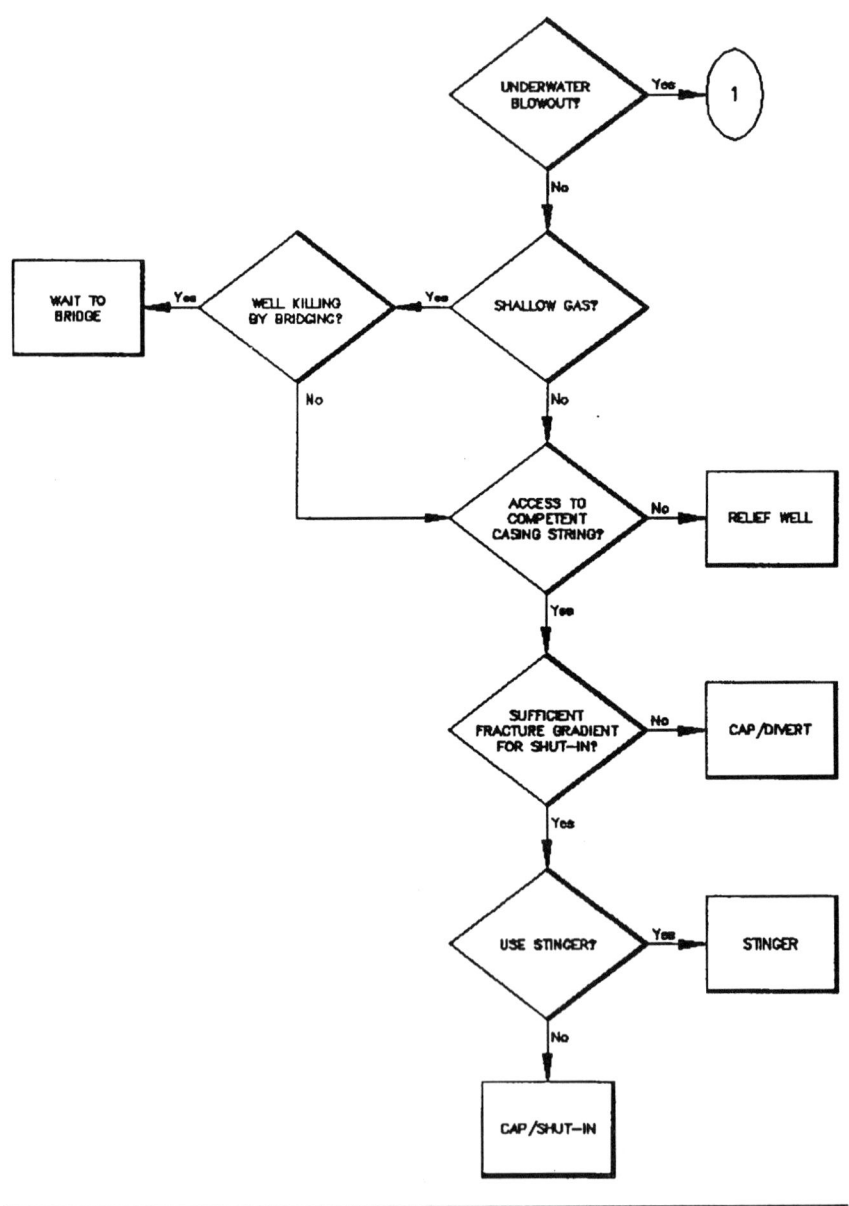

BLOWOUTS

FIGURE 4–12 (b) • Kill technique selection

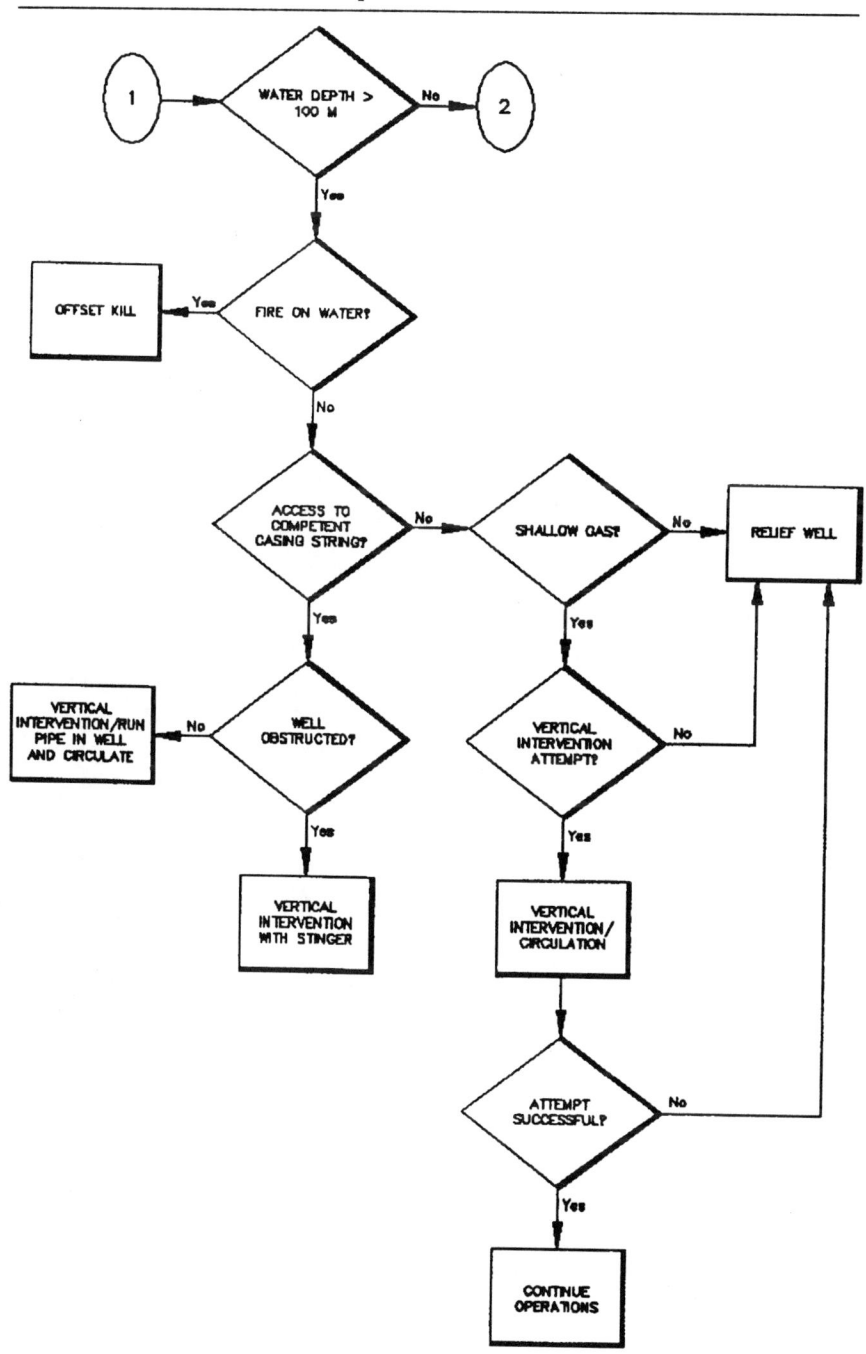

FIGURE 4-12 (c) • Kill technique selection

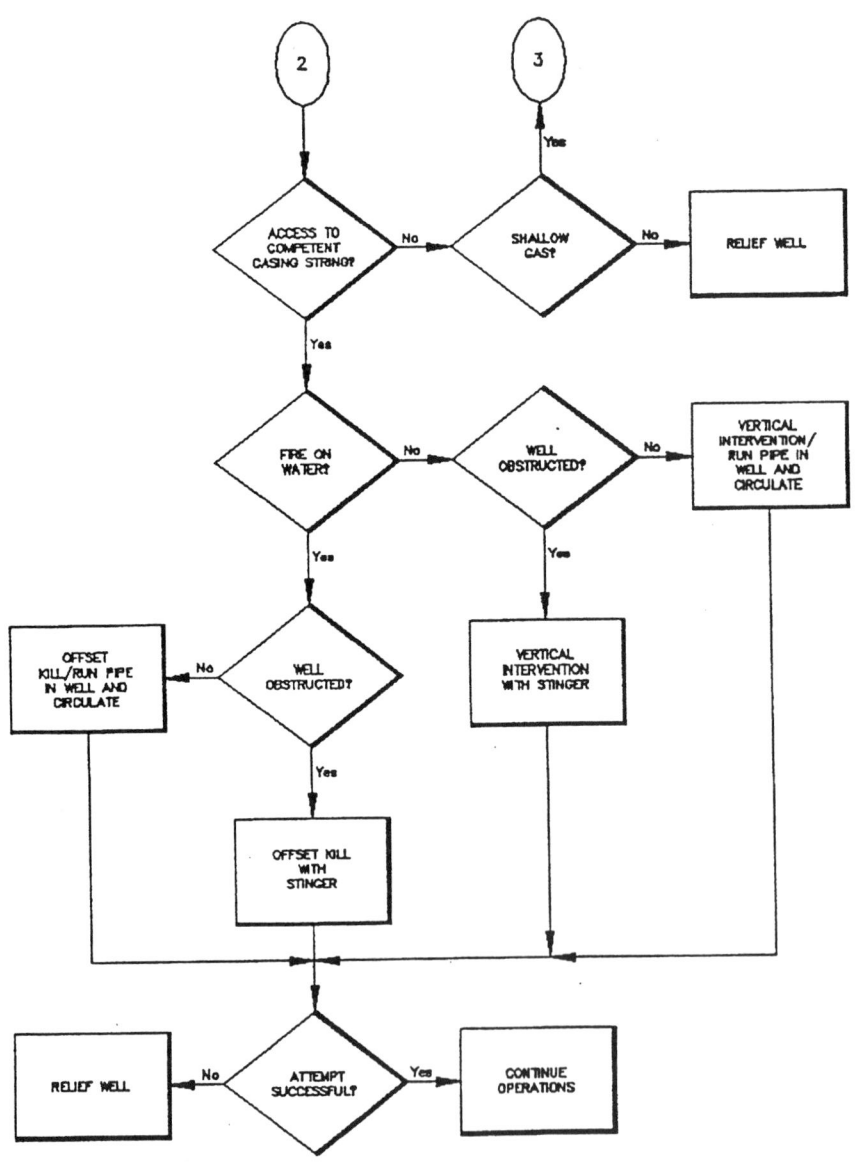

FIGURE 4–12 (d) • Kill technique selection

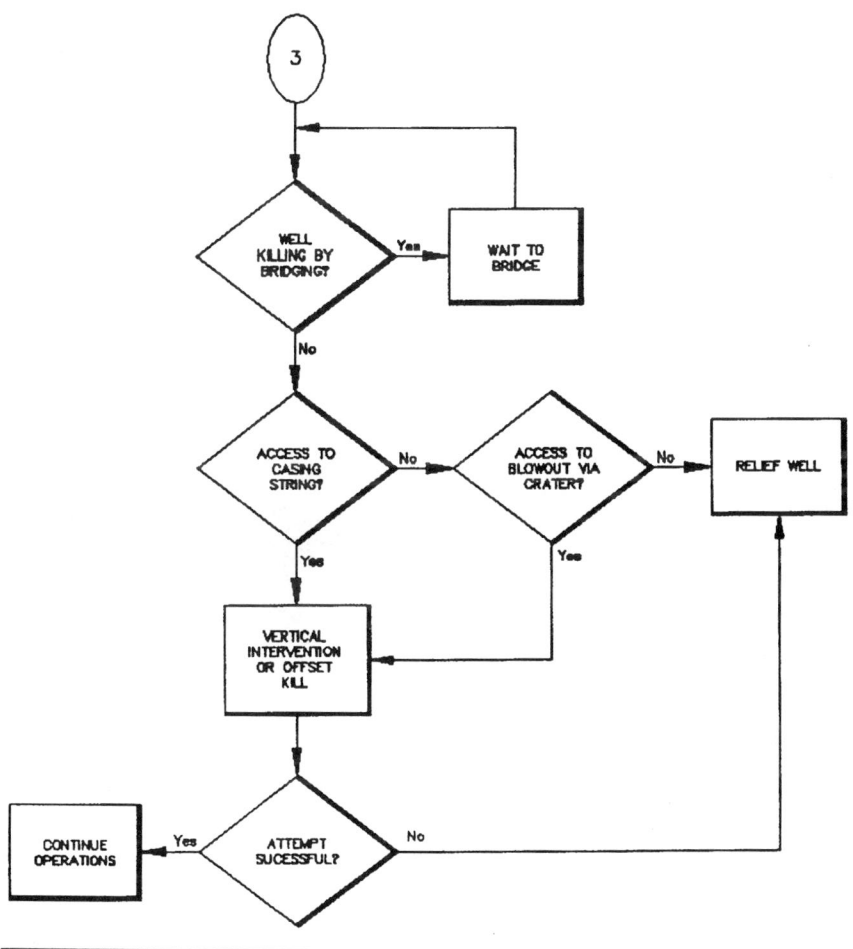

WELL CAPPING

A surface blowout is an uncontrolled flow of formation fluids observed at the surface. The most common case is a blowout that occurs through the casing and surface equipment, although many wells have blown out around the outside of the casing. Also, it is not uncommon for this problem to cause smaller blowouts in adjacent water wells as a result of formation fluids flowing through permeable zones common to the water well and the original blowout.

Planning steps must be taken before an actual killing plan can be implemented. These include organizing personnel, selecting the most feasible kill procedures for the circumstances involved, and gathering the equipment necessary for the selected kill procedure. Proper planning is essential to a quick, successful kill. Proper contingency planning, discussed in the first section in the chapter, is a key element in being properly prepared for a blowout.

The initial step in a kill operation is making the decision on whether to set fire to the blowout, if not previously ignited. It is generally preferable to avoid setting fire to the well either intentionally or by accident. Lack of fire aids the kill operation by reducing the temperatures under which personnel must work and minimizing equipment damage.

If the blowout fluid contains significant quantities of hydrogen sulfide and the well is on land, the well must be ignited immediately to burn the gas and minimize the associated dangers of personnel exposure. Wells drilled in suspected hydrogen sulfide environments should be equipped with flare guns.

After the well is ignited, personnel equipped with flare guns and protective breathing apparatus must be stationed near the well to reignite it if the fire is extinguished accidentally. Caution must be exercised even after the well is ignited because sulfur dioxide, a by-product of burning hydrogen sulfide, is toxic in sufficient quantities.

Capping Overview

Well capping can be separated into three distinct phases:

- Extinguishing the fire
- Capping the well
- Well killing

Although blowout specialists are often known as "firefighters," the most difficult task, or phase, is controlling the well. This requires experience and the ability to improvise quickly.

Extinguishing the Fire

If a well ignites, it will usually cause destruction to any overhead structures such as a drilling rig or platform. This results in debris that must be cleared.

The fire is typically extinguished prior to starting capping operations. Without the fire, the area surrounding the well is cooler. Approach is generally easier and safer.

If the well is blowing but not on fire, efforts should be made to prevent it from igniting. This includes shutting down all rig power and spraying the structure with water. Injection of water into the flow through the blowout preventers, casing valve(s), or other point will assist in preventing ignition.

Ignition sources include electrical motors, lighting, or sparks caused by well debris hitting a steel member of the rig.

The primary exception to the rule of extinguishing the fire is when the well effluent contains sour gas, H_2S. The gas is heavier than air and will settle in low places around the rig. Also, it is extremely toxic in small concentrations. A well fire converts the hydrogen sulfide to sulfur dioxide, which is less toxic. Also, the SO_2 is carried into the atmosphere with the thermal plume from the fire so the gas is more readily dispersed. Chapter 3 contains more details on sour gas.

Several techniques are used to extinguish fires. All rely on one of the three basic requirements to have a fire:

- Ignition source
- Fuel
- Oxygen

Common techniques to extinguish the fire include the following:

- Water
- Explosives
- Nitrogen
- Carbon dioxide
- Dry chemicals
- Foam
- Jet engines
- Smokestack
- Wellhead clearance

A combination of these is common. Water is usually needed with the other techniques.

Water. Water is sprayed on the fire to cool the fire. This allows the specialists to work in close proximity to the fire. With proper focusing of the spray, many well fires can be extinguished exclusively with water. It was estimated that 90% of the Kuwait well fires were handled with water.

Water is cheap and relatively easy to supply in large volumes in most situations. Water is needed to cool the general area to prevent reignition regardless of whether it is utilized as the well extinguishing method. Water sprays also provide personnel protection and equipment cooling.

Water management requires gaining access to a water source, transporting the water to the site, temporarily storing it at the well site, and spraying the fire. Oil company personnel should have a basic knowledge of these requirements so they can begin this task while awaiting arrival of the blowout specialists. Offshore events do not pose the same water management problems as land wells.

The heart of any water system is the pumps and piping. Pump output ranges from 2,000–6,000 gpm (450–1,400 m^3/hr). Some companies prefer

FIGURE 4-13 • High volume firefighting pumps

the flexibility and portability associated with smaller output pumps while other blowout specialists prefer the high volume output of bigger pumps. If a fire can be extinguished with water, it typically does not require large volumes. (See Figure 4–13.)

High capacity monitors are placed in several positions to direct water at the well fire and surrounding area. After cooling the area for a period of time, several water streams are directed at the well fire while continuing to cool the area.

The piping system transports water from the pumps to the monitors and nozzles. Old systems used 4 in. (102 mm) piping that required extensive handling to make up each connection. In some cases, the piping was welded. New systems employed by some companies use 6 in. and 8 in. (152 and 203 mm) piping with quick-lock connections. A long run of large diameter piping can be installed quickly.

Water usage has few negative aspects except a water source must be available. A lagoon (small pond) or steel tanks must be provided. The lagoon may require a plastic lining to prevent water loss in desert or dry, permeable soils.

Explosives. Explosives are often considered by the general public to be a standard part of the blowout specialist's tools. This is generally not the case. They are rarely used in today's industry, although older generation companies still use the explosives.

Explosives snuff the fire by consuming the oxygen and disrupting oxygen flow near the fire. If the area surrounding the well is cooled below the auto-ignition temperature with water, the fire will not reignite. In any case, water is a necessity.

Typically, a barrel is mounted on the end of an athey wagon and packed with dynamite. It is protected by insulation and water spray while moving it near the base of the fire. The explosion momentarily deprives the fire of oxygen.

Explosives have several disadvantages. An explosives specialist is usually required. They expose firefighters and support personnel to increased risks. Explosives usually cause heavy damage to the end of traditional athey wagons. It is an extra expense. In some areas, importing explosives has severe restrictions. Explosives have often been used on well fires that could have been easily extinguished with water. Some firefighters consider the explosives method outdated and more theatrical than practical.

Nitrogen. Nitrogen starves the fire of oxygen. The fire will not reignite if the surrounding area is cooled with water after the nitrogen is depleted. This technique has applications where quantities of water are not readily available.

Both gaseous and liquid nitrogen were employed in Kuwait in conjunction with long, large diameter tubes (smokestacks) placed over the burning well. Nitrogen, or a mixture of water and nitrogen, was injected through a hose attached to a pipe nipple on the side of the stack. Water was sprayed at the base near the wellhead for cooling and as an extinguishing aid.

Carbon Dioxide. Carbon dioxide can be applied with the same smokestack method described for nitrogen. Nitrogen is more commonly available in large volumes. Nitrogen is less risky for personnel.

Carbon dioxide has also been used to snuff fires located in low areas by blanketing the area with carbon dioxide. It is heavier than nitrogen and stays close to the ground.

Dry Chemicals. Dry chemicals, such as potassium bicarbonate (trade name–Purple K), were utilized by one company in Kuwait. Their use was mainly limited to small well fires and ground fires. The dry chemical units in Kuwait had a discharge rate of 200 lbs/sec (90 kg/sec). The material is propelled by nitrogen.

Dry chemicals can also be injected into smokestacks placed over the burning well. The chemical is directed at the base of the stack and is sucked into the flow stream. The induced flow assists to disperse the dry chemicals more uniformly.

Foam. Foam is widely used to handle oil fires in refineries and other similar places. It is particularly applicable to pit, tank, and ground fires. Foam does not work well on oil well blowouts because of the associated gas with the fire. The foam extinguishes the fire by separating the oxygen from the burning oil or by chemically locking liquid hydrocarbon molecules. However, the gas continues to burn. When the foam is exhausted, the oil will reignite.

Jet Engines. Jet engines became popular in Kuwait. The Hungarian team had a device called the "Big Wind." It consisted of two Mig-21 Russian jet engines mounted on a military tank body. The Russian team had a similar single engine device. The exhaust of the engines was turned on the fire. The fire was extinguished by flooding the area with carbon dioxide/monoxide. However, large quantities of water was also sprayed on the fire. It is doubtful that the jet engines would have functioned without the application of water. Reports suggested the jet engines were not successful on very large fires.

Smokestacks. Smokestacks, in combination with water/nitrogen injection, can be effective in extinguishing the fire. Smokestacks have other applications. Smokestacks can be placed over a well fire to elevate the flames. This is useful when working on hydrogen sulfide or hot wells. A smokestack is shown in prior Figure 4–14.

The well flow stream through the stack induces an air flow into the base of the stack. The induced flow picks up water, dry chemicals, and other material directed at the stack base. The induced air flow also cools the area.

Smokestacks are typically 20–36 in. (51–91 cm) diameter. The length varies from 14–30 ft (4.3–9.1 m). A swivel mount is attached to the athey wagon. Ears on the smokestack are secured in mount hangers to permit free rotation in one plane. Elevation changes are made by raising or lowering the athey wagon boom.

Smokestacks are frequently used to divert oil flows after the fire is extinguished. On calm days the oil goes up into the air and falls on workers. The stack placed as a diverter helps keep the site clear of raining oil.

Wellhead Clearance. Wellhead clearance is not actually an extinguishing technique. It simplifies the process.

Severely damaged xmas trees and wellheads often have multiple flow paths with fire shooting out in several directions. Access is limited. Extinguishing multiple-burning oil streams is difficult using any of the previous techniques. Sometimes the vertical flames can be snuffed, but the horizontal flames reignite the well.

Firefighters clear the wellhead area to attain a vertical flow. The term used for this is "to get the fire going straight up." This simplifies the extinguishing process, as well as making a safer workplace for the firefighters.

FIGURE 4–14 • Fire extinguishment-smokestack and water/nitrogen

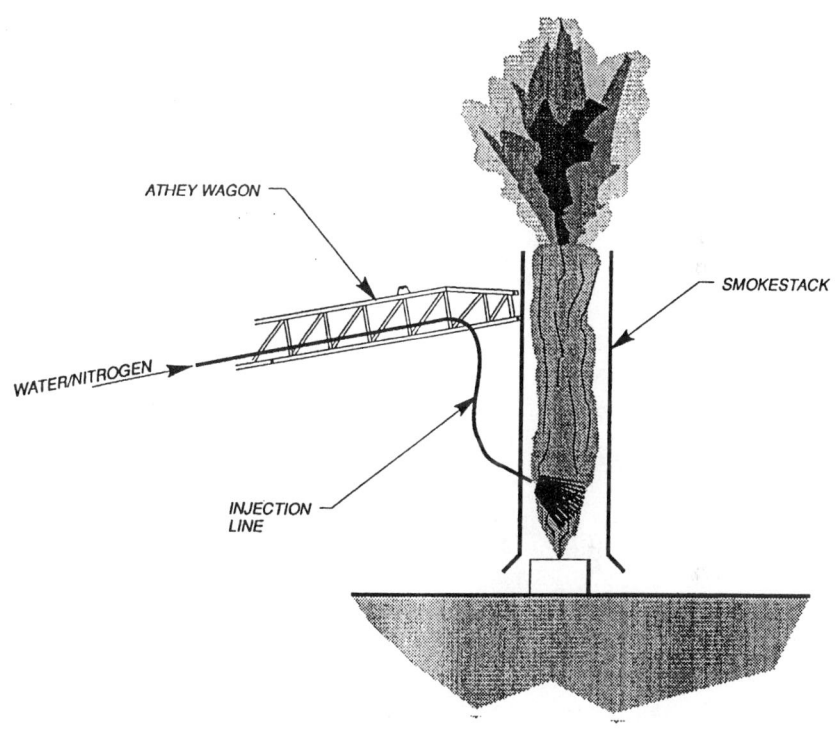

Capping the Well

The major objectives in well capping are as follows:

- Cool the area with water
- Remove the debris
- Extinguish the fire
- Replace existing equipment with new equipment or make serviceable the existing equipment on the well. (An alternative procedure is to use a stinger.)
- Shut in or divert the well
- Kill the well

Cooling the area with water has already been discussed in the previous section.

Debris Removal. Debris removal is accomplished with the most conveniently available equipment. This includes athey wagons, cranes, track hoes, wireline, and so forth.

The athey wagon is the only purpose-built device for debris removal. They are tracked units with a vertically moveable boom extending out from one end. They are generally operated by pin connecting the wagon to a large bulldozer. The operator's cab is protected by a corrugated metal enclosure. The dozer backs toward the well with the athey wagon. Various boom implements can be attached to lift equipment, snag and remove obstructions, and perform other tasks.

The boom is rigged with multisheave blocks. Bulldozers are used that have a heavy-duty hydraulic winch on the back. The winch line reeled onto the boom blocks is used to raise/lower the boom. Although separate hydraulic winches have been provided on some athey wagons, the traditional use of bulldozer winches has the major advantages of simplification and heat protection for the hydraulics. All hydraulic lines are enclosed inside the protection of the dozer body.

Repairing/Replacing Equipment. Repairing/replacing equipment is the next step in the capping operation. This step requires experience and the ability to improvise. Knowledge of wellheads and BOPs is important. The degree of difficulty varies widely. Capping can be as simple as changing out a valve to complete wellhead replacement.

The first step is to remove the damaged tree, BOP, or wellhead components. Damaged flowlines, chokes, valves, and other items are disassembled or cut with pneumatic saws and water abrasive cutters.

Flanged, stacked heads can sometimes be salvaged by removing cracked, warped, or otherwise damaged components down to a reasonably sound flange. A mating spool is prepared and installed. Capping spools have at least one outlet and a valve or BOP assembled to the upper flange. The assembly is set on the existing flange with the upper valve/BOP open to allow the well to flow while studs are placed and tightened.

Ring groove damage and warpage can be overcome in some cases with lead seals. Lead seals are only suitable for low pressure, temporary capping assemblies.

Severely damaged heads must be removed. This is done by complete cut off using one of the cutting techniques or by knocking off the wellhead. The outer casing strings are stripped down to the production or inner casing string. A capping stack and/or new wellhead is installed.

Several types of casing cutters are available. The old traditional method was to use two wireline units—one with a full reel and the other empty. The line is passed from one reel around the casing and attached to the empty reel. The line is pulled taut and then cycled back and forth

BLOWOUTS

between reels slowly sawing through the casing strings. The method is slow and the wire becomes jammed and breaks.

Linear shaped charges are used occasionally to make vertical and horizontal cuts for stripping casing and complete wellhead cutoffs. Care must be taken because the larger charges can penetrate outer and inner casing strings. Some cases of blowout reignition have occurred with this technique.

Water abrasive and pneumatic cutters are the favored tools of most groups. The portable lathe-type pneumatic cutters have a split frame that fits around tubulars. A lathe cutting tool is rotated around the pipe to produce a clean, even cut suitable for a capping stack or wellhead installation. The units make horizontal cuts for stripping outer casing strings to expose production casing.

Water abrasive cutters utilize a high pressure water jet with entrained abrasive material to cut tubulars or other components. They can be rigged on a circular track system for cutting tubulars or mounted on a tripod for cutting bolts, etc. The units are available with operating pressure from 5,000 to 30,000 psi (34–210 MPa).

Stingers. Stinger operations are shown in Figure 4–15. Metal and inflatable packer stingers have been used. Tapered cone metal stingers are

FIGURE 4–15 • Stinger operation

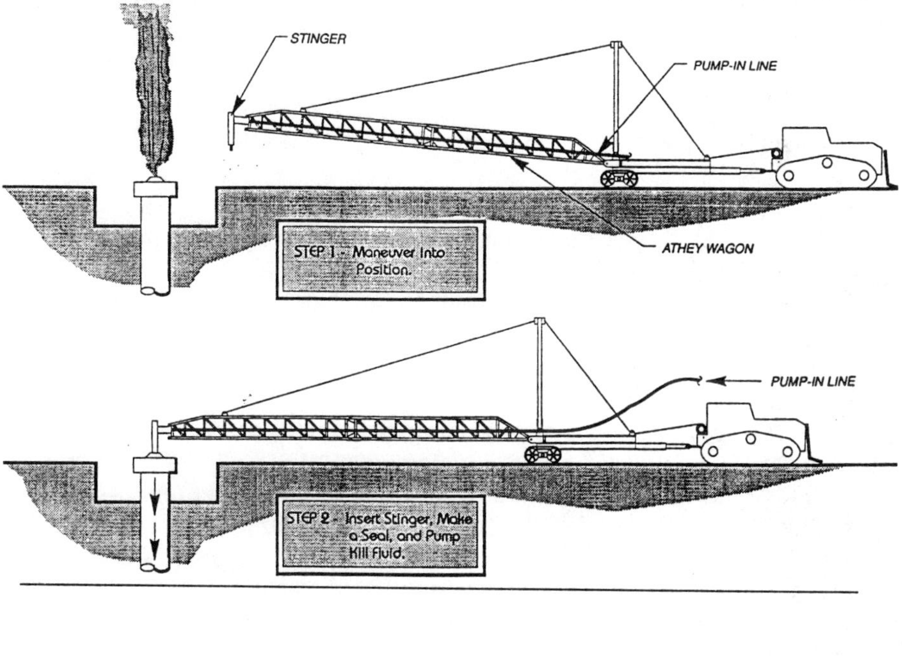

made of steel, brass or aluminum. Softer metals have been tested to find a better method to seal against deformed wellheads and tubulars.

Metal stingers are bolted or pinned to the end of the athey wagon. They have also been attached to the end of a mobile hydraulic crane. The stinger shown in Figure 4–15 has a quick connect stinger mount that attaches with four pins to the athey wagon. It is designed for adjustments in two planes.

Moderate wellhead deformities can be accommodated after stinger insertion. Various plugging materials are added to the kill fluid and pumped slowly. These materials include pieces of polypropylene or cotton rope, cut rubber, lost circulation material, and golf balls. They tend to seek out the leak path and bridge gaps to effect a seal.

Usually, the combined weight of the stinger assembly and the athey wagon boom contains low flow rates and well pressures up to 300–400 psi (2–3 MPa), depending on the tubing/casing cross-sectional area and the boom weight. Snubbing lines can be attached for higher flows and pressures. The snubbing configuration consists of dual wirelines attached to the stinger assembly and ran through snatch blocks to winches.

Baker has supplied Lynes inflatable packers used on some wells in Kuwait. Inflatable packers had limited success under restricted conditions—low flow rate, low well pressure, and no obstruction in the top 6–8 ft (1.8–2.4 m) of the well. Inflatable packers attached to a swivel mount on the athey wagon hang vertically. Free motion was a problem. As packer insertion was attempted, the well flow tended to kick the packer to one side.

Special Equipment

Explosives may be used in attempting to cap the well. The most common use is in extinguishing a fire as discussed previously. Also, explosives may be used to aid in debris removal, although this application often sets fire to the well. Explosives are normally those readily available in the area and may include dynamite or plastics, such as C-4.

Sheets of corrugated steel (roofing "tin") are used as individual heat shields to protect against radiated heat. The sheets are equipped with a handle and a small viewport. The tin is a size easily handled by the workers, yet sufficiently large to shield them from the blaze. Tractors, cranes, and bulldozers may also be fitted with tin to protect the equipment operators.

Fire retardant clothing is normally worn by firefighters. This can include coveralls or underclothing. A common brand name is NOMEX. Also, they are sprayed with water for cooling and fire protection.

Fire entry suits may be used when it is not possible to obtain sufficient heat protection from the water or tin sheets. Although the suits are used when necessary, they are considered a last resort because of the increased

Blowouts

difficulty in performing the manual operations necessary for the removal or installation of equipment at the wellhead.

Techniques to Kill an Annular Blowout

As discussed, the procedures involved in killing an annular blowout are based on installing new control equipment or repairing the existing blowout preventers. After the equipment is operational, the well is shut in or capped, and mud is pumped into the well. The steps required to prepare the control tools for service will vary with each application.

Whichever procedure is used to cap or divert the blowout, the next step in the kill process is to pump heavy mud into the annulus. Assuming no pipe is in the hole, pressure limitations on casing and preventers often prohibit bullheading or directly pumping mud into the well. It may be necessary to lubricate mud into the annulus by (1) bleeding small amounts of formation fluids from the well, (2) quickly pumping small amounts of mud into the annulus, (3) allowing time for the mud to fall, and (4) again bleeding formation fluids.

Although lubrication is a tedious process, it will eventually kill the well or reduce surface pressures to the point where other techniques can be employed. If drill pipe is in the well, it should be possible to pump mud directly into the hole and kill the well.

In cases where drill pipe is not in the well and mud lubrication is not possible, attempts should be made to snub pipe, tubing, or coiled tubing. The equipment necessary for the snubbing process can be rigged directly on top of the capping assembly. The well can be killed using conventional procedures after pipe is in the hole.

A suitable choice of kill mud weights must be made before the kill process begins. In some field cases, the well was successfully capped and mud circulation begun only to fracture the formation with excessive mud weights. The natural tendency at this time is to overcompensate and use high mud weights to prevent further trouble. The worst possible conditions exist before pumping begins, and the pressure should diminish thereafter. The kill mud weight will be only slightly greater than the weight used when the well blew out.

Techniques to Kill a Drill Pipe Blowout

Techniques used to kill drill pipe blowouts are based on the same principles used to kill annular blowouts. A diverter/capping assembly shuts in the well until heavy mud can be pumped down the drill pipe. A modified drill pipe connection or a pipe ram locking system can be used. (See Figure 4–16.)

FIGURE 4-16 • Typical diverter-capping assembly for drill pipe blowouts

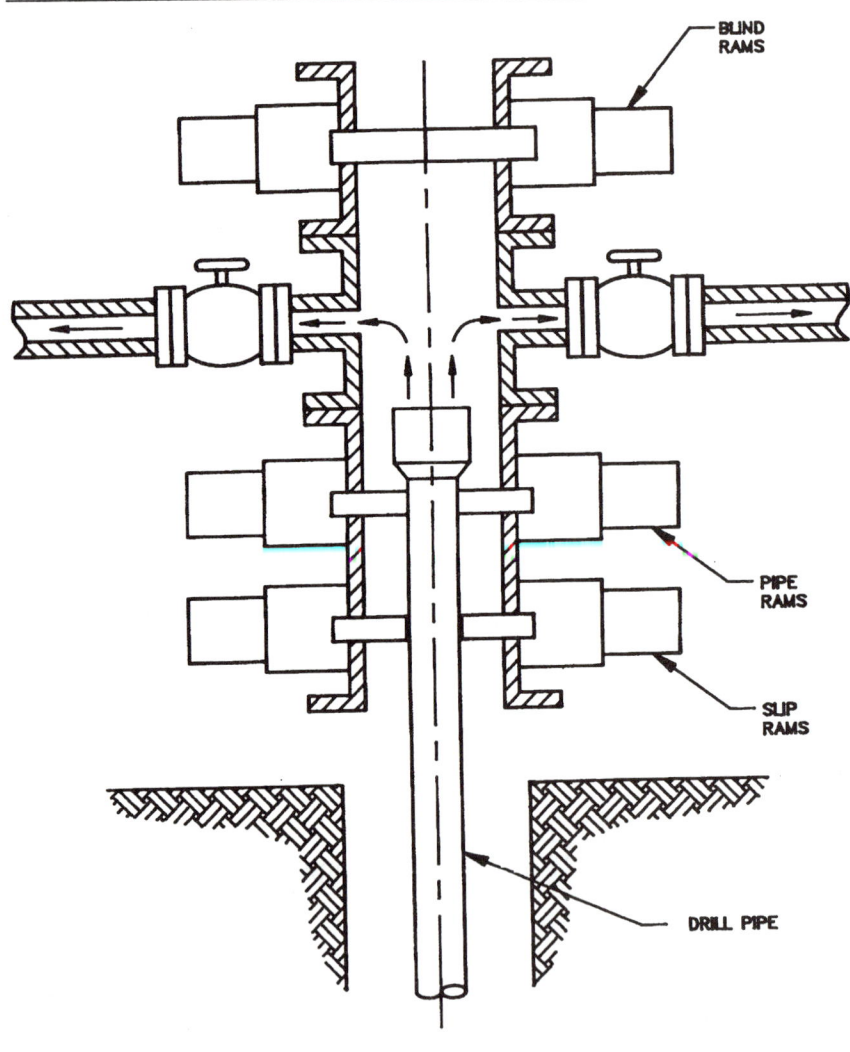

RELIEF WELLS

Relief wells fall into the category of bottom kills. A directional well is drilled to intersect the blowout well. Fluids in the form of water, acid, and/or mud are pumped into the well at specified rates and sequences until the well is dead.

Many relief wells have been drilled over the years. The first documented case was supervised by the legendary John Eastman in the Humble Field, Texas, in the early 1920s.

The original wells were given the name of "relief wells" because their purpose was to relieve reservoir pressure. The intent, or "hope" may be a better term, was that the reservoir pressure would draw down and the well would die.

This relief well discussion is primarily oriented toward the most difficult environment, which is offshore. However, much of the information will be applicable in all situations. This applies to land wells. Relief well technology can be used on most wells anywhere in the world. Obviously, the rig selection discussion presented herein pertains to floating rigs and not to onshore rigs.

In some areas, such as frontier Canada, the Arctic, and Northern North Sea, the operator may be required to demonstrate his capability to drill relief wells before being given approval to drill.

Overview of General Scenarios

Scenarios for offshore underwater blowouts vary, but they can be grouped into several general categories. The general assumption is the water depth is greater than typical jack-up capability for all scenarios discussed (i.e., the relief well requires a floater and cannot be drilled with a jack-up rig). To meet this constraint, a minimum water depth of 300 ft (91 m) has been selected.

All scenarios assume that the wells are drilled in a nonprotected environment (i.e., an open sea situation). This qualification has only minor impact for relief well drilling. It has a greater impact on pollution control. Land wells are a simpler case.

Key factors in the various general scenarios are as follows:

- Water depth
- Blowout depth
- Fluid type

Each will be discussed.

Water Depth

Relief well drilling has slightly different requirements with varying water depths. In general, requirements can be considered for water depth ranges of 300–600 ft (91–180 m), 600–1,500 ft (180–460 m), and >1,500 ft (>460 m).

Water depth has an impact on the blowout. Key parameters are as follows:

- Seawater hydrostatic acts as a choke and prevents gas expansion in the critical low pressure environments from 500 psi to atmospheric conditions.
- The water acts as a buffer, and it allows a safe vertical intervention as a kill option when evaluating the relief well as a control option.
- Entrained water in the blowout plume disperses the blowout effluent so it poses minimal risks to the relief well rig and crew.
- The water column reduces the effects of methane and H_2S release on the surface.
- Seawater back pressure reduces flow rates out of the well.
- Reduced flow rates inhibit bridging which, according to statistics, will increase the likelihood that a relief well will be required.
- Reduced flow rates mitigate reservoir drawdown, which equates to a higher reservoir pressure that must be controlled by the relief well.

The water depth range of 300–600 ft (91–180 m) has some interesting characteristics relative to relief wells. Blowout effluent release at the surface for a large blowout can impact the site location for the relief well rig. The farther removed the rig is from the blowout site, the more stringent the directional drilling requirements will be.

An H_2S blowout in the shallow end of this depth range may require special consideration. Longer contact with water in deeper environments may strip H_2S from the hydrocarbon gas as evidenced by field case histories.

An ignited blowout may continue to burn even if the rig sinks below the water line or is moved off location. It is not considered likely that the fire can be extinguished. Based on historical data, heat loading is not expected to be a controlling factor, but the heat must be evaluated at the time.

The depth range of 600–1,500 ft (180–460 m) allows a large flexibility without posing many constraints on the relief well. The water depth is sufficient to prevent any adverse effects from combustible gas, H_2S, fire on the water, or pollution. Water depths beyond 1,500 ft (460 m) do not provide any real additional benefits in terms of reducing the adverse effects of these parameters.

Site selection for the rig is not limited by any surface conditions. The rig could move a short distance from the center-line of the blowout well and drill a vertical well throughout most of the drilling program. It could be made to track the blowout well. This technical feature of deepwater environments is interesting, and it could be used on a future blowout to ease directional drilling tasks.

The water depth in this range may seem to be deep, but it is not deep enough to tax most equipment available on today's market. This includes mooring systems, risers, and control systems. Most relatively modern semisubmersibles can meet the requirements. Also, a drillship becomes a

Blowouts

viable option as a drilling vessel for blowout control as the water depth increases.

Water depths greater than 1,500 ft (460 m) for a relief well begin to pose equipment problems not related to the blowout itself. Riser design becomes more complicated, and many rigs will not have adequate capability. BOP control systems are reaching the edge of their limits as water depth increases, particularly if the well is blowing out underwater. Rigging a kill system to pump fluids at high rates into the annulus increases complexity and begins to eliminate some of the options discussed in later sections of this report.

Wells drilled in 5,000 ft (1,500 m) or greater water depths are unique. They are quite often a "one off" well, which means that many aspects of the well were special designed on a one-time basis. If a "one off" well blows out, the question arises as to how long will it take to rig up and kill the blowout when perhaps years went into preparation for drilling the initial well. This situation is analogous to a well drilled in severe Arctic conditions where the drilling season is very short. Fortunately, in these deeper environments, vertical intervention becomes an attractive kill option.

Blowout Depth

Depth of the blowout affects relief well drilling strategy. Shallow blowouts can be more complicated in many respects than medium depth or deep blowouts. For purposes of discussion, blowout depth ranges might be grouped as 0–3,000 ft (910 m), 3,000–10,000 ft (910–3,000 m), and >10,000 ft (>3,000 m).

As previously stated, shallow blowouts (i.e., 0–3,000 ft (910 m)) cause many difficulties not encountered in deeper blowouts. Some are:

- Shallow kick off depths
- High build and drop rates
- High drift angles
- Hole opening and underreaming difficulties in soft sediments in the shallow depths
- Possible charged sands from shallow gas blowouts
- Requirements for a special-built diverter unless the well is drilled riserless
- Greater than expected drill times due to directional control complexities
- Casing program modifications to account for bending in large diameter tubulars

The blowout depth range of 3,000–10,000 ft poses the least problems of all depths. Factors supporting this statement are as follows:

- Modest and manageable ellipses of uncertainty under most anticipated conditions
- Reasonable directional profiles and drilling requirements
- Drill times that are usually acceptable prior to reaching the target

The drill time issue is worth discussion. A typical well can be drilled to 10,000 ft (3,000 m) in reasonable and acceptable times for most situations. This avoids the decision to intersect deep at the bottom or at a shallower depth. This decision is important in deeper wells because of time required to drill the relief well.

Furthermore, the reverse situation can be a factor in the 3,000–10,000 ft (910–3,000 m) range. If the relief well can be drilled and made ready for the killing phase quickly, it may be necessary to wait until kill equipment can be located, assembled, tested, and mobilized to the well site.

This situation almost occurred on Piper Alpha's P-01 well. The TVD was ≈8,000 ft (2,400 m) with an 8,500 ft MD (2,600 m). Killing equipment was difficult to locate in sufficient quantities. It was a tight race to beat the deadlines; but at the end of the day, the equipment was located and assembled. This might not have been accomplished on a blowout in a remote location.

Blowout depths greater than 10,000 ft have advantages and disadvantages. The luxury of a deeper blowout is that sufficient time is available for planning and equipment procurement and mobilization.

Deeper blowouts have a number of disadvantages including the following:

- Higher formation pressures that place more stringent requirements on the relief well.
- Reduced casing sizes for deeper relief wells that consume more hydraulic horsepower in pumping kill fluids to the blowout well.
- Ellipses of uncertainty that may be unmanageable in deep situations unless bypasses and sidetracks are made.
- Long drilling times.

Again, the issue of drilling time is worth discussion. Consider an example of a 17,500 ft (5,330 m) blowout well. Drilling time required to drill to bottom under normal conditions can be considerable. It is increased by directional requirements to change hole angles in deep, hard sections. Ellipses of uncertainty can be unmanageable requiring sidetracks and bypasses.

A shallow intersect avoids many of these difficulties. Ellipses of uncertainty are smaller and probably manageable. Drilling times are reasonable. The key issue is whether the well can be killed at a shallower depth [i.e., 8,000–12,000 ft (2,400–3,700 m)]. These topics will be discussed in detail in later sections of this book.

Fluid Types

Blowout fluids have an impact on the relief well. A gas blowout does not cause significant environmental damage. The gas, if necessary, can be burned. Other than the routine expediency associated with the desire to kill a blowout quickly, additional urgency due to pollution is not imposed with a gas blowout, on land, offshore, or subsea.

A gas blowout, if not on fire, should not be ignited unless extenuating circumstances exist. The fire may collapse the rig or other equipment. It will cause a heat loading that increases the effort required for capping. An unignited gas well does not have the dramatic impact associated with a burning well. Public pressure is increased with a burning well.

Oil blowouts pose an obvious pollution problem. Oil does not burn cleanly in most cases, so ignition does not always provide a solution. An oil blowout in 1,000 ft (300 m) of water, for example, will be dispersed over a large area and may negate the effectiveness of surface spill containment efforts.

H_2S is toxic and must be burned on land blowouts. Underwater blowouts are different because it is difficult to ignite and maintain ignition on a blowout if it did not ignite initially. Fortunately, some case histories have shown that the water will strip H_2S from the gas and allow the release of sweet gas at the surface. One of these involved a sour gas blowout in 300 ft (90 m) of water. It is anticipated that greater water depths will completely sweeten the gas regardless of the toxic concentration. This matter needs further investigation.

Blowout Well Path Location

The blowout well path must be known with some degree of certainty before a directional plan for the relief well can be developed. Ranging tools have a reliable accuracy of 50–100 ft (15–30 m) under average conditions and, supposedly, 200 ft (60 m) under ideal conditions. If the wellbore path uncertainty is 300 ft (90 m) at the proposed point of intersection, for example, it is clear that the ranging tool limits have been exceeded. A shallower point of intersection with smaller uncertainties may be required.

Surface Site Evaluation

Finding the surface location of a land blowout without a crater is simple. Triangulation or other conventional surveying techniques can meet the requirement. It is recommended to use two independent surveyors and compare the results. A third survey should be taken if the initial surveys do not agree. A land well crater makes the job more difficult. An error of ±10 ft (3 m) is important relative to accuracy of the ranging tools.

Underwater blowouts can present a challenge. The surface blowout plume meanders randomly, similar to a cyclone, and cannot be used to suggest mudline location of the blowout. Conventional surveying techniques are not available to fix the relative positions of the two wells.

The approach for underwater blowouts is to assume the blowout site is at the original coordinates. The relief well is spotted at a location based on the assumed site for the blowout. Accurate satellite navigation and other surveying techniques used for spotting are essential for the relief well. The accuracy must be less than 3 ft (1 m) even if multiple satellite passes are required.

Survey Analysis

Original surveys on the blowout well should be obtained, if possible, and reanalyzed. This is particularly important if the well is more than 10 years old. All directional calculations, including site location data, should be checked thoroughly.

Ellipse of Uncertainty

The well path is seldom in the exact spot suggested by survey analysis. It could lie in an area known as the "ellipse of uncertainty" or "cone of uncertainty."

Wolff and de Wardt are credited with quantifying survey errors and developing an approach to calculate the uncertainties.

They found that systematic errors had a greater influence on inaccuracies than random errors. After an analysis of magnetic and gyrosurveying techniques, they found that five sources of inaccuracy contribute to borehole position uncertainty: compass reference, compass instrument, inclination, misalignment, and depth errors.

Relative lateral position uncertainties were determined by Wolff and de Wardt, for example, as a function of the average hole inclination and are presented in Figure 4–17. A comparison of vertical, radial, and lateral error magnitudes revealed that the latter is the greatest over the full inclination range. Hence, only this one is discussed. The graph demonstrates the increasing lateral uncertainty with inclination for all types of surveys. A 13,000 ft (4,000 m) deep well of 45° average inclination cannot be surveyed more accurately than ± 115 ft (35 m). The uncertainty can be larger.

Uncertainty calculations are done by computer because of the large number of calculations required. Some operators and survey companies have in-house computer programs, and several commercial programs are available.

FIGURE 4-17 • Typical lateral position uncertainties of inclined wells for poor and good, magnetic and conventional gyro surveys (Courtesy of JPT, © 1981.)

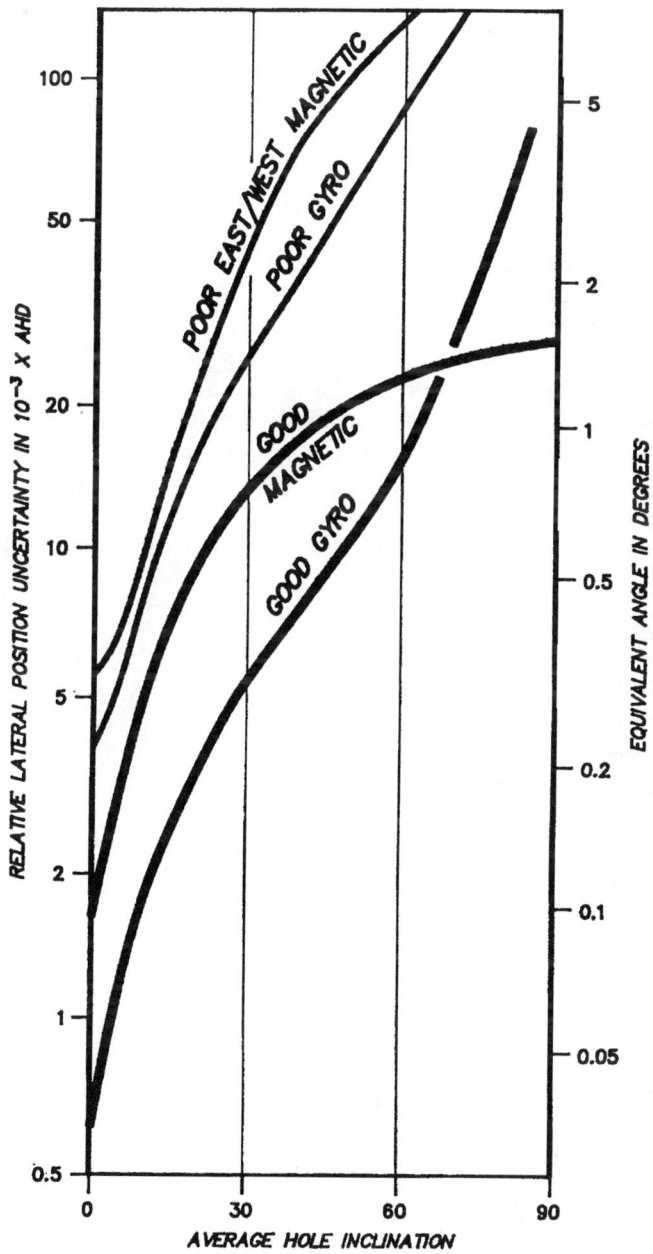

Relief Well Site Selection

Selecting a surface site, on land or water, for the relief well rig is usually simple. However, it can prove to be the most difficult aspect of the planning process for which no best solution exists.

Often, the site is selected hastily in an attempt to start drilling quickly. The site can prove to be a poor choice and result in a much longer time to drill the well than if more consideration had been given initially to proper selection of a well site.

Site selection is normally done by process of elimination. Over a dozen factors must be considered in some cases. These factors may eliminate certain sites or regions. After all factors have been considered, the remaining areas must be evaluated and a site selected from them. It is not uncommon that a good or desirable option for a site plan is not available.

If multiple wells are blowing out, an optimum site must be selected singularly for each well. A compromise site that allows hitting several wells is seldom the optimum site for hitting any single well.

Although a few factors may change at the time of a blowout, the optimum relief well location investigation should be carried out in advance in the blowout contingency plan. This would avoid compromise site selection difficulties.

Factors to be considered in relief well site selection include:

- Offset distance
- Optimum intercept/approach path
- Ellipse of uncertainty
- Proximity to other wells
- Shallow gas blowouts
- Debris
- Wind currents
- Water currents
- Heat
- Noise
- Bathymetry
- Localized gas seepage
- Insurance
- Regulatory agency requirements
- Mooring patterns

Each of these factors is discussed in the following sections.

Offset Distance

Often, a specific minimum offset distance is established between the relief and blowout well sites. It is commonly 3,000–4,500 ft (1,000–1,500 m). This offset distance seldom has any basis in fact. It is not a requirement of insurance underwriters or government agencies.

FIGURE 4-18 • Field layout and relief well site selection for Alaska blowout

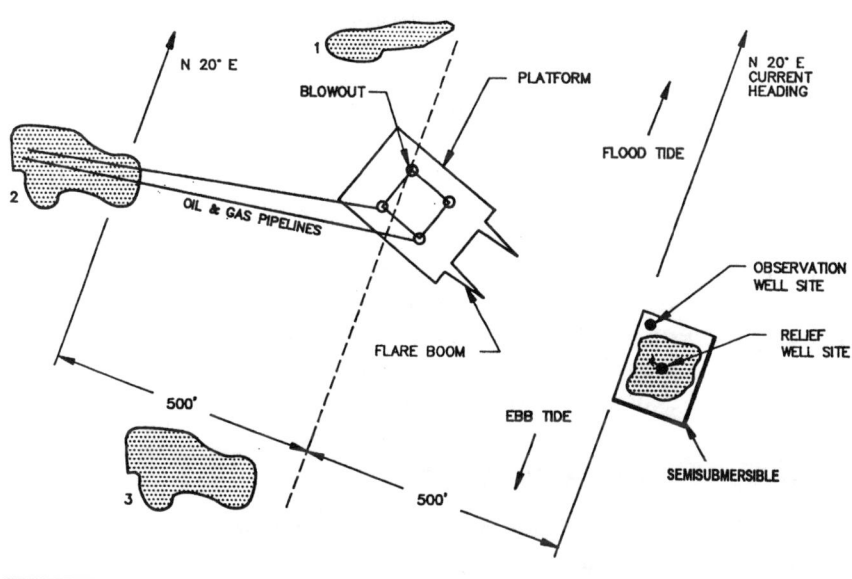

Technical reasons do not exist for arbitrarily establishing a minimum offset distance. Each situation must be established on its own set of technical facts. The existence of fire may be a prevailing factor.

Figure 4-18 is an example for an offset distance of 500 ft (150 m). Fires were not present.

Optimum Intercept/Approach Path

A primary consideration in selecting a site is the appropriate directional plan to be used. To some degree, this is an argument similar to the old question of "which came first, the chicken or the egg."

The key aspect of the directional plan is the approach angle between the relief well and the desired intersect point on the blowout well. Usually, the angle will be small and in the range of 5–15°. The selected surface site should provide the required horizontal displacement and a 5–15° relief well approach angle near target. Relief wells for shallow blowouts will require a site near the blowout well. The kickoff point (KOP) must be shallow, and usually the drift angles will be high so the hole can drop to the appropriate approach angle.

Deep blowouts allow more flexibility in site selection. However, a site as near as possible to the blowout is generally desirable to minimize the required horizontal displacement.

Ellipse of Uncertainty Consideration

Systematic survey errors create a cone or ellipse of uncertainty relative to the specific location of the blowout well and the relief well.

The depth of investigation for ranging tools should be considered with respect to the combined ellipses of uncertainty for the blowout and relief wells. The ellipse for the blowout well is fixed. However, the ellipse for the relief well depends to some degree on the directional plan. Consideration should be given to selecting a directional plan that minimizes the relief well ellipse of uncertainty if the combined ellipses for the two wells exceed the ranging tool's capability. The ranging tool's range of investigation is 200 ft (60 m) normal to the centerline of the tool under optimum conditions. It is often the case that the combined ellipse radius exceeds the 200 ft (60 m) range capability. If this is the case, even under optimum conditions for the relief well, multiple ranging runs will be required as the well is drilled near the blowout target.

Proximity to Other Wells

Site selection for relief wells must consider other wells in the area. The worst situation is shallow blowouts under or near a platform. A site and a directional plan must be chosen to avoid a collision with another well. More importantly, however, are ranging difficulties caused by interference from other wells. Since the wells are in close proximity under a platform, it can be difficult to select an appropriate site and directional plan to avoid well interference.

A field example is shown in Figure 4–19. For various reasons including water currents and pipeline restrictions, the rig position as shown was the only available site to drill the relief well. Magnetic ranging was hampered because of interference of other nonblowout wells. If the ranging was restricted until the relief well was near the blowout well, the ellipses of uncertainty between the two wells would have significant overlap. The relief well intersected the blowout well at the expected depth so ranging was not necessary. A post-intersect ranging run was made and the blowout well was 2 ft (60 cm) away from the relief well.

Shallow Gas Blowouts

Shallow gas blowouts cause significant site selection difficulties. A shallow gas blowout can deplete some shallow zones while charging others. The phenomenon of simultaneous depleting and charging from zones that were normally pressured in the same wellbore is difficult to explain but has been observed from field experience.

FIGURE 4–19 • Relief well proximity to other wells

To obtain reasonable directional programs for shallow blowouts, the relief well site should be near to the blowout well. However, sites near the blowout have a greater potential for being pressure-charged.

The following examples illustrate this phenomenon. The data are for known pressure charging at distances from the relief/observation well to the blowout.

Date	Event	Distance from Blowout to Known Gas Charging, ft
1983	Mobil/Banteng South China Sea	2,000 (610 m)
1984	Mobil/West Venture Sable Island, Canada	3,000 (910 m)
1985	Union/Grayling Plat. Cook Inlet, Alaska	2,000 (610 m)
1985	Shell/Patricia PA-5 Sarawak, Malaysia	1,500 (460 m)
1988	Cook Inlet, Alaska	500 (150 m)
1989	NFA16, Qatar	2,485 (757 m)

This does not appear to be an acute problem in deep blowout situations from field experiences.

Shallow seismic surveys should be run after a shallow gas blowout to evaluate possible pressure charging and direction of gas travel. The surveys should be compared against preblowout surveys. If gas flow is detected, the surveys should be rerun frequently to evaluate magnitude and direction.

Gas may flow preferentially according to fault orientation. This was observed on Mobil's West Venture event in 1984 as reported by Booth in *World Oil*, May 1990. If this is observed to be the case, relief well sites perpendicular to the fault orientation appear to be preferable.

Gas has traveled up vertically oriented faults in some field cases. The gas was observed at some distance from the blowout well. Shallow seismic surveys can usually identify this gas.

Debris. Debris is not typically a concern in relief well site selection. A significant amount of rig and platform debris at the blowout well site may negate the possibility of a vertical intervention and can be the controlling parameter in resorting to a relief well. However, it seldom has an impact on the relief well site.

Wind. Wind has both pro and con effects with respect to rigsite selection. The wind can carry any available gases to the relief well rig if it is located on the downwind side. The particular concerns are explosions related to flammable gases, such as methane, and for toxicity from hydrogen sulfide. The advantage to having a wind is obvious. Gases are diluted and dissipated.

The typical procedure is to evaluate the wind rosette for an area. Meteorological groups can define predominant wind directions and seasonal variations. The appropriate rig site would be upwind.

Offshore blowouts have not shown the degree of gas problems that theoretical models predict should occur. It appears that the gases are dissipated to a greater degree than predicted by the model before they reach the rig. The exact reason for this phenomenon is somewhat inexplicable.

Water Currents. The concern for water currents in an offshore environment relates to possible movement of an oil slick toward the rig. Similar to wind rosettes, guides are available for current predictions. The rig site should be evaluated for up-current positions if possible. If the blowout fluid is gas, water currents for pollution potential should not be a consideration.

Currents with respect to mooring considerations are a concern in some situations. If the currents are in a single direction, the problem is simply relegated to mooring analyses.

Heat. Heat loading from a blowout fire can be significant. However, it is often exaggerated. Very few fires have created heat loading that would require a rig to be positioned more than a few thousand feet from the blowout site.

Noise. Similar to the previous discussions on heat from a fire, the noise can be significant but is often exaggerated. Noise testing should be run at the time of the event.

Bathymetry. Minimum water depths required for support vessels must be evaluated if water depth varies significantly around the site. Minimum acceptable depths must be established for workboats, pump vessels, and the drilling rig. The seabed gradient must also be taken into consideration in selecting locations for jack-up operations. Bathymetry considerations do not apply to any degree in floating drilling except for anchor positioning.

Localized Gas Seepage. A preblowout localized gas seep will have an effect on site selection if the drilling rig is planned as a dynamically positioned drillship. Gas can interfere with the hole positioning/referencing system. The operator would not normally position his rig over a large gas seep.

Insurance. Contrary to the understanding of many groups concerning insurance requirements for relief well site location, very few regulations exist. The operative term in most insurance contracts is that the operator will act in a "prudent manner." This provides the necessary flexibility to make good engineering judgments based on facts relative to the current situation as opposed to arbitrary rules not applicable to the event.

Although not generally possible, it may prove beneficial to seek the advice of the adjuster in an unofficial manner. The adjuster is on-site to work as a representative for the underwriter and not the operator. They do not want to be seen as giving advice to the operator. However, the adjusters are typically very experienced and have seen many bad situations. If their input can be obtained, it will often prove valuable, but it should be considered in context with all other available facts.

Regulatory Agency Requirements

Government agencies are, in one sense, similar to insurance underwriters. They will not respond until a plan has been presented for review. Therefore, the site selection issue still remains with the operator. Government regulations relative to site selection are not known to exist at this time.

Many government agencies worldwide are taking a greater role in the technical review of proposed relief well plans and other activities. The agencies are better informed and very knowledgeable in many cases. Often experienced petroleum engineers and field operations personnel are staff members.

Mooring Patterns

Mooring patterns must be considered if multiple relief well rigs are used. The general guideline is to avoid crossing anchor lines. In some cases, the mooring spread on one or both rigs can be modified to achieve a noncrossing pattern.

Although the ideal situation is a mooring spread for both vessels that avoids line crossing, provisions can be made to safely cross the anchor lines if alternatives are not available. One vessel with chain mooring lines and the other with wireline can provide adequate separation due to the different line catenaries.

Intercept Point Selection

The "intercept point" is the depth at which the relief well establishes communication, or is about to establish communication, with the blowout well. It is not the same as a bypass made for location determination. Communications establishment is discussed in greater detail in later sections.

An off-bottom intercept seldom has been used. An apparent concern is that the pressure seen at a shallow depth in the blowout well could not be controlled with a shallow-intercept relief well. The fallacy with this thought process is it assumes that the blowout well does not have any depletion effects. Field cases clearly show that depletion occurs and that under blowout conditions the depletion is more extreme than could be expected from normal production rates. This applies to oil or gas blowouts. If the depletion effect is considered, an off-bottom intercept is worth consideration.

Advantages and disadvantages of a bottom and off-bottom intercept are described below.

Bottom Intercept

As stated previously, the bottom intercept approach has been used on most relief wells. It has functioned reasonably well. The advantages of the bottom intercept, as compared to an off-bottom or mid-range intercept, are as follows:

- A longer column of kill fluid exerts a greater hydrostatic pressure and, for a given pumping rate, the frictional back pressure will be higher in the blowout wellbore.
- The bottom-hole kill process minimizes the dilution of the kill fluids by the produced fluid influx; therefore, the buildup of the controlling pressure is achieved much faster.
- Required pumping capacity may be reduced because of the greater hydrostatic of the long kill column. However, it will generally tend to be greater because of frictional losses associated with the reduced hole geometry of a deep relief well.

The bottom intercept has distinct disadvantages that must be considered as well:

- The error of uncertainty increases with depth and may be quite large in a deep well.
- Directional control becomes more difficult as rock strength increases with depth.
- Directional control becomes more difficult as the size of the directional tool string is restricted by the progressively smaller strings of casing.
- High temperature gradients in deep wells can hamper logging and ranging surveys.
- Time to drill the relief well will be extended exponentially.

Intermediate (Off-Bottom) Intercept

The intermediate intercept has several advantages:

- The error of uncertainty for the relief and blowout wells will be reduced as compared to a deeper well with equivalent survey accuracy. The blowout well may be more easily located with ranging tools.
- Relief well casing size can be larger at the shallow intercept. This significantly reduces the pumping equipment requirements.
- The time required to drill the relief well to intermediate depth will be less than if drilled to total depth. Oil pollution is minimized due to reduced drilling time.

The intermediate intersect approach has few disadvantages assuming that a proper evaluation is completed with respect to depth of intersection, depletion in the blowout well, and fracture gradient at the intercept point. If the problem well is cased at the depth of the intercept, that casing must be perforated or milled prior to killing.

Figure 4–20 is an example of a blowout that required a bottom and mid-range intersect. The blowout was believed to be flowing from as many as six sands. An additional intersect at a shallower depth was planned but never required.

Observation Wells

An observation well is designed to perform the task suggested by its name. It is designed to provide a means to observe subsurface formations and evaluate several parameters that may be affected by a blowout.

Observation wells are becoming more common for shallow gas blowout control. It is recommended to drill an observation well(s) prior to commencing a relief well for a shallow gas blowout. This recommendation is based on case histories and field experiences. The mechanics of shallow gas blowouts have tended to cause subsurface soil disturbances.

The same situation has not been noticed, or at least it has not been reported for deeper blowouts. However, no reason exists for not drilling an observation well for deeper blowouts other than the time required to drill an observation well and then the relief well. Procedures have been designed for some wells where it can be used as an observation well and then a relief well.

Observation wells can monitor formations as they are drilled and logged. They are typically plugged after being drilled. Occasionally, wells have been drilled and used as a permanent monitoring station by placing downhole pressure sensors to provide a continuous record of transient pressures.

The first observation well that has been fully reported was in 1985 in Southeast Asia. It was instrumental in drilling a relief well for a shallow gas blowout.

FIGURE 4-20 • Hole and casing profile for dual intersect kill

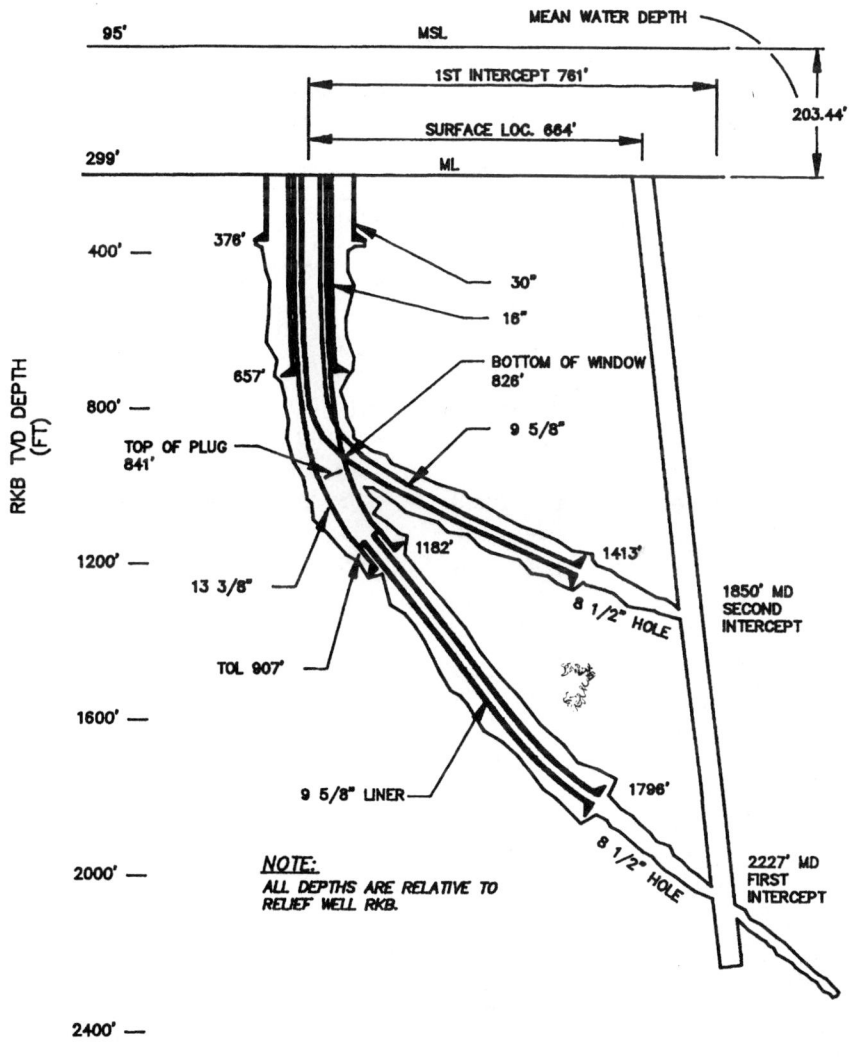

Purpose

The purpose of observation wells is simple. They provide monitoring or observation of the soil/formation and pressure situations. Typically, they are used for the following reasons:

- Identify zones of pressure charging and depletion
- Determine rate of pressure change
- Evaluate fluid movement near the well bore

- Identify the degree of subsurface soil disturbance that may impact drilling of the relief well

Information obtained from the observation well is used to plan the relief well. It can affect casing setting depth programs. This could affect casing sizing if additional unexpected strings of pipe are required. Muds may be required where seawater had been sufficient to drill the equivalent sections on the blowout well.

The typical procedure is to drill the observation well at least as deep as the depth of the first major casing string in the relief well. This could be the depth for structural or conductor pipe. Preferably, it will be the depth of the surface casing.

Drilling Guidelines

A key factor in drilling observation wells is to use caution. Penetrated formations may contain uncertainties that could cause problems. In fact, the purpose of drilling these wells is to eliminate the uncertainties.

The wells can be drilled vertically from the site. This will provide some valuable information. However, it should be remembered that zones in close proximity to the blowout could pose problems not encountered at a greater distance.

Another option, other than a vertical well, is to drill a directional profile toward the blowout well that does not intersect the well completely. This provides a better observation point without actually drilling into the blowout well.

Drilling riserless is recommended for offshore situations, if possible. The shallow charged zones can blow out. It is not desirable to bring this gas to the rig. Allowing it to flow subsea will not cause a problem to the rig if proper safeguards are implemented.

Some field cases have shown subsurface disturbances. These were evidenced by the fact that seawater had been used originally to drill the same sections in the blowout well. However, the observation well required mud or weighted gel water to drill the same intervals.

The well can be drilled as a completely expendable, low cost hole. It is drilled with a bit, motor, and MWD tool. The MWD tool provides a complete logging service except for RFTs. Guide bases are not used. RFTs can be run if the tool can be lowered into the well. After the well is drilled, it can be plugged. This option is clearly suited for floater drilling.

Another option is to drill the observation well in a manner similar to a normal well. Casing and cement are used. The well can be fully logged including RFTs. If desired, procedures can be established for using the well as a continuous pressure monitoring source.

Precautions

Most precautions for the observation wells should be obvious. Encountering pressure charged zones is possible. Hole stability may be a concern.

If the well starts to flow and casing/BOPs have not been used, procedures do not exist for controlling the flow. However, this may not be the major concern that it seems. If the flow is caused by the pressure charging from the blowout, the flow should cease after the blowout well is killed.

The benefit of the well flowing is that it serves as a vent for the gas. This should facilitate drilling the relief well. This matter must be addressed with the appropriate authorities.

Drilling precautions include the standard warnings. Drill at low rates. The rates should be sufficiently slow to obtain a good MWD log of the drilled section. Circulate the well clean. If casing is to be used, cement carefully and use gas blocking agents. Use subsea TV cameras or sonar units to detect gas leaving the hole if riserless mode is used.

Kill Hydraulics

The general objective of the relief well is to kill the blowout well with hydraulic control. This includes hydrostatic and friction pressure components. Early relief wells relied principally on hydrostatic control. Techniques developed in the late 1970s and the early 1980s combined friction and hydrostatic pressures to gain control of the well in two stages.

Various techniques have been proposed to kill blowouts via relief wells. These include the following:

- Overbalance kill
- Dynamic kill
- Reservoir flood (saturation kill)
- Momentum kill
- High rate production kill

They are summarized in the following paragraphs.

Overbalance Kill

Historically, most blowout kill attempts were based on the overbalance kill concept. After fluid flow communication has been established between the relief well and the blowout wellbore with water, drilling fluid of the required density is pumped at a rate sufficiently high to overcome flow and kill the well. The method requires a good understanding of the reservoir pressure in order to select the kill fluid density. Many wells have been killed by this method using heavy weight drilling mud or cement. The technique usually requires a significant number of high pressure pumps to achieve the required flow rates.

The major disadvantage of the technique lies in the potential for fracturing away from the problem wellbore. This would preclude the kill fluid from entering the blowing well bore (i.e., it would enter the formation instead of moving up the blowing well). The potential for high injection pressures can also mean that the required flow rates cannot be achieved and that the technique will fail to stop the flow.

This technique has been most successful where the blowout rate itself was relatively low. It will not be discussed further in this report. Other techniques that can provide quantitative results will be presented.

Dynamic Kill

This is a relatively new technique, developed in the late 1970s and early 1980s, which has been used successfully in controlling various high rate blowouts.

In this method the blowout is brought under control by initially pumping water or brine at a rate sufficient to overcome the blowout source's formation pressure. This occurs through a combination of the hydrostatic pressure of the water or brine in the wellbore supplemented by frictional pressure associated with the flow of kill fluid up the problem wellbore. After the formation flow is stopped, a drilling fluid of sufficient density to statically control formation pressure is pumped into the blowout well. The dynamic kill process must be continued until the higher density drilling fluid provides sufficient hydrostatic head to control the well under static conditions.

Water is pumped during the initial phase because of its wider availability. Communication is established with the water. Mud, which is more difficult to prepare and store, can be used for a final kill after the water has dynamically killed the flow.

During the pumping process, a monitoring string is used in the relief well to provide continuous pressure data. Frictional and hydrostatic pressure components in the blowout can be controlled through adjustment of the injection rate, and thus be related to a balance between the required kill pressures and formation fracturing pressures.

Reservoir Flooding

This process is occasionally called the "saturation method." It involves flooding the producing reservoir in the vicinity of the problem well by pumping water from a closely positioned relief well until production in the blowout well changes completely to water. If the water bank pressure has been maintained above the reservoir pressure, it will stop the gas or oil flow.

This technique is limited because higher volumes of water will be required as the distance between wells increases. With a high blowout rate, it may not be possible to develop sufficient pump rates to flood the producing formation around the problem wellbore. The reservoir parameters must be understood particularly well for this concept to be effective. Multi-layered zones cause complications.

Momentum Kill

This kill method supposedly utilizes the momentum of the kill fluid to overcome the momentum of the well fluids and reverse the flow. Although various technical papers have described successful field cases, the data in those papers do not seem to support the method. It appears that fluid friction actually killed the blowout, not momentum.

High Rate Production Kill

The high rate production kill concept is based on a relief well to produce fluids from the blowout source zone. The producing rate must be sufficient under controlled conditions to kill the blowout through a combination of near wellbore pressure drawdown and depletion of blowout zone fluids.

It has been used successfully to kill a dual zone producing well where both zones were blowing out concurrently. Design of a conventional relief well kill program was complicated by parted tubing strings in the original well and a rupture in the casing through which water was being produced. After commingled production was initiated from the relief well, both producing zones were killed within the calculated time frame.

Dynamic Kill Development

The dynamic kill method was developed by E.M. Blount of Mobil in the late 1970s as a response to a blowout in Arun, Indonesia. An article was published on the topic in *World Oil*, October 1981. The following description of the dynamic kill is based principally from the article.

Dynamic kill is an interim condition where a blowout is killed by injecting a fluid through a communication link and up the blowout annulus at such a rate that the static formation pressure is exceeded and the well ceases to produce. The flow is multiphase (produced fluid plus injected fluid) before the well is killed and single phase (injected fluid only) immediately after the well is killed.

Flow rate must be maintained so that the sum of frictional and hydrostatic pressure exceeds the static formation pressure until a heavier static kill fluid can replace the lighter dynamic kill fluid. The injection rate can be varied to control the bottom-hole pressure by adjusting the frictional component much in the same way the back pressure is controlled with an adjustable choke when conventionally circulating out a kick on a drilling rig. The basic approach to dynamic kill uses methods developed for analyzing performance of producing wells and considers the relief well and blowout well as a single system.

A communication link must be established between the two wells. Tubing is run in the relief well and filled with water to monitor pressure. Kill fluid, usually water, is injected down the annulus of the relief well and up the annulus of the blowout well along with produced fluids.

The object of the kill is to achieve a bottom-hole pressure (BHP) dynamically that exceeds the static formation pressure but does not fracture the formation. Controlling and monitoring BHP is the basis for success in a dynamic kill. BHP consists of the hydrostatic pressure exerted by the column of multiphase fluid, plus friction pressure exerted by the kill fluid being pumped up the blowing well's annulus.

BHP is controlled by altering the flow rate into the annulus of the relief well to adjust the frictional pressure, since there is no control on the blowing well as in kick control (i.e., a choke). BHP is monitored by observing surface pressure at the tubing in the relief well and adding hydrostatic pressure of the fluid inside the tubing. All injection must be down the relief well annulus, and the tubing must remain full of static fluid.

The system's "relief valve" is the fracture pressure of the formation. If the formation is fractured, some fluid injected into the relief well will go into the formation and not up the blowing well. Fracture pressure limitation is imposed on the relief well, upstream of the communication channel, rather than in the blowout well.

The kill procedure can be controlled precisely by observing the tubing pressure. The injection rate of the initial kill fluid can be increased until the static formation pressure is exceeded. This will permit the blowing well to be displaced with initial kill fluid by preventing formation fluids from entering the blowout well. The well should be dynamically killed at this point. Injection of the intermediate fluid can commence, and the rate reduced after the intermediate fluid enters the blowout well to keep the BHP below the fracture pressure but above the static formation pressure.

Design Parameters

In designing a dynamic kill operation, several parameters must be predetermined. Basic factors are:

- Kill fluid density
- Kill fluid injection rates

BLOWOUTS

- Size of relief well
- Hydraulic horsepower
- Maximum allowable surface pressure to prevent drill pipe from being ejected (pumped out of the hole at the relief well)

Each is discussed in the following sections.

Kill Fluid Density

The density of the ideal dynamic kill fluid can be determined by finding a fluid so that the introduction of a bubble of gas into the single phase stream flowing at the rate required to control the dead well will increase the frictional pressure component as much as the hydrostatic pressure component is reduced. The density of the initial kill fluid can be determined by the following condition: (Refer to end of section for nomenclature.)

$$\rho_f \leq \frac{12.836\ P_s}{TVD}$$

The derivation follows:
The frictional pressure,

$$\Delta P_f \leq \frac{CfL\ \rho_f V_f^2}{d_h}$$

Where V_f is the velocity of the fluid, d_h is the hydraulic diameter and C is a constant. Assume that gas bubbles entered the flow stream. In bubble flow regime, the continuous fluid is the liquid phase. Let ϕ_g be the fraction of gas volume to the total fluid volume, then,

$$V_f = \frac{V_1}{\left(1 - \phi_g\right)}$$

$$\rho_f = \rho_1\left(1 - \phi_g\right) + \rho_g \phi_g$$

Since

$$P = \Delta P_{hyd} + \Delta P_f$$

By taking the derivative of the above equation with respect to the gas fraction, then,

$$\frac{dP}{d\phi_g} = \frac{d\ \Delta P_{hyd}}{d\phi_g} + \frac{d\ \Delta P_f}{d\phi_g} \tag{4.1}$$

$$\Delta P_{hyd} = \frac{0.433}{8.337} \rho_f TVD$$

$$= \frac{0.433}{8.337} TVD \left[\rho_1\left(1 - \phi_g\right) + \rho_g \phi_g\right]$$

Then,

$$\frac{d \Delta P_{hyd}}{d\phi_g} = \frac{0.433}{8.337} TVD \left[-\rho_l + \rho_g\right] \quad (4.2)$$

$$= -\Delta P_{hyd} + \Delta P_{hyd_g}$$

$$\approx -\Delta P_{hyd}$$

$$\Delta P_f = \frac{CfL \rho_l V_f^2}{d_h} = \frac{CfL \rho_l V_l^2}{d_h (1-\phi_g)^2}$$

$$\frac{d\Delta P_f}{d\phi_g} = \frac{2CfL \rho_l V_f^2}{d_h (1-\phi_g)^3} = \Delta P_{f_1} \frac{2}{(1-\phi_g)^3} \quad (4.3)$$

Substitute eq. (4.2) and (4.3) into (4.1)

$$\frac{dP}{d\phi_g} = 2\Delta P_{f_1} \frac{1}{(1-\phi_g)^3} - \Delta P_{hyd}$$

The pressure should increase with the introduction of gas bubbles, i.e., the following should occur:

$$\frac{dP}{d\phi_g} \geq 0$$

Since ΔP_{f_1} and ΔP_{hyd} are always non-negative, and

$$\frac{1}{(1-\phi_g)^3} \geq 1 \quad (4.4)$$

then the condition

$$2 \Delta P_{f_1} \geq \Delta P_{hyd} \quad (4.5)$$

will ensure condition (4.4)
Since

$$P_s = \Delta P_{hyd} + \Delta P_{f_1}$$

or,

$$\Delta P_{f_1} = P_s - \Delta P_{hyd}$$

substitute into (4.5)

$$2(P_s - \Delta P_{hyd}) \geq \Delta P_{hyd}$$

or

$$P_s \geq 1.5 \Delta P_{hyd}$$

BLOWOUTS

but

$$\Delta P_{hyd} = \frac{0.433}{8.337} \rho_f \text{ TVD}$$

$$P_s \geq \frac{1.5 \times 0.433}{8.337} \rho_f \text{ TVD}$$

or,

$$\rho_f \geq \frac{12.836 \, P_s}{\text{TVD}}$$

Estimation of Flow Rate Requirement:

$$\Delta P_f = \frac{11.41 \, fL \, \rho q^2}{d_e^5}$$

In the blowout well, there should be,

$$\Delta P_{f_b} = P_s - \Delta P_{hyd}$$

Such that the BHP of blowout well is P_s
Then

$$P_s - \Delta P_{hyd} = 11.41 \left(\frac{fL}{d_e^5}\right)_b \rho_f q_b^2$$

$$\text{or } q_b^2 = \frac{(P_s - \Delta P_{hyd})}{11.41 \, \rho_f} \left(\frac{d_e^5}{fL}\right)_b$$

$$q_b = \left[\frac{(P_s - \Delta P_{hyd})}{11.41 \, \rho_f} \left(\frac{d_e^5}{fL}\right)_b\right]^{1/2}$$

Recall $k = \dfrac{q_b}{q_r}$

Injection rate required in relief well:

$$q_r = \frac{1}{k}\left[\frac{(P_s - \Delta P_{hyd})}{11.41 \, \rho_f} \left(\frac{d_e^5}{fL}\right)_b\right]^{1/2}$$

Size of the Relief Well

In this section a technique is derived for determining the size a relief well must be or how many relief wells are required to enable a blowout to be dynamically killed considering the poorest communication system and without exceeding the pressure limitations of the surface equipment.

Examine the frictional loss equation:

$$\Delta P_f = \frac{CfL\,\rho_f q^2}{d_e^5}$$

$$= C\left(\frac{fL}{d_e^5}\right)\rho_f q^2$$

Assuming complete turbulence, or

$$f = \frac{0.25}{\left(2\log\frac{d_h}{e} + 1.14\right)^2}$$

then the term

$$\left(\frac{fL}{d_e^5}\right)$$

is a casing-tubing characteristic, which does not depend on fluid properties. This term is called "flow resistance." The flow resistance of a well consisting of N multiple sections in series is the sum of each section flow resistance, i.e.,

$$\left(\frac{fL}{d_e^5}\right)_{total} = \sum_{i=1}^{N}\left(\frac{fL}{d_e^5}\right)_i \qquad (4.6)$$

For a well with N parallel flow resistances, the equivalent flow resistance is given by

$$\left(\frac{fL}{d_e^5}\right)_{equiv.} = \frac{1}{\left[\sum_{i=1}^{N}\left(\frac{d_e^5}{fL}\right)_i^{1/2}\right]^2} \qquad (4.7)$$

Consider the blowout well/communication channel/relief well system. Assuming the flow resistance of the blowout well is

$$\left(\frac{fL}{d_e^5}\right)_b$$

BLOWOUTS

then the question is to determine the flow resistance of relief well

$$\left(\frac{fL}{d_e^5}\right)_r$$

kill fluid, and the corresponding hydraulic horsepower required. Assume that the maximum surface equipment operating pressure $P_{an\text{-}max}$, formation fracture pressure P_{frac}, and the reservoir static pressure P_s are known.

In a dynamic kill, the BHP of blowout well should be kept above the static reservoir pressure and the BHP of the relief well below the formation fracture pressure. Therefore, the maximum allowable pressure drop across the communication channel still achieving dynamic kill is $P_{frac} - P_s$. When a single fluid is injected through the relief well and comes out from the blowout well with a WHP = 0 psig, the injection pressure of the relief well equals the total frictional pressure loss, i.e.,

$$P_{an} = \Delta P_{f_b} + \Delta P_{f_r} + \Delta P_{f_c}$$

Since the worst communication should be prepared for, the BHP of blowout well and relief well are assumed to be P_s and P_{frac}, respectively. Then,

$$\Delta P_{f_b} = P_s - \Delta P_{hyd}$$

and,

$$\Delta P_{f_b} = P_{an} - P_{frac} + \Delta P_{hyd}$$

Then the maximum allowable

$$\Delta P_{f_{r\,max}} = P_{an\text{-}max} - P_{frac} + \Delta P_{hyd}$$

The relief well should be designed so that

$$\Delta P_{f_r} \leq \Delta P_{f_{r\,max}}$$

$$\frac{\Delta P_{f_r}}{\Delta P_{f_b}} \leq \frac{\Delta P_{f_{r\,max}}}{\Delta P_{f_b}}$$

or,

$$\frac{\Delta P_{f_r}}{\Delta P_{f_b}} \leq \frac{P_{an\text{-}max} - P_{frac} + \Delta P_{hyd}}{P_s - \Delta P_{hyd}} \qquad (4.8)$$

Since

$$\Delta P_f = \frac{11.41\, fL\, \rho_f q^2}{d_e^5}$$

then

$$\Delta P_{f_r} = 11.41 \left(\frac{fL}{d_e^5}\right)_r \rho_f q_r^2$$

$$\Delta P_{f_b} = 11.41 \left(\frac{fL}{d_e^5}\right)_b \rho_f q_b^2$$

Then (4.8) becomes

$$\frac{\left(\dfrac{fL}{d_e^5}\right)_r q_r^2}{\left(\dfrac{fL}{d_e^5}\right)_b q_b^2} \leq \frac{P_{an-max} - P_{frac} + \Delta P_{hyd}}{P_s - P_{hyd}}$$

or

$$\frac{\left(\dfrac{fL}{d_e^5}\right)_r}{\left(\dfrac{fL}{d_e^5}\right)_b} \leq k^2 \left[\frac{P_{an-max} - P_{frac} + \Delta P_{hyd}}{P_s - P_{hyd}}\right] \quad (4.9)$$

where $k = \dfrac{q_b}{q_r}$ = fraction of flow entering blowout well and $1-k$ = leak off in fraction of q_r.

Equation (4.9) is the basic equation for designing the relief well. Precise calculation of rates is not required. Errors in assumptions of roughness factors, fluid properties, and so forth, cancel out. If a single relief well cannot be practically completed with large enough d_e, then multiple relief wells will be required. Equation (4.7) can be used to predict the effective d_e for different size wells.

Estimation of HHP Required

Assuming injection wellhead pressure at relief well is P_{an-max} then,

$$HHP = \frac{42 \, q_r P_{an-max}}{1714}$$

$$= \frac{q_r P_{an-max}}{40.81}$$

Derivation of Maximum Allowable BHP to Prevent Drill String Ejection

This section only considers ejection from a vertical hole. A force tending to eject the drill string is composed of the frictional drag and the hydraulic force acting on various cross-sections of the drill string. The hydraulic force is sometimes considered as two forces called "buoyancy" and form drag but is correctly handled as one force, which results from the hydraulic pressure acting on the cross-section of the drill string.

$$\text{Hydraulic Force } (F_H) = \frac{\pi}{4} d_i^2 P_{BH} \tag{4.10}$$

where,

d_i = OD of drill pipe

d_o = ID of casing or open hole

The total frictional drag can be calculated by determining the frictional pressure drop (ΔP_f) and applying this stress to the cross-section of flow (A_{an}):

$$\text{Total Drag} = \Delta P_f A_{an} \tag{4.11}$$

This total frictional drag is applied to both the inside surface of the casing and the outside surface of the drill string. The ratio (R) of the total frictional drag that applies to the inner string is determined by the ratio of the shear stresses.

Drag on drill string (F_{DS}):

$$F_{DS} = R\Delta P_f \frac{\pi}{4}\left(d_o^2 - d_i^2\right) \tag{4.12}$$

$$R = \frac{1}{2\ln\left(\frac{d_o}{d_i}\right)} = \frac{d_i^2}{\left(d_o^2 - d_i^2\right)} \tag{4.13}$$

The weight of the drill string (W_s) resists the ejection force. If the ejection force is greater than the weight, the pipe will be ejected. The air (vacuum) weight of the string is used as W_s because the buoyancy is included in the hydraulic force. If the bit is plugged and the drill pipe is full of mud, the total weight of the mud and drill pipe should be included in the weight of the drill string. If the bit is plugged and the drill pipe empty, only the weight of the steel is considered. If the bit is not plugged and flow goes up the inside drill pipe as well as outside the drill pipe, the drag on the inside must also be considered.

$$\frac{\pi}{4} d_i^2 P_{BH} + \frac{\pi}{4}\left(d_o^2 - d_i^2\right) R \Delta P_f \leq W_s \qquad (4.14)$$

P_f can be calculated from various flow equations, but since we are monitoring and controlling on bottom-hole pressure (P_{BH}), much of the potential inaccuracies of frictional calculation can be eliminated by calculating ΔP_f as follows:

$$\Delta P_f = P_{BH} - \Delta P_{hyd} \qquad (4.15)$$

$$\frac{\pi}{4} d_i^2 P_{BH} + \frac{\pi}{4}\left(d_o^2 - d_i^2\right) R\left(P_{BH} - \Delta P_{hyd}\right) \leq W_s$$

$$P_{BH}\left[\frac{\pi}{4} d_i^2 + \frac{\pi}{4}\left(d_o^2 - d_i^2\right) R\right] \leq W_s + \frac{\pi}{4}\left(d_o^2 - d_i^2\right) R \Delta P_{hyd}$$

$$P_{BH_{max}} = \frac{W_s + \frac{\pi}{4}\left(d_o^2 - d_i^2\right) R \Delta P_{hyd}}{\frac{\pi}{4} d_i^2 + \frac{\pi}{4}\left(d_o^2 - d_i^2\right) R} \qquad (4.16)$$

where

$$\Delta P_{hyd} = \frac{0.433}{8.33} \rho_f (TVD) = \frac{\rho_f(TVD)}{19.25} \qquad (4.17)$$

$$P_{BH} \leq \frac{W_s + A_{an} R \Delta P_{hyd}}{A_{d_p} + A_{an} R} \qquad (4.18)$$

Nomenclature

A_{an} = area of annulus, in²
A_{dp} = area of drill string O.D., in²
D = true vertical depth, feet (TVD)
d_r = equivalent diameter, in.
d_h = hydraulic diameter, in.
f = Fanning friction factor (0.25 Moody Friction Factor)
$1-k$ = fractional leak off, $k = q_b/q_r$
L = measured depth, feet (MD)
P_{frac} = fracture pressure of formation, psig
P_{BH} = bottom-hole pressure (BHP), psig
P_{an} = injection pressure in relief well annulus, psi
P_s = static formation pressure, psig
P_{tbg} = tubing pressure, relief well, psig
ΔP_f = frictional pressure loss, psi
ΔP_{f_b} = frictional pressure loss, blowout well ($P_s - P_{hyd}$), psi

ΔP_{f_c} = frictional pressure loss, communication channel between wells, psi
ΔP_{f_r} = frictional pressure loss, relief well ($P_{an} - [P_{frac} - P_{hyd}]$)
ΔP_{hyd} = component of BHP due to hydrostatic weight of fluid, psi
R = ratio of frictional drag on drill string, total friction
$\varnothing_g$ = gas fraction
q = flow rate, bpm
q_b = flow up blowout well (kill rate), bpm
q_s = injection down relief well, bpm
W_s = weight of drill string in air, lb
WHP = wellhead pressure, psi
ρ_f = density of fluid, ppg

Subscripts

b = blowout well
c = communication
f = fluid
r = relief well
g = gas

Example of Dynamic Kill Calculations

Designing a kill job using the dynamic approach is best suited to computers. The experience of Neal Adams Firefighters is that many sets of calculations will be necessary because of uncertainties associated with the kill operations. An example of a run for a recent 1990 blowout is presented. The calculations are run with DYNKIL.

Example 4.1.

A well blew out in the Gulf of Mexico in September 1990. The data used for calculating the dynamic kill procedure is included in the attached computer printouts. The well was difficult to kill because the blowout occurred after the tubing was out of the well. (See Figure 4–21 and Table 4–1.)

Workover/Production Kill Operations

The dynamic kill principle has applications in workovers and production operations to solve various problems. An example might be a hole in the tubing. The relatively small diameters of the tubulars enhance the dynamic killing operations. The kill string should use an inner diameter as large as reasonable to minimize the parasite friction pressures if pumping is down the tubing. An example is shown in Figure 4–22.

FIGURE 4-21 • Blowout configuration for example 4.1

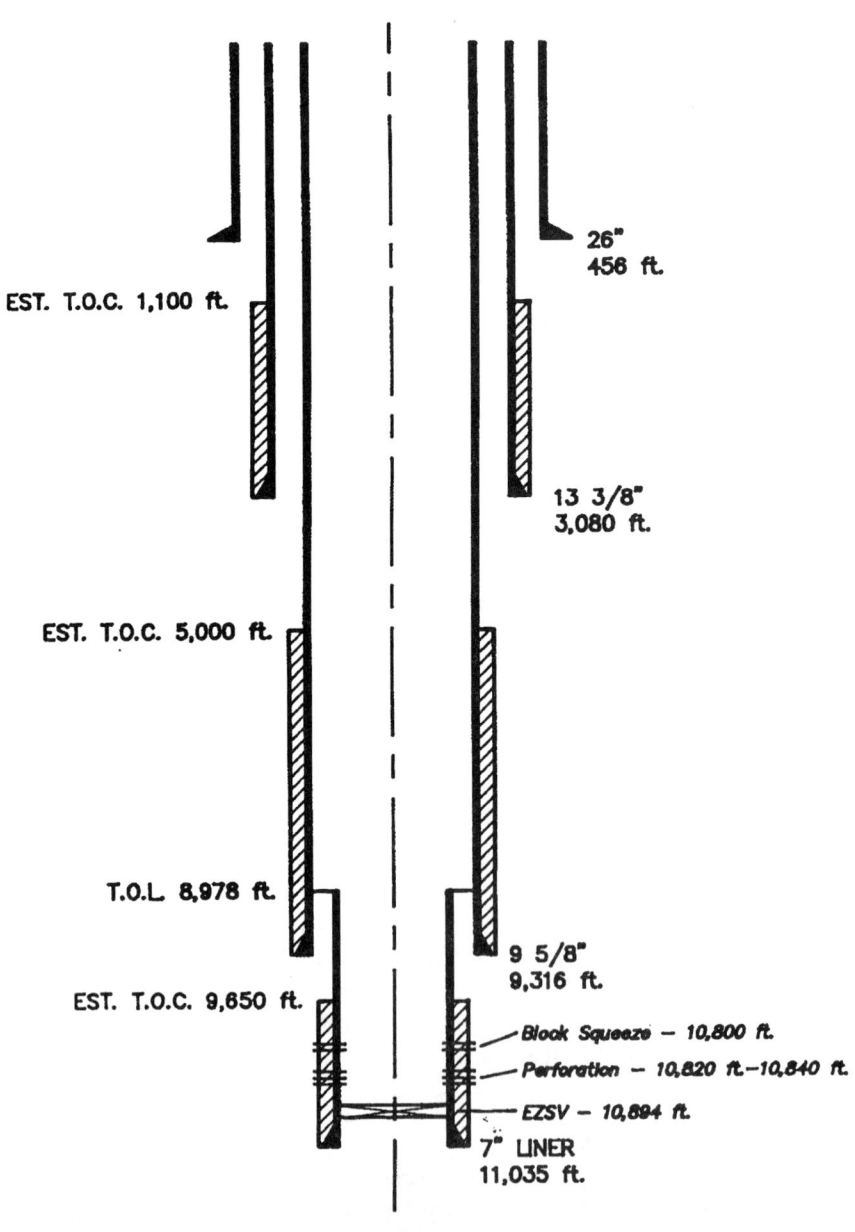

TABLE 4-1 • Dynamic kill 13 3/8" casing (BHP 6460 psi)

```
OPERATOR:                                              DATE:   12-SEP-90
LEASE:    Gulf of Mexico                               FIELD:  EUGENE ISLAND, BLOCK 2
SEC.         TWP.        RNG.         COUNTY:                  STATE:  LA
```

```
============================================================================
                         DYNAMIC KILL SUMMARY
============================================================================

     VOLUMES:
     ANNULAR VOLUME OF BLOWOUT WELL (BBLS) ............ =     723.049
     ANNULAR VOLUME OF RELIEF WELL (BBLS) ............. =    1425.938

     INITIAL KILL:
     WEIGHT OF INITIAL KILL FLUID (PPG) ............... =       8.600
     PUMPING RATE (BBLS/MIN) .......................... =     146.928

     PUMPING RATE TO EJECT EMPTY DRILLSTRING (BBLS/MIN) =      78.685
     CORRESPONDING BOTTOM-HOLE PRESSURE (PSI) ......... =    5342.662

     PUMPING RATE TO EJECT FULL DRILLSTRING (BBLS/MIN) =       81.017
     CORRESPONDING BOTTOM-HOLE PRESSURE (PSI) ......... =    5372.194

     FINAL KILL:
     WEIGHT OF FINAL KILL FLUID (PPG) ................. =      12.500
     RESERVOIR PRESSURE (PPG) ......................... =      11.624

     PUMPS:
     MAXIMUM PUMP PRESSURES (PSI) ..................... =    5491.963
     HYDRAULIC HORSEPOWER REQUIRED .................... =   19772.630
============================================================================
```

```
============================================================================
                 PUMPING SCHEDULE -- 8.60 PPG TO 12.50 PPG
============================================================================
```

| TIME | VOLUME | INJECTION | RELIEF WELL | | RELIEF WELL | |
| | PUMPED | RATE | ANNULAR PRESSURE | | TUBING PRESSURE | |
(MIN)	(BBLS)	(BBLS/MIN)	MIN (PSI)	MAX	MIN (PSI)	MAX
.00	.0	146.9	2226.	5492.	1680.	4946.
9.75	1418.2	146.9	737.	4003.	1680.	4946.
10.00	1486.1	102.7	186.	3452.	1680.	4946.
10.25	1510.9	95.4	105.	3371.	1680.	4946.
10.50	1534.3	91.8	64.	3330.	1680.	4946.
10.75	1556.8	88.4	26.	3291.	1680.	4946.
11.00	1578.5	85.0	0.	3256.	1680.	4946.
11.25	1599.3	81.8	0.	3223.	1680.	4946.
11.50	1619.4	78.6	0.	3192.	1680.	4946.
11.75	1638.6	75.6	0.	3163.	1680.	4946.
12.25	1675.0	69.9	0.	3111.	1680.	4946.
12.75	1708.6	64.4	0.	3066.	1680.	4946.
13.25	1739.5	59.3	0.	3026.	1680.	4946.
13.75	1768.0	54.5	0.	2991.	1680.	4946.
14.25	1794.1	50.0	0.	2961.	1680.	4946.
14.75	1818.1	45.9	0.	2935.	1680.	4946.
15.25	1840.0	41.9	0.	2913.	1680.	4946.
15.75	1860.1	38.2	0.	2892.	1680.	4946.
16.25	1878.2	34.5	0.	2874.	1680.	4946.
16.75	1894.6	31.0	0.	2858.	1680.	4946.
17.25	1909.3	27.6	0.	2844.	1680.	4946.
17.75	1922.3	24.4	0.	2831.	1680.	4946.
18.25	1933.7	21.2	0.	2820.	1680.	4946.
18.75	1943.5	18.2	0.	2810.	1680.	4946.
19.75	1959.1	13.0	0.	2802.	1680.	4946.
20.75	1969.2	7.2	0.	2795.	1680.	4946.
21.75	1973.6	1.4	0.	2788.	1680.	4946.
22.75	1974.5	.5	0.	2786.	1680.	4946.

```
============================================================================
```

FIGURE 4-22 • Dynamic killing of production well

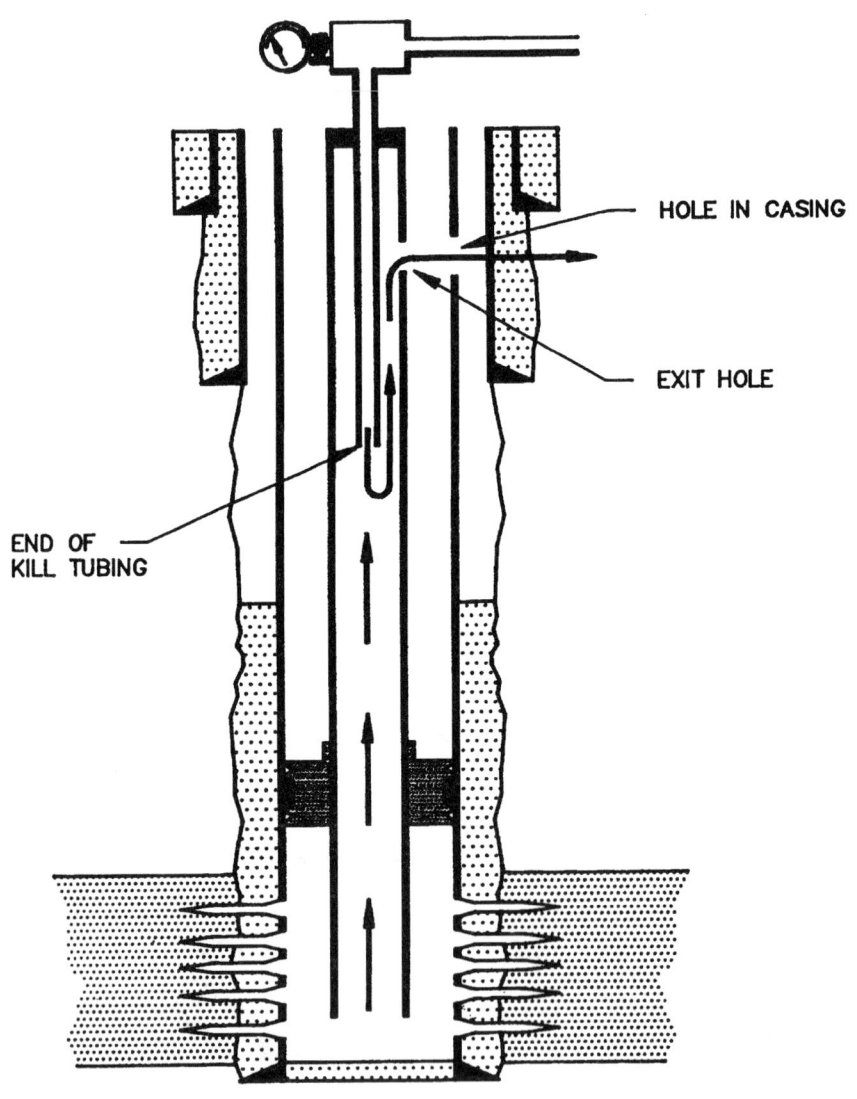

Reservoir Flooding Development

Reservoir flooding, or saturation flooding as it is occasionally termed, was perhaps the first formal technique developed for blowout kill operations. It was developed mathematically from basic reservoir equations. The key document used as the basis for the following discussion is "Reservoir

BLOWOUTS

Engineering Techniques To Predict Blowout Control During the Bay Marchand Fire" by Miller and Clements presented in *The Journal of Petroleum Technology*, March 1972.

The reservoir flooding technique was formally developed by Shell Oil Company in response to its Bay Marchand platform blowouts in 1970. Eleven of the 22 wells were burning and relief wells were required. Since ranging tools were not developed at that time, the selected approach was to drill into the reservoir as near as possible and then the reservoir was flooded via the relief well.

Reservoir simulation is an effective approach to predicting kill requirements. However, it is often impractical because of the number of uncertainties requiring many time-consuming runs. An easily run model was developed by Shell for the Bay Marchard incident.

The objective of the kill operation is simple (see Figure 4–23). A water bank is created by injecting water into the reservoir via the relief well(s). The injection rate must be sufficient so the pressure internal to the water bank exceeds reservoir pressure. If the situation is maintained, the oil or gas flow will be shutoff when the leading edge of the water bank surrounds the blowout well.

The method will probably fail, as it has in many case histories, if the pumping rate is not sufficient to maintain the internal water bank pressure in excess of the reservoir pressure. If the water pressure is less than reservoir pressure, the pumped fluids might be gas lifted up the blowout well. This gas lifting eventually allows the blowout to be killed if the water entrained in the flow stream increases the hydrostatic pressure and results in a lower flow rate. If this situation continues, it could "load up" the well and kill the flow, particularly if the reservoir pressure has been depleted via the flow.

Assuming a limiting bottom-hole injection pressure, Darcy's law of fluid flow yields the following expression for the maximum rate of water injection for a given size of water bank, r_b.

$$i_{wmax} = \frac{0.00707 \, k_w h \left(P_{iwfmax} - \overline{P} \right)}{\mu_w \ln \left(\frac{r_b}{r_{we}} \right)} \quad (4.19)$$

$$\text{where } r_{we} = r_{we}^{-s} \quad (4.20)$$

It is necessary to study the rise in bottom-hole injection pressure as the water bank volume increases while injecting at a constant rate, i_w. The above equations can be rewritten in terms of the bottom-hole injection pressure, P_{iwf}.

$$P_{iwf} = \overline{P} + \frac{141.4 \, i_w \, \mu_w \ln \frac{r_b}{r_{we}}}{k_w h} \quad (4.21)$$

FIGURE 4-23 • Reservior flood kill operations with two relief wells (Courtesy of JPT, © 1972.)

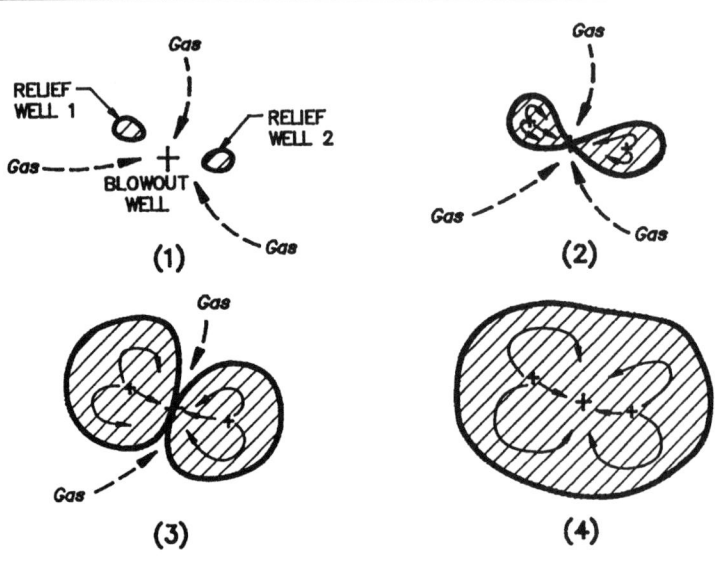

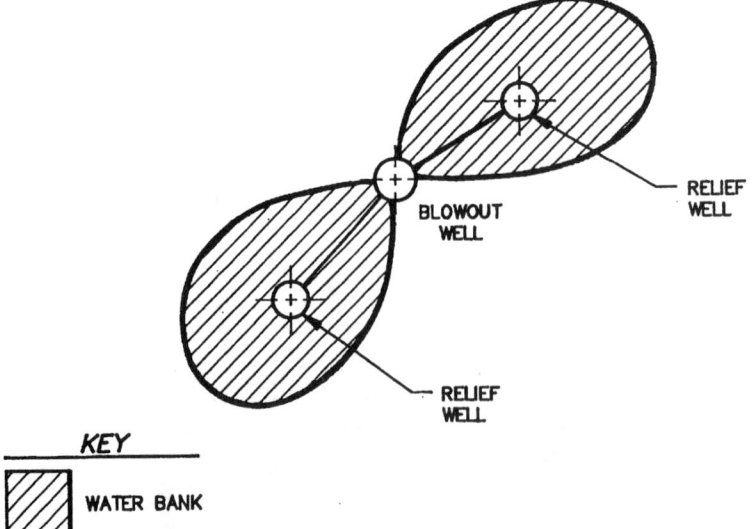

KEY

WATER BANK

*After Miller and Clements

The cumulative volume of water injected can be expressed as a function of the radius of the water bank, r_b.

$$W_i = \frac{\pi \phi h \left(1 - S_{wc} - S_{or}\right) \left(r_b^2 - r_w^2\right)}{5.615} \quad (4.22)$$

These equations permit the calculation of the apparent time of water breakthrough and volume of water because they ignore the way the water bank is distorted by flow into the producer. It is believed to be preferable to err on the conservative rather than risk underestimating the time and volumes to achieve breakthrough.

The dimensions for each variable are as follows:

h = net thickness, ft
i_w = water injection rate, bbl/day
k_w = effective permeability to water, md
$\overline{P}$ = average static reservoir pressure, psia
P_{iwf} = injection well bottom-hole pressure, flowing, psia
r_b = radius of injected water bank, ft
r_w = wellbore radius of injector, ft
r_{we} = effective wellbore radius of injector, ft
S = skin effect factor
S_{wc} = connate water saturation, fraction
S_{or} = residual oil saturation, fraction
W_i = cumulative water injected, bbl
μ_w = water viscosity, cp
ϕ = porosity, fraction

Although the dynamic kill method is considered preferable in most situations, the reservoir flooding method has distinct applications. An example involves shallow gas blowouts. The shallow formations often erode to a diameter that makes a dynamic kill virtually impossible. This situation lends itself to flooding.

Momentum Kill

The use of "engineered" fluid dynamics was first reported in 1977 as the momentum kill. The fluid dynamics kill concept utilizes the momentum of the kill fluid to overcome the momentum of the well blowout fluids and reverse the flow.

The momentum of the blowout fluids is shown by:

$$M_g = \frac{\rho_{sc} Q_{sc} U_i}{g_c} \quad (4.23)$$

where:

M_g = gas momentum
ρ_{sc} = gas density, standard conditions
Q_{sc} = gas flow rate at standard conditions
Z_i = gas compressibility factor
T_i = temperature
g_c = gravitational constant
R = gas constant
S = specific gravity of the gas
M_m = air molecular weight
P_i = pressure, point of interest
A_i = area, point of interest
U_i = velocity, point of interest

Units are in any basic system. As can be seen in Eq. 4.23, the momentum of the gas is primarily a function of its velocity.

The momentum of the kill fluid is given by Eq. 4.24.

$$M_m = \frac{Q^2 \rho}{g_c A} \qquad (4.24)$$

ρ = fluid density
Q = volume flow rate
g_c = gravitational constant
A = area, point of intersect

Again, the units must be consistent with any basic system. The momentum of the kill fluid is a function of both density and velocity. The density of the kill fluid is considered to be critical in keeping the well killed once the momentum of the kill fluid has overcome the flow from the blowout.

The momentum kill is considered to have unanswered questions. If the field cases quoted by the authors are examined closely, it appears that these wells were killed dynamically, (i.e., friction and hydrostatic pressure). It is not clear in the published papers what is offered by the momentum kill that is different than a dynamic kill. It is possible that it has differences that can be translated as advantage, but these differences are not self-evident. As such, the technique will not be further discussed in this book.

Number of Required Relief Wells

A common question in blowouts relates to the number of required kill wells. This is particularly true of big blowouts:

- Is one well sufficient for the kill?
- Will two relief wells be required?
- Should a second well be started as a standby?

Several factors affect this question. Each will be discussed. The issue should be decided on technical basis as opposed to an irrational decision.

Kill Hydraulics

The initial step to determine the required number of kill wells is to evaluate the blowout well for the following items:

- Estimated bottom-hole pressure
- Blowout fluid type
- Estimate of permeability ranges
- Zone thickness
- Wellbore geometry
- Water depth

Other factors not shown have a minor effect on the kill hydraulics.

Bottom-hole pressure can be estimated from reservoir production information or from offset well data. If the blowout occurs while tripping on an exploratory well, the formation pressure is assumed to be equal to or less than the original mud weight before starting the trip. If the blowout occurs while drilling and taking a kick, the formation pressure is known to be greater than the original mud weight. If no other data is available, an average kick value of 0.5 ppg (60 kg/m^3) can be used. This value is a statistical average from 3,800 well kicks.

With respect to blowout fluid types, gas and oil pose different kill situations. Gas has a lower hydrostatic pressure and higher blowout rates, but it draws down reservoir pressure more quickly. Oil blowouts are easier to kill from a hydrostatic view, but they have less drawdown in the reservoir.

A key factor is reservoir permeability. This value is seldom known with any degree of certainty. When making kill requirement estimates, it is important that the permeability be viewed with practicality. "What if" situations should be avoided. As an example for a blowout, "we believe the permeability to be 250 millidarcies but what if it is 500 millidarcies?" This "what if" can mean the difference between 5 and 15 kill pumps and 50 to 100 bbl/min (8–16 m^3/min) requirements. If our oil and gas reservoirs performed worldwide like we think they might in blowouts, we would never have an energy shortage.

Water depth has an impact on blowouts. Key effects are:

- The seawater hydrostatic acts as a choke and prevents gas expansion in the critical low pressure environments.
- The water acts as a buffer and allows a safe vertical intervention.

- The water masks the effects of methane and H_2S release on the surface.
- The back pressure reduces flow rates out of the well.
- Reduced flow rates inhibit bridging.
- Reduced flow rates mitigate reservoir drawdown.

Several of these factors relate to kill hydraulics.

After these factors have been evaluated, the kill system must be designed. The kill calculations on dynamic killing and reservoir flooding are commonly used. They are performed with computers and the results are given in horsepower (i.e., pressure and flow rates).

When converting from calculated horsepower based on pressures and flow rates to actual mechanical horsepower, an efficiency factor must be introduced. It is not appropriate to use exactly 10 × 400 hp (300 kW) pumps if the calculations show that 4,000 hp (3,000 kW) is required. Some pumps invariably will fail when pressed into service. The service duration is important:

- Intermittent service is for pump usage less than 4 hours. An efficiency factor of 1.2 is suggested.
- Expected kill time from 4–8 hours should use a factor of 1.3.
- Continuous service greater than 8 hours should use 1.5.

Calculated horsepower should be increased by the appropriate factor to determine the mechanical horsepower requirements.

A word of caution is extended. Realistic kill estimates should be used. Again, avoid "what if" situations. Most wells have been killed recently in 0.25–2 hours at low kill rates [e.g., Saga 2/4-14 (1989) and Ormat (1989)].

After mechanical horsepower is determined for the blowout well, the relief well must be addressed. It will consume horsepower due to friction pressures. This is the primary technical basis for using large casing strings and a small drill string when pumping on the blowout. The parasite horsepower for the relief well is added with the blowout well requirements to give total horsepower.

To establish the number of kill wells, an arbitrary pressure limit is established as the upper pressure limit on a given relief well. If the hydraulic calculations are such that the pressure limit is exceeded, several options exist:

- Increase the size of the casing to be used on the relief well.
- Use a smaller drill string in the relief well.
- Add friction reducers to the kill fluid.
- Drill a second or third relief well.

Assuming that the initial three options have been exercised and that the relief well pressure still exceeds the maximum pressure limit, a second or third well is required.

BLOWOUTS

The maximum pressure limit is arbitrary. Values from 2,500 to 7,500 psi (17–52 MPa) have been used on various jobs. However, the range of 2,500 to 5,000 psi is recommended. Equipment availability is much greater in the lower pressure ranges and equipment downtime during pumping is lower.

Calculated horsepower to kill the blowout is directly related to the reservoir pressure. This pressure can be considered as follows:

- Assume absolute open flow (AOF) with no reservoir depletion.
- Account for reservoir depletion but discount formation damage at high flow velocities.
- Account for reservoir depletion and formation damage.

Procedures that discount reservoir depletion are most common. However, reservoir depletion does occur and should be considered. It has not been done industry-wide until very recently. Quantitative procedures for evaluating formation damage at high flow rates are not available and thus are not considered.

Effect of Reservoir Depletion

A reservoir under blowout conditions will experience a rapid pressure drop. The phenomenon is factual, calculatable, and has been verified in blowouts where basic data exists. The pressure drop is important to kill calculations for the obvious reason that it makes the blowout well easier to kill. A field case will be used to illustrate the depletion occurrence.

Example 4.2

The SLB-5-4X well in Lake Maracaibo, Venezuela, blew out on 28 May 1986. Pertinent data are as follows:

Thickness	800 ft (240 m)
Permeability	Unknown
Porosity	15%
Fluid	Oil and gas
Gravity, API	38 deg (830 kg/m^3)
Flow rates	7,000 bbl/day(est.) (1,100 m^3/D)
	40 × 10^6 scfd (1.1 × 10^6 m^3/D)
BHT	400°F (200°C)
BHP	14,620 psi (101 MPa)

The well was capped and diverted on 24 October 1986 using an offset kill technique with a derrick barge. A snubbing unit was rigged on top of the well. A 3 1/2 in. (89 mm) fish at 3,642 ft (1,110 m) was latched with an overshot. The drill string was cleaned out with a 1 in. (25 mm) string of tubing. The well was killed on December 12, 1986, by pumping fresh water at 5 bbl/min (0.8 m^3/min). During the 5 1/2 month period from the time the well blew out until it was killed, the reservoir around the wellbore depleted to a level sufficient to allow a fresh water kill. The pressure may have been depleted much lower than a fresh water equivalent.

Reservoir depletion is not magic. It is calculated with basic reservoir and fluid flow equations and is well suited for PC applications. Reservoir simulation models are quite effective but probably are overkill. If they are used, it is recommended to consider a 2-D model with minimum vertical permeability.

Reservoir depletion as related to blowouts is affected by numerous factors. Key issues include:

- Low permeabilities create maximum early drawdown around the wellbore.
- Large blowout annuli allow faster depletion.
- Smaller drill strings allow faster depletion.
- Increasing water depth retards depletion.

Considering worst case scenarios in conjunction with reservoir depletion, the recommended approach for kill hydraulics design is:

- Develop the worst case scenario of absolute open flow with no drawdown.
- Evaluate reservoir depletion and the expected pressure at the kill time.
- Design hydraulics to handle the reservoir depletion case and add as much capability as reasonable toward spanning the gap between the reservoir depletion results and the worst case scenarios.

History and calculations show that the design objective should be the reservoir depletion approach. It is still conservative because it discounts the beneficial effect of formation damage from high flow rates.

Back-up Well

Legitimate reasons for starting a second well are as follows:

- The plan for the primary well has a high degree of complexity and/or will require a long time to implement.
- A simultaneous top kill effort has a high risk factor or a high degree of uncertainty.
- Public pressure or media response is intense and negative.
- The blowout fluid is oil.
- The blowout well has high pressure with a large casing string or the blowout occurred with pipe out of the hole.

The last item is based on the technical requirements for killing the specified blowout with a relief well.

An interesting application for a back-up well was the 1988 Enchova blowout. The initial well missed the target sand in the blowout well but hit below it. The flow and surface fire were diminished and the second well was successful. The second well was being drilled simultaneously with the first.

Casing Size Selection

A key issue in relief well planning is the selection of casing sizes. The well must have a kill string of sufficient size that will allow kill fluids to be pumped at appropriate rates to control the blowout well. If the kill string is too small, the pumping pressures can exceed pump capabilities or exceed some safe design working limits. The well may not be controllable with the originally selected design.

General Size Selection Criteria

The book *Drilling Engineering* has identified the factors to be considered in casing size selection for drilling wells. These include the following:

- Casing coupling clearance
- Bit clearance
- Annular hydraulics
- Cementing

These apply to relief wells, also.

Relief wells pose additional constraints on the size selection issue. Key factors include pumping pressure conditions and allowances for a backup casing string.

Larger casing strings are usually applicable only for "normal" blowout situations. The presence of shallow charged formations may result in reduced hole size at total depth—which effectively increases kill pump pressures.

Pumping Pressures. An important aspect of casing size selection for relief wells is the consideration for high pump pressures associated with kill rates. Most well killing is down the annulus. The annular geometry must be large enough so friction pressures do not pose any restriction on kill capability. This issue is not a concern in typical drilling operations.

Backup Strings. Relief wells often pose uncertainties because of the blowout well's effect on subsurface formations. Typical results can include pressure charging or depletion. If the charging or depletion is severe, an additional casing string may be required to drill the well safely. Thus, allowances should be made in selecting casing sizes for a backup string of pipe if required.

Casing Program

Essential elements of any drilling program include proper casing designs and setting depth selections. These designs, particularly with respect to setting depths, can create the difference between a successful, trouble-free well and a problem-plagued situation. These designs play an even more critical role in relief well drilling.

Setting Depth Guidelines

The initial design task in preparing the relief well plan is selecting the depths to which the casing will be run and cemented. Key factors worth consideration include formation pressures and fracture gradients, hole problems, pressure-charged zones, reservoir or zone depletion, internal company concerns, and, in some situations, possible government regulations.

Conventional setting depth design procedures have been described in considerable detail in other publications. They will not be reviewed in detail in this book.

Casing setting depth guidelines for relief wells have additional considerations. These include the following:

- Pressure-charged or depleted zones
- Top of the blowout zone
- Ranging tool design and operations
- Directional drilling requirements (below the kill string)
- Reservoir depletion (kill mud requirements)
- Hole stability as it relates to high volume pumping
- Back up casing string

Each will be discussed in the following sections.

Pressure-Charged Zones. Pressure charging implies that the pressure in a zone has been increased to a level greater than its original pressure. With respect to blowouts, the consideration is that the blowout zone may have flowed into a lower pressured environment and increased its pressure.

Although the matter should always receive consideration, the typical case is that pressure charging does not occur in blowouts where the fluid can exit the surface. The pressure under blowout conditions usually decreases in all zones that are exposed to the wellbore. Zones not originally involved in the blowout can begin to contribute to the well flow if the wellbore pressure drops to a level lower than the zone's fluid pressure.

Pressure charging occurs and must be considered. Field cases have shown that shallow gas blowouts can increase pressures in other shallow zones by a small margin. Also, underground flows can increase the pressure in shallow zones if the flow is not allowed to exit freely at the surface.

Abnormal pressure detection techniques do not account for pressure anomalies due to pressure charging. Thus, it is difficult to predict the location of zones that are subject to the pressure increases. Shallow seismic has been used successfully to detect charged sands.

Historically, casing setting depth calculations have been based on a worst case situation for pressure charging. If the charging is considered to be a possibility, the flowing zone is assumed to be transmitting its pressure to the suspect zone. An analysis of the mud weights required to drill the suspect zone with its charged pressures is compared to the formation fracture gradient to determine if it can be drilled without setting an additional string of casing. If the resulting formation pressure-fracture gradient relationship is not acceptable, a casing string may be required to be set on top of the zone. It is possible that the formation pressure-fracture gradient relationship may require another casing string below the zone to isolate it.

A rotating head may be required to drill through the charged zone. If this is the case, it may be necessary to use the rotating head to run and cement the next casing string.

Pressure Depletion. Partial pressure depletion of zones other than the blowout zone is more common than pressure charging. The blowout environment will generally lead to fluids flowing from other zones into the wellbore. This will result in some degree of pressure depletion.

Potential problems from pressure depletion include differential pressure sticking and lost circulation. Fortunately, field cases from relief wells do not indicate an unusually high frequency of these problems.

Identification of the pressure depleted zones suffers from the same difficulty as identifying the charged zones. Suspect zones may require additional casing strings above or below the zone. The worst case depletion can be estimated from an analysis of the depletion tendencies from the blowing zone.

Top of the Blowing Zone and Near the Blowing Well. The general approach to a setting depth for the kill string is that it will be set near the top of the blowing zone. The logic has been that it would be set as near as possible to the blowing well to maximize formation fracture gradient.

An underlying concern has been the potential of the blowing well to cause a problem in the relief well. Fortunately, the history of relief wells shows that they do not experience kick problems from the blowing well. In fact, the opposite is generally true; namely, that lost circulation occurs from the relief well to the blowing well.

Several drilling and magnetic ranging factors affect the proximity of the relief well casing seat to the blowout well. These will be discussed in the following sections.

Ranging Tool Design and Operations. The ranging tools are used to define the distance and direction from the relief well to the blowing well.

Active detection tools have an effect on casing setting depth programs. The active tools induce a magnetic field in the casing or drill string in the blowing well by injecting current into the formation. An electrode on the wireline tool injects the current. The electrode is placed several hundred feet above the tool in some cases. Therefore, the casing setting depth program must be adjusted so the electrode is in the open hole when it is run.

This consideration is applicable only when ranging must be done below the kill string. If a definite location fix has been made on the blowout well prior to running the casing, it may not be necessary to run the tool again, or it may be possible to use a short bridle for the electrode if the relief well is near to the blowing well.

Directional Drilling Requirements. Directional drilling requirements below the kill string affect the placement depth for the string. The desired situation is that the casing string and the wellbore are positioned so drilling into the blowout well will not require any directional modifications.

Reservoir Depletion (Kill Mud Requirements). Reservoir depletion is a known occurrence that has not been widely considered in relief well planning and killing. If the depletion is not considered, kill planning and equipment design requirements can be demanding for a worst case assumption.

A key aspect is the mud weight used in the kill operation. If depletion is considered, the actual kill weight may be much lower than the mud weight originally required to drill the well. This may have an impact on the casing setting depth program because of the kill mud weight-fracture gradient relationship. It is possible that the casing can be set higher in the relief well, since high kill mud weights may not be required. This can give more flexibility to the drilling program if the casing setting depth requirements are not so rigid.

Hole Stability and High Volume Pumping. A question arises as to the stability of the wellbore under high rate pumping conditions. Some erosion will certainly occur. However, the issue concerns the structural integrity of the rock. Will it become unstable under high rate pumping?

Historical relief well experience does not suggest that hole integrity is more of a problem than it would be under normal drilling circumstances. If this is the case, additional flexibility exists for a casing setting depth program if the goal is to set the kill string higher in the relief well. A kill string set at a shallower depth allows more directional flexibility.

Backup Casing String. The casing program must allow for the flexibility of running an additional casing string if unexpected hole problems occur. The setting depth for the string may be decided as the well is being drilled and when problems such as charging or depletion are encountered. The casing size selection must account for the additional string.

Impact of Deepwater Fracture Gradients on Casing Depth Selection

Deepwater environments have lower fracture gradients than equivalent depths on land. The interval from the rotary kelly bushing (RKB) to the mud line has a lower pressure gradient than the overburden stress over a similar interval on land. Thus, the fracture gradients are reduced.

One word of caution must be given. Fracture gradient calculations in deepwater environments are not as straightforward as in land situations. Some calculational techniques, such as Eaton, that are widely used for land fracture gradients do not appear to be valid for deep waters unless some type of correction factor modification is made.

Casing setting depths must account for the reduced fracture gradients. The approach for determining the setting depths would be similar to the technique used for the original blowing well. However, the original well must be analyzed very closely to determine if the source of its original problem was related to improper setting depth selection.

Casing Design

Design procedures for the various casing strings used in a relief well should be initially established as if the relief well does not pose any problems different from a standard drilling well for that environment. After the initial designs have been completed, the unusual problems that may be encountered in the relief well should be considered. If necessary, pertinent strings should be upgraded.

Pressure charged zones should be considered in the casing designs that will handle those zones. An estimate can be made of either worst case, virgin blowout zone pressures or depleted pressures by using some computer modeling routine. After the blowout zone pressures have been established, the estimated pressure in the charged zone can be determined. Casing burst design pressures are determined accordingly.

Depletion affects the collapse design. The worst case involves lost circulation of the drilling fluid into the depleted zone so the backup fluid inside the casing is reduced. The design procedure is to consider the heaviest mud weight to be used below the particular casing string and assume that it is lost into the depleted zone with a resultant fluid level drop inside the casing. The calculation procedure is described in more detail in the following section relating to design of the kill string.

Kill String Design. Several factors affect the design of the kill string. They are discussed in the separate burst and collapse designs.

Burst design must account for the fracture gradient, kill mud weight, and the friction pressure associated with the kill pump rates. An example is shown in Figure 4–24. Consider a kill string with a vertical setting depth of 13,000 ft (4,000 m) and a fracture gradient at the casing seat of 17.5 ppg (2,100 kg/m^3). Also, consider the original virgin blowout pressure to be 15.0 ppg (1,800 kg/m^3).

The controlling parameter is the fracture gradient at the casing seat. If a 1.0 ppg (120 kg/m^3) safety margin is applied, the maximum pressure at the bottom is defined as the "injection pressure" and is as follows:

$$\text{Inj. Pressure} = \text{Fracture Gradient} + \text{Safety Margin} \quad (4.25)$$

FIGURE 4–24 • Typical burst design configuration for a kill string

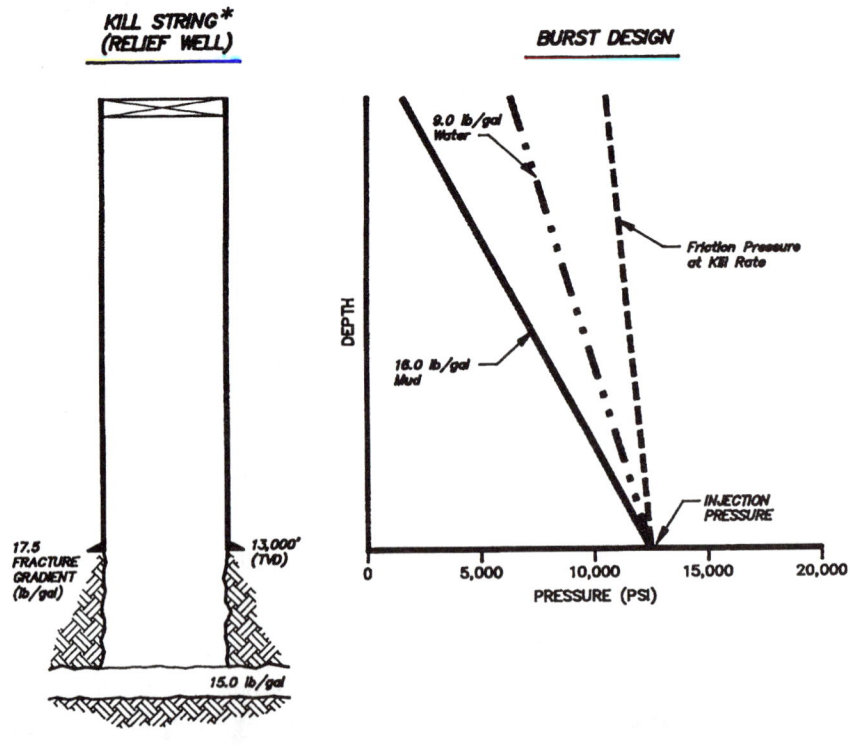

* Vertical depth shown for illustrative purposes.

BLOWOUTS

Maximum surface pressure is the injection pressure less a column of kill fluid. Options for the kill fluid include water as the first phase to be pumped or a mud weight that will exceed the original virgin blowout pressure. Assume 16.0 ppg (1,900 kg/m^3) for the purposes of this example. (See Figure 4–25.)

Pumping friction pressures can be added to the surface design values. They are calculated for the kill pump rates and the annular geometries.

FIGURE 4–25 • Effect of water depth on formation fracture mud weight for normal formation pressure

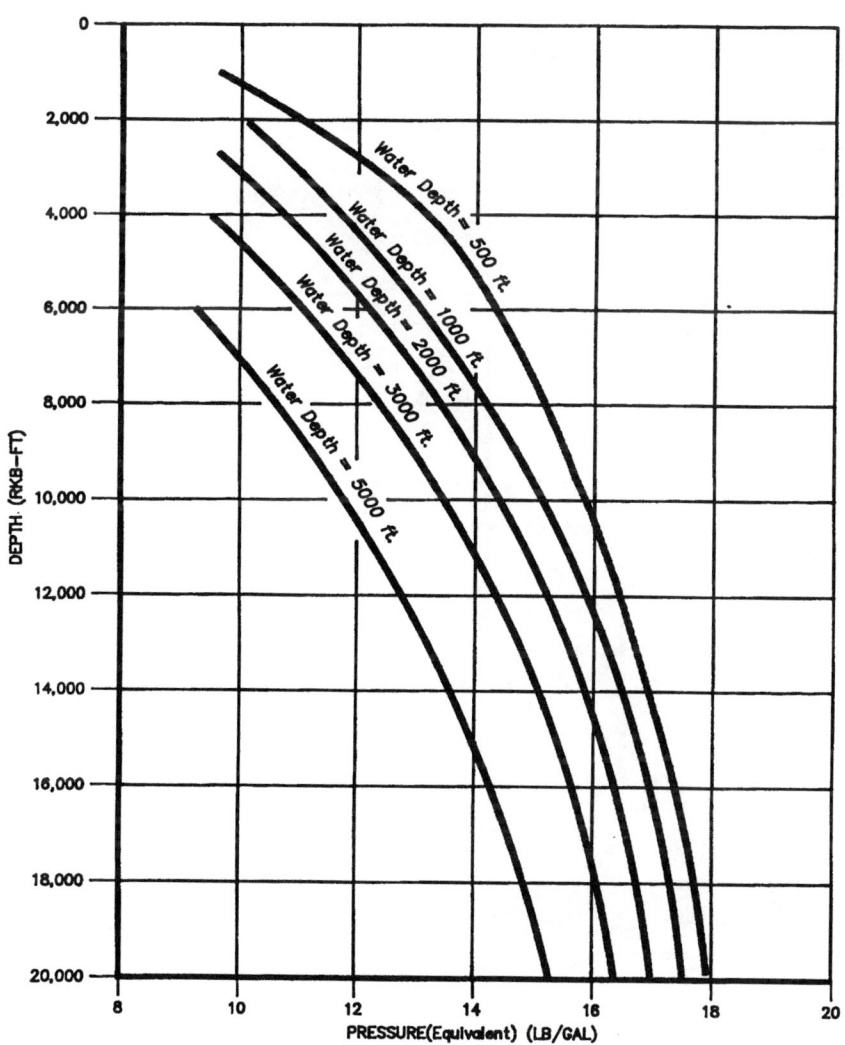

An important logic consideration is that the pressure at the bottom of the string will not exceed the injection pressure, even at kill pump rates. This is a reasonable assumption for most situations.

This approach to design for burst is a worst case scenario. For most situations, field experience has shown that the reservoir is depleted to some degree. The kill rates and required mud weights are much lower than originally anticipated.

Collapse design assumes the worst case that the blowout reservoir is depleted to some low level. Lost circulation occurs in the relief well when the zone is penetrated. The mud level falls in the annulus. The worst situation occurs if the heaviest mud to be used below the string is considered.

For the purposes of illustration, refer to Figure 4–26 and assume that the bottom-hole pressure has been reduced to an equivalent of 5.0 ppg

FIGURE 4–26 • Typical collapse design configuration for a kill string

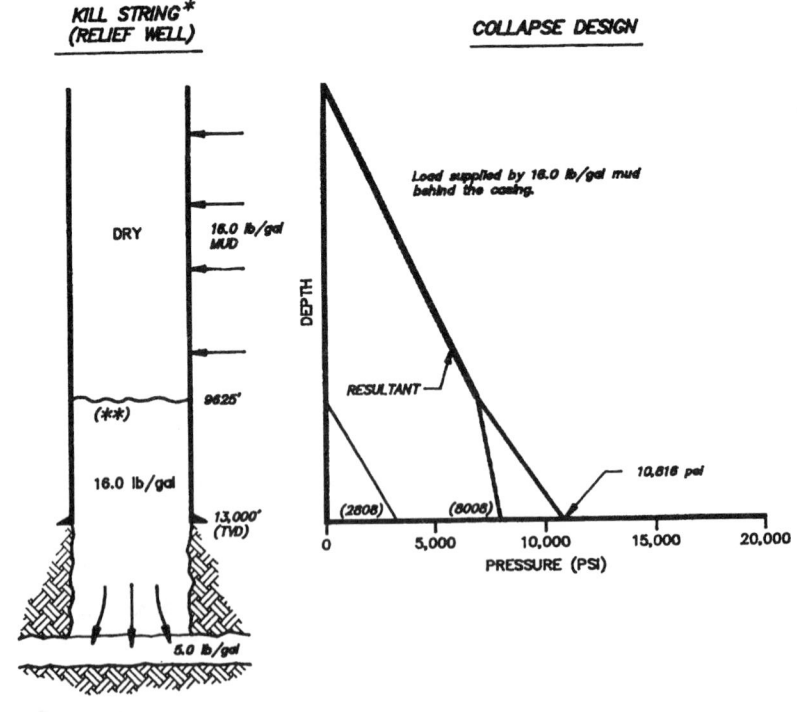

* Vertical depth shown for illustrative purposes.

** Worst Case

(600 kg/m³). If a 16.0 ppg (1,900 kg/m³) mud has been used to drill the zone, the mud level would drop in the annulus to a level of 9,625 ft (2,934 m). If the kill string was set in 16.0 ppg mud, the resultant would be as shown in Figure 4-26.

Due to the variables involved in the casing design for the kill string, considerable attention should be given to this important problem. However, it should be noted that available historical records do not indicate that kill string designs have ever hampered the kill process in any manner. Casing sizing is perhaps the key casing design concern.

Directional Planning

Directional planning for a relief well is similar in the general approach as directional planning for any directional well. The need for preciseness in drilling and surveying is more acute.

The directional plan for a relief well is bound by several constraints. Some of these are:

- Intersection at the bottom of the blowout well or at a shallower point
- Ellipse of uncertainty considerations
- Surface site selection
- Blowout depth (i.e., shallow vs. deep)
- Interference from other wells

Others will certainly enter the picture for specific cases of relief wells.

The blowout well intercept usually controls the lower "half" of the directional plan. Ranging tools function most effectively at low approach angles. Therefore, the relief well normally follows an "S" curve so the bottom section of the well approaches the blowout well at low angles. Straight-kick directional plans are seldom used.

The ellipses of uncertainty for the two wells affect the program. As the two ellipses overlap near the bottom of the well, the plan must proceed slowly to minimize inadvertent intersect. A typical directional well seldom considers the error associated with survey accuracy.

A shallow gas blowing zone or a shallow intersect will require a compressed plan (i.e., shallow kick off point, high build rates and hold angles, and high drop rates near the bottom). These requirements pose unique drilling difficulties for shallow gas relief wells. The difficulty is often coupled with gas charging of shallow zones.

Well interference is (more or less) a routine directional planning concern. It has proven to be an overriding concern in some situations of shallow blowouts under platforms.

Course Path Selection

Many directional course paths have been discussed over the years. Some of the common approaches are shown in Figure 4–27. Each has its basis on technical requirements.

The course paths are heavily influenced by ranging tool capability and ellipses of uncertainty. Since the exact location of the relief and blowout wells are uncertain, it is not a simple task of drilling directly toward the blowout well. Ranging tools must be used to define the relative locations of each well. If the area of blowout location uncertainty is large and beyond the ranging capability of the logging tool, a shallow by-pass may be required to reduce the cone of uncertainty. The relief well is then continued to the desired intersect point.

Course path selection is impacted by the type of blowout fluid, although its effect is nontechnically based. It is generally desirable to kill blowouts as quickly as possible. An added emphasis is placed on oil blowouts where pollution can be a major concern with respect to cleanup cost and public pressure. Igniting an oil blowout should be a consideration immediately after the event occurs, although ignition may not by desirable under some circumstances. Gas blowouts do not pose this element of emergency. The consequence is that a direct approach with the relief well has more advantages with oil blowouts than gas.

The controlling parameters for the directional plan selection are the required kill hydraulics and mud weights. The solution to the hydraulics and mud weight issues depend primarily on depth of intersect and amount of reservoir drawdown. The design procedure should be as follows:

- Run an appropriate blowout depletion model and determine the sand face pressure for various times, t, including the time required to drill a relief well to the deepest possible intersect.
- Determine kill mud requirements for various intersect depths and the associated time, t, to reach that depth.
- Select an intersect depth that has operationally acceptable kill mud and hydraulic requirements.

Purpose of Ranging Tools

In simple terms, ranging tools are designed to guide a relief well to a blowout well. They should determine (range) and direction to the blowout well. To be more specific, a ranging tool fixes the relative location of the two wells (i.e., where the blowout well is located relative to the relief well).

Ranging tools are sophisticated instruments that are only as good as the experience/knowledge of the individual operating the tool. Claims are commonly made of ranging distances up to 200 ft, although many oil

FIGURE 4-27 • Directional kill plans

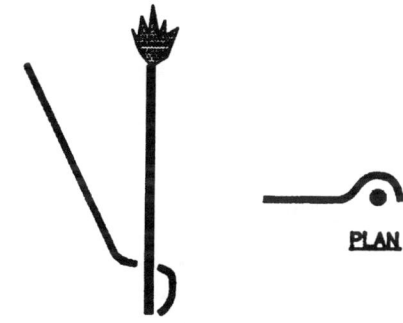

TRIANGULATION APPROACH, DEEP KILL [1]

DIRECT APPROACH, DEEP KILL [2]

ONE WELL, INTERMEDIATE SEARCH, DEEP KILL [4]

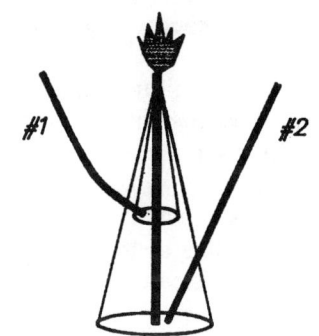

2 WELL, INTERMEDIATE SEARCH, DEEP KILL [3]

DIRECT APPROACH, INTERMEDIATE KILL [5]

companies suggest the actual effective range is much less [i.e., 50–125 ft (15–38 m)]. Tools usually employ magnetic detection sensors, known as "magnetometers," to investigate casing or drill pipe in the blowout well. Hopefully, future tool and technology development will increase the reliability and accuracy of ranging techniques.

The limited distance measuring capability of ranging tools restricts their usage until the relief well is near the blowout well. Thus, the directional program near the bottom of the relief well is dictated to some degree by the ranging tool. Multiple runs are often required.

Magrange. Tensor Corporation of Austin, Texas, offers the Magrange II as a ranging tool for relief wells. The tool was the first of its type to determine distance and direction of blowout wells from relief wells. It has been perhaps the most widely used tool until recently.

The Magrange II system consists of a downhole instrument, a winch and seven-conductor cable, a surface electronic unit, a programmable calculator, and plotter. The downhole instrument contains magnetic field sensors arranged in a noninterferring orthogonal configuration and in a gradiometric measurement configuration. The sensors, along with their associated electronics and signal condition circuitry, are housed in a nonmagnetic cylindrical container.

Experience has shown that under the optimum conditions, Magrange II can detect targets at a range of 100 ft (30 m). The direction from the relief well to a target well can be determined to within a few degrees.

The Magrange II system using the passive technique is the oldest such tool in the industry. It detects magnetic dipoles. Figure 4–28 shows a plot of Magrange data. The casing near-point in the blowout well is shown at 2,230 ft (680 m).

The passive tool can measure distance and direction in a homogeneous formation to approximately 70 ft (20 m) for 9 5/8 in. (244 mm), 47 ppf (70 kg/m) casing in the blowout well. This example is given by the manufacturer.

The present tool offers some advantages over active tools of any manufacturer's origin. It is not affected by oil muds in the relief well, whereas active tools, depending on manufacturer, can cause the effectiveness to be reduced by 50% for oil muds. Also, the tool functions as effectively near the bottom of the pipe string in the blowout well as up the hole. Again, this is contrary to active tools. The presence of hematite as a weighting material in the mud does not affect tool range or response according to Dr. Waters.

Magrange has recently introduced an active detection tool. Its initial field trial was on a Cook Inlet, Alaska, blowout in 1988. Recent verbal reports of its usage on a Corpoven relief well in Venezuela indicate that it can provide good measurements to 200 ft (60 m) separation between the relief and blowout wells. Consult the manufacturer for more information.

FIGURE 4-28 • Magrange plot (Courtesy Tensor Corp.)

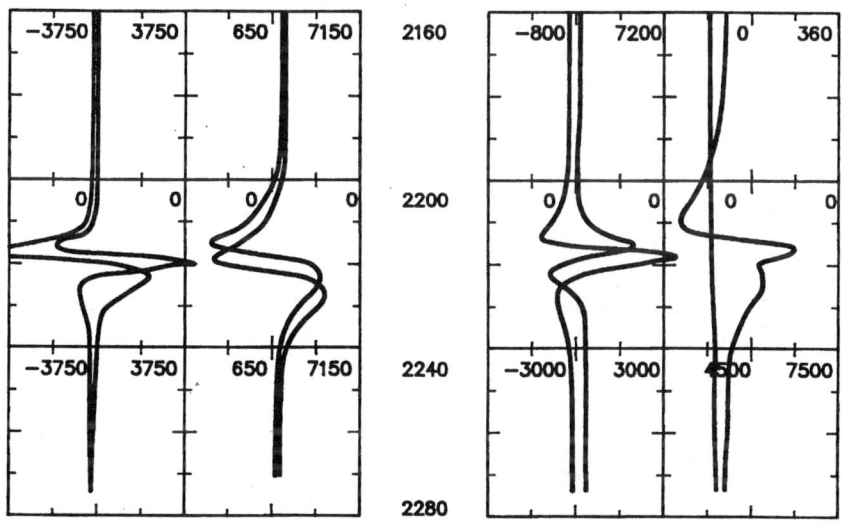

Vector Magnetics. Their Wellspot tool uses a low frequency alternating current flow in the blowout well's casing or drill string to develop a magnetic field. The current is injected from an electrode placed some distance above the Wellspot tool in the relief well, or by attaching an electrode directly to the blowout well's tubular at the surface.

Measurements are taken at selected depths to determine the magnitude and direction of the magnetic field resulting from the induced current. At the same time, measurements are made of the magnitude and direction of the earth's magnetic field as the orientation of the measurement device can be determined. From these measurements, the compass direction and the distance to the target well can be determined.

The Wellspot equipment consists of a sensor sonde, 2-inch diameter, 72 inches long, (51 mm diameter, 180 cm long) to which are attached sinker bars and a spring tip if needed. Attached to the top of the sonde is a bridle 150 to 400 ft (46–120 m) long that electrically insulates the sensor sonde from the electrode at the bottom of the wireline. The electrode is the torpedo connector that fastens the bridle to the conductor openhole wireline.

The procedure in oil-based muds is the same for water-based muds except for the bridle arrangement. The range of detection in oil-based operations is about 50% of water-based muds under ideal conditions.

Accuracy of Wellspot determination depends on the geometry of the wells. In ideal conditions, the range of detection is approximately 200 ft

(61 m) and inaccuracies exist. The range of detection is considerably smaller in less than ideal conditions. From 200 to about 100 ft (61–30 m), the direction can be determined to about 10 degrees and distances to +/– 20% of the distance, again under the qualification of ideal conditions. These accuracies generally improve as the target well is approached.

The resistivity of the surrounding formations can generate a background signal. If the formations are uniform and have no dip, the spurious signal is very small or virtually nonexistent. Lateral resistivity changes and dipping beds will generate a small bias signal. Within about 100 ft (30 m), the magnitude of the signal generated by the heterogeneities in the earth is small compared to the signal from the target well, and it will have only small bias effects on the results. This is one of the reasons for a greater uncertainty assigned to the distances and angles at ranges of greater than 100 ft (30 m).

Approach Angle Considerations

The approach angle is defined as the angle between the relief well and the blowout well. It is used to specify the closure or approach conditions between the relief well and the target in the blowout well. The target could be for a by-pass at a shallow depth or for the kill intersect at some deeper point.

The general tone of the directional planning for the relief well is controlled by the desired bottom positioning relative to the blowout. Likewise, the bottom positioning is controlled by the approach angle.

Factors affecting the approach angle are as follows:

- Ranging tool considerations
- Concern relative to a premature intersect
- Casing milling considerations

An approach angle less than 30° is desirable for relief wells.

Drilling Guidelines

Drilling mechanics for the relief well do not differ appreciably from that of a conventional high priority well. The differences typically rest in the killing operation and the required equipment for pumping and ranging. As such, a viable organization structure for relief well killing, described in Blowout Contingency Planning, involves the operator handling all routine drilling tasks while a blowout specialist coordinates killing functions.

The differences for relief well drilling can be grouped into general guidelines, considerations for the shallow section of the well if gas charging has occurred, and operations required for the deeper sections of the well. They will be discussed in the following sections. Conventional drilling operations such as running casing, tripping, and electric logging will not be discussed.

General Guidelines

General guidelines discussed herein are recommended for most wells, but they are not considered mandatory for all cases. In some cases, they are clearly not applicable (i.e., offshore related items may not apply to land wells).

An MWD system should be used to monitor drilling conditions near the bit. A system with a full complement of services is recommended including directional and formation logging capability. The logging system should have a gamma ray and a resistivity tool, as a minimum, so lithology can be easily correlated.

A computerized mud logging unit should be used. It should contain most currently available services. It is desirable in some situations to have the capability to transmit data to the operator's central district office to allow viewing of various logs as they are generated. A remote set of MWD printouts should be set up in the mud logging unit so all operations can be monitored by the supervisors from one site.

Additional gas detectors may be required in excess of the rig's normal complement. The situation where they may be required is for operations in a shallow gas blowout. If the water depth is greater than 500–600 ft (150–180 m), it is not anticipated that they will be needed. The gas monitors should be supplied by the mud logging company so the readouts can be observed by the mud logging crew if their unit is established as the control center.

Oil muds may be required to drill some relief wells. The only difficulty with oil muds relative to well killing is their adverse effect on ranging tools. If possible, consideration should be given to changing out the oil mud to a water-based mud near the bottom of the well to enhance ranging logging. Alternatives in the ranging logging program are available if the oil mud is necessary to drill the appropriate sections.

Accurate well surveying is obviously important. Past experiences of Neal Adams Firefighters have shown that supposedly "accurate state-of-the-art tools" may not give repeatable results and differ significantly with other tools that may be run. This situation has been observed by other operators in conventional well surveying practices. The difficulty is determining which surveys are most representative of the actual borehole location. In one field case performed by NAF, the directional surveys from the MWD tool proved more reliable and repeatable than "highly accurate" survey tools.

Furthermore, it is recommended that the operator obtain, as part of the organization team, a specialist in survey interpretation. The specialist must know the operational principles of each tool so decisions can be made about reliability under the actual relief well conditions. The specialist should come from within the operator's organization, if possible. Most blowout service companies, as a rule, do not have these specialists on staff. The specialist should also be consulted to reanalyze the surveys on the blowout well to determine if its position can be more accurately determined by a re-examination of the data.

Meetings should be held at the rigsite prior to each critical function. The meetings should be attended by the operator representatives, blowout specialists, and key members of the service companies involved on the rig. The meeting should include all service companies and not just those involved with the particular activity. Exclusion of noninvolved groups leads to miscommunications and rumors.

Crew psychology should be considered. The crew will typically be wary and perhaps nervous at the beginning of the well, but then they tend to become casual about their operations as the project continues. Unfortunately, they tend to become casual at the most critical part of the operations when the kill operations commence at the conclusion of the drilling. A meeting should be held with all crew members prior to the kill operations to gain their full attention and alertness.

The crew should be assured that relief wells have not historically blown out. The crew will obviously know that the original well blew out and, as such, the relief well has a high probability of blowing out. This is clearly not the case, and the crew should be advised as such. Crude and thoughtless jokes about a relief well blowout should be avoided.

Various rig modifications may be required for drilling the well.

Shallow Drilling Guidelines

Shallow drilling for relief wells does not pose any unusual requirements unless natural shallow gas problems exist at the relief well site or the blowout well has charged shallow gas zones.

It is recommended to drill riserless for offshore situations, if possible. A riser can be used after casing has been run to a depth sufficient to allow shut-in of a kick. If the relief well begins to flow and it cannot be killed dynamically, the rig can move off and let the well serve as a vent well. After the blowout well is killed, the vent well will soon die without further intervention.

If drilling riserless is not possible for any reason, a special purpose-built diverter system should be used. See Figure 4–29 for a typical system with an erosion resistant section. This diverter system is designed from new technology and has proven serviceable under stringent conditions.

An ROV with a sonar head or independent sonar units should be used to track any possible gas under the rig. The sonar has additional capability

FIGURE 4-29 • Purpose built diverter system

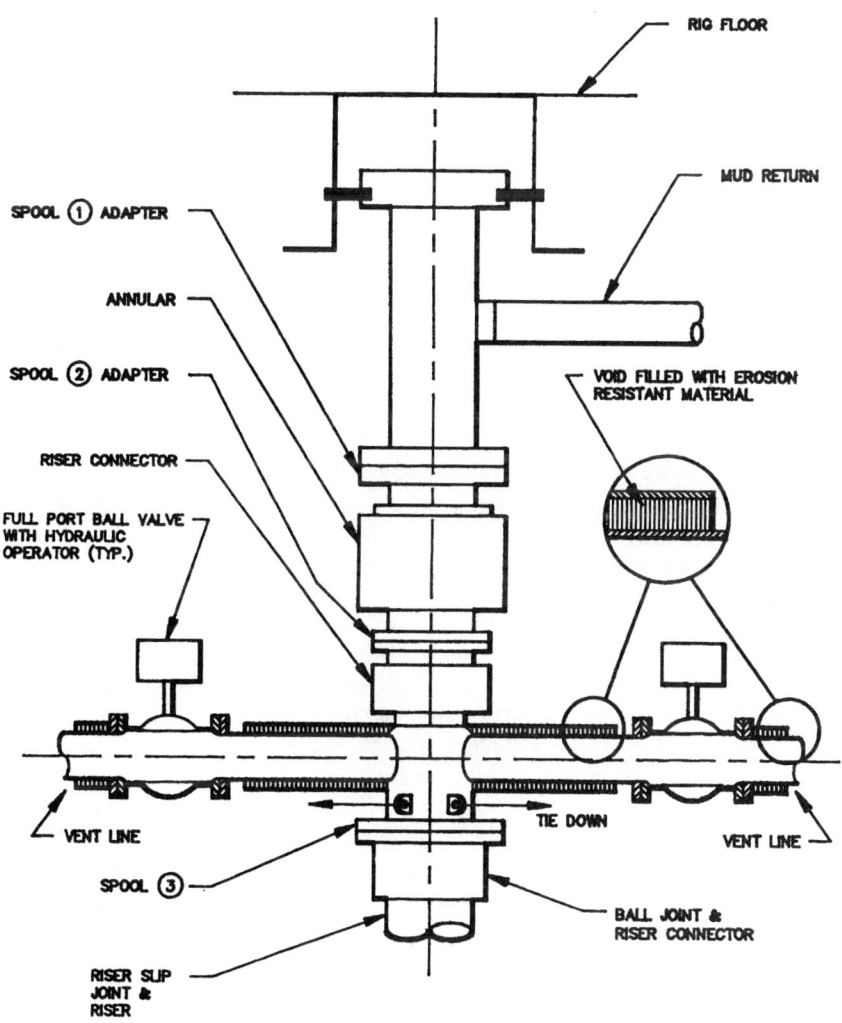

NOTE:
Outlet Lines Extend Outboard of the Rig

not provided with a TV picture. The sonar requirements may involve several sonar heads with varying frequencies that have proven useful for functions including running a BOP stack in murky environments and tracking gas at relatively long distances. The ROV should have observation capabilities and sufficient power to move at high velocity under adverse conditions. Manipulator and work package capability has a lower priority.

Drilling bits without jets should be used when drilling possible charged zones. Jets cause a pressure restriction that can be important when attempting to dynamically kill the well if flow should start. Drilling efficiency via optimum hydraulics is not a high priority consideration at this point.

Likewise, consideration should be given to using motors, turbines, and MWD tools with minimum internal restrictions. These restrictions can impede a dynamic kill. Also, pumping at high rates with turbine-driven MWD tools destroys some internal components of the tools. A sacrifice must be made on some occasions between MWD performance and dynamic killing capability.

Deep Drilling Guidelines

Drilling guidelines for deep drilling are not as critical as for shallow situations. However, this is true for most situations when comparing deep drilling versus drilling a shallow gas horizon. Conversion from the drilling process to the killing operations does require differences.

The kill system must be tested. The flow lines should be color-coded if the system is complex and confusion could exist when opening or closing valves. All key valves should be controlled from the operator's console so manual intervention is not required.

A small drill string should be used when drilling into the blowout zone. The small pipe allows optimum kill hydraulics in the event that the large drill string cannot be pulled from the well and changed out. The drill bit should not be equipped with jets when drilling in the zones where possible intersect could be made. Drilling efficiency is not the controlling priority at this point. Bits without cones (i.e., PDC bits, fish-tail bits, flat-bottom mills, etc.) are preferred to reduce the risk of a fishing job.

Killing Equipment

The equipment used in blowout killing operations is different to some degree than conventional drilling equipment. The general objective is to pump large volumes of kill fluids at high rates into the annulus. The annulus is used preferentially to the drill string for pumping kill fluids because of its large flow area and lower friction losses. The drill string is occasionally used for pumping, but it is more commonly used as a bottom-hole pressure monitoring device.

Pumping Equipment

Kill systems usually utilize auxiliary pumps instead of (or in addition to) the rig pumps. The auxiliary pump system requirements may be large (i.e., 5,000–10,000 hp (3,700–7,500 kW)]. Special considerations include pump type and liner sizing, number of required pumps, long-term pumping efficiency factors, and pump placement.

The initial step is to determine the number of required pumps. The flow rate and maximum pumping pressure controls the number of pumps. If the pressure is not excessive [i.e., 0–6,000 psi (40 MPa)], large liners can be used to increase output per pump. High pressures restrict the pump liner size.

An efficiency factor must be applied for long term pumping. It must be noted that very few operations require long term pumping so this efficiency consideration seldom is applied.

After the minimum number of required pumps is defined, they must be transported to the rig and organized in a manageable layout. Land jobs usually ease the difficulties because the pumps can be spotted on the area adjacent to the rig. Offshore sites pose more problems because of limited deck space and variable deck load capacity. Also, additional pits for kill mud are usually required, which further restricts the deck space.

Options for offshore pump equipment hookup are:

- Deck placement
- Barges
- Stimulation vessels
- Purpose-built vessel

Most offshore areas worldwide have access to stimulation vessels.

Deck Placement. Pumps can be placed on the deck of the drilling rig. The number and size of pumps are controlled by the deck loading limitation.

Figure 4–30 shows the pump arrangement used on one notable blowout in Cook Inlet, Alaska. Although the pumps were more than adequate, the upper limit was set by deck loading because of pits on the deck filled with kill mud.

Barges. Barges can be used for pump placement assuming sea conditions are moderate. Stacking and deck placement are similar to that for drilling rigs.

FIGURE 4-30 • Equipment and diverter system layout (relief well for platform blowout)

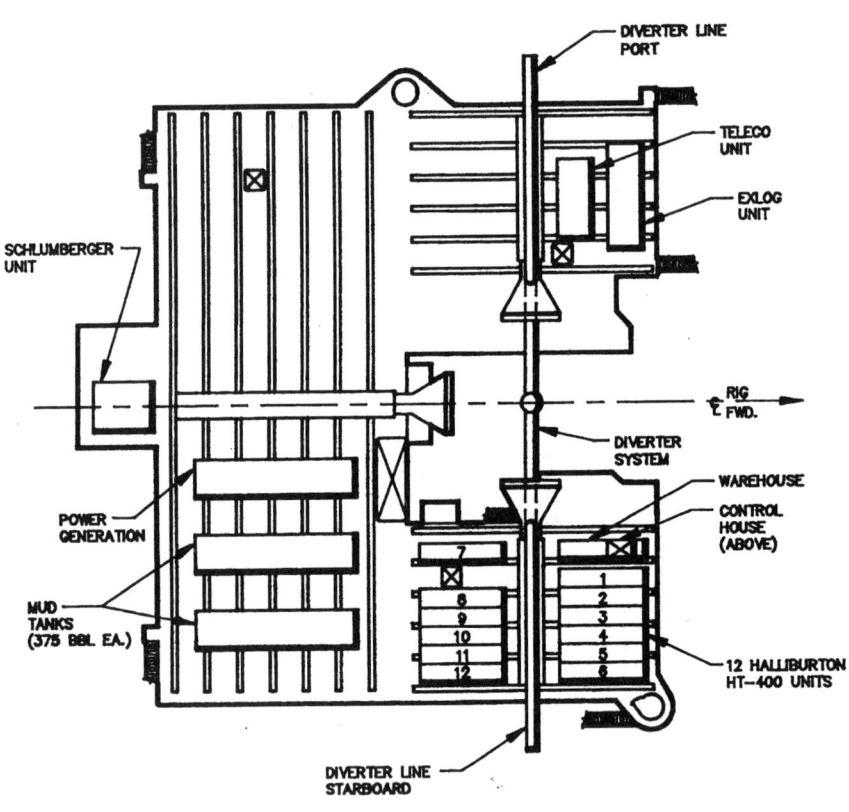

Supplying mud to the barge pumps may require several 3 in. or 4 in. (76–100 mm) ID hoses from the rig pits and centrifugals. This situation is different than pump placement on the rig where 6 in. or 8 in. (152–203 mm) ID hard piping can be installed. Pits can be placed on the barge, but it is more convenient to work with the mud on the rig with the rig's associated fluid handling/treatment equipment.

Stimulation Vessels. Most cementing companies operate stimulation vessels at various worldwide locations. As an example, Dowell has vessels located at offshore bases in Dubai, Aberdeen, Singapore, Venezuela, Brazil, Congo, Gabon, and Berwick, Louisiana, and the United States.

These vessels are almost ideally suited for blowout pumping requirements. Their basic function as a stimulation vessel is for high pressure pumping at relatively high rates. As such, they contain integrated pumps with manifolding, blending equipment, and computer control/monitoring. The vessels are self-propelled.

The only apparent limitation with these vessels for some blowouts is an upper pumping limit of 60–100 bpm (10–16 m^3/min). This rate will clearly handle almost every conceivable situation, particularly if reservoir drawdown is considered. Larger pump liners may increase pump output in some cases.

Purpose-built Pump Vessels. On occasion, purpose-built pump vessels have been outfitted for controlling a blowout. An example occurred during the Bay Marchand blowouts (1971). A jack-up rig was converted to a pump vessel and stationed alongside the rig used to drill one of the relief wells. The pump rig contained the mud handling equipment, pits, and pumps.

Design specifications for purpose-built vessels depend on the blowout scenario, logistics, and operations. Obviously, it is not possible at this time to prepare and present general specifications other than those in the previous paragraph.

Control Systems

The pumping system should be operated from a central control station. Key features should include:

- Operate all pumps including start-up
- Hydraulically or pneumatically operate key suction and discharge valves
- Control the mixture rate for key additives such as friction reducers
- Monitor and record pressures, flow rates, and volumes

Most stimulation companies offer this capability.

It is not desirable to have manually controlled systems involving individuals to control each operation. Communications can be interrupted at critical times. The noise level is a deterrent to efficient operations. Quick action from a central command/operations position may be required.

Manifolds

Rigging the kill equipment will require several special manifolds. Suction and discharge systems are required. Additional capability is necessary if the rig pumps are connected to the kill system.

Suction

Supplying mud or water to the kill system is not a small task. Water volume requirements can be high. Mud weights can exceed 18 ppg (2,200 kg/m^3). The initial step in designing a suction system requires an assessment of the kill requirements for the blowout well.

Water is often pumped as the initial fluid. It is usually available in large quantities. Plans must be made to move the water from its source to the kill pumps at the kill rate. The source may be a lake, river, or holding pit on land or the ocean if offshore. Centrifugal pumps are commonly used for this purpose. Maintenance problems with centrifugals prompt a recommendation to have at least a 150–200% excess capacity available.

Moving large volumes of water over long distances on land is usually done in a relay system. Aluminum irrigation pipe is used to transport the water from its source to a storage pit at the rig site. Other centrifugals pick up the water from the storage pit and feed the suction manifold for the kill system.

A kill system built on the deck of an offshore vessel will usually require installation of large feed line(s) from the centrifugals to the pumps. As an example, several 6 in. (152 mm) lines may be required to feed water at 100 bbl/min (16 m^3/min).

A suction system for high volume, high density kill operations requires large diameter lines to minimize pressure drops. Several additional centrifugal pumps may be required. Pumping heavy weight mud from pits on lower rig levels to pumps on upper decks must enter into the horsepower calculations.

Caution must be exercised to prevent barite settling. The mud should be mixed long before pumping. Mixing and simultaneous pumping (i.e., "mixing on the fly") is not recommended. The pits should be thoroughly agitated. Suction lines should be checked frequently for plugging. If initiation of the heavy mud is critical in a timing sequence, such as in a dynamic kill, the mud valves and centrifugal pumps should be activated from the operations center.

Discharge

Discharge from the kill side of the pump must be at the same rate as the feed side but at a higher pressure. As the required pressure output increases, equipment design complexity increases by an order of magnitude. It is desirable to design an overall system so the well can be killed with pressure less than 4,000–5,000 psi (28–34 MPa). If properly designed, it is not likely that higher pressures will be required.

Rig Pumps

Tying the rig pumps to the kill manifold is necessary in most cases to prevent interruptions in the kill operations. The initial well killing usually begins when the relief well intersects the blowout well and lost circulation occurs. A smooth transition between the kill package and the rig pumps can be made via the common manifold. Field experiences show the rig pumps will be satisfactory to kill the blowout in many cases.

Kill Spools

A kill spool is often installed in the stack to provide large flow-volume capability. It generally is designed with 4, 4 1/16 in. inlets (102 mm) (or 2 spools with 2, 4 1/16 in. inlets).

The spool is special-built for each job. This practice of special-building a spool is more historically oriented than based on actual requirements. It is usually more cost effective to build a spool than to transport a prefabricated spool and pay rental charges.

Kill spools are used predominantly on land jobs, jack-ups, or platforms. The spool is accessible for installation and operation of valves.

The spools have not been used on subsea BOP stacks. The additional valves would require significant planning for a satisfactory hydraulic control system. The common practice is to use the choke and kill lines, although this approach has pitfalls if high volume pumping is required.

The spool is designed to match the BOP flange sizes and pressure ratings. The inlets should have 4 1/16 in. (103 mm) inner diameters and flange connections. Two 4 1/16 in. hydraulic valves should be connected to each inlet. These guidelines can be reduced to two inlets or to one hydraulic valve and one manual valve per inlet if the service conditions are not demanding. The spool and all valves should be appropriately tested with the stack.

Annulus Injection Operations

Killing a blowout with a floater requires different pumping arrangements than with a jack-up. The general objective with any kill operation is to pump large volumes of kill fluid into the blowout well.

Fluids can be pumped down the drill string, but this approach is seldom used. The friction pressures in a drill string at the associated kill rates would be excessive.

Most kill pumping is down the annulus of the relief well. The annular space is larger than the drill pipe and the pumping pressures are reduced. Also, the drill pipe can be used as a bottom-hole pressure monitor.

Gaining access to the annulus is relatively easy on land, platform, or jack-up rigs. A kill spool is often used.

Floaters pose a more difficult situation because the BOP stack is subsea. The kill and choke lines present an apparent solution to pumping fluids into the annulus. However, the kill and choke lines are not recommended for pumping. High flow rates can be destructive. If the kill and choke lines are damaged, the only remaining means to pump is down the drill pipe, which would restrict the kill rate in most cases to a level that would be inadequate to kill the well. Kill and choke lines have been used successfully for annulus injection on some blowouts, but the flow rates were small [i.e., 5–10 bbl/min (0.8–1.6 m^3/min)].

Pressure and Flow Testing

The kill system must pass a testing program before beginning the kill operation. The tests are more varied than conventional BOP testing. Testing a system to 70–130 bbl/min (11–21 m^3/min) capacity is a significant operation. The recommended program is as follows:

Operation	Fluid
Low rate function test (10–20 bbl/min) (1.6–3.2 m^3/min)	Water
Low pressure BOP test	Water
High pressure BOP test	Water
Volume rate test to full kill rate for 15–30 minutes	Water
1/4–1/2 kill volume rate	Kill mud
Low pressure BOP test	Water
High pressure BOP test	Water

The static BOP testing is conducted before full dynamic testing to find leaks that could be dangerous if identified at full-rate test.

A static BOP test is again conducted after the volume rate testing. The high flow rate will often create numerous leaks that were not identified in the original pressure test. As leaks are repaired, flow testing is repeated. This procedure is time consuming and may require several days.

The pressure and flow testing program should also be used as an operational readiness practice. Organizational meetings should be held with all group leaders. Responses to various operational events should be discussed along with contingency responses.

The 1/4–1/2 kill volume rate test with mud is described. A full kill rate test with mud for 15–30 minutes is preferable, but it is often difficult to implement without large losses of mud. As an alternative, a short duration test of 2–3 minutes at full kill rate with mud is recommended even if the mud is sacrificed.

Kill Fluids

The ultimate objective of blowout control is to regain control of the well with hydrostatic pressure. Mud or water may be used for the final control. However, several fluids may be used on an intermediate basis to increase communication between the blowout and relief wells or for multi-stage pressure control. These fluids include acid, water, and/or mud.

Acid

Acid is used occasionally to increase communication between the two wells. The intent is to create or enlarge existing flow channels, or "worm holes."

An acid program for worm hole development is different than a normal acidizing program. A worm hole is a small flow path. The acid increases the size of the path to allow greater flow capacity.

The primary concern in relief wells is to create massive destruction between the two wells without regard for long-term production or clay stabilization. A typical program will delete clay stabilizers, alter the ratios and concentrations of HCL and HF, modify or enlarge the required volumes of acid and use action retarders.

Large volumes of acid may be required. The optimum approach, if possible, is to pump acid until an acceptable injection rate is obtained.

Short-term corrosion of the acid should be evaluated. Long-term consideration should not be important if the relief well is expendable.

Acid will not be needed on most wells. Limestone or dolomite formations are the most likely candidates because of low permeabilities. Also, their rock stability under drawdown conditions is greater than shale and may prevent lost circulation when the relief well intercepts the blowout zone. An injection test is essential in determining if an acid is required.

Water

Water is used in blowout operations for several reasons:

- Establish communications
- Intermediate kill fluid in dynamic control
- Final kill fluid for normal or subnormal pressured reservoirs

A key feature of water as a kill fluid is that it is typically available in large supplies, whereas mud has restricted volume availability. It is an easy task in most areas to build large earthen pits and fill them full of water, or, alternatively, tap into the ocean as on offshore wells.

Friction reducers can be used with water to either (1) reduce the pump pressure for a given rate or (2) allow greater rates for a given maximum pressure. A mixing and metering system for the polymer is necessary. Also, the shelf life of the polymer should be evaluated.

Mud

Case histories of blowout control operations in the 1960s and 1970s include large volumes of kill mud. Often, the fluid had very high densities. This situation has changed in the recent decade because of the development of the dynamic kill technique and the ability of ranging tools to place the relief well very near or in the blowout well bore. The dynamic kill technique has reduced the need for large volumes of mud because the initial step of the dynamic kill technique involves pumping water at high rates until the well is killed dynamically.

Mud is still required, however, and it may involve large quantities. The viscosity of the fluid must be sufficient to prevent barite settling in the pits, yet it should be as low as possible to reduce friction pressures. The mud may be in the pits for an extended time so agitation must be available.

Mud should never be mixed "on the fly." Another way of stating this is to avoid simultaneous mixing and pumping. It is not uncommon that bits plug from poorly mixed mud. Also, it is difficult to maintain a uniform mud weight for mixing and pumping at high rates.

Transfer capability among pits is necessary. Flexibility should be built into the design to allow quick suction changes in an emergency.

Kill Operations

Actual kill operations for a relief well require very little time in comparison to the planning and drilling. It is difficult to jeopardize the success if planning has been done correctly and reasonable, informed supervision takes place at the time of the kill. The key factor that can hinder success is to fracture a formation if natural fracturing does not occur at the time of the intersect.

Intersecting the Well

The point at which the relief well intersects the blowout well is always a major concern to the supervisory group. The concern is based somewhat on the possibility that the blowout well may, in some way, cause a well control problem to the relief well. The truth is that the intersect is typically anticlimactic and rather boring. However, attention should be maintained at a high level for any possible occurrence.

Prior Preparation

Several operations should be conducted prior to intersecting the well. On-site meetings should be held with all key supervisory personnel from the operator, blowout specialist, and service companies. Discussions should include planned operations and contingencies for unplanned events. Personnel responsibility assignments should be made and emphasized.

The kill equipment should have been installed and tested prior to intersecting the well.

The pumping equipment should be running at idle speed with pumps disengaged if an open hole intersect is planned. Immediate pumping can commence if dictated by the operating conditions. Immediate pumping is seldom required for deeper blowouts. A shallow gas blowout, particularly in a diverter operation, may warrant immediate pumping at the time that the diverter bag is closed. In this situation, the pumping supervisor should be instructed to start pumping at a certain rate if the diverter is closed.

First Warning Signs

The initial warning signs of an intersect can be subtle. A drilling rate change is not likely in most cases. The bit usually does not jump or increase in torque. In some cases, it may be difficult to identify the exact time of intersect, since clear signs may not exist.

The usual warning sign is that of partial or complete loss of circulation. The blowout well has experienced pressure drawdown that is overbalanced by the weight of the drill mud in the relief well. A fracture is formed from the relief well to the blowout well. If the loss is complete, drilling should be stopped and the well secured for the kill operation. If the loss is partial, drilling should continue for several feet to determine if the loss will become a complete loss. The complete loss of circulation is desirable because it indicates a good communications path between the two wells.

Establishing Communications

Several options exist to establish communications between the two wells. These are:

- Lost circulation
- Fracturing
- Acid (worm) holes
- Perforating
- Milling

Each has its own place in relief well drilling.

Lost Circulation. The typical reaction for an open hole intersect between a relief well and a blowout well is complete mud loss, as previously described. This situation probably occurs 70–80% of the time on actual field cases. If it does not occur, it suggests that the relief well is some distance away from the blowout well and that a sidetrack will be required.

Upon encountering losses, drilling should stop and the well secured for the kill operation. The mud pumps should be left running at idle to keep the hole full of mud until the kill system is started. Many recent blowouts were killed with the rig pumps in the idle mode. The well should be monitored closely at this point for indications of good communications and any influences the relief well may have on the blowout.

Fracturing. Fracturing between the two wells should be avoided if lost circulation did not occur naturally. The fracture direction cannot be controlled, and it is more likely that the fracture is not in the direction of the blowout well. This often is determined after a significant amount of ineffective pumping.

Acid Holes. Some formations such as limestone are not as susceptible to the lost circulation because of rock stability. Shales are more prone to natural fractures upon intersect.

Limestone may exhibit a partial loss of circulation that indicates close proximity to the blowout well. If the communications channels could be opened, the kill could be made successfully at the intersect point. These channels are opened with acid and are termed as "worm holes." Large volumes of acid should be pumped at modest rates until the desired injectivity is reached. The kill fluid is injected at this point.

Perforating. Cased hole intersects require a means to penetrate the casing in the blowout well before pumping can begin. If the relief well is cased also, it must be penetrated in addition to perforating the blowout well. Large perforating guns are usually used for this purpose.

Several companies offer large guns for blowout control work. Vann has the most widely used gun with perhaps the best features. It offers large diameter charges with reasonable penetration. (See Figure 4–31.) However, only a few charges are mounted on the gun, and they are usually mounted in a vertical column. This means the gun must be oriented so the charges are in the direction of the blowout well. An orienting sub is available for this purpose.

Vann makes the following recommendations for their gun when used in a blowout situation:

- 12 in. (305 mm) or less separation between wellbores
- 7 5/8 in. (194 mm) OD minimum casing size on relief well
- 6 in. (152 mm) OD guns
- 300 gram HMX charges
- 9 in. (229 cm) minimum vertical spacing between charges
- 28 shots minimum (2–11 ft (3.35 m) carriers)
- Shot phasing to cover angle of 10° minimum

One recent application in Venezuela had some difficulty in orienting the gun. Several unsuccessful firing attempts were made without hitting the blowout well. The situation was resolved by setting a packer and running the gun into the packer. It was oriented and then fired. The blowout casing was penetrated and the well was killed.

Milling. Although perforating is the preferred method for establishing communications in a cased intersect scenario, a few field cases have used

FIGURE 4–31 • Vanngun for perforating blowout casing (Courtesy of Halliburton)

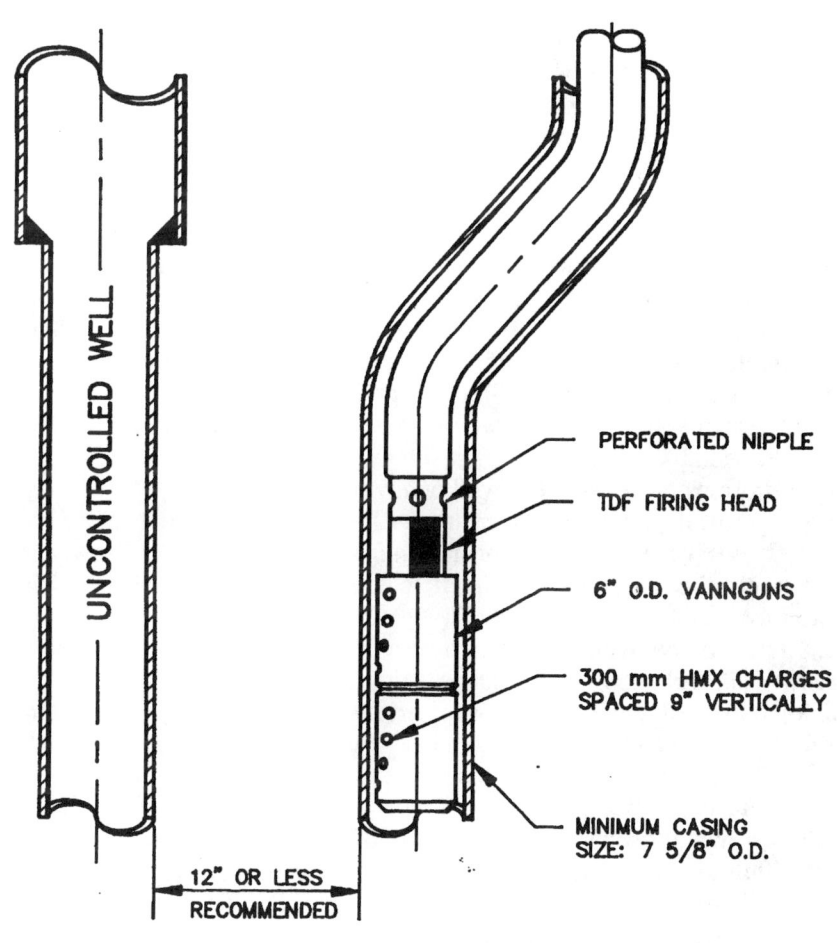

a mill to create a window in the blowout casing string. The only known guideline for the operation is that a milling angle of about 3.0° works most effectively.

Pumping. The pumping operations must be closely supervised by an individual or team. The expected well behavior must be anticipated prior to pumping. All actual pumping data should be recorded and monitored. A real time analysis should be done of the operations in progress.

An important guide is to have an understanding of the expected well behavior. Yet, an open mind must be maintained to interpret the blowout well behavior if it deviates from the expected trend. Blowout wells tend to have many variables that cannot be predefined and, as such, it is difficult to develop accurate pumping predictions. A key factor is the level of reservoir drawdown and the manner in which it affects the well killing.

Some type of "kill sheet" should be used. The term "kill sheet," with respect to blowouts, is analogous to the standard kill sheet used for kick killing. Of course, the parameters are different.

The kill sheet will contain data, or more likely it will be a computer printout showing expected pressures and the minimum and maximum acceptable values. The minimum allowable pressure is the pressure that must be maintained while pumping into the relief well to equal or exceed the bottom-hole pressure in the blowing well. The maximum pressure is equal to or less than the fracture gradient at the casing seat. Since several kill fluids may be used, these minimum and maximum values change throughout the kill operations. The pressures decrease almost linearly as heavy mud goes down the relief well and up the blowout well.

A reasonable medium between the minimum and maximum pressures should be maintained. This approach gives some latitude in the event of uncertainties or unexpected well behavior.

Operations must be controlled with a single individual making the decisions. He may receive information from several advisors. However, a single contact person has proven to work most effectively from field experiences.

The blowout well should be observed for signs of influence from the relief well. Indications include lessening of the observed flow and color change in the flame if the well was burning or on extinguishing the flame. An observer may be required as near as possible to the blowout well to notice any initial subtle changes in the blowout.

Continue to pump until the flow is apparently killed. After the kill, pump slowly for several hours until all of the gas or oil is worked out of the kill fluids. If the dynamic kill technique was used, the second stage of kill mud should be pumped according to the kill sheet prior to pumping at the slow rates to clean the annulus.

If the flow is not killed at the initial pump rates, increase the pump rates to the maximum allowable and observe the well. If good communications exist between the two wells and the kill calculations were done correctly, the blowout will be killed. If it does not die, the likely problem is that the relief well was not sufficiently close to the blowout. Some of the kill fluids are being lost to the formation and not directly into the blowout well. The relief well must be sidetracked to a more favorable location in better proximity to the blowout well.

After the well is dead and mud has been circulated for several hours, the next typical operation will be to prepare for cementing and the plug/abandonment. The cementing program should receive consideration if reservoir drawdown is observed in the blowout well. The cement may go directly into the reservoir and not up the blowout well. Several cement jobs may be required. Some thought may be given to sidetracking the relief well for a shallower intersect and setting an additional cement plug.

Also, if the top of the blowout well is accessible, consider cementing down the top of the blowout well. A snubbing unit or large diameter coil tubing unit must be used if a rig cannot be placed over the top of the well. This should be done after the initial cement plugs are set at the bottom by the relief well.

Rig Selection Guidelines

The selection of a suitable rig for a relief well and kill operation is a key decision during the planning phase. The choice of a rig directly affects the mobilization scheduling and logistics. It will subsequently impact the rig systems available for drilling and well kill activities and the operational safety and efficiency. A list of operating conditions and rig system requirements must be prepared based on the area of operations and preliminary kill plan.

Rig selection for land operations is relatively simple. If the rig can run the necessary casing strings for the relief well, then it is generally suitable for the work.

Rig Availability

The first step is to determine availability and state of readiness of rigs in or near the area of operations. For remote operations, long mobilization times can be involved. Computer data bases offered by several companies, such as Oceandril Data Services (Houston, Texas) division of PennWell Publishing Company, can provide a list of likely candidates. This is then followed up by direct contact with the respective contractors.

Operating Conditions

The operator will need to provide comprehensive operating conditions data. The data should include:

- General weather summary for the expected period of operations—prevailing weather, mean wind speeds and directions, etc.
- Storm conditions
- Currents
- Tides
- Unusual considerations—icing, etc.
- Sea bed soil conditions and mooring data
- Water depth

- Blowout well records—maximum bottom-hole temperature and pressure, location surveys, casing depths, logs, daily reports, etc.
- Blowout effluent and estimated flow rate
- Local regulatory restrictions
- Logistics situation—local supply of equipment, materials, and services; support vessels; ground and air transportation facilities

Rig Type Evaluation

Rig type recommendations are based on actual field experience in operating rigs over or near blowouts and a knowledge of case histories from other blowouts. Jack-up rigs provide a stable platform for operations and adequate load carrying capacity. The depth limitation for jack-ups is approximately 300 ft (100 m). Jack-ups are applicable in shallow water depths where floaters cannot operate. In the case of a secondary blowout on a vent well or relief well, a jack-up cannot be easily moved off. A particular danger exists on shallow gas blowouts that the soil may be disturbed to the extent that a jack-up can overturn from soil instability. A hard bottom with the possibility of boulders is not suitable for placement of jack-up legs/spud cans/mats. Soft, unconsolidated soil or soft soil containing sand lenses can also pose a problem.

Drillships have a high load carrying capacity. Drillships generally have a higher degree of motion than a semisubmersible for the same sea conditions. In calm areas, this will not be a deciding factor. Drillships have a relatively low freeboard and no air gap. Gas accumulations at the sea surface are a greater risk than on a semisubmersible. Due to radial outward surface flow during a subsea blowout, drillships tend to set off to one side of the boil. For a moored ship, this results in mooring line tightening on the boil side and load relaxation on the opposite side, which induces an overturning moment. Moored drillships are generally more difficult to move off of location during a blowout emergency. Dynamic positioning eliminates this particular risk.

Semisubmersibles are very stable and have a high load carrying capacity. The large air gap and open construction reduce the risk of gas reaching the main deck and ignition sources. Semis can be operated in a blowout boil under most conditions and, in fact, have been used for vertical re-entry into blowing wells. Therefore, they have advantages if a secondary blowout occurs while drilling the relief well. Propulsion/thrusters can assist in maintaining position and reducing the loads on the mooring system. Semis can be equipped to move off readily in an emergency.

Suitable rig availability becomes very limited for water depths over 1,500 ft (460 m). Relatively few floaters are fitted with mooring systems or dynamic positioning for operations in depths greater than 1,500 ft. Serious consideration should be given to relief well contingency planning in general, but particularly for deepwater drilling programs.

BOP Stack, Riser, and Subsea Equipment

The BOP stack, riser, and subsea equipment rated working pressure and depth capability must be suitable for the intended relief well operation. It is recommended to use a proven design, high release angle wellhead and LMRP connectors for floating operations. Risers, tensioners, and associated equipment have depth limitations. Rig selection can be restricted for deepwater operations. For ultra-deep water, guidelineless re-entry is utilized. It would be expensive and time-consuming to reoutfit a rig for deepwater operations if it is not already fitted out with adequate tensioners and riser. Occasionally, it is possible to locate some of this equipment that can be borrowed or rented from a contractor or operator.

Diverter Systems

A high capacity, state-of-the-art diverter system is necessary for shallow gas blowout kill operations. Most rigs are not equipped with suitable diverter systems. However, it is possible to prepare and install the necessary components in a relatively short period of time. The system should be sized conservatively, particularly if it is to be used for operations on a known shallow gas location.

An annular blowout preventer is preferred for the diverter unit itself. Porting and control lines can be provided that will give the desired response time of 20 seconds or less. Be aware that there are inherent advantages and disadvantages to the purpose-built diverter units generally found on most rigs.

The configuration should be simple and straight. Bends of any type should be avoided, if at all possible. Special designs should be utilized for the outlet area from the diverter annular and at any sections that will cause flow disturbances.

The control system, like the configuration, should be as simple as possible. Problems have arisen in the past where sophisticated control options were selected.

It may not be necessary to upgrade a relief well rig's diverter system if the drilling is going to be in an area with a low shallow gas risk.

Load Capacity and Deck Layout

The load capacity and layout must be suitable to handle the kill pumping system, supplementary mud tanks, additional supplies, and other equipment required for the planned operations. Many of these items are in addition to the materials and equipment normally required in a drilling operation. A typical high volume pumping system layout on a semisubmersible is shown in Figure 4–30.

Weights and sizes must be defined early in the planning. Information from the drilling contractor, service companies, and suppliers must be efficiently coordinated. The drilling contractor or consultants must confirm that the proposed loads are within the vessel's stability limits.

Gas Detection Systems

The existing rig's gas detection systems will need to be inspected, repaired (if necessary), and calibrated. In most cases, the existing systems will have to be supplemented. The rig may be potentially working near or in a live gas boil. Gas detectors must be located in strategically selected positions to provide early warning in case gas starts drifting.

Mud System and Bulk Storage

The drilling and well kill operations will be defined at an early stage. Additional tanks, mix pumps, and associated piping are sometimes necessary for the well killing activities. (See Figure 4–30.)

Living Quarters

Blowout kill operations generally necessitate manning levels significantly higher than for normal drilling operations. The various service contractors will need personnel on board for well monitoring, kill pump systems, directional drilling, cementing, ranging tools, or other special services. The blowout and firefighting specialists will have a team of two to four people. The drilling contractor may be required to supply extra personnel for support of various operations. Typical manning levels will range from 75 to 85 people. For remote operations where shuttling is not possible, it would be advisable to have sufficient facilities for 90 to 100 people.

Mooring and Stationkeeping

Mooring and stationkeeping are particularly important evaluation factors when the relief well will be drilled in deep water or where special conditions exist, such as fast currents, extreme tides, sea ice, or severe weather. Dynamic positioning is required for some deepwater sites. Special mooring arrangements and additional anchors may be necessary. Directional drilling will be used. The hole position indicating system needs to be a reliable design and fully operational.

Rig Maintenance and Warehouse Stock On Board

The candidate rig should have a planned maintenance system in place. All key systems must be operationally tested prior to commencing the drilling operations. The warehouse inventory should be adequate to support the rig in the area of operations. Any deficiencies in stock, parts, and so forth, should be purchased and in hand before drilling starts.

Additional Considerations

Rig selection, depending on the particular situation, will be influenced to some extent by many factors. The rig design, existing equipment, level of maintenance, and proven performance record must be considered. Personnel is the key to a successful operation. Experienced, knowledgeable personnel can overcome some deficiencies, but no amount of sophisticated equipment or controls can overcome poorly trained or inexperienced personnel.

Rig Inspection

The data for available rigs must be compared to the requirements described for the relief well rig. Contractors need to be contacted about various questions that arise during the evaluation process. A short list will be prepared composed of those rigs that most closely meet the requirements. Inspection visits should be made to those rigs on the short list. Sufficient time should be allocated to check each rig thoroughly and preferably visit with key rig personnel that will be assigned to the job. These visits allow time to confirm layout limitations and other points. A spread sheet should be prepared to compare all key factors for the short-listed rigs. A report should be prepared by the inspection team giving general impressions and pertinent comments about each rig.

Organization, Planning, and Logistics

Seldom considered topics for blowout control discussions are organization, planning, and logistics. However, operators seem to be giving more attention to these topics recently. Perhaps the importance of the issues are becoming more evident.

These topics are discussed in more detail in the earlier sections on contingency planning.

UNDERGROUND BLOWOUTS

An underground blowout occurs when formation fluids flow from one zone to another. The receiving zone could be a permeable/porous interval, a fractured formation, or behind ruptured casing which exposes a weaker formation. Occasionally, these blowouts occur without detection at the surface, although this case is rare.

Fluid flow direction is an important concern when choosing a control procedure. The cause of the underground transfer often indicates direction of the flow. Since most underground blowouts occur after the blowout preventers have been closed on a kick taken while drilling, the flow will normally be from the zone being drilled to some upper exposed zone. This is based on the assumption that shallow zones will fracture prior to deeper zones, and the initial kicking zone will be the primary source of formation fluid flow. (See Figure 4–32.)

The flow can be from shallow to deeper zones if the lost circulation is near the bottom. This is common when lost circulation is encountered at the bit while drilling. The zone may be naturally fractured, structurally weak, or an unsealed fault plane. The fluid level may fall and decrease the hydrostatic pressure sufficiently to allow an upper zone to flow. The common case is for the fluid to flow downward although the reverse occasionally occurs. (See Figure 4–33.)

Underground blowouts can cause problems by pressure charging exposed formations. Cases have been reported in which water wells have blown out with pressure charged zones from underground blowouts in adjacent oil and gas wells. These high pressures can be a serious problem in drilling offset relief and/or new wells. Conventional abnormal pressure

FIGURE 4–32 • Typical location of the flowing and fractured zones in an underground blowout

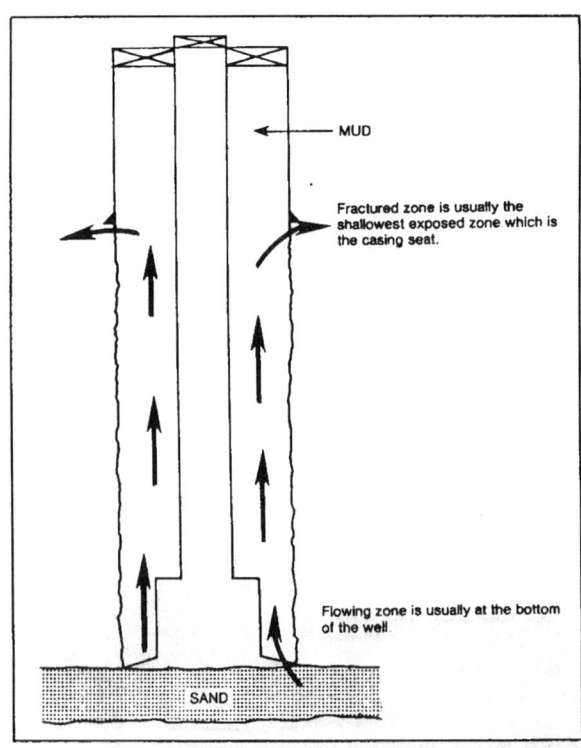

FIGURE 4-33 • Effect of encountering the lost circulation zone at the bottom of the well

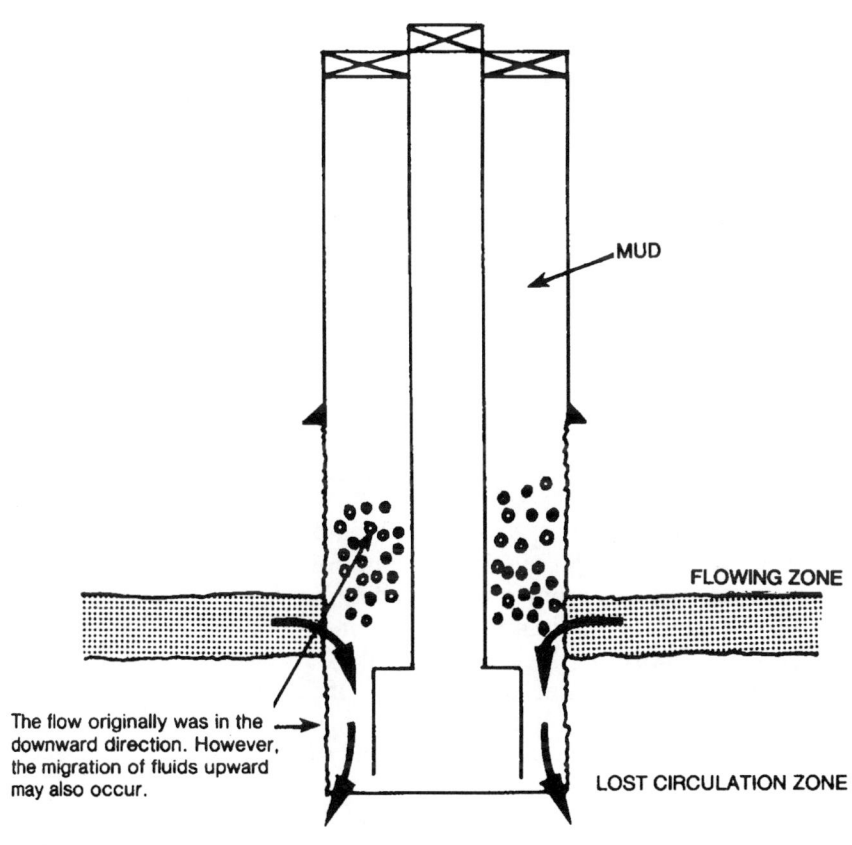

detection techniques and principles cannot be applied. Drilling offset wells often requires rotating heads to drill and deplete the intervals.

Underground Blowout Indicators

Several indicators identify underground blowouts. One, none, or several of these may be observed. Close observation may aid in determining the appropriate kill procedure.

A primary indicator is an initial drill pipe and casing pressure buildup and subsequent reduction. When the kick is detected, the preventers are closed and surface pressures begin building to balance bottom-hole pressures. If the casing pressures create equivalent mud weights greater than formation fracture gradients, a fracture will be formed. It relieves wellbore pressures and reduces the casing pressure. The reduced pressure may be stable or fluctuating. See Example 4.3.

Example 4.3

A well was shut in after warning signs of a kick were observed. These pressure data were observed.

Time	SIDPP, psi (MPa)	SICP, psi (MPa)
3:15	350 (2.41)	1,100 (7.58)
3:18	475 (3.28)	1,300 (8.96)
3:20	510 (3.52)	1,360 (9.38)
3:22	525 (3.62)	1,380 (9.51)
3:24	475 (3.28)	1,340 (9.24)
3:26	475 (3.28)	1,110 (7.65)
3:28	425 (2.93)	1,090 (7.52)
3:30	350 (2.41)	1,090 (7.52)
3:40	0 (0)	1,100 (7.58)
3:50	125 (0.86)	1,250 (8.62)
4:00	140 (0.97)	1,200 (8.27)
5:00	130 (0.90)	1,120 (7.72)

Monitoring initial pressures on the drill pipe and casing may be an important factor when killing the blowout. Although the initial pressure on the drill pipe does not give a reading to determine bottom-hole pressure, it indicates minimum pressures on the formation that must be achieved to control the kick. In Example 4.3, the minimum kill mud weight would be calculated with a SIDPP of 525 psi (3.62 MPa).

Fluctuating or unstable pressure readings indicate an underground blowout. Fluctuations result from unsteady flow from one or several formations or from the fractured formation tending to open and close as pressures change. Both drill pipe and casing pressures may fluctuate uniformly or independently of each other. If the annular formation collapses or bridges on the drill string, the casing pressure may stabilize while the drill pipe pressure continues to change.

Drill pipe pressure may be higher than casing pressure. This is usually the result of formation fluids entering the drill pipe after the blowout has been initiated. The drill pipe mud may overbalance the cut annular fluid and U-tube out of the pipe. This leaves a void that can be occupied by gas. Although this can occur in the annulus, the larger feet-per-barrel ratio in the drill pipe gives rise to higher pressures.

Low pressures may be noticed in some cases. If the mud falls out of the pipe, lower or even zero pressure may be recorded if no fluids enter the pipe due to bit jet plugging or severe flocculation in the pipe.

In most cases, there will be little or no direct communications between drill pipe and annulus. The casing pressure may change without affecting the drill pipe, or the drill pipe pressure may change with no uniform reflection on the casing pressure. Lack of communication is from loss of integrity in the U-tube or borehole.

Kill Procedures for Underground Blowouts

Unlike conventional well control, established kill procedures that work in most situations do not exist for underground blowouts. Although some work more effectively, these are based on adequate knowledge of the cause of the underground blowout, the location of the thief zone(s) and flowing zone(s), formation pressures, and limitations of the proposed procedures.

The most common case of underground blowouts is a kick occurring in a deep zone when surface casing is set in the well. This situation exposes large sections of hole to high equivalent mud weights resulting in formation fracture. (See Figure 4–34.)

FIGURE 4–34 • Effect of equivalent mud weight vs. depth during a typical pre-underground blowout situation

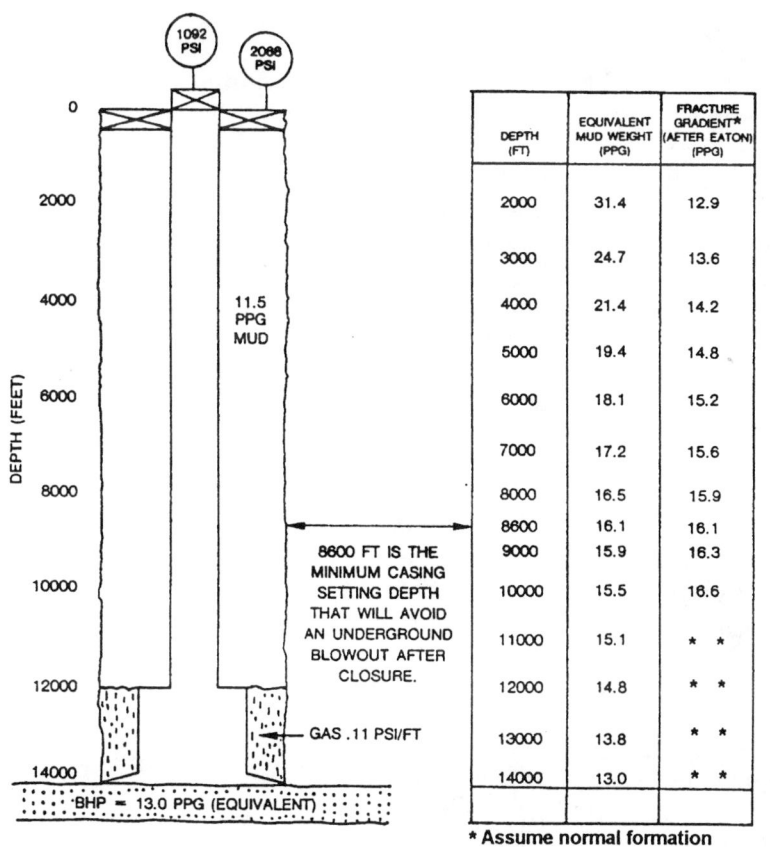

Although it is advisable to set the intermediate casing string as deep as possible to achieve maximum fracture gradients (if this is in accordance with the well design), precautions must be taken when selecting the surface casing setting depth to account for the deep intermediate casing seat and prevent the situation in Figure 4–34.

Proper setting depth selection will vary for every well that exhibits different formation pressure characteristics. If a casing depth of 8,600 ft (2,600 m) had been chosen in Figure 4–34, the underground blowout would have been avoided at initial closure.

Heavy Slug. The most successful kill procedure for this type of blowout is spotting a heavy slug of mud in the open hole below the point of lost circulation. The objective is to generate high bottom-hole pressures from (1) hydrostatic pressure of high and low density muds in the well and (2) any existing casing pressure. Although the heavy slug may have a density greater than the equivalent fracture gradient at the casing seat, this does not present lost circulation problems, since mud will not be circulated above this point.

The required kill density can be estimated from initial shut-in pressures and drilling and penetration rate data immediately prior to the kick. The volume of heavy mud required to build necessary hydrostatic pressure depends on mud density, hole geometry (washout), and rate of formation influx into the wellbore. A rule of thumb used successfully in many applications is to build a volume of mud three times the calculated volume to allow for mud cutting and hole washout.

When spotting the slug, pump as fast as reasonable. This is contrary to conventional kick killing where pumping is at a controlled rate to allow proper pressure control, mud handling, and degassing. The heavy slug is pumped at a faster rate to minimize influx cutting.

Example 4.4 illustrates this heavy slug kill procedure.

Example 4.4

An underground blowout is occurring on a particular well. An attempt is made to kill the well using the heavy slug method. With the following known data, what amount of mud should be used? Use the pressure data from Example 4.3.

Well depth = 10,000 ft

Casing seat = 3,500 ft

Original mud weight = 11.0 ppg

Hole size = 8 1/2 in.

Drill pipe = 4 1/2 in.

Drill collars = 6 1/2 in.

Solution

1. Bottom-hole pressure (minimum)

   $(0.052 \times 11.0 \text{ ppg} \times 10{,}000 \text{ ft}) + 525 \geq 6{,}245 \text{ psi}$

2. Assume the total annular pressure (mud + SICP) is equivalent to 11.0 ppg mud. The heavy slug must exert at least 525 psi more than an equal height column of 11.0 ppg mud. Any of the following options accomplish this objective.

Mud weight, ppg	Column height, ft
15	2,524
16	2,019
17	1,682
18	1,440

3. The operator elects to mix and pump a volume of 15.0 ppg mud that gives a minimum annular height of 2,500 ft. Drill collars were disregarded in the calculation.

   (Fluid height) (annular capacity) = volume

   $(2{,}500 \text{ ft}) (0.05 \text{ bbl/ft}) = 125 \text{ bbl}$

4. Using a safety factor multiple of 3 to reduce the effects of mud cutting, the operator pumped 375 bbl of 15.0 ppg mud to kill the well.

After the blowout is killed, steps must be taken to restore the well to a drillable condition. The fractured zone(s) must be cemented to resolve the lost circulation problem. The original mud in the annulus must be conditioned to a density that controls formation pressures but less than the fracture gradient of exposed zones. Heavy mud used to kill the well must be circulated out in several stages to avoid refracturing the lost circulation zone. (See Figure 4–35.)

Barite Plug. Another procedure to kill underground blowouts is spotting a barite plug. This technique forms a barite bridge that seals the blowout and allows a heavier mud to be circulated above the plug. The technique is not based on well control through a hydrostatic pressure increase as was the heavy slug procedure. Instead, it uses a bridging effect.

The density and composition of the barite plug is the key to its success. A typical mix composition is given in Table 4–2. Recent field applications show a density of 18–22 ppg (2,200–2,600 kg/m^3) yields optimum settling and bridging characteristics. A small amount of SAPP (sodium acid pyrophosphate) is often added to reduce the viscosity and channeling effects of flocculated mud in the annulus, increase the barite settling rate by thinning contamination of the plug, and by forming a tight, low

FIGURE 4-35 • Steps required to complete the kill process for the situation developed in Example 4.4

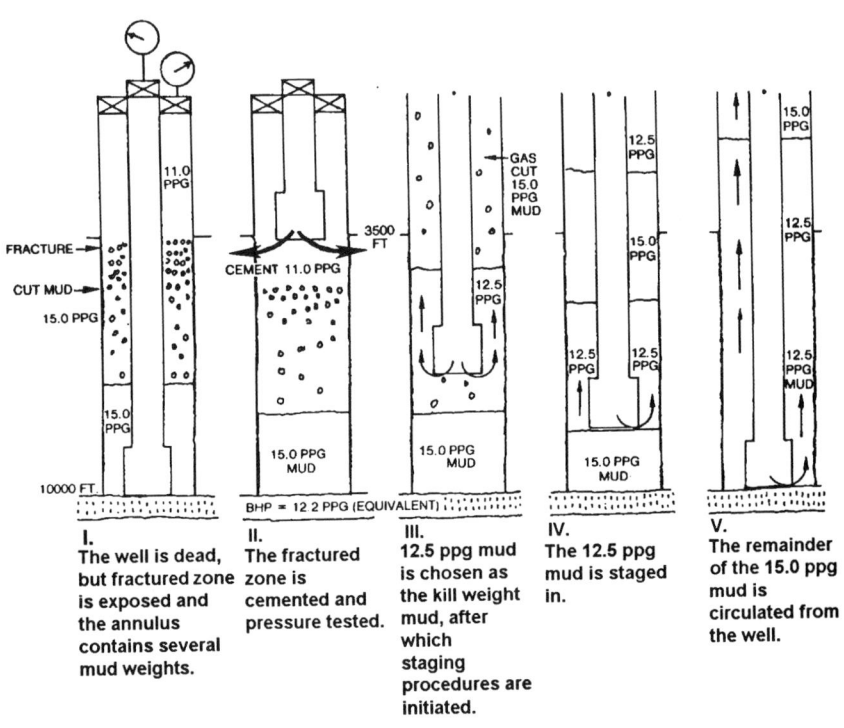

TABLE 4-2 • Typical barite plug composition

Formula for a Barrel of 22 ppg Slurry
• 750 lb of barite
• 21 gal of fresh water
• 1/2 lb SAPP
• 1/4 lb caustic soda

permeability filter cake. The plug volume should normally give at least a 500 ft (150 m) column. Barite plugs are more effective in smaller hole sizes.

The barite plug contains no suspension agent for the barite. (See Table 4–2.) Surface mixing facilities and plug placement must be continuous and rapid. If the mixing or pumping is halted for a short time, settling in the pits or drill pipe plugging may occur. Plug mixing will generally require a cementing pump and agitation system. Rig facilities are not suitable due to low mixing rates.

A barite plug has been found effective in hole sizes of 9 7/8 in. (251 mm) or smaller. Large hole sizes make it difficult to achieve sufficient annular velocity to remove old viscous mud. The plug mixes with the mud and the high viscosity prevents barite settling.

Squeezes with Diesel Oil. Squeezes with diesel oil, called "gunk" squeezes, as the base fluid have proven effective in solving the circulation problems when the thief zone is not caused by low fracture gradients. Applications include porous and permeable zones, unsealed fault planes, or poor cement jobs.

The diesel oil is used as a transport agent for water reactive additives such as bentonite or cement. The mixture is pumped into the formation where it mixes with mud or formation water and hydrates, or sets. The slurry, often called a "gunk squeeze," forms a high viscosity pill in the formation that restricts or retards the fluid flow. (See Table 4–3.)

A diesel oil squeeze contains additives not soluble in diesel oil. The mixture has an extended pumping time unless it becomes contaminated with mud. This feature reduces the possibility of premature setting before placement is complete. Also, if it becomes necessary to drill the set plug, sidetrack tendencies are reduced. The final plug does not form a firm bridge.

Remedial procedures, other than the diesel squeeze, can be used to re-establish circulation. These include lost circulation materials, cement, polymer pills, and numerous commercial specialty products. When these products are used, it may be advisable to remove or blast the bit jets with primer cord on an electric wireline.

Time may be a factor in killing an underground blowout. The hole may bridge from chemical sloughing or mechanical heaving if the flow is allowed to continue. The kicking reservoir may deplete or reduce its pressure to a point so lower mud weights can regain control of the well. The operator cannot rely solely on time because many underground blowouts have shown the capability of flowing for extended periods.

Certain techniques can be used in conjunction with kill procedures to aid in controlling the blowout. Bullheading injects material into the

TABLE 4–3 • Typical composition for a diesel oil-bentonite squeeze

- 300 pounds of bentonite/bbl of diesel
- 15 pounds of mica (or walnut hulls)/bbl of diesel for additional plug strength
- Diesel oil as required

formation by pumping fluid into the annulus with the drill pipe closed or pumping mud into the drill pipe with the blowout preventers closed. Although this technique is not normally recommended for conventional well control, the procedure is applicable in underground blowouts because the zone is already fractured.

Bullheading may be quicker and more economical in cases such as wells with shallow surface casing. The mud lost by pumping down the casing into the formation is considerably less than pumping down the drill string and up the annulus. Bullheading must be evaluated at the rig site to determine its applicability.

Another procedure that aids in implementing blowout control procedures is stripping out of the open hole. This avoids hole bridging on the pipe and allows the operator to apply lost circulation remedies directly to the thief zone if it is located at the casing seat. If stripping procedures are followed and a float is not included in the drill string, fill the pipe with heavy, viscous mud to prevent a drill pipe blowout after the kelly is removed.

Lost Circulation Zone Detection Techniques. It is important to determine the thief zone's location in order to calculate volumes and densities of kill fluids and the position at which they should be spotted. The history of the well or field may supply the information necessary to locate the zone. A structurally weak zone or an unsealed fault plane may be present. The field may contain one of more depleted sands causing the loss. Also, the conditions under which the blowout occurred may indicate the zone is at the bottom of the well or at the casing seat.

Common logging tools include the following:

- Temperature
- Radioactive tracer
- Noise

The most common tool used to define the interval is the temperature log. It is generally not used to record absolute temperature; instead, it's used to record differential temperature. As the logging tool is lowered down the drill pipe, it will read an abnormal change at the loss zone if the underground flow is continuous. (See Figure 4–36.)

The tool senses the heat from the fluid greater than it should be for the depth at which it is encountered. In some cases, the temperature change has been reported as a cooling effect supposedly due to gas expansion. Nonetheless, a temperature change is the key. Figure 4–37 shows a section from an actual temperature log indicating a fluid exit at approximately 1,225 ft (373 m).

Figure 4–36 showed the case in which the formation fluids were moving. If the log gave the results shown in Figure 4–38, the indications

FIGURE 4-36 • Illustration of the expected results from a temperature log used to locate the loss zone during an underground blowout

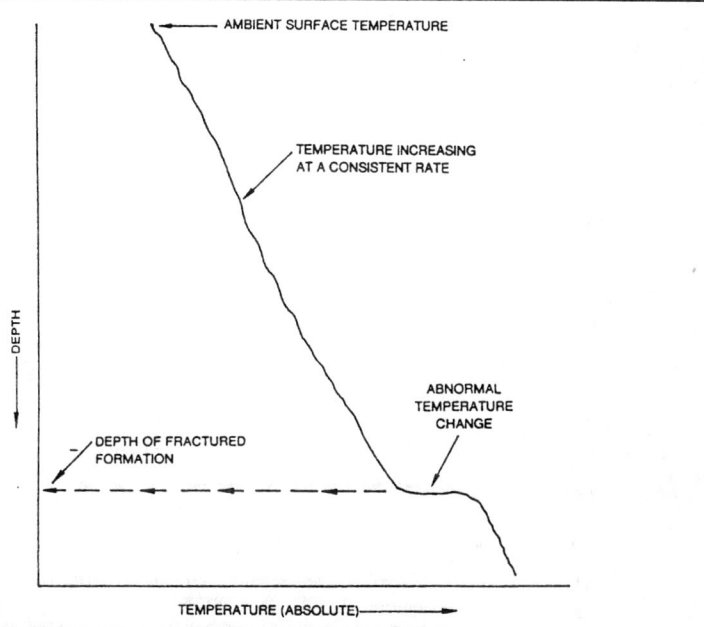

FIGURE 4-37 • Section of an actual temperature log

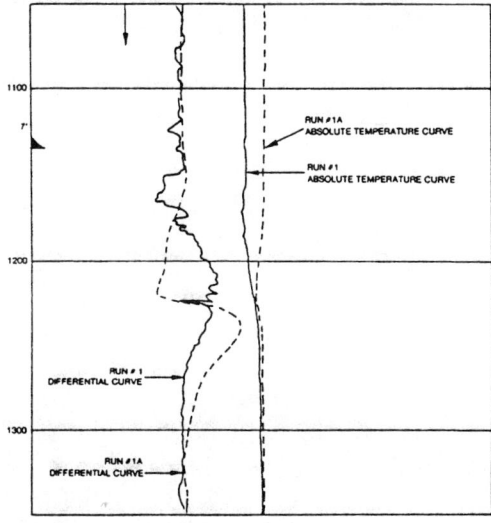

FIGURE 4-38 • Readings from a temperature log when the well bore fluids are static

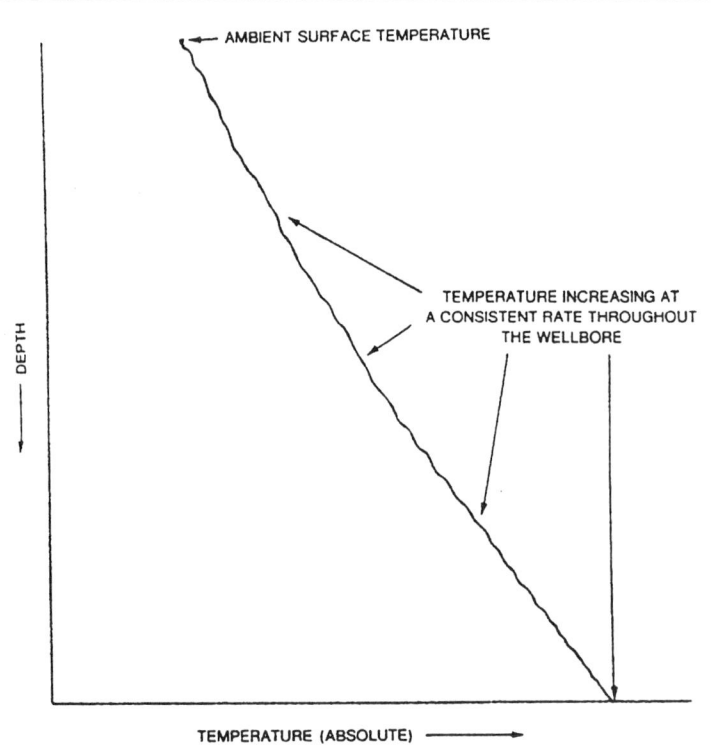

are that the thief zone is at the bottom of the well or that a static situation exists in which the fluid is no longer flowing.

The temperature log can be used in this case by pulling the tool up the drill pipe, pumping a volume of mud into the annulus, and then running the logging tool again. The tool will read a fluid with a lower than normal temperature until it reaches the thief zone where it will record normal temperatures. (See Figure 4-39.) If the results on the second attempt are similar to those shown in Figure 4-38, it is assumed that the lost circulation zone is at the bottom of the well.

A radioactive tracer tool is often used to isolate lost circulation problems. A radioactive material is pumped into the well through the mud system. A logging tool, usually a gamma ray detector, is used to determine areas of high radioactive concentration. The depths of high concentrations is assumed to be the point at which the fluid entered the formation.

FIGURE 4-39 • Illustration of the readings from a temperature log after mud has been pumped down the annulus

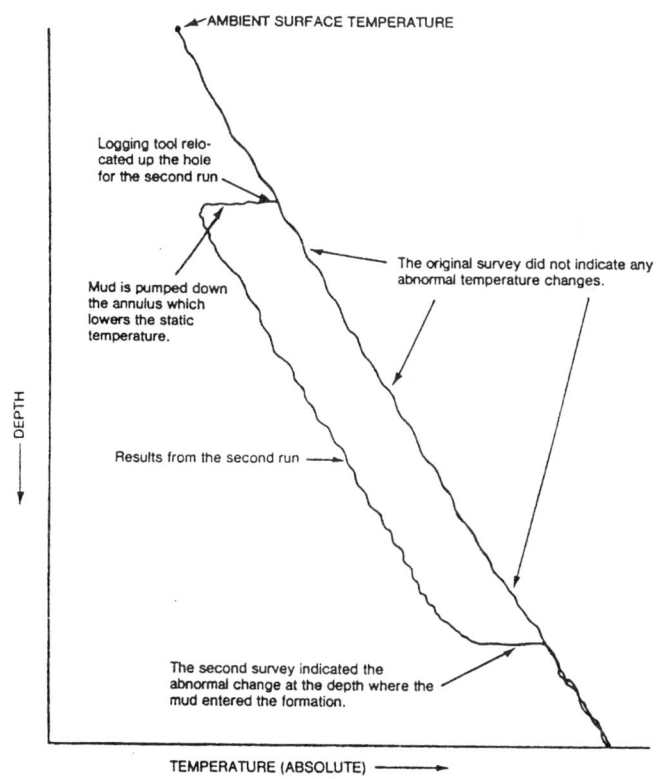

Figure 4-40 shows an actual case in which a radioactive tracer was used to determine the lost circulation interval.

Typically, radioactive tracers are not effective. The common result is that the material contaminates the annulus, or it is lost and never tracked.

A noise log can be used to determine the lost circulation zone. The tool is a sonic detector that records sounds created by fluid movement. The tool delineates the static fluid column above the loss zone from the moving fluid below. The sensitivity of the tool may be a deterrent to its usage.

Field cases report an interference from surface rig vibrations transported through the tubular goods in the well. Regardless, the tool can be used as a qualitative indicator.

FIGURE 4-40 • Results of a radioactive tracer log

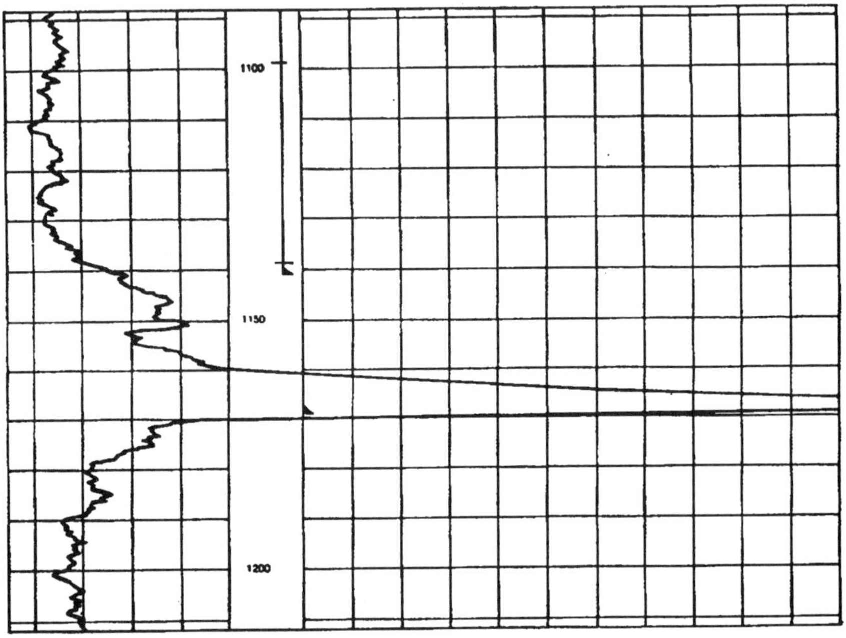

WHAT CAUSES BLOWOUTS

Kato and Adams have presented a paper that studied many aspects of blowouts. The paper was based on a study of a large blowout database containing nearly 1,000 blowouts.

Figure 4-41 shows common operations occurring at the time of the blowout. Blowouts seem to occur about as frequently while tripping out of the hole as while drilling. If kick prevention while coming out of the hole can be improved, it should decrease the number of blowouts. As discussed in Chapter 3, handling kicks off bottom are more difficult than drilling kicks because mud cannot be circulated to bottom.

Figure 4-42 shows blowouts causes. High pressures or abnormal and unusual circumstances seldom cause blowouts. It is more common that routine situations are involved during the blowout. Industry attention in these areas can reduce the number of blowouts.

Figure 4-43 shows drilling blowout frequency in the United States for a certain time period. Note the decrease in blowouts in 1978, which is contrary to the increase in drilling activity. This is associated with the implementation of well control training standards and mandatory training. This clearly shows that training pays dividends.

FIGURE 4-41 • Operation when blowout occurred (All areas except for Alberta) (Courtesy of SPE, 23289, © 1987.)

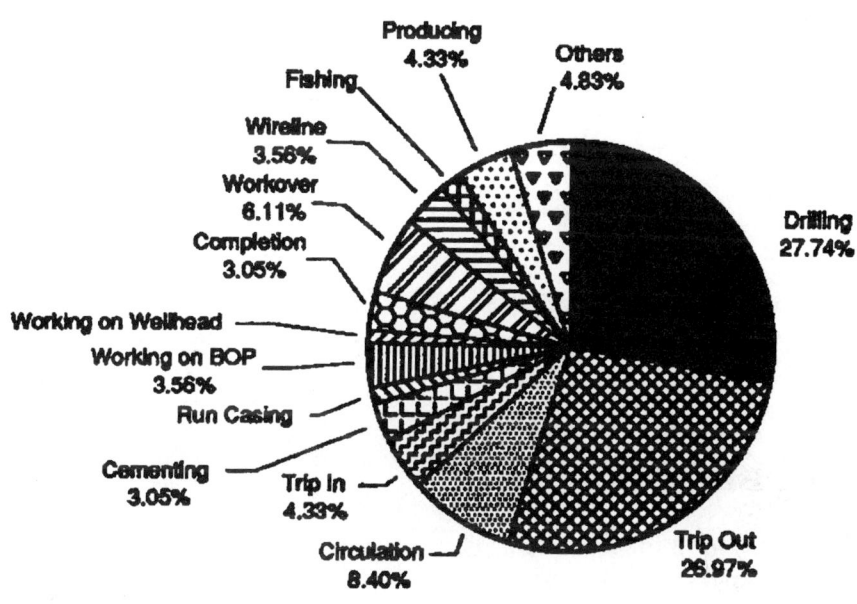

FIGURE 4-42 • Blowout causes (Courtesy of SPE, 23289, © 1987.)

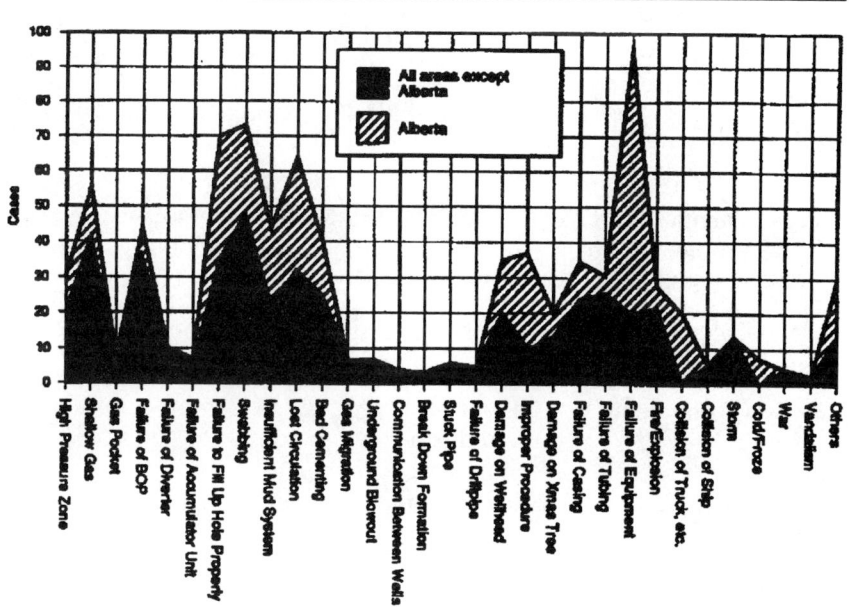

FIGURE 4-43 • Blowouts and drilled wells in the U.S. (Courtesy of SPE, 23289, © 1987.)

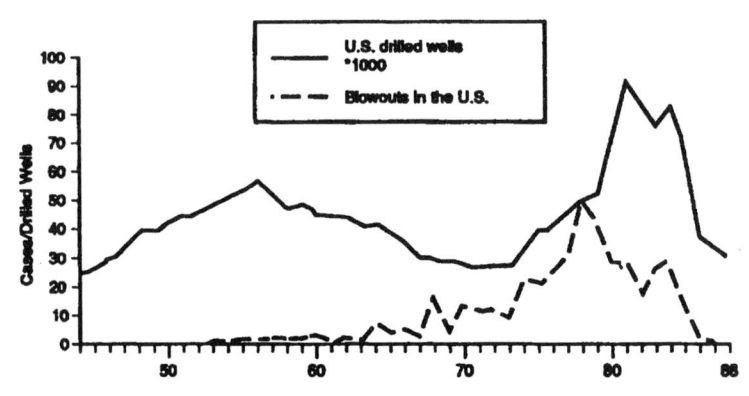

PROBLEMS

4.1 What is the maximum allowable injection pressure for the following circumstances?

Depth of casing = 8,500 ft

Fracture mud weight (at 8,500 ft) = 15.8 ppg

Mud weight = 9.0 ppg

Friction pressure during pump-in (calculated) = 400 psi

4.2 Calculate and plot the maximum allowable injection pressures for Problem 4.1 using mud weights of 9.5 ppg, 10.0 ppg, 10.5 ppg, and 11.0 ppg.

4.3 Calculate and plot the maximum allowable injection pressures for Problem 4.1 using friction pressures of 300 psi, 500 psi, 700 psi, and 900 psi.

4.4* Calculate and plot the maximum allowable injection pressures for the following well.

Casing depth = 13,000 ft

Casing ID = 7.0 in.

Fracture mud weight at the casing seat = 15.9 ppg

Kill mud weight = 12.7 ppg

BLOWOUTS

Plastic viscosity = 25 cp

Flow rate = 30 bbl/min

4.5* Using the data from Problem 4.4, what are the friction pressures for pumping rates of 40 bbl/min; 50, 60, and 75 bbl/min? Calculate the maximum injection pressures for these rates.

4.6 Use the following data to determine the required hydraulic horsepower for a pump-in operation.

Flow rate = 40 bbl/min

Maximum pressure = 4,000 psi

Design factor = 1.3

4.7* Calculate and plot the required hydraulic horsepower for Problem 4.5. Use a design factor of 1.5.

4.8 Use the following data to calculate the maximum rate of water injection during a pump-in operation.

Radius of injected water bank = 9 ft

Effective wellbore radius = 0.75 ft

Water viscosity = 1.1 cp

Permeability = 600 md

Net formation thickness = 25 ft

Average reservoir pressure = 2,000 psia

Flowing injection bottom-hole pressure = 3,000 psia

4.9* In Problem 4.8, what is the maximum rate if the injected water bank radius is 29 ft?

4.10* What would be the expected cumulative volume of injected water at the front of the water bank under the following circumstances?

Connate water saturation = 0.5

Residual oil saturation = 0.2

Radius of injected water bank = 25 ft

Radius of injector well = 0.5 ft

Net formation thickness = 40 ft

Formation porosity = 0.3

4.11* Calculate and plot the expected cumulative injected volumes for the following water bank radii: 20 ft, 30, 40, 60, and 70 ft. Use the data from Problem 4.10.

4.12 The following pressure and hole data were observed on a kick that preceded an underground blowout. What is the minimum bottom-hole pressure? If a heavy mud slug is 3 ppg (equivalent) greater than the bottom-hole pressure, what vertical column height is necessary? Assume a 2.5 volume safety margin to reduce the effects of mud cutting.

Well depth = 13,000 ft

Casing seat = 2,000 ft

Original mud weight = 12.9 ppg

Hole size = 12 1/4 in.

Drill pipe = 5.0 in.

Drill collars = 7 1/2 in., 1,200 ft

Total annular pressure = 12.9 ppg

Time	SIDPP, psi	SICP, psi
1:00	650	930
1:05	670	950
1:10	690	970
1:15	700	910
1:25	700	880
1:40	660	750
2:00	650	750
3:00	650	760

SOLUTIONS

Note: All solutions are approximate depending on user's round off procedures.

4.1   3,405 psi

4.2   9.5; 3,185 psi

        10.0; 2,964 psi

        10.5; 2,743 psi

        11.0; 2,522 psi

4.3   300; 3,305 psi
      500; 3,505 psi
      700; 3,705 psi
      900: 3,905 psi
4.4   584 psi
      2,747 psi
4.5   966 psi; 3,129 psi
      1,428 psi; 3,591 psi
      1,965 psi; 4,128 psi
      2,904 psi; 5,067 psi
4.6   5,096 hhp
4.7   40; 1,420 hhp
      50; 2,624 hhp
      60; 4,333 hhp
      75; 8,005 hhp
4.8   38,846 bpd
4.9   26,380 bpd
4.10 4,196 bbls
4.11 20; 2,685 bbls
      30; 6,043 bbls
      40; 10,742 bbls
      60; 24,170 bbls
      70; 32,898 bbls
4.12 1,420 bbls of 17.4 ppg

REFERENCES

Aadnøy, B.S. and P. Bakøy. "Relief Well Breakthrough at the Problem Well 2/4-14 in the North Sea," SPE/IADC 20915, SPE/IADC European Petroleum Conference, 22–24 October 1990.

"Accidents Connected with Federal Oil and Gas Operations on the Outer Continental Shelf Gulf of Mexico," Vol. 1 1956–1979, U.S. Geological Survey Conservation Division, December 1979.

"Accidents Connected with Federal Oil and Gas Operations on the Outer Continental Shelf Gulf of Mexico OCS Region," Vol. 2 January 1980–December 1984, U.S. Department of the Interior/Minerals Management Service Gulf of Mexico OCS Region, June 1985.

Adams, N.J. *Drilling Engineering: A Well Planning Approach*, Tulsa: PennWell Publishing Co., 1985.

Adams, N.J. "How to drill a relief well," *Oil & Gas Journal*, September 1980.

Adams, N.J. "Underground Blowouts," *Oil & Gas Journal*, October 1980.

Adams, N.J. *Well Control Problems and Solutions*, Tulsa, Oklahoma: The Petroleum Publishing Co., 1980.

Adams, N.J. and D. Carter. "Hydrogen Sulfide in the Drilling Industry," Deep Drilling Symposium, Amarillo, Texas, 1979.

Adams, N.J., B. Hansen, A.D. Stone, J. Voisin, and S. Clement. "A Case History of Underwater Wild Well Capping–Successful Implementation of New Technology on the SLB-5-4X Blowout in Lake Maracaibo, Venezuela," SPE 16673, 62nd Annual Technical Conference, Dallas, 27–30 September 1987.

Adams, N.J. and L.G. Kuhlman. "A Discussion on Casing Settling During Shallow Gas Blowouts," SPE/IADC 27502, SPE/IADC Drilling Conference, Houston, Texas, 15–18 February 1994.

Adams, N.J. and L.G. Kuhlman. "Shallow Gas Blowout Kill Operations," SPE/IADC 21455, SPE Middle East Oil Show, Bahrain, 16–19 November 1991.

Adams, N.J. and L.G. Kuhlman. "How to prevent or minimize shallow gas blowouts–Part 1," *World Oil*, May 1991, pp. 51–58.

Adams, N.J. and L.G. Kuhlman. "How to prevent or minimize shallow gas blowouts–Part 2," *World Oil*, June 1991, pp. 66–71.

Adams, N.J. and L.G. Kuhlman. "Case History Analyses of Shallow Gas Blowouts," SPE/IADC 19917, SPE/IADC Drilling Conference, Houston, Texas, 27 February–2 March 1990.

Barnett, R.D. "A Logical Approach to Killing an Offshore Blowout, West Cameron 165 Well No. 3, Offshore Louisiana," SPE/IADC 6903, SPE/IADC Fall Technical Conference, 1977.

Blount, E.M. and E. Soeiinah. "Dynamic Kill: Controlling Wild Wells in a New Way," *World Oil*, October 1981.

Blowout Database Offshore Blowouts, Part (1), 1955–1974, Trondheim, Norway: Marintek Sintef-Gruppen.

Blowout Database Offshore Blowouts, Part (2), 1975–1985, Trondheim, Norway: Marintek Sintef-Gruppen.

Booth, J.E. "Use of Shallow Seismic Data in Relief Well Planning," *World Oil*, May 1990.

Britt, E.L. "Theory and Applications of the Borehole Audio Tracer Survey," SPWLA Seventeenth Annual Logging Symposium, June 1976.

Bruist E.H. "A New Approach in Relief Well Drilling," *Journal of Petroleum Technology*, June 1972, pp. 713–722.

Davenport, H.H., B.J. Bulpard, and J.A. Cashman. "How Shell Controlled its Gulf of Mexico Blowouts," *World Oil,* November 1971, pp. 71–73.

DeWardt, J.P. and C.M. Wolff. "Borehole Position Uncertainty Analysis of Measuring Methods and Deviation of Systematic Error Model," *Journal of Petroleum Technology,* December 1981.

Eaton, B.A. "Fracture Gradient Prediction and Its Application in Oilfield Operations," *SPE Journal,* October 1969, p. 1,353.

Ely, J.W. and S.A. Holditch. "Conventional and Unconventional Kill Techniques for Wild Wells," SPE/IADC 16674, SPE/IADC 62nd Annual Technical Conference, 27–30 September 1987.

Fagerjord, O. "Worldwide Offshore Accident Databank," *Veritec Marine Technology Journal,* Noroil Publishing House, Ltd. A/S, 1986.

Fanneløp, T.K. and K. Sjøen. "Hydrodynamics of Underwater Blowouts," *Norwegian Maritime Research,* no. 4, 1980, pp. 17–34.

Flak, L.H. and W.C. Goins. "New Relief Well Technology is Improving Blowout Control," *World Oil,* December 1983.

Flak, L.H. and W.C. Goins. "New Techniques and Equipment Improve Relief Well Success," *World Oil,* January 1984.

Fosdick, M.R. "Compilation of Blowout Data from Southeast U.S. Gulf of Mexico Area Wells," The University of Texas at Austin, College of Engineering, August 1980.

Grace, R.D. "Case History of Texas' Largest Blowout Shows Successful Techniques on Deepest Relief Well," *Oil & Gas Journal,* 20 May 1985.

Haanschoten, G.W. and B.V. Gaatschappij. "ULSEL System Performs on Brunei Blowout Under Tough Conditions," *Oil & Gas Journal,* 17 January 1977.

Hughes, Virginia M. "Reducing Blowout Incidents Through a Computer Assisted Analysis of Trends Among Gulf Coast Blowouts," Report No. UT 86-3, Texas Petroleum Research Committee, University Division, August 1986.

"Joint Industry Program for Subsea Blowout Control Capability," Phase I Report, managed by Adams Engineering Inc., Houston, 1986.

Jones, L.B., E.M. Blount, and O.H. Glaze. "Use of Short-Term Multiple Rate Tests to Predict Performance of Wells Having Turbulence," SPE/IADC 6133, 1976 SPE Fall Technical Conference, 3–6 October 1976.

Koederitz, W.L., F.E. Beck, J.P. Langlinais, and A.T. Bourgoyne, Jr. "Method for Determining the Feasibility of Dynamic Kill of Shallow Gas Flows," SPE 16691, SPE Fall Technical Conference, 27–30 September 1987.

Kuckes, A.F. "An Electromagnetic Survey Method for Directionally Drilling a Relief Well into a Blown Out Oil or Gas Well," *SPE,* June 1984.

Lewis, J.B., G.J. Mabie, J.Z. Harris, and R.D. Barnett. "New Innovations for Fighting Blowouts," OTC 2766, Offshore Technology Conference, 1977.

Lewis, J.B. "New Uses of Existing Technology for Controlling Blowouts; Chronology of a Blowout Offshore Louisiana," *Journal of Petroleum Technology,* October 1978, pp. 1,473–1,480.

Lewis, J.B. "The Use of the Computer and Other Special Tools for Monitoring a Gas Well Blowout During the Kill Operation–Offshore Louisiana," SPE/IADC 6836, SPE/IADC Fall Technical Conference, 1977.

Matthews, W.R., and J. Kelly. "How to Predict Formation Pressure and Fracture Gradient," *Oil & Gas Journal,* 20 February 1967.

McLamore, R.T. and G.O. Suman, Jr. "Explosive Termination of a Wild Well–Evaluation of a Concept," SPE/IADC 3591, SPE/IADC Fall Technical Conference, 1971.

McManus, C., A. Da Mota, and H. Strand. "A Continuous Flexible 15,000 psi Choke and Kill Line System From the BOP Stack to Deck; First Experience in the North Sea," OTC 4352, Offshore Technology Conference, 3–6 May 1982.

Milgram, H.H. and R.J. Van Houten. "Plumes from Subsea Well Blowouts," Behavior of Off-shore Structures (BOSS), Third International Conference, August 1982, pp. 659–677.

Miller, R.T. and R.L. Clements. "Reservoir Engineering Techniques Used to Predict Blowout Control During the Bay Marchand Fire," *Journal of Petroleum Technology,* March 1972, pp. 234–240.

Morris, F.J., R.L. Waters, G.F. Roberts, and J.P. Costa. "A New Method of Determining Range and Direction From a Relief Well to a Blowout Well," SPE/IADC 6836, SPE/IADC Fall Technical Conference, 1977.

Nelson, R.F. "The Bay Marchand Fire," *Journal of Petroleum Technology,* March 1972.

Pidcock, G.A. and D.R. Fowler. "Relief Well Contingency Drilling Plans For Remote Areas," SPE/IADC 21997, SPE/IADC Drilling Conference, 11–14 March 1991.

Podio, A.L., R. Fosdick, and J. Mills. "Analysis Gives Blowout Causes, Trends, Cost," *Oil & Gas Journal,* 7 November 1983.

Prentice, C.M. "Maximum Load Casing Design," *Journal of Petroleum Technology,* July 1970, pp. 805–811.

Risnes, R. and P. Horstud. "Drilling of Relief Wells," Rogalandsforskning, 1980.

Robinson, J.D. and J.P. Vogiatzis. "Magnetostatic Methods for Estimating Distance and Direction from a Relief Well to a Cased Wellbore," *Journal of Petroleum Technology,* June 1972, pp. 741–749.

Rygg, O.B. and T. Gilhus. "Use of Two-Phase Pipe Flow Simulator in Blowout Kill Planning," SPE/IADC 20433, SPE/IADC 65th Annual Conference, 23–26 October 1990.

Schlumberger Log Interpretation/Applications, pp. 110–116.

Stephensen, M. "Program Challenges Directional Survey Accuracy Claims," *Oil & Gas Journal,* August 1984.

Tannich, J.D., C.W. Sandlin, and A.N. Gist. "Well Killing: Possibilities and Limitations," Shallow Gas Seminar, Norwegian Petroleum Directorate, 27–28 August 1987.

Tarr, B.A., A.F. Kuckes, and M. Ac. "Use of New Ranging Tool to Position a Vertical Well Adjacent to a Horizontal Well," SPE/IADC 20446, SPE/IADC 65th Annual Conference, 23–26 September 1990.

Tucker, J., L. Nunenmacher, and H. Williamson. "Shallow Gas Events," United States Department of the Interior/Minerals Management Service, 1985.

"Veritas Claims Diverters Fail in 50% of Blowouts," *Offshore,* November 1986, p. 54.

Voisin, J.A., G.A. Quiroz, J. Wright, R. Pounds, and K. Bierman. "Relief Well Planning and Drilling for SLB-5-4X Blowout, Lake Maracaibo, Venezuela," SPE/IADC 16677, SPE/IADC 62nd Annual Technical Conference, 27-30 September 1987.

West, C.L., A.F. Kuckes, and H.J. Ritch. "Successful ELREC Logging for Casing Proximity in an Offshore Louisiana Blowout," Technical Conference Exhibit, SPE - AIME, October 1983.

Westergaard, R.H. *All About Blowouts, Norwegian Oil Review,* 1985.

Wiryodiarjo, S., K. Sumatra, and L. Pertamina. "Formation Fracturing Kills Indonesian Blowout," *Oil & Gas Journal,* November 1982.

Wright, J.W. "Directional Drilling Azimuth Reference Systems," SPE/IADC 17212, SPE/IADC Conference, 28 February-2 March 1988.

APPENDIX

TABLE OF CONTENTS

Capacity and Displacement of Drill Pipe	432
Capacity and Displacement of Drill Collars	433
Capacities of Pipe and Hole	435
Tubing Sizes and Capacities	441
Casing Sizes and Capacities	443
Extreme Line Casing Sizes and Capacities	448
Aluminum Liner Sizes and Capacities	450
Pump Output	451
Blowout Preventer and Hydraulic Valve Data	454

CAPACITY AND DISPLACEMENT OF DRILL PIPE*

DRILL PIPE		DRILL PIPE			DISPLACEMENT		CAPACITY	
O.D. Size (in.)	I.U.	Weight (Lbs/Ft) E.U.	I.U. & E.U.	I.D. (in.)	Bbl/Ft	Bbl/Ft 93' Stand	Bbl/Ft	Bbl/Ft 93' Stand
2 3/8	4.85	4.85		1.995	.0019	.174	.00386	0.359
	6.65	6.65		1.815	.0025	.237	.00320	0.292
2 7/8	6.45			2.469	.002	.225	.00592	0.551
	6.85	6.85		2.441	.0027	.248	.00579	0.538
	8.35			2.323	.0032	.299	.00524	0.487
	10.40	10.40	10.40	2.151	.0041	.378	.00449	0.418
3 1/2	8.50			3.063	.0032	.304	.00911	0.847
	9.50	9.50		2.992	.0038	.352	.00870	0.811
	11.20			2.900	.0043	.399	.00817	0.760
	13.30	13.30	13.30	2.764	.0054	.505	.00742	0.690
	15.50	15.50	15.50	2.602	.0061	.571	.00658	0.612
4	11.85	11.85		3.476	.0044	.414	.01174	1.092
	14.00	14.00	14.00	3.340	.0056	.526	.01084	1.008
	15.70		15.30	3.24	.0063	.589	.01020	0.949
4 1/2	12.75			4.00	.0048	.448	.01554	1.445
	13.75	13.75		3.958	.0055	.515	.01522	1.415
	16.60	16.60	16.60	3.826	.00648	.602	.01422	1.322
	20.00	20.00	20.00	3.640	.0081	.754	.01287	1.197
5	16.25		16.25	4.408	.0064	.592	.01888	1.756
	19.50		19.50	4.267	.0075	.707	.01776	1.652
	20.50	20.50		4.214	.0078	.727	.01730	1.609
5 1/2	21.90		21.90	4.778	.0094	.874	.02218	2.063
	24.70		24.70	4.670	.0105	.977	.02119	1.971
5 9/16	19.00			4.975	.0084	.777	.02404	2.236
	22.20		22.20	4.859	.0090	.845	.02294	2.133
	25.25			4.733	.0112	1.047	.02176	2.024
6 5/8**	22.20			6.065	.008	.750	.03573	3.323
	25.20		25.20	5.965	.009	.852	.03456	3.214
	31.90			5.761	.012	1.07	.03224	2.998
7 5/8**	29.25			6.969	.011	.988	.04718	4.388

* - These figures do not include the effects of upsets or tool joints
** - Conventionally used in mining operatons

Capacity and Displacement of Drill Collars*

O.D. Size (in.)	Drill Collar			Displacement		Capacity	
	Weight (Lbs/Ft)	I.D. (in.)		Bbl/Ft	Bbl per 93' Stand	Bbl/Ft	Bbl per 93' Stand
4½	51	1		.0187	1.74	.0009	.0837
	48	1½		.0175	1.62	.0022	.2046
	43	2		.0158	1.47	.0039	.3627
4¾	54	1½		.0197	1.83	.0022	.2046
	52	1¾		.0189	1.757	.0030	.279
	50	2		.018	1.67	.0039	.3627
5	61	1½		.0221	2.055	.0022	.2046
	59	1¾		.0213	1.98	.0030	.279
	56	2		.0204	1.89	.0039	.3627
5¼	68	1½		.0246	2.29	.0022	.2046
	65	1¾		.0238	2.21	.0030	.279
	63	2		.0229	2.13	.0039	.3627
5½	75	1½		.0272	2.53	.0022	.2046
	73	1¾		.0264	2.46	.0030	.279
	70	2		.0255	2.37	.0039	.3627
5¾	82	1½		.0299	2.78	.0022	.2046
	80	1¾		.0291	2.71	.0030	.279
	78	2		.0282	2.62	.0039	.3627
6	88	1¾		.032	2.98	.0030	.279
	85	2		.0311	2.89	.0039	.3627
	83	2¼		.0301	2.80	.0049	.4557
6¼	96	1¾		.0349	3.24	.0030	.279
	94	2		.034	3.16	.0039	.3627
	91	2¼		.033	3.07	.0049	.4557
6½	105	1¾		.038	3.53	.0030	.279
	102	2		.0371	3.45	.0039	.3627
	99	2¼		.0361	3.36	.0049	.4557
6¾	114	1¾		.0413	3.84	.0030	.279
	111	2		.0404	3.76	.0039	.3627
	108	2¼		.0394	3.66	.0049	.4557

Drill Collar			Displacement		Capacity	
O.D. Size (in.)	Weight (Lbs/Ft)	I.D. (in.)	Bbl/Ft	Bbl per 93' Stand	Bbl/Ft	Bbl per 93' Stand
7	120	2	.0437	4.06	.0039	.3627
	114	2½	.0415	3.86	.0061	.5673
	107	3	.0388	3.61	.0088	.8184
7¼	130	2	.0472	4.39	.0039	.3627
	124	2½	.045	4.18	.0061	.5673
	116	3	.0423	3.93	.0088	.8184
7½	139	2	.0507	4.72	.0039	.3627
	133	2½	.0485	4.51	.0061	.5673
	126	3	.0458	4.26	.0088	.8184
7¾	144	2½	.0522	4.85	.0061	.5673
	136	3	.0495	4.6	.0088	.8184
	128	3½	.0464	4.32	.0119	1.107
8	147	3	.0534	4.97	.0088	.8184
	143	3¼	.0519	4.83	.0103	.958
	138	3½	.0503	4.68	.0119	1.107

Capacities of Pipe and Hole

CAPACITIES OF VARIOUS DIAMETERS

HOLE DIAM (IN)	CU FT PER LIN FT	LIN FT PER CU FT	BARRELS PER LIN FT	LIN FT PER BARREL	GALLONS PER LIN FT
1	0.0055	183.3465	0.0010	1029.4142	0.0408
1/8	0.0069	144.8664	0.0012	813.3643	0.0516
1/4	0.0085	117.3418	0.0015	658.8251	0.0637
3/8	0.0103	96.9767	0.0018	544.4835	0.0771
1/2	0.0123	81.4873	0.0022	457.5174	0.0918
5/8	0.0144	69.4330	0.0026	389.8373	0.1077
3/4	0.0167	59.8682	0.0030	336.1352	0.1249
7/8	0.0192	52.1519	0.0034	292.8111	0.1434
2	0.0218	45.8366	0.0039	257.3535	0.1632
1/8	0.0246	40.6027	0.0044	227.9672	0.1842
1/4	0.0276	36.2166	0.0049	203.3411	0.2065
3/8	0.0308	32.5046	0.0055	182.5000	0.2301
1/2	0.0341	29.3354	0.0061	164.7063	0.2550
5/8	0.0376	26.6081	0.0067	149.3934	0.2811
3/4	0.0412	24.2442	0.0073	136.1209	0.3085
7/8	0.0451	22.1818	0.0080	124.5416	0.3372
3	0.0491	20.3718	0.0087	114.3794	0.3672
1/8	0.0533	18.7747	0.0095	105.4120	0.3984
1/4	0.0576	17.3582	0.0103	97.4593	0.4309
3/8	0.0621	16.0963	0.0111	90.3738	0.4647
1/2	0.0668	14.9671	0.0119	8.40338	0.4998
5/8	0.0717	13.9526	0.0128	78.3383	0.5361
3/4	0.0767	13.0380	0.0137	73.2028	0.5737
7/8	0.0819	12.2104	0.0146	68.5562	0.6126
4	0.0873	11.4592	0.0155	64.3384	0.6528
1/8	0.0928	10.7752	0.0165	60.4982	0.6942
1/4	0.0985	10.1507	0.0175	56.9918	0.7369
3/8	0.1044	9.5789	0.0186	53.7816	0.7809
1/2	0.1104	9.0541	0.0197	50.8353	0.8262
5/8	0.1167	8.5713	0.0208	48.1245	0.8727
3/4	0.1231	8.1262	0.0219	45.6250	0.9205
7/8	0.1296	7.7148	0.0231	43.3153	0.9696
5	0.1364	7.3339	0.0243	41.1766	1.0200
1/8	0.1433	6.9805	0.0255	39.1924	1.0716
1/4	0.1503	6.6520	0.0268	37.3484	1.1245
3/8	0.1576	6.3462	0.0281	35.6314	1.1787
1/2	0.1650	6.0610	0.0294	34.0302	1.2342
5/8	0.1726	5.7947	0.0307	32.5346	1.2909
3/4	0.1803	5.5455	0.0321	31.1354	1.3489
7/8	0.1833	5.3120	0.0335	29.8246	1.4082

NOTE: Some diameters and weights are non-API.
No allowance has been made for couplings or upsets.
*Plain end weights are indicated by asterisk.

CAPACITIES OF VARIOUS DIAMETERS

HOLE DIAM (IN)		CU FT PER LIN FT	LIN FT PER CU FT	BARRELS PER LIN FT	LIN FT PER BARREL	GALLONS PER LIN FT
6		0.1963	5.0930	0.0350	28.5948	1.4688
	1/8	0.2046	4.8872	0.0364	27.4396	1.5306
	1/4	0.2131	4.6937	0.0379	26.3530	1.5937
	3/8	0.2217	4.5114	0.0395	25.3297	1.6581
	1/2	0.2304	4.3396	0.0410	24.3648	1.7238
	5/8	0.2394	4.1773	0.0426	23.4541	1.7907
	3/4	0.2485	4.0241	0.0443	22.5935	1.8589
	7/8	0.2578	3.8791	0.0459	21.7793	1.9284
7		0.2673	3.7418	0.0476	21.0085	1.9992
	1/8	0.2769	3.6116	0.0493	20.2778	2.0712
	1/4	0.2867	3.4882	0.0511	19.5846	2.1445
	3/8	0.2967	3.3709	0.0528	18.9263	2.2191
	1/2	0.3068	3.2595	0.0546	18.3007	2.2950
	5/8	0.3171	3.1535	0.0565	17.7056	2.3721
	3/4	0.3276	3.0526	0.0583	17.1390	2.4505
	7/8	0.3382	2.9565	0.0602	16.5993	2.5302
8		0.3491	2.8648	0.0622	16.0846	2.6112
	1/8	0.3601	2.7773	0.0641	15.5935	2.6934
	1/4	0.3712	2.6938	0.0661	15.1245	2.7769
	3/8	0.3826	2.6140	0.0681	14.6764	2.8617
	1/2	0.3941	2.5377	0.0702	14.2479	2.9478
	5/8	0.4057	2.4646	0.0723	13.8380	3.0351
	3/4	0.4176	2.3947	0.0744	13.4454	3.1237
	7/8	0.4296	2.3277	0.0765	13.0693	3.2136
9		0.4418	2.2635	0.0787	12.7088	3.3048
	1/8	0.4541	2.2019	0.0809	12.3630	3.3972
	1/4	0.4667	2.1428	0.0831	12.0311	3.4909
	3/8	0.4794	2.0861	0.0854	11.7124	3.5859
	1/2	0.4922	2.0315	0.0877	11.4063	3.6822
	5/8	0.5053	1.9791	0.0900	11.1119	3.7797
	3/4	0.5185	1.9287	0.0923	10.8288	3.8785
	7/8	0.5319	1.8802	0.0947	10.5564	3.9786
10		0.5454	1.8335	0.0971	10.2941	4.0800
	1/8	0.5591	1.7885	0.0996	10.0415	4.1826
	1/4	0.5730	1.7451	0.1021	9.7981	4.2865
	3/8	0.5871	1.7033	0.1046	9.5634	4.3917
	1/2	0.6013	1.6630	0.1071	9.3371	4.4982
	5/8	0.6157	1.6241	0.1097	9.1187	4.6059
	3/4	0.6303	1.5866	0.1123	8.9079	4.7149
	7/8	0.6450	1.5503	0.1149	8.7043	4.8252
11		0.6600	1.5153	0.1175	8.5078	4.9368
	1/8	0.6750	1.4814	0.1202	8.3174	5.0496
	1/4	0.6903	1.4487	0.1229	8.1336	5.1637
	3/8	0.7057	1.4170	0.1257	7.9559	5.2791
	1/2	0.7213	1.3864	0.1285	7.7839	5.3958
	5/8	0.7371	1.3567	0.1313	7.6174	5.5137
	3/4	0.7530	1.3280	0.1341	7.4561	5.6329
	7/8	0.7691	1.3002	0.1370	7.3000	5.7534

CAPACITIES OF VARIOUS DIAMETERS

HOLE DIAM (IN)		CU FT PER LIN FT	LIN FT PER CU FT	BARRELS PER LIN FT	LIN FT PER BARREL	GALLONS PER LIN FT
12		0.7854	1.2732	0.1399	7.1487	5.8752
	⅛	0.8018	1.2471	0.1428	7.0021	5.9982
	¼	0.8185	1.2218	0.1458	6.8599	6.1225
	⅜	0.8353	1.1972	0.1488	6.7220	6.2481
	½	0.8522	1.1734	0.1518	6.5883	6.3750
	⅝	0.8693	1.1503	0.1548	6.4564	6.5031
	¾	0.8866	1.1279	0.1579	6.3324	6.6325
	⅞	0.9041	1.1061	0.1610	6.2101	6.7632
13		0.9218	1.0849	0.1642	6.0912	6.8952
	⅛	0.9396	1.0643	0.1673	5.9757	7.0284
	¼	0.9575	1.0443	0.1705	5.8635	7.1629
	⅜	0.9757	1.0249	0.1738	5.7544	7.2987
	½	0.9940	1.0060	0.1770	5.6484	7.4358
	⅝	1.0125	0.9876	0.1803	5.5452	7.5741
	¾	1.0312	0.9698	0.1837	5.4448	7.7137
	⅞	1.0500	0.9524	0.1870	5.3472	7.8546
14		1.0690	0.9354	0.1904	5.2521	7.9968
	⅛	1.0882	0.9190	0.1938	5.1596	8.1402
	¼	1.1075	0.9029	0.1973	5.0694	8.2849
	⅜	1.1270	0.8873	0.2007	4.9817	8.4309
	½	1.1467	0.8720	0.2042	4.8961	8.5782
	⅝	1.1666	0.8572	0.2078	4.8128	8.7267
	¾	1.1866	0.8427	0.2113	4.7316	8.8765
	⅞	1.2068	0.8286	0.2149	4.6524	9.0276
15		1.2272	0.8149	0.2186	4.5752	9.1800
	⅛	1.2477	0.8015	0.2222	4.4999	9.3336
	¼	1.2684	0.7884	0.2259	4.4264	9.4885
	⅜	1.2893	0.7756	0.2296	4.3547	9.6447
	½	1.3104	0.7631	0.2334	4.2848	9.8022
	⅝	1.3316	0.7510	0.2372	4.2165	9.9609
	¾	1.3530	0.7391	0.2410	4.1498	10.1209
	⅞	1.3745	0.7275	0.2448	4.0847	10.2822
16		1.3963	0.7162	0.2487	4.0211	10.4448
	¼	1.4402	0.6943	0.2565	3.8984	10.7737
	½	1.4849	0.6734	0.2645	3.7811	11.1078
	¾	1.5302	0.6535	0.2725	3.6691	11.4469
17		1.5763	0.6344	0.2807	3.5620	11.7912
	¼	1.6230	0.6162	0.2891	3.4595	12.1405
	½	1.6703	0.5987	0.2975	3.3614	12.4950
	¾	1.7184	0.5819	0.3061	3.2673	12.8545
18		1.7671	0.5659	0.3147	3.1772	13.2192
	¼	1.8166	0.5505	0.3235	3.0908	13.5889
	½	1.8667	0.5357	0.3325	3.0078	13.9638
	¾	1.9175	0.5215	0.3415	2.9281	14.3437
19		1.9689	0.5079	0.3507	2.8516	14.7288
	¼	2.0211	0.4948	0.3600	2.7780	15.1189
	½	2.0739	0.4822	0.3694	2.7072	15.5142
	¾	2.1275	0.4700	0.3789	2.6391	15.9145

CAPACITIES OF VARIOUS DIAMETERS

HOLE DIAM (IN)		CU FT PER LIN FT	LIN FT PER CU FT	BARRELS PER LIN FT	LIN FT PER BARREL	GALLONS PER LIN FT
20		2.1817	0.4584	0.3886	2.5735	16.3200
	¼	2.2365	0.4471	0.3983	2.5104	16.7305
	½	2.2921	0.4363	0.4082	2.4495	17.1462
	¾	2.3484	0.4258	0.4183	2.3909	17.5669
21		2.4053	0.4158	0.4284	2.3343	17.9928
	¼	2.4629	0.4060	0.4387	2.2797	18.4237
	½	2.5212	0.3966	0.4490	2.2270	18.8598
	¾	2.5802	0.3876	0.4595	2.1761	19.3009
22		2.6398	0.3788	0.4702	2.1269	19.7472
	¼	2.7001	0.3704	0.4809	2.0794	20.1985
	½	2.7612	0.3622	0.4918	2.0334	20.6550
	¾	2.8229	0.3542	0.5028	1.9890	21.1165
23		2.8852	0.3466	0.5139	1.9460	21.5831
	¼	2.9483	0.3392	0.5251	1.9043	22.0549
	½	3.0121	0.3320	0.5365	1.8640	22.5317
	¾	3.0765	0.3250	0.5479	1.8250	23.0137
24		3.1416	0.3183	0.5595	1.7872	23.5007
	¼	3.2074	0.3118	0.5713	1.7505	23.9929
	½	3.2739	0.3055	0.5831	1.7150	24.4901
	¾	3.3410	0.2993	0.5951	1.6805	24.9925
25		3.4088	0.2934	0.6071	1.6471	25.4999
	¼	3.4774	0.2876	0.6193	1.6146	26.0125
	½	3.5466	0.2820	0.6317	1.5831	26.5301
	¾	3.6164	0.2765	0.6441	1.5525	27.0529
26		3.6870	0.2712	0.6567	1.5228	27.5807
	¼	3.7583	0.2661	0.6694	1.4939	28.1137
	½	3.8302	0.2611	0.6822	1.4659	28.6517
	¾	3.9028	0.2562	0.6951	1.4386	29.1949
27		3.9761	0.2515	0.7082	1.4121	29.7431
	¼	4.0501	0.2469	0.7213	1.3863	30.2965
	½	4.1247	0.2424	0.7346	1.3612	30.8549
	¾	4.2000	0.2381	0.7481	1.3368	31.4185
28		4.2761	0.2339	0.7616	1.3130	31.9871
	¼	4.3528	0.2297	0.7753	1.2899	32.5609
	½	4.4301	0.2257	0.7890	1.2674	33.1397
	¾	4.5082	0.2218	0.8029	1.2454	33.7237
29		4.5869	0.2180	0.8170	1.2240	34.3127
	¼	4.6664	0.2143	0.8311	1.2032	34.9069
	½	4.7465	0.2107	0.8454	1.1829	35.5061
	¾	4.8273	0.2072	0.8598	1.1631	36.1105
30		4.9087	0.2037	0.8743	1.1438	36.7199
	¼	4.9909	0.2004	0.8889	1.1250	37.3345
	½	5.0737	0.1971	0.9037	1.1066	37.9541
	¾	5.1572	0.1939	0.9185	1.0887	38.5789

CAPACITIES OF VARIOUS DIAMETERS

HOLE DIAM (IN)		CU FT PER LIN FT	LIN FT PER CU FT	BARRELS PER LIN FT	LIN FT PER BARREL	GALLONS PER LIN FT
31		5.2414	0.1908	0.9335	1.0712	39.2087
	¼	5.3263	0.1877	0.9487	1.0541	39.8437
	½	5.4119	0.1848	0.9639	1.0375	40.4837
	¾	5.4981	0.1819	0.9793	1.0212	41.1289
32		5.5851	0.1790	0.9947	1.0053	41.7791
	¼	5.6727	0.1763	1.0103	0.9898	42.4345
	½	5.7610	0.1736	1.0261	0.9746	43.0949
	¾	5.8499	0.1709	1.0419	0.9598	43.7604
33		5.9396	0.1684	1.0579	0.9453	44.4311
	¼	6.0299	0.1658	1.0740	0.9311	45.1068
	½	6.1209	0.1634	1.0902	0.9173	45.7877
	¾	6.2126	0.1610	1.1065	0.9037	46.4736
34		6.3050	0.1586	1.1230	0.8905	47.1647
	¼	6.3981	0.1563	1.1395	0.8775	47.8608
	½	6.4918	0.1540	1.1562	0.8649	48.5621
	¾	6.5862	0.1518	1.1731	0.8525	49.2684
35		6.6813	0.1497	1.1900	0.8403	49.9799
	¼	6.7771	0.1476	1.2071	0.8285	50.6964
	½	6.8736	0.1455	1.2242	0.8168	51.4181
	¾	6.9707	0.1435	1.2415	0.8054	52.1448
36		7.0686	0.1415	1.2590	0.7943	52.8767
	¼	7.1671	0.1395	1.2765	0.7834	53.6136
	½	7.2663	0.1376	1.2942	0.7727	54.3557
	¾	7.3662	0.1358	1.3120	0.7622	55.1028
37		7.4667	0.1339	1.3299	0.7519	55.8551
	¼	7.5680	0.1321	1.3479	0.7419	56.6124
	½	7.6699	0.1304	1.3661	0.7320	57.3749
	¾	7.7725	0.1287	1.3843	0.7224	58.1424
38		7.8758	0.1270	1.4027	0.7129	58.9151
	¼	7.9798	0.1253	1.4213	0.7036	59.6928
	½	8.0844	0.1237	1.4399	0.6945	60.4757
	¾	8.1898	0.1221	1.4587	0.6856	61.2636
39		8.2958	0.1205	1.4775	0.6768	62.0567
	¼	8.4025	0.1190	1.4965	0.6682	62.8548
	½	8.5098	0.1175	1.5157	0.6598	63.6581
	¾	8.6179	0.1160	1.5349	0.6515	64.4664
40		8.7266	0.1146	1.5543	0.6434	65.2798
	¼	8.8361	0.1132	1.5738	0.6354	66.0984
	½	8.9462	0.1118	1.5934	0.6276	66.9220
	¾	9.0570	0.1104	1.6131	0.6199	67.7508

CAPACITIES OF VARIOUS DIAMETERS

HOLE DIAM (IN)		CU FT PER LIN FT	LIN FT PER CU FT	BARRELS PER LIN FT	LIN FT PER BARREL	GALLONS PER LIN FT
41		9.1684	0.1091	1.6330	0.6124	68.5846
	¼	9.2806	0.1078	1.6529	0.6050	69.4236
	½	9.3934	0.1065	1.6730	0.5977	70.2676
	¾	9.5069	0.1052	1.6933	0.5906	71.1168
42		9.6211	0.1039	1.7136	0.5836	71.9710
	¼	9.7360	0.1027	1.7341	0.5767	72.8304
	½	9.8516	0.1015	1.7546	0.5699	73.6948
	¾	9.9678	0.1003	1.7753	0.5633	74.5644
43		10.0847	0.0992	1.7962	0.5567	75.4390
	¼	10.2023	0.0980	1.8171	0.5503	76.3188
	½	10.3206	0.0969	1.8382	0.5440	77.2036
	¾	10.4396	0.0958	1.8594	0.5378	78.0936
44		10.5592	0.0947	1.8807	0.5317	78.9886
	¼	10.6796	0.0936	1.9021	0.5257	79.8888
	½	10.8006	0.0926	1.9237	0.5198	80.7940
	¾	10.9223	0.0916	1.9453	0.5140	81.7044
45		11.0447	0.0905	1.9671	0.5084	82.6198
	¼	11.1677	0.0895	1.9891	0.5028	83.5404
	½	11.2915	0.0886	2.0111	0.4972	84.4660
	¾	11.4159	0.0876	2.0333	0.4918	85.3968
46		11.5410	0.0866	2.0555	0.4865	86.3326
	¼	11.6668	0.0857	2.0779	0.4812	87.2735
	½	11.7932	0.0848	2.1005	0.4761	88.2196
	¾	11.9204	0.0839	2.1231	0.4710	89.1707
47		12.0482	0.0830	2.1459	0.4660	90.1270
	¼	12.1767	0.0821	2.1688	0.4611	91.0883
	½	12.3059	0.0813	2.1918	0.4563	92.0548
	¾	12.4358	0.0804	2.2149	0.4515	93.0263
48		12.5664	0.0796	2.2382	0.4468	94.0030
	¼	12.6976	0.0788	2.2615	0.4422	94.9847
	½	12.8295	0.0779	2.2850	0.4376	95.9716
	¾	12.9621	0.0771	2.3087	0.4332	96.9635

TUBING SIZES AND CAPACITIES

SIZE OD INCHES	WEIGHT PER LIN FT	ID	CU FT PER LIN FT	LIN FT PER CU FT	BARRELS PER LIN FT	LIN FT PER BARREL	GALLONS PER LIN FT
0.750	0.42	0.636	.0022	453.2715	.0004	2544.9304	.0165
1.000	0.67	0.866	.0041	244.4763	.0007	1372.6328	.0306
1.050	1.14	0.824	.0037	270.0338	.0007	1516.1275	.0277
	1.20	0.824	.0037	270.0338	.0007	1516.1275	.0277
	1.55	0.742	.0030	333.0158	.0005	1869.7448	.0225
1.315	1.30	1.125	.0069	144.8664	.0012	813.3643	.0516
	1.43	1.097	.0066	152.3559	.0012	855.4150	.0491
	1.63	1.065	.0062	161.6491	.0011	907.5926	.0463
	1.70	1.049	.0060	166.6179	.0011	935.4900	.0449
	1.72	1.049	.0060	166.6179	.0011	935.4900	.0449
	1.80	1.049	.0060	166.6179	.0011	935.4900	.0449
	1.90	1.049	.0060	166.6179	.0011	935.4900	.0449
	2.25	0.957	.0050	200.1929	.0009	1123.9999	.0374
	2.30	0.957	.0050	200.1929	.0009	1123.9999	.0374
1.660	2.10	1.410	.0108	92.2220	.0019	517.7879	.0811
	2.30	1.380	.0104	96.2752	.0018	540.5451	.0777
	2.40	1.380	.0104	96.2752	.0018	540.5451	.0777
	3.02	1.278	.0089	112.2563	.0016	630.2726	.0666
	3.24	1.264	.0087	114.7568	.0016	644.3117	.0652
	3.29	1.264	.0087	114.7568	.0016	644.3117	.0652
1.900	2.40	1.650	.0148	67.3449	.0026	378.1136	.1111
	2.75	1.610	.0141	70.7328	.0025	397.1352	.1058
	2.90	1.610	.0141	70.7328	.0025	397.1352	.1058
	3.64	1.500	.0123	81.4873	.0022	457.5174	.0918
	4.19	1.462	.0117	85.7784	.0021	481.6099	.0872
2	3.30	1.670	.0152	65.7415	.0027	369.1112	.1138
	3.40	1.670	.0152	65.7415	.0027	369.1112	.1138
2¹⁄₁₆	2.66	1.813	.0179	55.7798	.0032	313.1804	.1341
	3.25	1.750	.0167	59.8682	.0030	336.1352	.1249
	4.50	1.613	.0142	70.4699	.0025	395.6593	.1062
2⅜	3.10	2.125	.0246	40.6027	.0044	227.9672	.1842
	3.32	2.107	.0242	41.2994	.0043	231.8788	.1811
	4.00	2.041	.0227	44.0136	.0040	247.1179	.1700
	4.60	1.995	.0217	46.0667	.0039	258.6451	.1624
	4.70	1.995	.0217	46.0667	.0039	258.6451	.1624
	5.00	1.947	.0207	48.3661	.0037	271.5553	.1547
	5.30	1.939	.0205	48.7660	.0037	273.8007	.1534
	5.80	1.867	.0190	52.5998	.0034	295.3259	.1422
	5.95	1.867	.0190	52.5998	.0034	295.3259	.1422
	6.20	1.853	.0187	53.3976	.0033	299.8053	.1401
	7.70	1.073	.0158	63.2184	.0028	354.9448	.1183

TUBING SIZES AND CAPACITIES

SIZE OD INCHES	WEIGHT PER LIN FT	ID	CU FT PER LIN FT	LIN FT PER CU FT	BARRELS PER LIN FT	LIN FT PER BARREL	GALLONS PER LIN FT
2⅞	4.36	2.579	.0363	27.5658	.0065	154.7702	.2714
	4.64	2.563	.0358	27.9110	.0064	156.7086	.2680
	5.90	2.469	.0332	30.0767	.0059	168.8682	.2487
	6.40	2.441	.0325	30.7707	.0058	172.7645	.2431
	6.50	2.441	.0325	30.7707	.0058	172.7645	.2431
	7.90	2.323	.0294	33.9762	.0052	190.7619	.2202
	8.60	2.259	.0278	35.9286	.0050	201.7240	.2082
	8.70	2.259	.0278	35.9286	.0050	201.7240	.2082
	9.50	2.195	.0263	38.0543	.0047	213.6590	.1966
	10.40	2.151	.0252	39.6271	.0045	222.4894	.1888
	10.70	2.091	.0238	41.9338	.0042	235.4410	.1784
	11.00	2.065	.0233	4299964	.0041	241.4071	.1740
	11.65	1.995	.0217	46.0667	.0039	258.6451	.1624
3½	5.63	3.188	.0554	18.0400	.0099	101.2870	.4147
	7.70	3.068	.0513	19.4788	.0091	109.3653	.3840
	8.50	3.018	.0497	20.1296	.0088	113.0191	.3716
	8.90	3.018	.0497	20.1296	.0088	113.0191	.3716
	9.20	2.992	.0488	20.4809	.0087	114.9918	.3652
	9.30	2.992	.0488	20.4809	.0087	114.9918	.3652
	10.20	2.992	.0488	20.4809	.0087	114.9918	.3652
	10.30	2.922	.0466	21.4740	.0083	120.5674	.3484
	11.20	2.900	.0459	21.8010	.0082	122.4036	.3431
	12.70	2.750	.0412	24.2442	.0073	136.1209	.3085
	12.80	2.764	.0417	23.9992	.0074	134.7454	.3117
	12.95	2.750	.0412	24.2442	.0073	136.1209	.3085
	13.30	2.764	.0417	23.9992	.0074	134.7454	.3117
	15.50	2.602	.0369	27.0806	.0066	152.0462	.2762
	15.80	2.548	.0354	28.2406	.0063	158.5591	.2649
	16.70	2.480	.0335	29.8105	.0060	167.3735	.2509
	17.05	2.440	.0325	30.7959	.0058	172.9062	.2429
4	9.25	3.548	.0687	14.5648	.0122	81.7754	.5136
	9.40	3.548	.0687	14.5648	.0122	81.7754	.5136
	9.50	3.548	.0687	14.5648	.0122	81.7754	.5136
	10.80	3.476	.0659	15.1745	.0117	85.1982	.4930
	10.90	3.476	.0659	15.1745	.0117	85.1982	.4930
	11.00	3.476	.0659	15.1745	.0117	85.1982	.4930
	11.60	3.428	.0641	15.6024	.0114	87.6009	.4794
	13.30	3.340	.0608	16.4354	.0108	92.2778	.4551
	13.40	3.340	.0608	16.4354	.0108	92.2778	.4551
	19.00	3.000	.0491	20.3718	.0087	114.3794	.3672
	22.50	2.780	.0422	23.7237	.0075	133.1989	.3153
4½	11.00	4.026	.0884	11.3116	.0157	63.5101	.6613
	11.80	3.990	.0868	11.5167	.0155	64.6613	.6495
	12.60	3.958	.0854	11.7036	.0152	65.7111	.6392
	12.75	3.958	.0854	11.7036	.0152	65.7111	.6392
	13.50	3.920	.0838	11.9316	.0149	66.9912	.6269
	15.40	3.826	.0798	12.5251	.0142	70.3235	.5972
	15.50	3.826	.0798	12.5251	.0142	70.3235	.5972
	16.90	3.754	.0769	13.0102	.0137	73.0469	.5750
	19.20	3.640	.0723	13.8379	.0129	77.6940	.5406
	21.60	3.500	.0668	14.9671	.0119	84.0338	.4998
	24.60	3.380	.0623	16.0487	.0111	90.1066	.4661
	26.50	3.240	.0573	17.4656	.0102	98.0619	.4283

CASING SIZES AND CAPACITIES

SIZE OD INCHES	WEIGHT PER LIN FT	ID	CU FT PER LIN FT	LIN FT PER CU FT	BARRELS PER LIN FT	LIN FT PER BARREL	GALLONS PER LIN FT
4	9.26	3.550	0.0687	14.5484	.0122	81.6833	0.5142
	9.50	3.550	0.0687	14.5484	.0122	81.6833	0.5142
	11.00	3.480	0.0661	15.1396	.0118	85.0025	0.4941
	11.60	3.430	0.0642	15.5842	.0114	87.4988	0.4800
	12.60	3.364	0.0617	16.2017	.0110	90.9658	0.4617
4½	9.50	4.090	0.0912	10.9604	.0163	61.5380	0.6825
	10.50	4.052	0.0896	11.1669	.0159	62.6977	0.6699
	10.98	4.030	0.0886	11.2892	.0158	63.3841	0.6626
	11.00	4.026	0.0884	11.3116	.0157	63.5101	0.6613
	11.60	4.000	0.0873	11.4592	.0155	64.3384	0.6528
	11.75	3.990	0.0868	11.5167	.0155	64.6613	0.6495
	12.60	3.958	0.0854	11.7036	.0152	65.7111	0.6392
	12.75	3.960	0.0855	11.6918	.0152	65.6447	0.6398
	13.50	3.920	0.0838	11.9316	.0149	66.9912	0.6269
	15.10	3.826	0.0798	12.5251	.0142	70.3235	0.5972
	16.60	3.826	0.0798	12.5251	.0142	70.3235	0.5972
	18.80	3.640	0.0723	13.8379	.0129	77.6940	0.5406
	21.60	3.500	0.0668	14.9671	.0119	84.0338	0.4998
	24.60	3.380	0.0623	16.0487	.0111	90.1066	0.4661
	26.50	3.240	0.0573	17.4656	.0102	98.0619	0.4283
4¾	16.00	4.082	0.0909	11.0034	.0162	61.7795	0.6798
	16.50	4.070	0.0903	11.0684	.0161	62.1443	0.6758
	18.00	4.000	0.0873	11.4592	.0155	64.3384	0.6528
	20.00	3.910	0.0834	11.9928	.0149	67.3343	0.6238
	21.00	3.850	0.0808	12.3695	.0144	69.4494	0.6048
5	11.50	4.560	0.1134	8.8174	.0202	49.5063	0.8484
	12.85	4.500	0.1104	9.0541	.0197	50.8353	0.8262
	13.00	4.494	0.1102	9.0783	.0196	50.9711	0.8240
	14.00	4.450	0.1080	9.2588	.0192	51.9841	0.8079
	15.00	4.408	0.1060	9.4360	.0189	52.9794	0.7928
	18.00	4.276	0.0997	10.0276	.0178	56.3008	0.7460
	20.30	4.184	0.0955	10.4734	.0170	58.8040	0.7142
	21.00	4.154	0.0941	10.6253	.0168	59.6564	0.7040
	23.20	4.044	0.0892	11.2112	.0159	62.9460	0.6672
	24.20	4.000	0.0873	11.4592	.0155	64.3384	0.6528
5¼	16.00	4.650	0.1179	8.4794	.0210	47.6085	0.8822
5½	13.00	5.044	0.1388	7.2065	.0247	40.4613	1.0380
	14.00	5.012	0.1370	7.2988	.0244	40.9796	1.0249
	15.00	4.974	0.1349	7.4107	.0240	41.6082	1.0094
	15.50	4.950	0.1336	7.4828	.0238	42.0126	0.9997
	17.00	4.892	0.1305	7.6613	.0232	43.0147	0.9764
	20.00	4.778	0.1245	8.0312	.0222	45.0918	0.9314
	23.00	4.670	0.1189	8.4070	.0212	47.2016	0.8898
	25.00	4.580	0.1144	8.7406	.0204	49.0749	0.8558
	26.00	4.548	0.1128	8.8640	.0201	49.7679	0.8439
	32.30	4.276	0.0997	10.0276	.0178	56.3008	0.7460
	36.40	4.090	0.0912	10.9604	.0163	61.5380	0.6825

CASING SIZES AND CAPACITIES

SIZE OD INCHES	WEIGHT PER LIN FT	ID	CU FT PER LIN FT	LIN FT PER CU FT	BARRELS PER LIN FT	LIN FT PER BARREL	GALLONS PER LIN FT
5¼	14.00	5.290	0.1526	6.5518	.0272	36.7857	1.1417
	17.00	5.190	0.1469	6.8067	.0262	38.2169	1.0990
	19.50	5.090	0.1413	7.0768	.0252	39.7333	1.0570
	20.00	5.090	0.1413	7.0768	.0252	39.7333	1.0570
	22.50	4.990	0.1358	7.3633	.0242	41.3418	1.0159
	23.00	4.990	0.1358	7.3633	.0242	41.3418	1.0159
	25.20	4.890	0.1304	7.6675	.0232	43.0499	0.9756
6	15.00	5.524	0.1664	6.0085	.0296	33.7352	1.2450
	16.00	5.500	0.1650	6.0610	.0294	34.0302	1.2342
	18.00	5.424	0.1605	6.2321	.0286	34.9906	1.2003
	20.00	5.352	0.1562	6.4009	.0278	35.9383	1.1687
	23.00	5.240	0.1498	6.6774	.0267	37.4910	1.1203
	26.00	5.140	0.1441	6.9398	.0257	38.9640	1.0779
6⅝	13.00	6.260	0.2137	4.6787	.0381	26.2689	1.5989
	17.00	6.135	0.2053	4.8713	.0366	27.3502	1.5356
	20.00	6.049	0.1996	5.0108	.0355	28.1334	1.4929
	22.00	5.980	0.1950	5.1271	.0347	28.7864	1.4590
	24.00	5.921	0.1912	5.2298	.0341	29.3630	1.4304
	25.00	5.880	0.1886	5.3030	.0336	29.7739	1.4106
	26.00	5.855	0.1870	5.3483	.0333	30.0287	1.3987
	26.80	5.837	0.1858	5.3814	.0331	30.2142	1.3901
	28.00	5.791	0.1829	5.4672	.0326	30.6961	1.3683
	29.00	5.761	0.1810	5.5243	.0322	31.0166	1.3541
	31.80	5.675	0.1757	5.6930	.0313	31.9638	1.3140
	32.00	5.675	0.1757	5.6930	.0313	31.9638	1.3140
	34.00	5.595	0.1707	5.8570	.0304	32.8844	1.2772
7	17.00	6.538	0.2331	4.2893	.0415	24.0824	1.7440
	20.00	6.456	0.2273	4.3989	.0405	24.6981	1.7005
	22.00	6.398	0.2233	4.4790	.0398	25.1479	1.6701
	23.00	6.366	0.2210	4.5242	.0394	25.4014	1.6535
	24.00	6.336	0.2190	4.5671	.0390	25.6425	1.6379
	26.00	6.276	0.2148	4.6549	.0383	26.1351	1.6070
	28.00	6.214	0.2106	4.7482	.0375	26.6592	1.5754
	29.00	6.184	0.2086	4.7944	.0371	26.9185	1.5603
	29.80	6.168	0.2075	4.8193	.0370	27.0584	1.5522
	30.00	6.154	0.2066	4.8413	.0368	27.1816	1.5452
	32.00	6.094	0.2026	4.9371	.0361	27.7195	1.5152
	35.00	6.004	0.1966	5.0862	.0350	28.5567	1.4708
	38.00	5.920	0.1911	5.2315	.0340	29.3729	1.4299
	40.20	5.836	0.1858	5.3832	.0331	30.2245	1.3896
	41.00	5.820	0.1847	5.4129	.0329	30.3909	1.3820
	43.00	5.736	0.1795	5.5726	.0320	31.2876	1.3424
	44.00	5.720	0.1785	5.6038	.0318	31.4629	1.3349
	49.50	5.540	0.1674	5.9738	.0298	33.5406	1.2522

CASING SIZES AND CAPACITIES

SIZE OD INCHES	WEIGHT PER LIN FT	ID	CU FT PER LIN FT	LIN FT PER CU FT	BARRELS PER LIN FT	LIN FT PER BARREL	GALLONS PER LIN FT
7⅝	20.00	7.125	0.2769	3.6116	.0493	20.2778	2.0712
	24.00	7.025	0.2692	3.7152	.0479	20.8592	2.0135
	26.40	6.969	0.2649	3.7751	.0472	21.1958	1.9815
	29.70	6.875	0.2578	3.8791	.0459	21.7793	1.9284
	33.70	6.765	0.2496	4.0062	.0445	22.4934	1.8672
	34.00	6.760	0.2492	4.0122	.0444	22.5267	1.8645
	35.50	6.710	0.2456	4.0722	.0437	22.8636	1.8370
	38.00	6.655	0.2416	4.1398	.0430	23.2431	1.8070
	39.00	6.625	0.2394	4.1773	.0426	23.4541	1.7907
	45.30	6.435	0.2259	4.4277	.0402	24.8595	1.6895
7¾	46.10	6.560	0.2347	4.2605	.0418	23.9212	1.7558
8	26.00	7.386	0.2975	3.3609	.0530	18.8700	2.2258
8⅛	28.00	7.485	0.3056	3.2726	.0544	18.3741	2.2858
	32.00	7.385	0.2975	3.3618	.0530	18.8751	2.2252
	35.50	7.285	0.2895	3.4547	.0516	19.3968	2.1653
	36.00	7.285	0.2895	3.4547	.0516	19.3968	2.1653
	39.50	7.185	0.2816	3.5516	.0501	19.9405	2.1063
	40.00	7.185	0.2816	3.5516	.0501	19.9405	2.1063
	42.00	7.125	0.2769	3.6116	.0493	20.2778	2.0712
8⅝	24.00	8.097	0.3576	2.7966	.0637	15.7015	2.6749
	28.00	8.017	0.3506	2.8527	.0624	16.0165	2.6223
	32.00	7.921	0.3422	2.9222	.0609	16.4070	2.5599
	36.00	7.825	0.3340	2.9944	.0595	16.8121	2.4982
	38.00	7.775	0.3297	3.0330	.0587	17.0290	2.4664
	40.00	7.725	0.3255	3.0724	.0580	17.2502	2.4348
	43.00	7.651	0.3193	3.1321	.0569	17.5855	2.3883
	44.00	7.625	0.3171	3.1535	.0565	17.7056	2.3721
	49.00	7.511	0.3077	392500	.0548	18.2471	2.3017
	52.00	7.435	0.3015	3.3167	.0537	18.6221	2.2554
9	34.00	8.290	0.3748	2.6679	.0668	14.9789	2.8039
	38.00	8.196	0.3664	2.7294	.0653	15.3245	2.7407
	40.00	8.150	0.3623	2.7603	.0645	15.4980	2.7100
	41.20	8.150	0.3623	2.7603	.0645	15.4980	2.7100
	45.00	8.032	0.3519	2.8420	.0627	15.9567	2.6321
	46.10	8.032	0.3519	2.8420	.0627	15.9567	2.6321
	54.00	7.810	0.3327	3.0059	.0593	16.8767	2.4886
	55.20	7.812	0.3329	3.0043	.0593	16.8681	2.4899

CASING SIZES AND CAPACITIES

SIZE OD INCHES	WEIGHT PER LIN FT	ID	CU FT PER LIN FT	LIN FT PER CU FT	BARRELS PER LIN FT	LIN FT PER BARREL	GALLONS PER LIN FT
9⅝	29.30	9.063	0.4480	2.2322	.0798	12.5327	3.3512
	32.30	9.001	0.4419	2.2630	.0787	12.7060	3.3055
	36.00	8.921	0.4341	2.3038	.0773	12.9349	3.2470
	38.00	8.885	0.4306	2.3225	.0767	13.0399	3.2209
	40.00	8.835	0.4257	2.3489	.0758	13.1879	3.1847
	42.00	8.799	0.4223	2.3681	.0752	13.2961	3.1588
	43.50	8.755	0.4181	2.3920	.0745	13.4301	3.1273
	44.30	8.750	0.4176	2.3947	.0744	13.4454	3.1237
	47.00	8.681	0.4110	2.4329	.0732	13.6600	3.0747
	47.20	8.680	0.4109	2.4335	.0732	13.6631	3.0740
	53.50	8.535	0.3973	2.5169	.0708	14.1313	2.9721
	57.40	8.450	0.3894	2.5678	.0694	14.4171	2.9132
	58.40	8.435	0.3881	2.5769	.0691	14.4684	2.9029
	61.10	8.375	0.3826	2.6140	.0681	14.6764	2.8617
10	33.00	9.384	0.4803	2.0821	.0855	11.6900	3.5928
	60.00	8.780	0.4205	2.3784	.0749	13.3537	3.1452
10¾	32.75	10.192	0.5666	1.7650	.1009	9.9099	4.2382
	35.75	10.140	0.5608	1.7832	.0999	10.0118	4.1950
	40.50	10.050	0.5509	1.8153	.0981	10.1920	4.1209
	45.50	9.950	0.5400	1.8519	.0962	10.3979	4.0393
	46.20	9.950	0.5400	1.8519	.0962	10.3979	4.0393
	48.00	9.902	0.5348	1.8699	.0952	10.4989	4.0004
	49.50	9.850	0.5292	1.8897	.0943	10.6101	3.9585
	51.00	9.850	0.5292	1.8897	.0943	10.6101	3.9585
	54.00	9.784	0.5221	1.9153	.0930	10.7537	3.9056
	55.50	9.760	0.5195	1.9247	.0925	10.8066	3.8865
	60.70	9.660	0.5090	1.9648	.0906	11.0315	3.8073
	65.70	9.560	0.4985	2.0061	.0888	11.2635	3.7289
	71.10	9.450	0.4871	2.0531	.0868	11.5273	3.6435
	76.00	9.350	0.4768	2.0972	.0849	11.7752	3.5668
	81.00	9.250	0.4667	2.1428	.0831	12.0311	3.4909
11¾	38.00	11.150	0.6781	1.4748	.1208	8.2802	5.0723
	42.00	11.084	0.6701	1.4924	.1193	8.3791	5.0125
	47.00	11.000	0.6600	1.5153	.1175	8.5076	4.9368
	50.00	10.956	0.6540	1.5291	.1165	8.5854	4.8920
	54.00	10.880	0.6456	1.5489	.1150	8.6963	4.8297
	60.00	10.772	0.6329	1.5801	.1127	8.8715	4.7343
	61.00	10.770	0.6326	1.5807	.1127	8.8748	4.7325
	65.00	10.682	0.6223	1.6068	.1108	9.0216	4.6555
12	40.00	11.384	0.7068	1.4148	.1259	7.9433	5.2875
12¾	33.38*	12.250	0.8185	1.2218	.1458	6.8599	6.1225
	37.42*	12.188	0.8102	1.2343	.1443	6.9299	6.0607
	41.45*	12.126	0.8020	1.2469	.1428	7.0009	5.9992
	43.77*	12.090	0.7972	1.2544	.1420	7.0427	5.9636
	45.58*	12.062	0.7935	1.2602	.1413	7.0754	5.9361
	49.56*	12.000	0.7854	1.2732	.1399	7.1487	5.8752
	53.00	11.970	0.7815	1.2796	.1392	7.1846	5.8458

CASING SIZES AND CAPACITIES

SIZE OD INCHES	WEIGHT PER LIN FT	ID	CU FT PER LIN FT	LIN FT PER CU FT	BARRELS PER LIN FT	LIN FT PER BARREL	GALLONS PER LIN FT
13	40.00	12.438	0.8438	1.1851	.1503	6.6541	6.3119
	45.00	12.360	0.8332	1.2002	.1484	6.7383	6.2330
	50.00	12.282	0.8227	1.2154	.1465	6.8242	6.1546
	54.00	12.200	0.8118	1.2318	.1446	6.9162	6.0727
13⅜	48.00	12.715	0.8818	1.1341	.1571	6.3673	6.5962
	54.50	12.615	0.8680	1.1521	.1546	6.4687	6.4928
	61.00	12.515	0.8543	1.1706	.1521	6.5725	6.3903
	68.00	12.415	0.8407	1.1895	.1497	6.6788	6.2886
	72.00	12.347	0.8315	1.2027	.1481	6.7525	6.2199
	77.00	12.275	0.8218	1.2168	.1464	6.8320	6.1476
	83.50	12.175	0.8085	1.2369	.1440	6.9447	6.0478
	85.00	12.159	0.8063	1.2402	.1436	6.9630	6.0319
	92.00	12.031	0.7895	1.2667	.1406	7.1119	5.9056
	98.00	11.937	0.7772	1.2867	.1384	7.2244	5.8137
14	50.00	13.344	0.9712	1.0297	.1730	5.7812	7.2649
16	55.00	15.375	1.2893	0.7756	.2296	4.3547	9.6447
	65.00	15.250	1.2684	0.7884	.2259	4.4264	9.4885
	70.00	15.198	1.2598	0.7938	.2244	4.4567	9.4239
	75.00	15.124	1.2476	0.8016	.2222	4.5005	9.3324
	84.00	15.010	1.2288	0.8138	.2189	4.5691	9.1922
	109.00	14.688	1.1767	0.8499	.2096	4.7716	8.8021
	118.00	14.570	1.1578	0.8637	.2062	4.8492	8.6612
18	80.00	17.180	1.6098	0.6212	.2867	3.4877	12.0422
18⅝	78.00	17.855	1.7388	0.5751	.3097	3.2290	13.0071
	87.50	17.755	1.7194	0.5816	.3062	3.2655	12.8618
	96.50	17.655	1.7001	0.5882	.3028	3.3026	12.7173
20	90.00	19.190	2.0085	0.4979	.3577	2.7954	15.0248
	94.00	19.124	1.9947	0.5013	.3553	2.8147	14.9216
	106.50	19.000	1.9689	0.5079	.3507	2.8516	14.7288
	133.00	18.730	1.9134	0.5226	.3408	2.9344	14.3131
	169.00	18.376	1.8417	0.5430	.3280	3.0485	13.7772
21½	92.50	20.710	2.3393	0.4275	.4166	2.4001	17.4992
	103.00	20.610	2.3168	0.4316	.4126	2.4235	17.3307
	114.00	20.510	2.2943	0.4359	.4086	2.4471	17.1629
24½	88.00	23.850	3.1024	0.3223	.5526	1.8097	23.2079
	100.50	23.750	3.0765	0.3250	.5479	1.8250	23.0137
	113.00	23.650	3.0506	0.3278	.5433	1.8405	22.8203
30	98.93*	29.376	4.7067	0.2125	.8383	1.1929	35.2083
	118.65*	29.250	4.6664	0.2143	.8311	1.2032	34.9069
	157.53*	29.000	4.5869	0.2180	.8170	1.2240	34.3127
	196.08*	28.750	4.5082	0.2218	.8029	1.2454	33.7237
	234.29*	28.500	4.4301	0.2257	.7890	1.2674	33.1397
	309.72*	28.000	4.2761	0.2339	.7616	1.3130	31.9871
	346.93*	27.750	4.2000	0.2381	.7481	1.3368	31.4185
	383.81*	27.500	4.1247	0.2424	.7346	1.3612	30.8549
	546.57*	27.000	3.9761	0.2515	.7082	1.4121	29.7431

EXTREME LINE CASING SIZES AND CAPACITIES

SIZE OD INCHES	WEIGHT PER LIN FT	ID	CU FT PER LIN FT	LIN FT PER CU FT	BARRELS PER LIN FT	LIN FT PER BARREL	GALLONS PER LIN FT
4½	11.60	4.000	.0873	11.4592	.0155	64.3384	0.6528
	13.50	3.920	.0838	11.9316	.0149	66.9912	0.6269
	15.10	3.826	.0798	12.5251	.0142	70.3235	0.5972
4¾	16.00	4.082	.0909	11.0034	.0162	61.7795	0.6798
	18.00	4.000	.0873	11.4592	.0155	64.3384	0.6528
5	15.00	4.408	.1060	9.4360	.0189	52.9794	0.7928
	18.00	4.276	.0997	10.0276	.0178	56.3008	0.7460
	21.00	4.154	.0941	10.6253	.0168	59.6564	0.7040
5½	15.50	4.950	.1336	7.4828	.0238	42.0126	0.9997
	17.00	4.892	.1305	7.6613	.0232	43.0147	0.9764
	20.00	4.778	.1245	8.0312	.0222	45.0918	0.9314
	23.00	4.670	.1189	8.4070	.0212	47.2016	0.8898
	25.00	4.580	.1144	8.7406	.0204	49.0749	0.8558
5¾	19.50	5.090	.1413	7.0768	.0252	39.7333	1.0570
	22.50	4.990	.1358	7.3633	.0242	41.3418	1.0159
	25.20	4.890	.1304	7.6675	.0232	43.0499	0.9756
6	18.00	5.424	.1605	6.2321	.0286	34.9906	1.2003
	20.00	5.352	.1562	6.4009	.0278	35.9383	1.1687
	23.00	5.240	.1498	6.6774	.0267	37.4910	1.1203
	26.00	5.140	.1441	6.9398	.0257	38.9640	1.0779
6⅝	24.00	5.921	.1912	5.2298	.0341	29.3630	1.4304
	26.00	5.855	.1870	5.3483	.0333	30.0287	1.3987
	28.00	5.791	.1829	5.4672	.0326	30.6961	1.3683
	29.00	5.791	.1829	5.4672	.0326	30.6961	1.3683
	32.00	5.675	.1757	5.6930	.0313	31.9638	1.3140
	34.00	5.595	.1707	5.8570	.0304	32.8844	1.2772
7	23.00	6.366	.2210	4.5242	.0394	25.4014	1.6535
	24.00	6.336	.2190	4.5671	.0390	25.6425	1.6379
	26.00	6.276	.2148	4.6549	.0383	26.1351	1.6070
	28.00	6.214	.2106	4.7482	.0375	26.6592	1.5754
	29.00	6.184	.2086	4.7944	.0371	26.9185	1.5603
	30.00	6.154	.2066	4.8413	.0368	27.1816	1.5452
	32.00	6.094	.2026	4.9371	.0361	27.7195	1.5152
	33.70	6.048	.1995	5.0124	.0355	28.1428	1.4924
	35.00	6.004	.1966	5.0862	.0350	28.5567	1.4708
	35.30	6.000	.1963	5.0930	.0350	28.5948	1.4688
	38.00	5.920	.1911	5.2315	.0340	29.3729	1.4299
	40.00	5.836	.1858	5.3832	.0331	30.2245	1.3896

EXTREME LINE CASING SIZES AND CAPACITIES

SIZE OD INCHES	WEIGHT PER LIN FT	ID	CU FT PER LIN FT	LIN FT PER CU FT	BARRELS PER LIN FT	LIN FT PER BARREL	GALLONS PER LIN FT
7⅝	26.40	6.969	.2649	3.7751	.0472	21.1958	1.9815
	29.70	6.875	.2578	3.8791	.0459	21.7793	1.9284
	33.70	6.765	.2496	4.0062	.0445	22.4934	1.8672
	36.00	6.705	.2452	4.0783	.0437	22.8977	1.8342
	38.70	6.625	.2394	4.1773	.0426	23.4541	1.7907
	39.00	6.625	.2394	4.1773	.0426	23.4541	1.7907
	45.00	6.445	.2266	4.4139	.0404	24.7825	1.6947
	45.30	6.435	.2259	4.4277	.0402	24.8595	1.6895
8⅝	32.00	7.921	.3422	2.9222	.0609	16.4070	2.5599
	36.00	7.825	.3340	2.9944	.0595	16.8121	2.4982
	38.00	7.775	.3297	3.0330	.0587	17.0290	2.4664
	40.00	7.725	.3255	3.0724	.0580	17.2502	2.4348
	43.00	7.651	.3193	3.1321	.0569	17.5855	2.3883
	44.00	7.625	.3171	3.1535	.0565	17.7056	2.3721
	48.00	7.537	.3098	3.2276	.0552	18.1215	2.3177
	49.00	7.511	.3077	3.2500	.0548	18.2471	2.3017
9	34.00	8.290	.3748	2.6679	.0668	14.9789	2.8039
	38.00	8.196	.3664	2.7294	.0653	15.3245	2.7407
	40.00	8.150	.3623	2.7603	.0645	15.4980	2.7100
	45.00	8.032	.3519	2.8420	.0627	15.9567	2.6321
	50.20	7.910	.3413	2.9304	.0608	16.4527	2.5528
9⅝	40.00	8.835	.4257	2.3489	.0758	13.1879	3.1849
	43.50	8.755	.4181	2.3920	.0745	13.4301	3.1273
	47.00	8.681	.4110	2.4329	.0732	13.6600	3.0747
	53.50	8.535	.3973	2.5169	.0708	14.1313	2.9721
	58.00	8.435	.3881	2.5769	.0691	14.4684	2.9029
	58.40	8.435	.3881	2.5769	.0691	14.4684	2.9029
	61.10	8.375	.3826	2.6140	.0681	14.6764	2.8617
10	41.50	9.200	.4616	2.1662	.0822	12.1623	3.4533
	45.50	9.120	.4536	2.2044	.0808	12.3766	3.3935
	50.50	9.016	.4434	2.2555	.0790	12.6638	3.3166
	55.50	8.908	.4328	2.3105	.0771	12.9727	3.2376
	61.20	8.790	.4214	2.3730	.0751	13.3233	3.1524
10¾	45.50	9.950	.5400	1.8519	.0962	10.3979	4.0393
	51.00	9.850	.5292	1.8897	.0943	10.6101	3.9585
	55.50	9.760	.5195	1.9247	.0925	10.8066	3.8865
	60.70	9.660	.5090	1.9648	.0906	11.0315	3.8073
	65.70	9.560	.4985	2.0061	.0888	11.2635	3.7289
	71.10	9.450	.4871	2.0531	.0868	11.5273	3.6435
	76.00	9.350	.4768	2.0972	.0849	11.7752	3.5668
	81.00	9.250	.4667	2.1428	.0831	12.0311	3.4909

ALUMINUM LINER SIZES AND CAPACITIES

SIZE OD INCHES	WEIGHT PER LIN FT	ID	CU FT PER LIN FT	LIN FT PER CU FT	BARRELS PER LIN FT	LIN FT PER BARREL	GALLONS PER LIN FT
2⅞	2.65	2.323	.0294	33.9762	.0052	190.7619	0.2202
3½	3.55	2.900	.0459	21.8010	.0082	122.4036	0.3431
4	4.35	3.364	.0617	16.2017	.0110	90.9658	0.4617
4½	5.18	3.826	.0798	12.5251	.0142	70.3235	0.5972
4¾	5.41	4.082	.0909	11.0034	.0162	61.7795	0.6798
5	6.08	4.290	.1004	9.9623	.0179	55.9340	0.7509
5½	6.80	4.778	.1245	8.0312	.0222	45.0918	0.9314
5¾	7.43	5.000	.1364	7.3339	.0243	41.1766	1.0200
6⅝	10.00	5.761	.1810	5.5243	.0322	31.0166	1.3541
7	10.25	6.154	.2066	4.8413	.0368	27.1816	1.5452
8⅝	14.50	7.651	.3193	3.1321	.0569	17.5855	2.3883

Pump Output Tables

Notes: 1. Volume shown are for one complete cycle or revolution.
2. To get output in volume/minute, multiply output/cycle by pump rpm.

DOUBLE ACTING DUPLEX PUMP

Note: For triplex double acting pump multiply output by 1.5.

Stroke, in.	Bore, in.	Rod D, in.	100% Efficiency		90% Efficiency	
			cu ft	bbl	cu ft	bbl
6.	4.00	1.5	0.1623	0.0289	0.1460	0.0260
8.	4.00	1.5	0.2163	0.0385	0.1947	0.0347
8.	4.50	1.5	0.2782	0.0495	0.2503	0.0446
8.	5.00	1.5	0.3472	0.0618	0.3125	0.0557
10.	4.00	1.5	0.2704	0.0482	0.2434	0.0433
10.	4.50	1.5	0.3477	0.0619	0.3129	0.0557
10.	5.00	2.0	0.4182	0.0745	0.3763	0.0670
12.	4.00	1.5	0.3245	0.0578	0.2921	0.0520
12.	4.50	1.5	0.4172	0.0743	0.3755	0.0669
12.	5.00	2.0	0.5018	0.0894	0.4516	0.0804
12.	5.50	2.0	0.6163	0.1098	0.5547	0.0988
14.	4.50	1.5	0.4868	0.0867	0.4381	0.0780
14.	5.00	2.0	0.5854	0.1043	0.5269	0.0938
14.	5.50	2.0	0.7190	0.1281	0.6471	0.1153
14.	6.00	2.0	0.8654	0.1541	0.7789	0.1387
14.	6.25	2.0	0.9433	0.1680	0.8490	0.1512
14.	6.50	2.0	1.0245	0.1825	0.9220	0.1642
14.	6.75	2.0	1.1088	0.1975	0.9979	0.1777
14.	7.00	2.0	1.1963	0.2131	1.0766	0.1918
14.	7.25	2.5	1.2583	0.2241	1.1325	0.2017
14.	7.50	2.5	1.3522	0.2408	1.2170	0.2167
14.	7.75	2.5	1.4492	0.2581	1.3043	0.2323
16.	5.00	2.5	0.6363	0.1133	0.5727	0.1020
16.	5.50	2.5	0.7890	0.1405	0.7101	0.1265
16.	6.00	2.5	0.9563	0.1703	0.8607	0.1533
16.	6.25	2.5	1.0454	0.1862	0.9408	0.1676
16.	6.50	2.5	1.1381	0.2027	1.0243	0.1824
16.	6.75	2.5	1.2345	0.2199	1.1110	0.1979
16.	7.00	2.5	1.3344	0.2377	1.2010	0.2139
16.	7.25	2.5	1.4381	0.2561	1.2943	0.2305
16.	7.50	2.5	1.5453	0.2752	1.3908	0.2477
16.	7.75	2.5	1.6562	0.2950	1.4906	0.2655
18.	5.00	2.5	0.7159	0.1275	0.6443	0.1147
18.	5.50	2.5	0.8877	0.1581	0.7989	0.1423
18.	6.00	2.5	1.0758	0.1916	0.9682	0.1725
18.	6.25	2.5	1.1761	0.2095	1.0584	0.1885
18.	6.50	2.5	1.2804	0.2280	1.1523	0.2052
18.	6.75	2.5	1.3888	0.2473	1.2499	0.2226
18.	7.00	2.5	1.5013	0.2674	1.3511	0.2406
18.	7.25	2.5	1.6178	0.2881	1.4561	0.2593
18.	7.50	2.5	1.7385	0.3096	1.5647	0.2787
18.	7.75	2.5	1.8633	0.3319	1.6769	0.2987

DOUBLE ACTING DUPLEX PUMP (Continued)

Stroke, in.	Bore, in.	Rod D, in.	100% Efficiency		90% Efficiency	
			cu ft	bbl	cu ft	bbl
20.	6.50	2.5	1.4226	0.2534	1.2804	0.2280
20.	6.75	2.5	1.5431	0.2748	1.3888	0.2473
20.	7.00	2.5	1.6681	0.2971	1.5013	0.2674
20.	7.25	2.5	1.7976	0.3202	1.6178	0.2881
20.	7.50	2.5	1.9317	0.3440	1.7385	0.3096
20.	7.75	2.5	2.0703	0.3687	1.8633	0.3319
20.	8.00	2.5	2.2135	0.3942	1.9921	0.3548

Appendix

SINGLE ACTING TRIPLEX PUMP

Note: For single acting quintuplex pump, multiply output by 1.67.

Stroke, in.	Bore, in.	100% Efficiency		90% Efficiency	
		cu ft	bbl	cu ft	bbl
4.	3.00	0.0491	0.0087	0.0442	0.0078
4.	3.75	0.0576	0.0103	0.0518	0.0093
4.	3.50	0.0668	0.0119	0.0601	0.0107
4.	3.75	0.0767	0.0137	0.0690	0.0123
4.	4.00	0.0873	0.0155	0.0786	0.0140
4.	4.50	0.1104	0.0197	0.0994	0.0177
4.	5.00	0.1364	0.0243	0.1228	0.0219
4.	6.00	0.1963	0.0350	0.1767	0.0315
4.	8.00	0.3491	0.0622	0.3142	0.0560
6.	3.00	0.0737	0.0131	0.0663	0.0117
6.	3.25	0.0864	0.0155	0.0777	0.0140
6.	3.50	0.1002	0.0179	0.0902	0.0161
6.	3.75	0.1151	0.0206	0.1035	0.0185
6.	4.00	0.1310	0.0233	0.1179	0.0210
6.	4.50	0.1656	0.0296	0.1491	0.0266
6.	5.00	0.2046	0.0365	0.1842	0.0329
6.	6.00	0.2945	0.0525	0.2651	0.0473
6.	8.00	0.5237	0.0933	0.4713	0.0840
8.	3.00	0.0982	0.0174	0.0884	0.0156
8.	3.25	0.1152	0.0206	0.1036	0.0186
8.	3.50	0.1336	0.0238	0.1202	0.0214
8.	3.75	0.1534	0.0274	0.1380	0.0246
8.	4.00	0.1746	0.0310	0.1572	0.0280
8.	4.50	0.2208	0.0394	0.1988	0.0354
8.	5.00	0.2728	0.0486	0.2456	0.0438
8.	6.00	0.3926	0.0700	0.3534	0.0630
8.	8.00	0.6982	0.1244	0.6284	0.1120
10.	3.00	0.1228	0.0218	0.1105	0.0195
10.	3.25	0.1440	0.0258	0.1295	0.0233
10.	3.50	0.1670	0.0298	0.1503	0.0268
10.	3.75	0.1918	0.0343	0.1725	0.0308
10.	4.00	0.2183	0.0388	0.1965	0.0350
10.	4.50	0.2760	0.0493	0.2485	0.0443
10.	5.00	0.3410	0.0608	0.3070	0.0548
10.	6.00	0.4908	0.0875	0.4418	0.0788
10.	8.00	0.8728	0.1555	0.7855	0.1400

BLOWOUT PREVENTER AND HYDRAULIC VALVE DATA

RAM-TYPE BLOWOUT PREVENTERS

Model or type	BOP size, in.	Working pressure max. psi	Vert. bore in.	Hydraulic operator psi*	Gal. to close	Gal. to open	Close ratio	Open ration
\multicolumn{9}{c}{Hydril Co., Houston, Texas}
V	6	3000	7 1/16	750	1.50	1.30	5.32:1	
V	6	5000	7 1/16	1175	1.50	1.30	5.32:1	
X	7 1/16	10000	7 1/16	1350/3000	1.90	1.80	7.7:1	1.7:1
X	7 1/16	15000	7 1/16	2200/3000	3.70	3.40	7.1:1	6.6:1
V	10	3000	11	550	3.30	3.20	6:1	N/A
V	10	5000	11	850	3.30	3.20	6:1	
X	11	10000	11	1050	12.90	11.80	10.56:1	3.8:1
	12	3000	13 5/8	700	5.90	4.90	5.2:1	
V	13 5/8	5000	13 5/8	1050	5.90	4.90	5.2:1	
X	13 5/8	10000	13 5/8	1050/3000	12.60	11.40	10.56:1	3.8:1
X	16 3/4	10000	16 3/4	1050/3000	15.60	14.10	10.56:1	2.4:1
X	18 3/4	10000	18 3/4	1050/3000	17.10	15.60	10.56:1	1.9:1
V	20	2000	21 1/4	500/3000	8.10	7.20	4.74:1	0.9:1
V	20	2000	21 1/4	1050/3000	18.00	16.30	10.6:1	2.2:1
V	20	3000	20 3/4	500/3000	8.10	7.20	4.8:1	0.9:1
V	20	3000	20 3/4	1050/3000	18.00	16.30	10.56:1	2.2:1
V	21 1/4	2000	21 1/4	500	17.20	16.30	10.14:1	
ML	7 1/16	3000	7 1/16	3000	1.00	0.93	4.8:1	1.5:1
MPL	7 1/16	3000	7 1/16	3000	1.20	0.93	5.4:1	1.5:1
ML	7 1/16	5000	7 1/16	3000	1.00	0.93	4.8:1	1.5:1
MPL	7 1/16	5000	7 1/16	3000	1.20	0.93	5.4:1	1.5:1
ML	7 1/16	10000	7 1/16	3000	1.90	1.80	7.7:1	1.7:1
MPL	7 1/16	10000	7 1/16	3000	2.00	1.80	8.2:1	1.7:1
\multicolumn{9}{c}{Bowen Tools, Houston, Texas}
51922	2 1/2 Single	6000	2 1/2	780	0.17	0.16	7.9:1	
51923	2 1/2 Single	10000	2 1/2	1300	0.19	0.19	7.9:1	
51924	2 1/2 Twin	5000	2 1/2	692	0.36	0.28	7.9:1	
60701	2 1/2 Twin	10000	2 1/2	1001	0.43	0.35	7.9:1	
50460	2 9/16 Single	15000	2 9/16	1000	.3	.3	8.18:1	
70051	2 9/16	20000	2 9/16	800	.87	.93	23.8:1	
51926	3 Single	5000	3	369	.3	0.22	13.2:1	
51927	3 Single	10000	3	738	.3	0.22	13.2:1	
51928	3 Twin	5000	3	369	0.54	0.49	13.2:1	
51929	3 Twin	10000	3	738	0.54	0.49	13.2:1	
61040	4 Single	5000	4	555	0.91	0.81	15.3:1	
61044	4 Single	10000	4	1110	0.91	0.81	15.3:1	

* Lower pressures for normal use, higher pressures for emergency use only

APPENDIX

Ram-type blowout preventers (cont.)

Model or type	BOP size, in.	Working pressure max. psi	Vert. bore in.	Hydraulic operator psi*	Gal. to close	Gal. to open	Close ratio	Open ration
\multicolumn{9}{c}{Bowen Tools, Houston, Texas (cont.)}
61048	4 Twin	5000	4	555	1.81	1.62	15.3:1	
61050	4 Twin	10000	4	1110	1.81	1.62	15.3:1	
79358	4 Twin	10000	4	1110	1.81	1.62	15.3:1	
47034	4 1/16 Single	10000	4 1/16	1000	0.43	0.34	13.6:1	
60467	4 1/16 Single	15000	4 1/16	1250	0.69	0.74	16.2:1	
70630	4 1/16 Twin	15000	4 1/16	1250	1.38	1.48	16.2:1	
61053	4 1/2 Single	3000	4 1/2	370	0.91	0.81	15.3:1	
66174	4 1/2 Single	5000	4 1/2	555	1.83	1.64	15.3:1	
79363	4 1/2 Single	10000	4 1/2	1110	0.91	0.81	15.3:1	
61055	4 1/2 Single	10000	4 1/2	1110	0.91	0.81	15.3:1	
61057	4 1/2 Twin	5000	4 1/2	555	1.83	1.64	15.3:1	
61060	4 1/2 Twin	10000	4 1/2	1110	1.83	1.64	15.3:1	
79362	4 1/2 Twin	10000	4 1/2	1110	1.83	1.64	15.3:1	
51938	5 1/2 Single	3000	5 1/2	240	1.51	1.37	20.8:1	
63642	7 1/16 Single	10000	7 1/16	900	1.02	1.10	16.2:1	
70466	7 1/16 Twin	10000	7 1/16	900	2.04	2.20	16.2:1	
60615	7 5/8 Single	5000	6 1/2	900	1.75	1.74	10.9:1	
70399	7 5/8 Twin	5000	6 1/2	1800	3.50	3.48	10.9:1	
\multicolumn{9}{c}{Cooper Oil Tool, Houston, Texas}
T	13 5/8	10000	13 5/8	1500/3000	13.90	12.90	8.6:1	4.3:1
T	18 3/4	15000	18 3/4	1500/3000	24.30	22.40	6.7:1	3.1:1
U	6	3000	7 1/16	1500/3000	1.30	1.30	6.9:1	2.3:1
U	6	5000	7 1/16	1500/3000	1.30	1.30	6.9:1	2.3:1
U	7 1/16	3000	7 1/16	1500/3000	1.30	1.30	6.9:1	2.2:1
U	7 1/16	5000	7 1/16	1500/3000	1.30	1.30	6.9:1	2.2:1
U	7 1/16	10000	7 1/16	1500/3000	1.30	1.30	6.9:1	2.2:1
U	7 1/16	15000	7 1/16	1500/3000	1.30	1.30	6.9:1	2.2:1
U	10	3000	11	1500/3000	3.31	3.16	7:1	2.5:1
U	10	5000	11	1500/3000	3.31	3.16	7:1	2.5:1
U	11	3000	11	1500/3000	3.50	3.40	7.3:1	2.5:1
U	11	5000	11	1500/3000	3.50	3.40	7.3:1	2.5:1
U	11	10000	11	1500/3000	3.50	3.40	7.3:1	2.5:1
U	11	15000	11	1500/3000	6.20	6.10	9.8:1	2.2:1
U	12	3000	13 5/8	1500/3000	5.80	5.40	7:1	2.3:1
U	13 5/8	3000	13 5/8	1500/3000	5.80	5.40	7:1	2.3:1
U	13 5/8	5000	13 5/8	1500/3000	5.80	5.40	7:1	2.3:1
U	13 5/8	10000	13 5/8	1500/3000	5.80	5.40	7:1	2.3:1
U	13 5/8	15000	13 5/8	1500/3000	10.60	10.40	10.6:1	3.6:1
U	16 3/4	3000	16 3/4	1500/3000	10.60	9.80	6.8:1	2.3:1
U	16 3/4	5000	16 3/4	1500/3000	10.60	9.80	6.8:1	2.3:1
U	16 3/4	10000	16 3/4	1500/3000	12.40	11.60	6.8:1	2.3:1

Ram-type blowout preventers (cont.)

Model or type	BOP size, in.	Working pressure max. psi	Vert. bore in.	Hydraulic operator psi*	Gal. to close	Gal. to open	Close ratio	Open ration
			Cooper Oil Tool, Houston, Texas (cont.)					
U	18 3/4	10000	18 3/4	1500/3000	23.10	21.20	7.4:1	3.7:1
U	20	2000	20 3/4	1500/3000	8.40	7.85	7:1	1.2:1
U	20	3000	20 3/4	1500/3000	8.40	7.85	7:1	1.2:1
U	20 3/4	3000	20 3/4	1500/3000	8.40	7.90	7:1	1.3:1
U	21 1/4	2000	21 1/4	1500/3000	8.40	7.90	7:1	1.3:1
U	21 1/4	5000	21 1/4	1500/3000	29.90	27.20	6.2:1	4.0:1
U	21 1/4	10000	21 1/4	1500/3000	26.90	24.50	7.2:1	4.0:1
U	26 3/4	2000	26 3/4	1500/3000	10.40	9.85	7:1	1.0:1
U	26 3/4	3000	26 3/4	1500/3000	10.80	10.10	7:1	1.0:1
U With	11	3000	11	3000	7.60	7.40	12.0:1	4.8:1
Large	11	5000	11	3000	7.60	7.40	12.0:1	4.8:1
Bore	11	10000	11	3000	7.60	7.40	12.0:1	4.8:1
Shear	11	15000	11	3000	9.00	8.90	15.2:1	3.7:1
Bonnet	13 5/8	3000	13 5/8	3000	10.90	10.50	10.8:1	4.5:1
	13 5/8	5000	13 5/8	3000	10.90	10.50	10.8:1	4.5:1
	13 5/8	10000	13 5/8	3000	10.90	10.50	10.8:1	4.5:1
"B"	13 5/8	15000	13 5/8	3000	16.20	16.00	16.2:1	16.0:1
"B"	16 3/4	3000	16 3/4	3000	19.00	18.10	10.4:1	4.4:1
"B"	16 3/4	5000	16 3/4	3000	19.00	18.10	10.4:1	4.4:1
	16 3/4	10000	16 3/4	3000	19.10	18.20	10.4:1	4.4:1
	20 3/4	3000	20 3/4	3000	14.90	14.30	10.8:1	1.7:1
	21 3/4	2000	21 3/4	3000	14.90	14.30	10.8:1	1.7:1
U II	18 3/4	10000	18 3/4	1500/3000	24.70	22.30	6.7:1	2.5:1
U II	18 3/4	15000	18 3/4	1500/3000	34.70	32.30	9.3:1	3.5:1
U II W/CYL	18 3/4	15000	18 3/4	1500/3000	28.10	25.70	7.6:1	2.9:1
U-Blind	13 5/8	5000	13 5/8	1500/2500	11.6	10.90	14:1	2.3:1
ram with	13 5/8	10000	13 5/8	1500/2500	11.6	10.90	14:1	2.3:1
shear	16 3/4	3000	16 3/4	1500/2500	10.8	11.70	9:1	1.4:1
booster	16 3/4	5000	16 3/4	1500/2500	10.8	11.70	9:1	1.4:1
	20	2000	20 3/4	1500/2500	16.8	15.70	14:1	1.2:1
	20	3000	20 3/4	1500/3000	16.8	15.70	14:1	1.2:1
QRC	6	3000	7 1/16	1500/3000	0.81	0.95	7.75:1	1.5:1
QRC	6	5000	7 1/16	1500/3000	0.81	0.95	7.75:1	1.5:1
QRC	8	3000	9	1500/3000	2.36	2.70	9.05:1	1.83:1
QRC	8	5000	9	1500/3000	2.36	2.70	9.05:1	1.83:1
QRC	10	3000	11	1500/3000	2.77	3.18	9.05:1	1.21:1
QRC	10	5000	11	1500/3000	2.77	3.18	9.05:1	1.21:1
QRC	12	3000	13 5/8	1500/3000	4.42	5.10	8.64:1	1.07:1

APPENDIX

Ram-type blowout preventers (cont.)

Model or type	BOP size, in.	Working pressure max. psi	Vert. bore in.	Hydraulic operator psi*	Gal. to close	Gal. to open	Close ratio	Open ration
Cooper Oil Tool, Houston, Texas (cont.)								
QRC	18	2000	17 3/4	1500/3000	6.00	7.05	8.64:1	0.62:1
QRC	20	2000	17 3/4	1500/3000	6.00	7.05	8.64:1	0.62:1
S/QRC WITHOUT SHUTTLE VALVE	3 1/16	20000	3 1/16	1500/3000	0.50	0.55	16.0:1	6.87:1
S/QRC WITHOUT SHUTTLE VALVE	4 1/16	10000	4 1/16	1500/3000	0.89	1.00	10.74:1	8.56:1
S/QRC WITHOUT SHUTTLE VALVE	4 1/16	15000	4 1/16	1500/3000	0.89	1.00	10.74:1	8.56:1
S/QRC WITHOUT SHUTTLE VALVE	4 1/16	25000	4 1/16	1500/3000	2.28	2.46	17.17:1	23.40:1
S/QRC WITHOUT SHUTTLE VALVE	7 1/16	20000	4 1/16	1500/3000	4.95	5.44	15.98:1	7.38:1
S/QRC WITH SHUTTLE VALVE	3 1/16	20000	3 1/16	1500/3000	0.50	0.60	16.0:1	0.76:1
S/QRC WITH SHUTTLE VALVE	4 1/16	10000	4 1/16	1500/3000	0.89	0.10	10.74:1	0.87:1
S/QRC WITH SHUTTLE VALVE	4 1/16	15000	4 1/16	1500/3000	0.89	0.10	10.74:1	0.87:1

Ram-type blowout preventers (cont.)

Model or type	BOP size, in.	Working pressure max. psi	Vert. bore in.	Hydraulic operator psi*	Gal. to close	Gal. to open	Close ratio	Open ration
			Cooper Oil Tool, Houston, Texas (cont.)					
WITH SHUTTLE VALVE								
S/QRC WITH SHUTTLE VALVE	7 1/16	20000	7 1/16	1500/3000	4.95	0.50	15.98:1	0.66:1
G-2	4 1/16	5000	4 1/16	1500/3000	0.65	0.71	7.9:1	6.2:1
G-2	4 1/16	10000	4 1/16	1500/3000	0.65	0.71	7.9:1	6.2:1
SS	6	3000	7 1/16	1500/3000	0.80	0.70	3.8:1	1:1
SS	6	5000	7 1/16	1500/3000	0.80	0.70	3.8:1	1:1
SS	8	3000	9	1500/3000	1.50	1.30	3.9:1	1:1
SS	8	5000	9	1500/3000	1.50	1.30	3.9:1	1:1
SS	10	3000	11	1500/3000	1.50	1.30	3.9:1	1:1
SS	10	5000	11	1500/3000	1.50	1.30	3.9:1	1:1
SS	12	3000	13 5/8	1500/3000	2.90	2.50	3.7:1	1:1
SS	14	5000	13 5/8	1500/3000	2.90	2.50	3.7:1	1:1
Type F with type W2 opr.	6	3000	7 1/16	500/1500	1.50	2.30	VARIABLE	4.5:1
	6	5000	7 1/16	500/1500	1.50	2.30	VARIABLE	4.5:1
	7	10000	7 1/16	500/1500	1.50	2.30	VARIABLE	4.5:1
	7	15000	7 1/16	500/1500	1.50	2.30	VARIABLE	4.5:1
	8	3000	9	500/1500	2.80	3.70	VARIABLE	2.5:1
	8	5000	9	500/1500	2.80	3.70	VARIABLE	2.5:1
	10	3000	11	500/1500	2.80	3.70	VARIABLE	2.5:1
	10	5000	11	500/1500	2.80	3.70	VARIABLE	2.5:1
	11	10000	11	500/1500	2.80	3.70	VARIABLE	2.5:1
	12	3000	13 5/8	500/1500	4.10	5.30	VARIABLE	2:1
	14	5000	13 5/8	500/1500	4.20	5.30	VARIABLE	2:1
	16	2000	16 3/4	500/1500	5.00	6.00	VARIABLE	2:1
	16	3000	16 3/4	500/1500	5.00	6.00	VARIABLE	2:1
	20	2000	20 1/4	500/1500	5.00	6.00	VARIABLE	2:1
	20	3000	20 1/4	500/1500	5.00	6.00	VARIABLE	2:1
Type F with type W opr.	6	3000	7 1/16	500/1500	2.30	3.05	VARIABLE	4.5:1
	6	5000	7 1/16	500/1500	2.30	3.05	VARIABLE	4.5:1
	7	10000	7 1/16	500/1500	2.30	3.05	VARIABLE	4.5:1
	7	15000	7 1/16	500/1500	2.30	3.05	VARIABLE	4.5:1
	8	3000	9	500/1500	3.70	4.60	VARIABLE	2.5:1
	8	5000	9	500/1500	3.70	4.60	VARIABLE	2.5:1

APPENDIX

Ram-type blowout preventers (cont.)

Model or type	BOP size, in.	Working pressure max. psi	Vert. bore in.	Hydraulic operator psi*	Gal. to close	Gal. to open	Close ratio	Open ration
			Cooper Oil Tool, Houston, Texas (cont.)					
	10	3000	11	500/1500	3.70	4.60	VARIABLE	2.5:1
	10	5000	11	500/1500	3.70	4.60	VARIABLE	2.5:1
	11	10000	11	500/1500	3.70	4.60	VARIABLE	2.5:1
	12	3000	13 5/8	500/1500	6.80	8.10	VARIABLE	2:1
	14	5000	13 5/8	500/1500	6.80	8.10	VARIABLE	2:1
	16	2000	16 3/4	500/1500	7.60	9.10	VARIABLE	2:1
	16	3000	16 3/4	500/1500	7.60	9.10	VARIABLE	2:1
	20	2000	20 1/4	500/1500	7.60	9.10	VARIABLE	2:1
	20	3000	20 1/4	500/1500	7.60	9.10	VARIABLE	2:1
Type F with type L opr.	6	3000	7 1/16	250/1500	3.97	3.46	VARIABLE	4.9:1
	6	5000	7 1/16	250/1500	3.97	3.46	VARIABLE	4.9:1
	7	10000	7 1/16	250/1500	3.97	3.46	VARIABLE	4.9:1
	7	15000	7 1/16	250/1500	3.97	3.46	VARIABLE	4.9:1
	8	3000	9	250/1500	6.85	6.19	VARIABLE	3.44:1
	8	5000	9	250/1500	6.85	6.19	VARIABLE	3.44:1
	10	3000	11	250/1500	6.85	6.19	VARIABLE	3.44:1
	10	5000	11	250/1500	6.85	6.19	VARIABLE	3.44:1
	11	10000	11	250/1500	6.85	9.38	VARIABLE	3.44:1
	12	3000	13 5/8	250/1500	10.30	9.38	VARIABLE	2.3:1
	14	5000	13 5/8	250/1500	10.30	10.66	VARIABLE	2.3:1
	16	2000	16 3/4	250/1500	11.71	10.66	VARIABLE	2.3:1
	16	3000	16 3/4	250/1500	11.71	10.66	VARIABLE	2.3:1
	20	2000	20 1/4	250/1500	11.71	10.66	VARIABLE	2.3:1
	20	3000	20 1/4	250/1500	11.71	10.66	VARIABLE	2.3:1
Type F with type H opr.	6	3000	7 1/16	1000/5000	0.52	1.05	VARIABLE	1.5:1
	6	5000	7 1/16	1000/5000	0.52	1.05	VARIABLE	1.5:1
	7	10000	7 1/16	1000/5000	0.52	1.05	VARIABLE	1.5:1
	7	15000	7 1/16	1000/5000	0.52	1.05	VARIABLE	1.5:1
	8	3000	9	1000/5000	0.90	1.80	VARIABLE	1:1
	8	5000	9	1000/5000	0.90	1.80	VARIABLE	1:1
	10	3000	11	1000/5000	0.90	1.80	VARIABLE	1:1
	10	5000	11	1000/5000	0.90	1.80	VARIABLE	1:1
	11	10000	11	1000/5000	0.90	1.80	VARIABLE	1:1
	12	3000	13 5/8	1000/5000	1.52	2.70	VARIABLE	2.3:1
	14	5000	13 5/8	1000/5000	1.52	2.70	VARIABLE	2.3:1
	16	2000	16 3/4	1000/5000	1.73	3.08	VARIABLE	2.3:1
	16	3000	16 3/4	1000/5000	1.73	3.08	VARIABLE	2.3:1
	20	2000	20 1/4	1000/5000	1.73	3.08	VARIABLE	2.3:1
	20	3000	20 1/4	1000/5000	1.73	3.08	VARIABLE	2.3:1

Ram-type blowout preventers (cont.)

Model or type	BOP size, in.	Working pressure max. psi	Vert. bore in.	Hydraulic operator psi*	Gal. to close	Gal. to open	Close ratio	Open ration
			Dresser OME (Guiberson), Dallas, Texas					
Type H	6	3000	N/A	2000	1.10	0.94	6.5:1	1.1
Hyd. Cyl.	8	2000	N/A	2000	1.10	0.94	6.5:1	1.1
			Shaffer (Varco), Houston, Texas					
LWS	4 1/16	10000	4 1/6	1500/3000	0.50	0.47	8.45:1	4.74:1
with	6	3000	7 1/16	1500/3000	1.20	1.00	4.44:1	1.82:1
locking	6	5000	7 1/16	1500/3000	1.20	1.00	4.45:1	1.82:1
manual	7 1/16	10000	7 1/16	1500/3000	6.35	5.89	10.63:1	19.40:1
screw	7 1/16	15000	7 1/16	1500/3000	6.35	5.89	10.63:1	19.40:1
	8	3000	9	1500/3000	2.58	2.26	5.58:1	3.00:1
	8	5000	9	1500/3000	2.58	2.26	5.58:1	3.00:1
	9	10000	9	1500/3000	2.44	2.44	5.58:1	1.69:1
	10	3000	11	1500/3000	1.75	1.45	4.45:1	1.16:1
	10	5000	11	1500/3000	2.98	2.62	5.58:1	2.10:1
	11	10000	11	1500/3000	3.62	3.31	7.83:1	2.20:1
	12	3000	13 5/8	1500/3000	3.36	2.95	5.58:1	1.75:1
	13 5/8	5000	13 5/8	1500/3000	3.36	2.95	5.58:1	1.75:1
	13 5/8	10000	13 5/8	1500/3000	10.59	9.82	10.63:1	3.47:1
	16	3000	16 3/4	1500/3000	4.69	4.13	5.58:1	1.40:1
	16 3/4	5000	16 3/4	1500/3000	6.6	6.03	7.85:1	1.59:1
	20	2000	21 1/4	1500/3000	5.07	4.46	5.58:1	0.78:1
	20	3000	21 1/4	1500/3000	5.07	4.46	5.58:1	0.78:1
LWP	6	3000	7 1/16	1500/3000	0.55	0.51	4.49:1	2.5:1
Type	8	3000	9	1500/3000	0.77	0.78	4.49:1	1.81:1
LWS	7 1/16	15000	7 1/16	1500/3000	7.24	6.60	10.85:1	19.44:1
WITH	10	5000	11	1500/3000	4.75	4.18	8.16:1	3.07:1
POSLOCK	10	5000	11	1500/3000	9.31	8.48	10.85:1	7.82:1
	11	10000	11	1500/3000	4.2	3.70	8.16:1	2.21:1
	+ 11	10000	11	1500/3000	8.23	7.50	10.85:1	5.24:1
	12	3000	13 5/8	1500/3000	5.34	4.70	8.16:1	2.56:1
	+ 12	3000	13 5/8	1500/3000	10.56	9.62	10.85:1	6.25:1
	13 5/8	5000	13 5/8	1500/3000	5.3	4.67	8.16:1	2.56:1
	+ 13 5/8	5000	13 5/8	1500/3000	10.56	9.62	10.85:1	6.25:1
	+ 13 5/8	10000	13 5/8	1500/3000	11.56	10.52	10.85:1	3.47:1
	16 3/4	3000	16 3/4	1500/3000	7.25	6.38	8.16:1	2.05:1
	16 3/4	5000	16 3/4	1500/3000	7.25	6.38	8.16:1	1.59:1
	+ 16 3/4	5000	16 3/4	1500/3000	13.97	12.71	10.85:1	3.61:1
	+ 18 3/4	10000	18 3/4	1500/3000	15.3	13.21	7.11:1	1.83:1
	20	2000	18 3/4	1500/3000	7.8	6.86	8.16:1	1.15:1
	+ 20	2000	21 1/4	1500/3000	16.88	15.35	10.85:1	2.52:1

Appendix

Ram-type blowout preventers (cont.)

Model or type	BOP size, in.	Working pressure max. psi	Vert. bore in.	Hydraulic operator psi*	Gal. to close	Gal. to open	Close ratio	Open ration
				Shaffer (Varco), Houston, Texas (cont.)				
	20	3000	21 1/4	1500/3000	7.8	6.86	8.16:1	1.15:1
	+ 20	3000	21 1/4	1500/3000	16.88	15.35	10.85:1	2.52:1
	+ 21 1/4	7500	21 1/4	1500/3000	16.05	13.86	7.11:1	1.63:1
	+ 21 1/4	10000	21 1/4	1500/3000	16.05	13.86	7.11:1	1.63:1
SL	7 1/16	15000	7 1/16	2200/3000	6.00	5.70	13.94:1	7.14:1
SL	7 1/16	15000	7 1/16	2200/3000	2.72	2.34	7.11:1	3.37:1
SL	11	15000	11	2200/3000	9.4	8.10	7.11:1	2.80:1
SL	13 5/8	15000	13 5/8	2200/3000	11.56	10.52	7.11:1	2.14:1
SL	18 3/4	15000	18 3/4	1500/3000	14.62	13.33	10.85:1	1.68:1
SL	7 1/16	10000	7 1/16	1500/3000	6.00	5.57	13.94:1	7.14:1
SL	7 1/16	10000	7 1/16	1500/3000	2.72	2.34	7.11:1	3.37:1
SL	11	10000	11	1500/3000	9.45	7.00	7.11:1	7.62:1
SL	13 5/8	10000	13 5/8	1500/3000	10.58	10.52	7.11:1	4.29:1
SL	16 3/4	10000	16 3/4	1500/3000	14.47	12.50	7.11:1	2.06:1
SL	18 3/4	10000	18 3/4	1500/3000	14.55	13.21	7.11:1	1.83:1
SL	21 1/4	10000	21 1/4	1500/3000	16.05	13.86	7.11:1	1.63:1
SL	13 5/8	5000	13 5/8	1500/3000	11.00	10.50	10.85:1	10.02:1
SL	13 5/8	5000	13 5/8	1500/3000	5.44	4.46	5.54:1	3.00:1
SL	16 3/4	5000	16 3/4	1500/3000	11.76	10.67	10.85:1	5.77:1
SL	16 3/4	5000	16 3/4	1500/3000	6.07	4.97	5.54:1	2.03:1
SL	13 5/8	3000	13 5/8	1500/3000	5.44	4.46	5.54:1	3.00:1
SENTINEL	7 1/16	3000	7 1/16	1500	0.29	0.28	4.00:1	2.50:1
SENTINEL II	7 1/16	3000	7 1/16	1500	0.29	0.28	4.00:1	2.50:1
	9	3000	9	1500	0.36	0.34	4.00:1	2.50:1

ANNULAR PREVENTERS, DIVERTERS, STRIPPERS

Model or type	BOP size, in.	Working pressure max. PSI	Vert. bore in.	Hydraulic operator PSI*	Gal. to close	Gal. to open	Packoff open hole min. psi
				Shaffer (Varco), Houston, Texas			
SPHERICAL BOP							
Bolted	30	1000	30	1500	122.00	55.00	VARIABLE
Bolted	21 1/4	2000	21 1/4	1500	32.59	16.92	VARIABLE
Bolted	7 1/16	3000	7 1/16	1500	4.92	3.43	VARIABLE
Bolted	9	3000	9	1500	7.23	5.03	VARIABLE
Bolted	11	3000	11	1500	11.00	6.78	VARIABLE
Bolted	13 5/8	3000	13 5/8	1500	23.50	14.67	VARIABLE
Bolted	20 3/4	3000	20 3/4	1500	43.40	26.90	VARIABLE

Annular preventer, diverters, stripper (cont.)

Model or type	BOP size, in.	Working pressure max. PSI	Vert. bore in.	Hydraulic operator PSI*	Gal. to close	Gal. to open	Packoff open hole min. psi
Shaffer (Varco), Houston, Texas (cont.)							
Bolted	7 1/16	5000	7 1/16	1500	4.57	3.21	VARIABLE
Bolted	9	5000	9	1500	11.05	8.72	VARIABLE
Bolted	11	5000	11	1500	18.67	14.59	VARIABLE
Bolted	13 5/8	5000	13 5/8	1500	23.58	17.41	VARIABLE
Wedge	16 3/4	5000	16 3/4	1500	33.26	25.61	VARIABLE
Wedge	18 3/4	5000	18 3/4	1500	48.16	37.61	VARIABLE
Wedge	21 1/4	5000	21 1/4	1500	61.37	47.76	VARIABLE
Bolted	4 1/16	10000	4 1/16	1500	2.38	1.95	VARIABLE
Bolted	7 1/16	10000	7 1/16	1500	17.11	13.95	VARIABLE
Wedge	11	10000	11	1500	30.58	24.67	VARIABLE
Wedge	13 5/8	10000	13 5/8	1500	40.16	32.64	VARIABLE
Wedge	18 3/4	10000	18 3/4	1500	85.00	66.00	VARIABLE
Diverter	21 1/4	2000	21 1/4	1500	32.59	16.92	VARIABLE
Hydril Company, Houston, Texas							
GK	6	3000	7 1/16	1500	2.85	2.24	1000
GK	6	5000	7 1/16	1500	3.86	3.30	1000
GK	7 1/16	15000	7 1/16	1500	11.20	7.50	1050
GK	7 1/16	20000	7 1/16	1500	10.90	7.20	1150
GK	8	3000	8 15/16	1500	4.33	3.41	1150
GK	8	5000	8 15/16	1500	6.84	5.80	1150
GK	10	3000	11	1500	7.43	5.54	1150
GK	10	5000	11	1500	9.81	7.97	1150
GK	12	3000	13 5/8	1500	11.36	8.94	1150
GK	13 5/8	5000	13 5/8	1500	17.98	14.15	1150
GK	16	2000	16 3/4	1500	17.42	12.53	1150
GK	16	3000	16 3/4	1500	21.02	15.8	1150
GK	16 3/4	5000	16 3/4	1500	28.7	19.93	1150
GK	18	2000	17 7/8	1500	21.09	14.44	1150
GK	7 1/16	10000	7 1/16	1500	9.42	7.08	1150
GK	9	10000	9	1500	15.90	11.95	1150
GK	11	10000	11	1500	25.10	18.87	1150
GK	13 5/8	10000	13 5/8	1500	34.53	24.66	1150
GL	13 5/8	5000	13 5/8	1500	19.70	19.70	1300
GL	16 3/4	5000	16 3/4	1500	33.80	33.80	1300
GL	18 3/4	5000	18 3/4	1500	44.00	44.00	1300
GL	21 3/4	5000	21 1/4	1500	58.00	58.00	1300
GS	4 1/16	10000	4 1/16		1.88	1.23	
GS	4 1/16	15000	4 1/16		1.95	1.30	

Annular preventer, diverters, stripper (cont.)

Model or type	BOP size, in.	Working pressure max. PSI	Vert. bore in.	Hydraulic operator PSI*	Gal. to close	Gal. to open	Packoff open hole min. psi
	Hydril Company, Houston, Texas (cont.)
GX	11	10000	11		17.88	17.88	
GX	13 5/8	5000	13 5/8		15.50	15.50	
GX	13 5/8	10000	13 5/8		24.14	24.14	
GX	18 3/4	10000	18 3/4		58.00	58.00	
GX	11	15000	11		24.14	24.14	
GX	13 5/8	15000	13 5/8		34.00	34.00	
MSP	6	2000	7 1/16	1500	2.85	1.98	1000
MSP	8	2000	8 15/16	1500	4.57	2.95	1050
MSP	10	2000	11	1500	7.43	5.23	1150
MSP	20	2000	20 3/4	1500	31.05	18.93	1100
MSP	20	2000	21 1/4	1500	31.05	18.93	1100
MSP	29 1/2	500	29 1/2	1500	60.00	0	1500
MSP	30	1000	30	1500	87.60	27.80	
	Cooper Oil Tool, Houston, Texas
A	6	5000	7 1/16	1500	2.20	1.90	
A	6	10000	7 1/16	1500	4.00	3.10	
A	6	15000	7 1/16	N/A	N/A	N/A	
A	11	5000	11	1500	7.80	6.50	
A	11	10000	11	1500	12.10	10.50	
A	11	15000	11	N/A	N/A	N/A	
A	13 5/8	5000	13 5/8	1500	15.50	13.90	
A	13 5/8	10000	13 5/8	1500	21.50	18.70	
A	16 3/4	5000	16 3/4	1500	33.00	29.00	
DL	7 1/16	3000	7 1/16	1500/3000	1.69	1.39	3000
DL	7 1/16	5000	7 1/16	1500/3000	1.69	1.39	3000
DL	7 1/16	10000	7 1/16	1500/3000	2.94	2.55	3000
DL	7 1/16	15000	7 1/16	1500/3000	6.94	6.12	3000
DL	7 1/16	20000	7 1/16	1500/3000	8.38	7.56	3000
DL	11	3000	11	1500/3000	5.65	4.69	3000
DL	11	5000	11	1500/3000	5.65	4.69	3000
DL	11	10000	11	1500/3000	10.15	9.06	3000
DL	11	15000	11	1500/3000	23.5	21.3	3000
DL	13 5/8	3000	13 5/8	1500/3000	12.12	10.34	3000
DL	13 5/8	5000	13 5/8	1500/3000	12.12	10.34	3000
DL	13 5/8	10000	13 5/8	1500/3000	18.10	16.15	3000
DL	16 3/4	3000	16 3/4	1500/3000	22.30	19.00	3000
DL	16 3/4	5000	16 3/4	1500/3000	22.30	19.00	3000
DL	18 3/4	5000	18 3/4	1500/3000	35.60	29.00	3000

Annular preventer, diverters, stripper (cont.)

Model or type	BOP size, in.	Working pressure max. PSI	Vert. bore in.	Hydraulic operator PSI*	Gal. to close	Gal. to open	Packoff open hole min. psi
		Cooper Oil Tool, Houston, Texas (cont.)					
DL	18 3/4	10000	18 3/4	1500/3000	51.00	45.10	3000
DL	20 3/4	3000	20 3/4	1500/3000	40.50	28.40	3000
DL	21 1/4	2000	21 1/4	1500/3000	40.50	28.40	3000

Model "D" has same specifications as the Model "DL". The model "D" does not have a locking ring, while the Model "DL" does.

Model or type	BOP & Vert. bore size, in.	Working pressure max. psi	Gal. to close	Gal. to open	Close ratio	Open ratio	Packoff open hole min. psi
One piece "D" type annular and "U" ram-type preventer							
D	7 1/16	3000	1.69	1.30	N/A	N/A	3000
U	7 1/16	3000	1.30	1.30	6.9:1	2.2:1	N/A
D	7 1/16	5000	1.69	1.30	N/A	N/A	3000
U	7 1/16	5000	1.30	1.30	6.9:1	2.2:1	N/A
D	7 1/16	10000	1.69	1.69	N/A	N/A	3000
U	7 1/16	10000	1.69	1.69	6.9:1	2.2:1	N/A
D	11	3000	5.65	4.69	N/A	N/A	3000
U	11	3000	3.50	3.40	7.3:1	2.5:1	N/A
D	11	5000	5.65	4.69	N/A	N/A	3000
U	11	5000	3.50	3.40	7.3:1	2.5:1	N/A

Model or type	BOP size, in.	Working pressure max. PSI	Vert. bore in.	Hydraulic operator PSI*	Gal. to close	Gal. to open	Packoff open hole min. psi
			ABB Vetco Gray, Houston, Texas				
DIVERTER UNITS							
KFDS	27 1/2	2000	10	2500	3	-	VARIABLE
KFDS	49 1/2	500	23/10	1500	2.7(1)	-	VARIABLE
KFDS-CSO	21 1/4	1000	20	3000	31.0	15.0	VARIABLE
KFDJ	36 1/2	500	22/10	1500	2.7(2)	-	VARIABLE
KFDJ	36 1/2	500	22/15	1500	5.9(3)	-	VARIABLE

APPENDIX

Annular preventers, diverters, strippers (cont.)

Model or type	BOP size, in.	Working pressure max. PSI	Vert. bore in.	Hydraulic operator PSI*	Gal. to close	Gal. to open	Packoff open hole min. psi
		ABB Vetco Gray, Houston, Texas (cont.)					
KFDJ	36 1/2	500	22/17.75	1500	9.2(4)	-	VARIABLE
KFDJ	36 1/2	500	22	1500	8.3(5)	-	VARIABLE

(1) 10" packer insert, 5" pipe
(2) 10" packer insert, 5" pipe, same for 2000 psi wp unit
(3) 15" insert, 8" casing; same for 2000 psi wp unit
(4) 17.75" packer insert, 16" casing; same for 2000 psi wp unit
(5) 22" packer insert, 16" casing; same for 2000 psi wp unit

HYDRAULICALLY OPERATED VALVES

Model or type	Line size, in.	Working pressure max. psi	Vert. bore, in.	Hydraulic operator psi*	Gal. to close	Gal. to open
		Shaffer (Varco), Houston, Texas				
Flo-seal	2	2000	1 11/16 2 1/16	3000	0.20	0.20
	2	3000	1 11/16 2 1/16	3000	0.20	0.20
	2	5000	1 11/16	3000	0.20	0.20
	2 1/16	10000	2 1/16	3000	0.40	0.40
	2 1/16	15000	2 1/16	3000	0.40	0.40
	2 1/2	2000	2 9/16	3000	0.30	0.30
	2 1/2	3000	2 9/16	3000	0.30	0.30
	2 1/2	5000	2 9/16	3000	0.30	0.30
	3	2000	3 1/8	3000	0.30	0.30
	3	3000	3 1/8	3000	0.30	0.30
	3	5000	3 1/8	3000	0.30	0.30
	3 1/16	10000	3 1/16	3000	0.60	0.60
	4	3000	4 1/16	3000	0.80	0.80
	4	5000	4 1/16	3000	0.80	0.80
	4 1/16	10000	4 1/16	3000	1.30	1.30
	6	3000	7 1/16	3000	2.00	2.00
Flo-seal with Ramlock	2	2000	1 11/16 2 1/16	3000	0.30	0.30
	2	3000	1 1/16 2 1/16	3000	0.30	0.30
	2	5000	1 1/16 2 1/16	3000	0.30	0.30

Hydraulically operated valves (cont.)

Model or type	Line size, in.	Working pressure max. psi	Vert. bore, in.	Hydraulic operator psi*	Gal. to close	Gal. to open
		Shaffer (Varco), Houston, Texas (cont.)				
	2 1/16	10000	2 1/16	3000	0.40	0.40
	2 1/16	15000	2 1/16	3000	0.40	0.40
	2 1/2	2000	2 9/16	3000	0.30	0.30
	2 1/2	3000	2 9/16	3000	0.30	0.30
	2 1/2	5000	2 9/16	3000	0.30	0.30
	3	2000	3 1/18	3000	0.40	0.40
	3	3000	3 1/18	3000	0.40	0.40
	3	5000	3 1/18	3000	0.40	0.40
	3 1/16	10000	3 1/16	3000	0.60	0.60
	4	3000	4 1/16	3000	0.80	0.80
	4	5000	4 1/16	3000	0.80	0.80
	4 1/16	10000	4 1/16	3000	0.80	0.80
	6	3000	7 1/16	3000	2.00	2.00
CB SUBSEA LONG SEA CHEST	3	5000	3 1/8	3000	0.45	-
	3 1/16	10000	3 1/16	3000	0.50	-
CB SUBSEA SHORT SEA CHEST	3	5000	3 1/8	3000	0.45	-
	3 1/16	10000	3 1/16	3000	0.50	-
DB	2	5000	2 1/16	3000	0.20	0.15
DB	2	10000	2 1/16	3000	0.20	0.15
DB	2	15000	2 1/16	3000	0.20	0.15
DB	3	3000	3 1/8	3000	0.30	0.30
DB	3	5000	3 1/8	3000	0.25	0.20
DB	3	10000	3 1/16	3000	0.40	0.35
DB	3	15000	3 1/16	3000	0.40	0.35
DB	4	3000	4 1/16	3000	0.80	0.80
DB	4	5000	4 1/16	3000	0.40	0.35
DB	4	10000	4 1/16	3000	0.50	0.45
DB	4	15000	4 1/16	3000	0.50	0.45
DB	6			3000	0.50	0.45
DB	6	3000	7 1/16	3000	0.50	0.00
HB	3	5000	3 1/16	3000	0.45	0.40
HB	3	10000	3 1/16	3000	0.50	0.45

APPENDIX

Hydraulically operated valves (cont.)

Model or type	Line size, in.	Working pressure max. psi	Vert. bore, in.	Hydraulic operator psi*	Gal. to close	Gal. to open
		Cooper Oil Tools, Houston, Texas				
Cameron F	2	960	1 13/16	1500	0.10	0.10
	2	2000	1 13/16	1500	0.10	0.10
	2	3000	1 13/16	1500	0.10	0.10
	2	5000	1 13/16	1500	0.16	0.16
	2	10000	1 13/16	1500	0.16	0.16
	2	15000	1 13/16	1500	0.16	0.16
	2	960	2 1/16	1500	0.10	0.10
	2	2000	2 1/16	1500	0.10	0.10
	2	3000	2 1/16	1500	0.10	0.10
	2	5000	2 1/16	1500	0.16	0.16
	2	10000	2 1/16	1500	0.16	0.16
	2	15000	2 1/16	1500	0.16	0.16
	2 1/12	960	2 9/16	1500	0.13	0.13
	2 1/12	2000	2 9/16	1500	0.13	0.13
	2 1/12	3000	2 9/16	1500	0.13	0.13
	2 1/12	5000	2 9/16	1500	0.20	0.20
	2 1/12	10000	2 9/16	1500	0.20	0.20
	2 1/12	15000	2 9/16	1500	0.40	0.40
	3	960	3 1/18	1500	0.15	0.15
	3	2000	3 1/18	1500	0.15	0.15
	3	3000	3 1/18	1500	0.24	0.24
	3	5000	3 1/18	1500	0.24	0.24
	3	10000	3 1/18	1500	0.28	0.28
	4	2000	4 1/18	1500	0.30	0.30
	4	3000	4 1/18	1500	0.30	0.30
	4	5000	4 1/18	1500	0.30	0.30
	4	10000	4 1/18	1500	0.59	0.59
	6	2000	6 1/8	1500	0.84	0.84
	6	3000	6 1/8	1500	0.84	0.84
	6	5000	6 1/8	1500	0.84	0.84
Cameron HCR	4	3000	4	1500	0.61	0.52
HCR	4	5000	4	1500	0.61	0.52
HCR	6	3000	7	1500	2.25	1.95
HCR	6	5000	7	1500	2.25	1.95
Cameron DV	4	3000	4	1500	0.80	1.10
DV	4	5000	4	1500	0.80	1.10
DV	6	3000	7	1500	2.10	3.60

Hydraulically operated valves (cont.)

Model or type	Line size, in.	Working pressure max. psi	Vert. bore, in.	Hydraulic operator psi*	Gal. to close	Gal. to open
Cooper Oil Tools, Houston, Texas (cont.)						
DV	8	3000	9	1500	2.40	5.60
DV	10	3000	11	1500	5.70	11.40
DV	10	5000	11	1500	5.70	11.40
DV	12	3000	13 5/8	1500	11.80	22.70
McEvoy AC Valve with U-1 Hyd. Opr.	2	2000	2 1/16	2500	0.13	0.11
	2	3000	2 1/16	2500	0.13	0.11
	2	5000	2 1/16	2500	0.13	0.11
	2	10000	2 1/16	2500	0.21	0.20
	2 1/2	2000	2 9/16	2500	0.26	0.23
	2 1/2	3000	2 9/16	2500	0.26	0.23
	2 1/2	5000	2 9/16	2500	0.26	0.23
	2 1/2	10000	2 9/16	2500	0.45	0.42
	3	2000	3 1/16	2500	0.30	0.25
	3	3000	3 1/16	2500	0.51	0.46
	3	5000	3 1/16	2500	0.51	0.46
	4	2000	4 1/16	2500	0.69	0.62
	4	3000	4 1/16	2500	0.69	0.62
	4	5000	4 1/16	2500	1.04	0.98
McEvoy EDU and EU Valve with U-1 actuator	3	5000	3 1/16	2500	0.52	0.47
	3 1/16	10000	3 1/16	2500	0.52	0.47

Index

Abnormal Formation Pressure, 93–94
Accumulators
  Bypass Valve, 75
  Design, 77
  Position, 7
  Pressure, 76
  Preventer Testing, 67
  Purpose, 74–75
API (American Petroleum Institute)
  BOP Ratings, 55
  Casing Head, 31
  Seal Rings, 44–46
Athey Wagon, 328

Barite, 3–4, 80
Barite Plug, 413–415
Blowout Preventers, 5
  Annular Group, 5–6, 56–58
  Blind Rams, 5, 20
  Bolts, 41–42
  Cooper D, 15
  Cooper F, 22
  Cooper QRC, 22
  Cooper SS, 22
  Cooper U, 23
  Cooper U-II, 23
  Cooper T, 24
  Chokes Manifolds, 63–65
  Downhole, 50
  Hydril GX, 8–9
  Hydril GK, 8, 10
  Hydril GL, 8, 11, 13
  Hydril MSP, 8, 13
  Hydril GS, 8
  Hydril RS, 8
  Kelly Cock, 47–48
  Lines, 37

Pipe Rams, 7, 17
Pressure Ratings, 55
Ram, 16–17
Rotating, 35
Seal Rings, 44
Shaffer LWP, 26
Shaffer LWS, 25
Shaffer SL, 27
Shaffer Rams, 25–28
Spherical, 13
Spools, 30, 37
Stack Design, 5, 17, 53
Testing, 66
Test Plugs, 68
Test Reports, 73
Blowouts
  Capping, 305, 319–329
  Causes, 420–422
  Contingency Planning, 289–319
  Definition, 287
  Diversion, 329
  Dynamic Killing, 351–364
  Environmental Protection, 287
  Hole Bridging, 303–305
  Igniting, 320–325
  Kill Procedures, 303–313
  Momentum Kill, 367–368
  Relief Wells, 330–407
  Reservoir Flooding, 364–367
  Underground Blowouts, 407–420
Buna, 14
Bullheading, 174, 192–193, 235–242

Calcium Carbonate, 3
Capacities
  Casing, 443–450
  Drill Collars, 433–434
  Drill Pipe, 432

469

Capacities, *continued*
　Pipe and Hole, 435–450
　Tubing, 441–442
Casing
　Burst, 53
Casing Head, 31, 37, 70
Cement
　Kicks Following, 175
　Choke Manifold, 63–65
Chokes
　Adjustable, 59–60
　Cooper, 61
　Manual, 59
Choke Line, 37, 56–57, 63–65, 167–171
Choke Manifolds, 63–64
Circulation
　One, 117–118
　Two, 117, 118
　Concurrent, 117, 119–120
Clamps, 41–44
Connectors
　Clamp, 41–44
　Flanged, 41–44
　Torque, 43
Contingency Plans, 289–319
Cooper Oil Tools
　Choke Control Panel, 64
　D, 15
　Drilling Choke, 59–63
　F, 22
　F Test Plug, 71
　QRC, 22
　SS, 22
　U, 23
　UII, 23
　T, 24
　VBR, 20–21

Degasser
　Atmospheric, 82
　Pressurized, 82–83
　Vacuum, 82–83
Displacement
　Choke Line, 167–171
　Riser, 172–173
　Pipe, 432
Diverters, 32
　ABB Vetco Gray, 35
　Hydril, 33–34

Shaffer, 34
System Failures 265–268
Driller's Method, 117, 118
Drilling Spool, 5, 30
Drill Pipe, 5, 46
　Aluminum, 17
　Capacity, 432–435
　Displacement, 186, 432–435
　Floats, 111
　Freezing, 204–205
　H_2S Design, 246–247
　Impairment, 207–210
　Plug, 209–210
　Washout, 207–209
Drill Stem Testing, 205–207
Dynamic Kill, 272–273, 351–364

Explosives, 321–323
Engineer's Method, 117

Flanges
　API, 42
　Bolt, 41–43
　Clamp, 41, 43
　Flange Bolts, 41–43
Flapper, 48–49
Float, 49
Floating Vessel, 163
Fluid Density, 2
Formation
　Fluids, 5
　Fracture Gradient, 164–166
　Permeability, 92
　Porosity, 92
Fracture Gradients, 164–166
Freeze Process, 204–205

Galena, 3
Gas
　Shallow Sands, 165
　Expansion, 121
　Counting, 84
　Detectors, 85
　Degassers, 82
Gaskets, 44

Horizontal Well, 227–235
Hydrates, 218–227
Hydril
　GK, 8, 10

Index

GL, 8, 11, 13
MSP, 8, 13
Rams, 28–30
Hydrogen Sulfide
  General, 9, 14, 26–27, 44–46, 60, 63, 242–258
Hydrostatic Pressure, 1, 92–98, 109

Joint, Test, 71

Kelly, 46
Kelly Cock 46–48
Kicks, 92
  Causes, 93–98
  Defintion, 92
  Identification, 115
  Warning Signs, 98–101, 163–164
Kick Killing Procedures, 4
  Bullhead, 235–242
  Constant Bottom-Hole Pressure, 108, 120
  Constant Pit Level, 152
  Concurrent Method, 119
  Deepwater, 163–173
  Diverter, 106, 265–268
  Drillers's Method (Two Circulation), 119
  Heavy Slug, 273–274, 412–413
  Horizontal Wells, 227–235
  Low Choke Pressure, 141–143
  Outrunning, 151–152
  Overkill, 130
  Slim Holes, 278–280
  Variables Affecting, 124–132
  Wait and Weight (One Circulation), 118, 132
  While Running Casing, 173–175
  While Cementing, 175–179
  While Pipe is Out of the Hole, 190–203
  With MWD In Pipe, 210–211
  With Downhole Motor, 211–215
  With Hydrates, 218–227
  With Oil-based Muds, 277–278

Locking Screws, 46
Logging
  Magrange, 384–385
  Noise, 419
  Radioactive, 418–419
  Temperature, 416–418
  Vector Magnetics, 385–386
Low Choke Pressure Method, 141–143

Manifolds
  Blowouts, 393–395
  Choke, 63–65
Mixing Pumps, 78
Mud
  Additives, 1–5
  Bulk, 80
  Corrosion, 252–256
  Density, 1
  Gas-Cut, 82
  Gradient, 2
  Hoppers, 78
  Mixing Pumps, 78
  Monitors, 83
  Pumps, 80

Neoprene, 9
Nitrile, 9, 14
Nitrogen, 66

Offset Kill, 309–310
Oil (Damage to BOP), 9, 14
Orifice, 60
Overburn, 165

Panels
  Choke, 63
Perforating, 210, 400–401
Permeability, 92
Pipe (see Casing, Drill Pipe)
Porosity, 92
Pressure
  Abnormal, 93
  Bottom-Hole, 107
  Build-up, 410
  Casing, 6, 107, 152
  Drill Pipe, 107–114
  Fracture, 92, 164–165
  Friction, 167–169
  Hydraulic, 5–6, 8
  Hydrostatic, 1
  Kill Rate, 111
  Spherical Closing, 9–11, 15
  Test, 66

Pressure, *continued*
  Trapped, 109–110
  Working, 44–46, 53
Procedures
  Blowout Kill (*see* Blowouts)
  Bullheading, 235–242
  Lubrication, 307
  Shut-in, 102–105
  Underground Blowout Kill, 407–420
Pump
  Accumulator, 74
  Centrifugal, 78
  Crippling, 80
  High Pressure, 67
  Output, 451–454
PVT, 84–85, 164

Rams
  Blind, 17, 20
  Bodies, 21
  Pipe, 5, 7, 17
  Self Feeding Action, 17
  Shear, 17, 21, 215
  Variable Bore, 17, 20
Relief Well Drilling
  Blowout Well Intersection, 344–346
  Casing Design, 373–377
  Communications, 399–403
  Directional Planning, 381–386
  Ellipse of Uncertainty, 336–338
  General, 310–313
  Kill Fluids, 396–398
  Kill Hydraulics, 349–368
  Kill Operations, 398–403
  Killing Equipment, 390–396
  Location Selection, 338–344
  Logging, 382–386
  Observation Wells, 346–349
  Number of Required Wells, 368–373
  Reasons for Drilling, 310–312
  Well Path Location, 335–338
Resistivity
  Cuttings, 94
  Mud, 94
Reverse Out, 172–173
Rings, 44
Ring Gaskets, 44
Risers, 58–59
Rotating BOP, 35–36

Rubber
  Choke, 59
  Packer Elements, 5, 14
  Self-feeding, 17–18

Sealing Elements
Shaffer
  B, 25
  E, 25
  LWP, 26
  LWS, 25
  SL, 27
  XHP, 225
Shale Density, 94
Shallow Gas Handling, 258–277
Shearing, 215–218
Snubbing
  Blowouts, 179–195
  Hydraulic Units, 184–186
  Rig Assisted, 181–184
Stingers, 327–328
Stripping
  Procedures, 8, 75, 186–190
  Rams, 184
  Operations, 179–181
  Equipment, 181–186

Time
  Effect on Pressures, 109–110
  Circulation of Kicks, 121

Underground Blowouts, 407–420

Valves
  Ball, 37
  Bypass, 75
  Check, 40
  Drilling, 203–204
  Flapper, 48–49
  Float, 48
  Gate, 38
  Hydraulic, 40
  Kelly Cock, 46–48
  Manual, 39
  Safety, 49
Vertical Intervention, 309
Volumetric Kill, 192

Well Planning, 4
Working Pressure, 44–46, 53